Handbook of
Data Mining for Design and Manufacturing

Handbook of
Data Mining for Design and Manufacturing

Contributors :
Fernando Sánchez Lasheras,
Paulino José García Nieto, *et al.*

Volume 1

AURIS REFERENCE LTD.
London, UK

Handbook of Data Mining for Design and Manufacturing
Contributors : Fernando Sánchez Lasheras *and* Paulino José García Nieto, *et al.*

Auris Reference Ltd., UK

www.aurisreference.com

United Kingdom

Printed in 2017 for Sale in the Indian Subcontinent

Handbook of Data Mining for Design and Manufacturing

ISBN: 978-1-78154-484-6

British Library Cataloguing in Publication Data
A CIP record for this book is available from the British Library

Exclusively distributed by CBS Publishers & Distributors Pvt. Ltd.

Sales & Distribution Rights only for India, Pakistan, Bangladesh, Sri Lanka, Nepal and Bhutan.This book is not to be sold outside these territories.

PREFACE

Data mining uses a relatively large amount of computing power operating on a large set of data to determine regularities and connections between data points. Algorithms that employ techniques from statistics, machine learning and pattern recognition are used to search large databases automatically. Data mining is also known as Knowledge-Discovery in Databases (KDD).

Like the term artificial intelligence, data mining is an umbrella term that can be applied to a number of varying activities. In the corporate world, data mining is used most frequently to determine the direction of trends and predict the future. It is employed to build models and decision support systems that give people information they can use. Data mining takes a front-line role in the battle against terrorism. It was supposedly used to determine the leader of the 9/11 attacks.

Data miners are statisticians who use techniques with names like near-neighbor models, k-means clustering, holdout method, k-fold cross validation, the leave-one-out method, and so on. Regression techniques are used to subtract irrelevant patterns, leaving only useful information. The term Bayesian is seen frequently in the field, referring to a class of inference techniques that predict the likelihood of future events by combining prior probabilities and probabilities based on conditional events. Spam filtering is arguably a form of data mining, which automatically brings relevant messages to the surface from a chaotic sea of phishing attempts and Viagra pitches.

Decision trees are used to filter mountains of data. In a decision tree, all data passes through an entrance node, where it faces a filter that separates the data into streams depending on its characteristics. For example, data about consumer behavior is likely to be filtered based on demographic factors. Data mining is not primarily about fancy graphs and visualization techniques, but it does employ them to show what it has found. It is known that we can absorb more statistical information visually than verbally and this format for presentation can be very persuasive and powerful if used in the right context.

As our civilization becomes increasingly data-saturated and sensors are distributed en masse into our local environments, we will inadvertently discover things that might be missed on the first pass over. Data mining will let us correct these mistakes and discover new insights based on past data, giving us more bang for our data storage buck.

Contents

This page left intentionally blank.

List of Contributors

Fernando Sánchez Lasheras

Department of Construction and Manufacturing Engineering, University of Oviedo, Gijón 33204, Spain

Paulino José García Nieto

Department of Mathematics, University of Oviedo, Oviedo 33007, Spain; E-Mail: lato@orion.ciencias.uniovi.es

Francisco Javier de Cos Juez

Department of Mining Engineering and Exploitation, University of Oviedo, Oviedo 33004, Spain; E-Mail: fjcos@uniovi.es

Ricardo Mayo Bayón

Department of Electrical Engineering, University of Oviedo, Gijón 33204, Spain; E-Mails: rmayo@uniovi.es (R.M.B.); vmsuarez@uniovi.es (V.M.G.S.)

Victor Manuel González Suárez

Department of Electrical Engineering, University of Oviedo, Gijón 33204, Spain; E-Mails: rmayo@uniovi.es (R.M.B.); vmsuarez@uniovi.es (V.M.G.S.)

This page left intentionally blank.

Chapter 1

DATA MINING

Data mining (the analysis step of the "Knowledge Discovery in Databases" process, or KDD), an interdisciplinary subfield of computer science, is the computational process of discovering patterns in large data sets involving methods at the intersection of artificial intelligence, machine learning, statistics, and database systems. The overall goal of the data mining process is to extract information from a data set and transform it into an understandable structure for further use. Aside from the raw analysis step, it involves database and data management aspects, data pre-processing, model and inference considerations, interestingness metrics, complexity considerations, post-processing of discovered structures, visualization, and online updating.

The term is a buzzword, and is frequently misused to mean any form of large-scale data or information processing (collection, extraction, warehousing, analysis, and statistics) but is also generalized to any kind of computer decision support system, including artificial intelligence, machine learning, and business intelligence. In the proper use of the word, the key term is *discovery*, commonly defined as "detecting something new". Even the popular book "Data mining : Practical machine learning tools and techniques with Java" (which covers mostly machine learning material) was originally to be named just "Practical machine learning", and the term "data mining" was only added for marketing reasons. Often the more general terms "(large scale) data analysis", or "analytics"-or when referring to actual methods, artificial intelligence and machine learning-are more appropriate.

The actual data mining task is the automatic or semi-automatic analysis of large quantities of data to extract previously unknown interesting patterns such as groups of data records (cluster analysis), unusual records (anomaly detection) and dependencies (association rule mining). Thi s usually involves using database techniques such as spatial indices. These patterns can then be seen as a kind of summary of the input data, and may be used in further analysis or, for example, in machine learning and predictive analytics. For example, the data mining step might identify multiple groups in the data, which can then be used to obtain more accurate prediction results by a decision support system. Neither the data collection, data preparation, nor result interpretation and reporting are part of the data mining step, but do belong to the overall KDD process as additional steps.

The related terms *data dredging, data fishing,* and *data snooping* refer to the use of data mining methods to sample parts of a larger population data set that are (or may be) too small for reliable statistical inferences to be made about the validity of any patterns discovered. These methods can, however, be used in creating new hypotheses to test against the larger data populations.

ETYMOLOGY

In the 1960s, statisticians used terms like "Data Fishing" or "Data Dredging" to refer to what they considered the bad practice of analyzing data without an a-priori hypothesis. The term "Data Mining" appeared around 1990 in the database community. At the beginning of the century, there was a phrase "database mining"™, trademarked by HNC, a San Diego-based company (now merged into *FICO*), to pitch their Data Mining Workstation; researchers consequently turned to "data mining". Other terms used include Data Archaeology, Information Harvesting, Information Discovery, Knowledge Extraction, etc. Gregory Piatetsky-Shapiro coined the term "Knowledge Discovery in Databases" for the first workshop on the same topic and this term became more popular in AI and Machine Learning Community. However, the term data mining became more popular in the business and press communities. Currently, Data Mining and Knowledge Discovery are used interchangeably.

BACKGROUND

The manual extraction of patterns from *data* has occurred for centuries. Early methods of identifying patterns in data include *Bayes' theorem* (1700s) and *regression analysis* (1800s). The proliferation, ubiquity and increasing power of computer technology has dramatically increased data collection, storage, and manipulation ability. As *data sets* have grown in size and complexity, direct "hands-on" data analysis has increasingly been augmented with indirect, automated data processing, aided by other discoveries in computer science, such as *neural networks, cluster analysis, genetic algorithms* (1950s), *decision trees* (1960s), and *support vector machines* (1990s). Data mining is the process of applying these methods with the intention of uncovering hidden patterns in large data sets. It bridges the gap from applied statistics and artificial intelligence (which usually provide the mathematical background) to database management by exploiting the way data is stored and indexed in databases to execute the actual learning and discovery algorithms more efficiently, allowing such methods to be applied to ever larger data sets.

RESEARCH AND EVOLUTION

The premier professional body in the field is the *Association for Computing Machinery*'s (ACM) Special Interest Group (SIG) on *Knowledge Discovery* and Data

Mining (*SIGKDD*). Since 1989 this ACM SIG has hosted an annual international conference and published its proceedings, and since 1999 it has published a biannual *academic journal* titled "SIGKDD Explorations".

Computer science conferences on data mining include :

- *CIKM Conference* — ACM *Conference on Information and Knowledge Management*
- *DMIN Conference* — International Conference on Data Mining
- *DMKD Conference* — Research Issues on Data Mining and Knowledge Discovery
- *ECDM Conference* — European Conference on Data Mining
- *ECML-PKDD Conference — European Conference on Machine Learning and Principles and Practice of Knowledge Discovery in Databases*
- *EDM Conference* — International Conference on Educational Data Mining
- *ICDM Conference — IEEEInternational Conference on Data Mining*
- *KDD Conference* — ACM SIGKDD *Conference on Knowledge Discovery and Data Mining*
- *MLDM Conference* — Machine Learning and Data Mining in Pattern Recognition
- *PAKDD Conference* — The annual *Pacific – Asia Conference on Knowledge Discovery and Data Mining*
- *PAW Conference* — Predictive Analytics World
- *SDM Conference — SIAM International Conference on Data Mining* (*SIAM*)
- *SSTD Symposium — Symposium on Spatial and Temporal Databases*
- *WSDM Conference* — ACM *Conference on Web Search and Data Mining*

Data mining topics are also present on many *data management/database conferences* such as the *ICDE Conference, SIGMOD Conference* and *International Conference on Very Large Data Bases.*

PROCESS

The **Knowledge Discovery in Databases (KDD) process** is commonly defined with the stages :

1. Selection
2. Pre-processing
3. Transformation
4. *Data Mining*
5. Interpretation/Evaluation.

It exists, however, in many variations on this theme, such as the *Cross Industry Standard Process for Data Mining* (CRISP-DM) which defines six phases :

1. Business Understanding
2. Data Understanding
3. Data Preparation
4. Modelling

5. Evaluation
6. Deployment.

or a simplified process such as (1) pre-processing, (2) data mining, and (3) results validation.

Polls conducted in 2002, 2004, and 2007 show that the CRISP-DM methodology is the leading methodology used by data miners. The only other data mining standard named in these polls was *SEMMA*. However, 3-4 times as many people reported using CRISP-DM. Several teams of researchers have published reviews of data mining process models, and Azevedo and Santos conducted a comparison of CRISP-DM and SEMMA in 2008.

PRE-PROCESSING

Before data mining algorithms can be used, a target data set must be assembled. As data mining can only uncover patterns actually present in the data, the target data set must be large enough to contain these patterns while remaining concise enough to be mined within an acceptable time limit. A common source for data is a *data mart* or *data warehouse*. Pre-processing is essential to analyze the *multivariate* data sets before data mining. The target set is then cleaned. *Data cleaning* removes the observations containing *noise* and those with *missing data.*

DATA MINING

Data mining involves six common classes of tasks :

- *Anomaly detection* (Outlier/change/deviation detection)—The identification of unusual data records, that might be interesting or data errors that require further investigation.
- *Association rule learning* (Dependency modelling)—Searches for relationships between variables. For example, a supermarket might gather data on customer purchasing habits. Using association rule learning, the supermarket can determine which products are frequently bought together and use this information for marketing purposes. This is sometimes referred to as market basket analysis.
- *Clustering*—is the task of discovering groups and structures in the data that are in some way or another "similar", without using known structures in the data.
- *Classification*—is the task of generalizing known structure to apply to new data. For example, an e—mail program might attempt to classify an e—mail as "legitimate" or as "spam".
- *Regression*—attempts to find a function which models the data with the least error.
- *Summarization*—providing a more compact representation of the data set, including visualization and report generation.

RESULTS VALIDATION

Data mining can unintentionally be misused, and can then produce results which appear to be significant; but which do not actually predict future behaviour and cannot be *reproduced* on a new sample of data and bear little use. Often this results from investigating too many hypotheses and not performing proper *statistical hypothesis testing*. A simple version of this problem in *machine learning* is known as *overfitting*, but the same problem can arise at different phases of the process and thus a train/test split-when applicable at all—may not be sufficient to prevent this from happening.

The final step of knowledge discovery from data is to verify that the patterns produced by the data mining algorithms occur in the wider data set. Not all patterns found by the data mining algorithms are necessarily valid. It is common for the data mining algorithms to find patterns in the training set which are not present in the general data set. This is called over fitting. To overcome this, the evaluation uses a test set of data on which the data mining algorithm was not trained. The learned patterns are applied to this test set, and the resulting output is compared to the desired output. For example, a data mining algorithm trying to distinguish "spam" from "legitimate" emails would be trained on a training set of sample e-mails. Once trained, the learned patterns would be applied to the test set of e-mails on which it had *not* been trained. The accuracy of the patterns can then be measured from how many e-mails they correctly classify. A number of statistical methods may be used to evaluate the algorithm, such as ROC curves.

If the learned patterns do not meet the desired standards, subsequently it is necessary to re-evaluate and change the pre-processing and data mining steps. If the learned patterns do meet the desired standards, then the final step is to interpret the learned patterns and turn them into knowledge.

STANDARDS

There have been some efforts to define standards for the data mining process, for example the 1999 European Cross Industry Standard Process for Data Mining and the 2004 Java Data Mining standard (JDM 1.0). Development on successors to these processes (CRISP-DM 2.0 and JDM 2.0) was active in 2006, but has stalled since. JDM 2.0 was withdrawn without reaching a final draft.

For exchanging the extracted models-in particular for use in predictive analytics—the key standard is the Predictive Model Markup Language (PMML), which is an XML—based language developed by the Data Mining Group (DMG) and supported as exchange format by many data mining applications. As the name suggests, it only covers prediction models, a particular data mining task of high importance to business applications. However, extensions to cover (for example) sub-space clustering have been proposed independently of the DMG.

NOTABLE USES

Games

Since the early 1960s, with the availability of *oracles* for certain *combinatorial games*, also called *tablebases* (*e.g.* for 3x3-chess) with any beginning configuration, small-board *dots-and-boxes*, small-board-hex, and certain endgames in chess, dots-and-boxes, and hex; a new area for data mining has been opened. This is the extraction of human-usable strategies from these oracles. Current pattern recognition approaches do not seem to fully acquire the high level of abstraction required to be applied successfully. Instead, extensive experimentation with the tablebases—combined with an intensive study of tablebase-answers to well designed problems, and with knowledge of prior art (*i.e.*, pre-tablebase knowledge)—is used to yield insightful patterns. *Berlekamp* (in dots-and-boxes, etc.) and *John Nunn* (in *chessendgames*) are notable examples of researchers doing this work, though they were not—and are not—involved in tablebase generation.

Business

Data mining is the analysis of historical business activities, stored as static data in data warehouse databases, to reveal hidden patterns and trends. Data mining software uses advanced pattern recognition algorithms to sift through large amounts of data to assist in discovering previously unknown strategic business information. Examples of what businesses use data mining for include performing market analysis to identify new product bundles, finding the root cause of manufacturing problems, to prevent customer attrition and acquire new customers, cross-sell to existing customers, and profile customers with more accuracy.

- In today's world raw data is being collected by companies at an exploding rate. For example, Walmart processes over 20 million point-of-sale transactions every day. This information is stored in a centralized database, but would be useless without some type of data mining software to analyse it. If Walmart analyzed their point-of-sale data with data mining techniques they would be able to determine sales trends, develop marketing campaigns, and more accurately predict customer loyalty.
- Every time a credit card or a store loyalty card is being used, or a warranty card is being filled, data is being collected about the users behaviour. Many people find the amount of information stored about us from companies, such as Google, Facebook, and Amazon, disturbing and are concerned about privacy. Although there is the potential for our personal data to be used in harmful, or unwanted, ways it is also being used to make our lives better. For example, Ford and Audi hope to one day collect information about customer driving patterns so they can recommend safer routes and warn drivers about dangerous road conditions.
- Data mining in *customer relationship management* applications can contribute significantly to the bottom line. Rather than randomly contacting a prospect or

customer through a call center or sending mail, a company can concentrate its efforts on prospects that are predicted to have a high likelihood of responding to an offer. More sophisticated methods may be used to optimize resources across campaigns so that one may predict to which channel and to which offer an individual is most likely to respond (across all potential offers). Additionally, sophisticated applications could be used to automate mailing. Once the results from data mining (potential prospect/customer and channel/offer) are determined, this "sophisticated application" can either automatically send an e-mail or a regular mail. Finally, in cases where many people will take an action without an offer, "*uplift modelling*" can be used to determine which people have the greatest increase in response if given an offer. Uplift modelling thereby enables marketers to focus mailings and offers on persuadable people, and not to send offers to people who will buy the product without an offer. *Data clustering* can also be used to automatically discover the segments or groups within a customer data set.

- Businesses employing data mining may see a return on investment, but also they recognize that the number of predictive models can quickly become very large. For example, rather than using one model to predict how many customers will *churn*, a business may choose to build a separate model for each region and customer type. In situations where a large number of models need to be maintained, some businesses turn to more automated data mining methodologies.
- Data mining can be helpful to human resources (HR) departments in identifying the characteristics of their most successful employees. Information obtained—such as universities attended by highly successful employees—can help HR focus recruiting efforts accordingly. Additionally, Strategic Enterprise Management applications help a company translate corporate-level goals, such as profit and margin share targets, into operational decisions, such as production plans and workforce levels.
- *Market basket analysis*, relates to data-mining use in retail sales. If a clothing store records the purchases of customers, a data mining system could identify those customers who favour silk shirts over cotton ones. Although some explanations of relationships may be difficult, taking advantage of it is easier. The example deals with *association rules* within transaction-based data. Not all data are transaction based and logical, or inexact *rules* may also be present within a *database*.
- Market basket analysis has been used to identify the purchase patterns of the *Alpha Consumer*. Analyzing the data collected on this type of user has allowed companies to predict future buying trends and forecast supply demands.
- Data mining is a highly effective tool in the catalog marketing industry. Catalogers have a rich database of history of their customer transactions for millions of customers dating back a number of years. Data mining tools can identify patterns among customers and help identify the most likely customers to respond to upcoming mailing campaigns.

- Data mining for business applications can be integrated into a complex modelling and decision making process. *Reactive business intelligence* (RBI) advocates a "holistic" approach that integrates data mining, *modelling*, and *interactive visualization* into an end-to-end discovery and continuous innovation process powered by human and automated learning.
- In the area of *decision making*, the *RBI* approach has been used to mine knowledge that is progressively acquired from the decision maker, and then self-tune the decision method accordingly. The relation between the quality of a data mining system and the amount of investment that the decision-maker is willing to make was formalized by providing an economic perspective on the value of "extracted knowledge" in terms of its payoff to the organization This decision-theoretic classification framework was applied to a real-world semi-conductor wafer manufacturing line, where decision rules for effectively monitoring and controlling the semi-conductor wafer fabrication line were developed.
- An example of data mining related to an integrated-circuit (IC) production line is described in the paper "Mining IC Test Data to Optimize VLSI Testing." In this paper, the application of data mining and decision analysis to the problem of die-level functional testing is described. Experiments mentioned demonstrate the ability to apply a system of mining historical die-test data to create a probabilistic model of patterns of die failure. These patterns are then utilized to decide, in real time, which die to test next and when to stop testing. This system has been shown, based on experiments with historical test data, to have the potential to improve profits on mature IC products. Other examples of the application of data mining methodologies in semi-conductor manufacturing environments suggest that data mining methodologies may be particularly useful when data is scarce, and the various physical and chemical parameters that affect the process exhibit highly complex interactions. Another implication is that on-line monitoring of the semi-conductor manufacturing process using data mining may be highly effective.
- Ford and Audi hope to one day collect information about customer driving patterns so they can recommend safer routes and warn drivers about dangerous road conditions.

Science and Engineering

In recent years, data mining has been used widely in the areas of science and engineering, such as *bioinformatics*, *genetics*, *medicine*, *education* and *electrical power* engineering.

- In the study of human genetics, *sequence mining* helps address the important goal of understanding the mapping relationship between the inter-individual variations in human *DNA* sequence and the variability in disease susceptibility. In simple terms, it aims to find out how the changes in an individual's DNA sequence affects the risks of developing common diseases such as *cancer*, which is of great importance to improving methods of diagnosing, preventing, and treating these diseases. One data mining method that is used to perform this task is known as *multifactor dimensionality reduction*.

- In the area of electrical power engineering, data mining methods have been widely used for *condition monitoring* of high voltage electrical equipment. The purpose of condition monitoring is to obtain valuable information on, for example, the status of the *insulation* (or other important safety-related parameters). *Data clustering* techniques—such as the *self-organizing map* (SOM), have been applied to vibration monitoring and analysis of transformer on-load tap-changers (OLTCS). Using vibration monitoring, it can be observed that each tap change operation generates a signal that contains information about the condition of the tap changer contacts and the drive mechanisms. Obviously, different tap positions will generate different signals. However, there was considerable variability amongst normal condition signals for exactly the same tap position. SOM has been applied to detect abnormal conditions and to hypothesize about the nature of the abnormalities.
- Data mining methods have been applied to *dissolved gas analysis* (DGA) in *power transformers*. DGA, as a diagnostics for power transformers, has been available for many years. Methods such as SOM has been applied to analyze generated data and to determine trends which are not obvious to the standard DGA ratio methods (such as Duval Triangle).
- In educational research, where data mining has been used to study the factors leading students to choose to engage in behaviours which reduce their learning, and to understand factors influencing university student retention. A similar example of social application of data mining is its use in *expertise finding systems*, whereby descriptors of human expertise are extracted, normalized, and classified so as to facilitate the finding of experts, particularly in scientific and technical fields. In this way, data mining can facilitate *institutional memory*.
- Data mining methods of *biomedical* data facilitated by domain *ontologies*, mining clinical trial data, and *traffic analysis* using SOM.
- In adverse drug reaction surveillance, the *Uppsala Monitoring Centre* has, since 1998, used data mining methods to routinely screen for reporting patterns indicative of emerging drug safety issues in the WHO global database of 4.6 million suspected *adverse drug reaction* incidents. Recently, similar methodology has been developed to mine large collections of *electronic health records* for temporal patterns associating drug prescriptions to medical diagnoses.
- Data mining has been applied *software* artifacts within the realm of *software engineering* : *Mining Software Repositories*.

Human Rights

Data mining of government records—particularly records of the justice system (*i.e.*, courts, prisons)—enables the discovery of systemic *human rights* violations in connection to generation and publication of invalid or fraudulent legal records by various government agencies.

Medical Data Mining

In 2011, the case of *Sorrell v. IMS Health, Inc.*, decided by the *Supreme Court of the United States*, ruled that *pharmacies* may share information with outside companies. This practice was authorized under the *1st Amendment of the Constitution*, protecting the "freedom of speech." However, the passage of the Health Information Technology for Economic and Clinical Health Act (HITECH Act) helped to initiate the adoption of the electronic health record (HER) and supporting technology in the United States. The HITECH Act was signed into law on February 17, 2009 as part of the American Recovery and Reinvestment Act and helped to open the door to medical data mining. Prior to the signing of this law, estimates of only 20% of United States based physician were utilizing electronic patient records. Søren Brunak notes that "the patient record becomes as information-rich as possible" and thereby "maximizes the data mining opportunities." Hence, electronic patient records further expands the possibilities regarding medical data mining thereby opening the door to a vast source of medical data analysis.

Spatial Data Mining

Spatial data mining is the application of data mining methods to spatial data. The end objective of spatial data mining is to find patterns in data with respect to geography. So far, data mining and *Geographic Information Systems* (GIS) have existed as two separate technologies, each with its own methods, traditions, and approaches to visualization and data analysis. Particularly, most contemporary GIS have only very basic spatial analysis functionality. The immense explosion in geographically referenced data occasioned by developments in IT, digital mapping, remote sensing, and the global diffusion of GIS emphasizes the importance of developing data-driven inductive approaches to geographical analysis and modelling.

Data mining offers great potential benefits for GIS-based applied decision-making. Recently, the task of integrating these two technologies has become of critical importance, especially as various public and private sector organizations possessing huge databases with thematic and geographically referenced data begin to realize the huge potential of the information contained therein. Among those organizations are :

- Offices requiring analysis or dissemination of geo-referenced statistical data
- Public health services searching for explanations of disease clustering
- Environmental agencies assessing the impact of changing land-use patterns on climate change
- Geo-marketing companies doing customer segmentation based on spatial location.

Challenges in Spatial Mining : Geospatial data repositories tend to be very large. Moreover, existing GIS data-sets are often splintered into feature and attribute components that are conventionally archived in hybrid data management systems.

Algorithmic requirements differ substantially for relational (attribute) data management and for topological (feature) data management. Related to this is the range and diversity of geographic data formats, which present unique challenges. The digital geographic data revolution is creating new types of data formats beyond the traditional "vector" and "raster" formats. Geographic data repositories increasingly include ill-structured data, such as imagery and geo-referenced multi-media.

There are several critical research challenges in geographic knowledge discovery and data mining. Miller and Han offer the following list of emerging research topics in the field :

- *Developing and supporting geographic data warehouses (GDW's)* : Spatial properties are often reduced to simple *aspatial* attributes in mainstream data warehouses. Creating an integrated GDW requires solving issues of spatial and temporal data interoperability—including differences in semantics, referencing systems, geometry, accuracy, and position.
- *Better spatio-temporal representations in geographic knowledge discovery* : Current geographic knowledge discovery (GKD) methods generally use very simple representations of geographic objects and spatial relationships. Geographic data mining methods should recognize more complex geographic objects (*i.e.*, lines and polygons) and relationships (*i.e.*, non-Euclidean distances, direction, connectivity, and interaction through attributed geographic space such as terrain). Furthermore, the time dimension needs to be more fully integrated into these geographic representations and relationships.
- *Geographic knowledge discovery using diverse data types* : GKD methods should be developed that can handle diverse data types beyond the traditional raster and vector models, including imagery and geo-referenced multimedia, as well as dynamic data types (video streams, animation).

Sensor Data Mining

Wireless sensor networks can be used for facilitating the collection of data for spatial data mining for a variety of applications such as air pollution monitoring. A characteristic of such networks is that nearby sensor nodes monitoring an environmental feature typically register similar values. This kind of data redundancy due to the spatial correlation between sensor observations inspires the techniques for in-network data aggregation and mining. By measuring the spatial correlation between data sampled by different sensors, a wide class of specialized algorithms can be developed to develop more efficient spatial data mining algorithms.

Visual Data Mining

In the process of turning from analogical into digital, large data sets have been generated, collected, and stored discovering statistical patterns, trends and information which is hidden in data, in order to build predictive patterns. Studies suggest visual data mining is faster and much more intuitive than is traditional data mining.

Music Data Mining

Data mining techniques, and in particular *co-occurrence* analysis, has been used to discover relevant similarities among music corpora (radio lists, CD databases) for purposes including classifying music into *genres* in a more objective manner.

Surveillance

Data mining has been used by the U.S. government. Programs include the *Total Information Awareness* (TIA) program, Secure Flight (formerly known as Computer-Assisted Passenger Prescreening System (*CAPPS II*)), Analysis, Dissemination, Visualization, Insight, Semantic Enhancement (*ADVISE*), and the Multi-state Anti-Terrorism Information Exchange (*MATRIX*). These programs have been discontinued due to controversy over whether they violate the 4th Amendment to the United States Constitution, although many programs that were formed under them continue to be funded by different organizations or under different names.

In the context of combating terrorism, two particularly plausible methods of data mining are "pattern mining" and "subject-based data mining".

Pattern Mining

"Pattern mining" is a data mining method that involves finding existing *patterns* in data. In this context *patterns* often means *association rules*. The original motivation for searching association rules came from the desire to analyze supermarket transaction data, that is, to examine customer behaviour in terms of the purchased products. For example, an association rule "beer ⇒ potato chips (80%)" states that four out of five customers that bought beer also bought potato chips.

In the context of pattern mining as a tool to identify terrorist activity, the *National Research Council* provides the following definition : "Pattern-based data mining looks for patterns (including anomalous data patterns) that might be associated with terrorist activity — these patterns might be regarded as small signals in a large ocean of noise." Pattern Mining includes new areas such a *Music Information Retrieval* (MIR) where patterns seen both in the temporal and non-temporal domains are imported to classical knowledge discovery search methods.

Subject-based Data Mining

"Subject-based data mining" is a data mining method involving the search for associations between individuals in data. In the context of combating terrorism, the *National Research Council* provides the following definition : "Subject-based data mining uses an initiating individual or other datum that is considered, based on other information, to be of high interest, and the goal is to determine what other persons or financial transactions or movements, etc., are related to that initiating datum."

Knowledge Grid

Knowledge discovery "On the Grid" generally refers to conducting knowledge discovery in an open environment using *grid computing* concepts, allowing users to integrate data from various online data sources, as well make use of remote resources, for executing their data mining tasks. The earliest example was the *Discovery Net*, developed at *Imperial College London*, which won the "Most Innovative Data-Intensive Application Award" at the ACM SC02 conference and exhibition, based on a demonstration of a fully interactive distributed knowledge discovery application for a bioinformatics application. Other examples include work conducted by researchers at the *University of Calabria*, who developed a Knowledge Grid architecture for distributed knowledge discovery, based on *grid computing*.

PRIVACY CONCERNS AND ETHICS

While the term "data mining" itself has no ethical implications, it is often associated with the mining of information in relation to peoples' behaviour (ethical and otherwise).

The ways in which data mining can be used can in some cases and contexts raise questions regarding privacy, legality, and ethics. In particular, data mining government or commercial data sets for national security or law enforcement purposes, such as in the *Total Information Awareness* Program or in *ADVISE*, has raised privacy concerns.

Data mining requires data preparation which can uncover information or patterns which may compromise confidentiality and privacy obligations. A common way for this to occur is through *data aggregation*. Data aggregation involves combining data together (possibly from various sources) in a way that facilitates analysis (but that also might make identification of private, individual-level data deducible or otherwise apparent). This is not data mining *per se*, but a result of the preparation of data before-and for the purposes of—the analysis. The threat to an individual's privacy comes into play when the data, once compiled, cause the data miner, or anyone who has access to the newly compiled data set, to be able to identify specific individuals, especially when the data were originally anonymous.

It is recommended that an individual is made aware of the following ***before*** *data are collected :*

- The purpose of the data collection and any (known) data mining projects;
- How the data will be used;
- Who will be able to mine the data and use the data and their derivatives;
- The status of security surrounding access to the data;
- How collected data can be updated.

Data may also be modified so as to *become* anonymous, so that individuals may not readily be identified. However, even "de-identified"/"anonymized" data sets can potentially contain enough information to allow identification of individuals, as occurred when journalists were able to find several individuals based on a set of search histories that were inadvertently released by AOL.

Situation in the United States

In the United States, privacy concerns have been addressed to some extent by the *US Congress* via the passage of regulatory controls such as the *Health Insurance Portability and Accountability Act* (HIPAA). The HIPAA requires individuals to give their "informed consent" regarding information they provide and its intended present and future uses. According to an article in *Biotech Business Week*', *"'[i]n practice, HIPAA may not offer any greater protection than the longstanding regulations in the research arena,' says the AAHC. More importantly, the rule's goal of protection through informed consent is undermined by the complexity of consent forms that are required of patients and participants, which approach a level of incomprehensibility to average individuals." This underscores the necessity for data anonymity in data aggregation and mining practices.*

Situation in Europe

Europe has rather strong privacy laws, and efforts are underway to further strengthen the rights of the consumers. However, the *U.S.-E.U. Safe Harbor Principles* currently effectively expose European users to privacy exploitation by U.S. companies. As a consequence of *Edward Snowden*'s *Global surveillance disclosure*, there has been increased discussion to revoke this agreement, as in particular the data will be fully exposed to the *National Security Agency*, and attempts to reach an agreement have failed.

ANOMALY DETECTION

In data mining, **anomaly detection** (or **outlier detection**) is the identification of items, events or observations which do not conform to an expected pattern or other items in a data-set. Typically the anomalous items will translate to some kind of problem such as bank fraud, a structural defect, medical problems or finding errors in text. Anomalies are also referred to as outliers, novelties, noise, deviations and exceptions.

In particular in the context of abuse and network intrusion detection, the interesting objects are often not *rare* objects, but unexpected *bursts* in activity. This pattern does not adhere to the common statistical definition of an outlier as a rare object, and many outlier detection methods (in particular unsupervised methods) will fail on such data, unless it has been aggregated appropriately. Instead, a cluster analysis algorithm may be able to detect the micro clusters formed by these patterns.

Three broad categories of anomaly detection techniques exist. **Unsupervised anomaly detection** techniques detect anomalies in an unlabelled test data set under the assumption that the majority of the instances in the data set are normal by looking for instances that seem to fit least to the remainder of the data set. **Supervised anomaly detection** techniques require a data set that has been labelled as "normal" and "abnormal" and involves training a classifier (the key difference to many other statistical classification problems is the inherent unbalanced nature of outlier detection). **Semi-supervised anomaly detection** techniques construct a model representing normal behaviour from a given *normal* training data set, and then testing the likelihood of a test instance to be generated by the learnt model.

Applications

Anomaly detection is applicable in a variety of domains, such as *intrusion detection, fraud detection,* fault detection, system health monitoring, event detection in sensor networks, and detecting Eco-system disturbances. It is often used in preprocessing to remove anomalous data from the data-set. In *supervised learning,* removing the anomalous data from the data-set often results in a statistically significant increase in accuracy.

Popular Techniques

Several anomaly detection techniques have been proposed in literature. Some of the popular techniques are :

- Distance based techniques (*k-nearest neighbour, local outlier factor*).
- One class *support vector machines.*
- Replicator *neural networks.*
- *Cluster analysis* based outlier detection.
- Pointing at records that deviate from *learned association rules.*

Fuzzy Logic Based Method

In recent years *Fuzzy Logic* has been adopted in several outlier detection approaches in order to improve results coming from popular *outlier detection* techniques. Yousri *et. al.* proposed an approach based on *fuzzy logic* in order to merge results obtained with an *outlier detection* method, which establishes if a pattern is an *outlier* and a *clustering* algorithm which provides results in allocating patterns to clusters. The two results, provided by the two different approaches, are then combined to give a meaure of outlierness. Another fuzzy based approach is proposed by Xue *et. al.* This approach, called Fuzzy Rough Semi Supervised Outlier Detection (FRSSOD), is the combination of two approaches : the Semi-Supervised Outlier Detection method (SSOD) and the Fuzzy Rough C-means clustering (FRCM). The objective of the FRSSOD method is to establish if a pattern under consideration can be considered as an *outlier.* Finally another fuzzy-based method, called Fuzzy Combination of Outlier Detection Techniques (FUCOT) combines different popular *outlier* detection methods exploiting the advantages of each of them while overcoming their drawbacks.

Application to Data Security

Anomaly detection was proposed for *Intrusion detection systems* (IDS) by *Dorothy Denning* in 1986. Anomaly detection for IDS is normally accomplished with thresholds and statistics, but can also be done with *Soft computing,* and inductive learning. Types of statistics proposed by 1999 included profiles of users, workstations, networks, remote hosts, groups of users, and programs based on frequencies, means, variances, covariances, and standard deviations. The counterpart of Anomaly detection in *Intrusion detection* is *Misuse Detection.*

ASSOCIATION RULE LEARNING

Association rule learning is a popular and well researched method for discovering interesting relations between variables in large databases. It is intended to identify strong rules discovered in databases using different measures of interestingness. Based on the concept of strong rules, Rakesh Agrawal *et. al.* introduced association rules for discovering regularities between products in large-scale transaction data recorded by point-of-sale (POS) systems in supermarkets. For example, the rule {onions, potatoes} = {burger} found in the sales data of a supermarket would indicate that if a customer buys onions and potatoes together, he or she is likely to also buy hamburger meat. Such information can be used as the basis for decisions about marketing activities such as, *e.g.,* promotional pricing or product placements. In addition to the above example from market basket analysis association rules are employed today in many application areas including Web usage mining, intrusion detection, Continuous production, and bioinformatics. As opposed to sequence mining, association rule learning typically does not consider the order of items either within a transaction or across transactions.

Definition

Table : Example database with 4 items and 5 transactions.

Transaction ID	*Milk*	*Bread*	*Butter*	*Beer*
1	1	1	0	0
2	0	0	1	0
3	0	0	0	1
4	1	1	1	0
5	0	1	0	0

Following the original definition by Agrawal *et. al.* the problem of association rule mining is defined as : Let $I = \{i_1, i_2, \ldots, i_n\}$ be a set of n binary attributes called *items*. Let $D = \{t_1, t_2, \ldots, t_m\}$ be a set of transactions called the *database*. Each transaction in D has a unique transaction ID and contains a subset of the items in I. A *rule* is defined as an implication of the form $X \Rightarrow Y$ where $X, Y \subseteq I$ and $X \cap Y = \emptyset$. The sets of items (for short *item-sets*) X and Y are called *antecedent* (left-hand-side or LHS) and *consequent* (right-hand-side or RHS) of the rule respectively.

To illustrate the concepts, we use a small example from the supermarket domain. The set of items is I = {milk, bread, butter, beer} and a small database containing the items (1 codes presence and 0 absence of an item in a transaction) is shown in the table. An example rule for the supermarket could be {butter, bread} ⇒ {milk} meaning that if butter and bread are bought, customers also buy milk.

Note : This example is extremely small. In practical applications, a rule needs a support of several hundred transactions before it can be considered statistically significant, and data-sets often contain thousands or millions of transactions.

Useful Concepts

To select interesting rules from the set of all possible rules, constraints on various measures of significance and interest can be used. The best-known constraints are minimum thresholds on support and confidence.

- The *support* supp(X) of an item-set X is defined as the proportion of transactions in the data set which contain the item-set. In the example database, the item-set {milk, bread, butter} has a support of 1/5 = 0.2 since it occurs in 20% of all transactions (1 out of 5 transactions).
- The *confidence* of a rule is defined $\text{conf}(X \Rightarrow Y) = \text{supp}(X \cup Y)/\text{supp}(X)$. For example, the rule {butter, bread} ⇒ {milk} has a confidence of 0.2/0.2 = 1.0 in the database, which means that for 100% of the transactions containing butter and bread the rule is correct (100% of the times a customer buys butter and bread, milk is bought as well). Be careful when reading the expression : here supp(X ∪ Y) means "*support for occurrences of transactions where* ***X and Y both appear***", not "*support for occurrences of transactions where* ***either X or Y appears***", the latter interpretation arising because set union is equivalent to logical disjunction. The argument of supp() is a set of preconditions, and thus becomes more restrictive as it grows (instead of more inclusive).
- Confidence can be interpreted as an estimate of the probability $P(Y \mid X)$, the probability of finding the RHS of the rule in transactions under the condition that these transactions also contain the LHS.
- The *lift* of a rule is defined as $\text{lift}(X \Rightarrow Y) = \dfrac{\text{supp}(X \cup Y)}{\text{supp}(X) \times \text{supp}(Y)}$ or the

ratio of the observed support to that expected if X and Y were independent. The rule {milk, bread} ⇒ {butter} has a lift of $\dfrac{0.2}{0.4 \times 0.4} = 1.25$.

- The *conviction* of a rule is defined as $\text{conv}(X \Rightarrow Y)\ \dfrac{1-\text{supp}(Y)}{1-\text{conf}(X \Rightarrow Y)}$. The rule

{milk, bread} ⇒ {butter} has a conviction of $\dfrac{1-0.4}{1-.5} = 1.2$, and can be interpreted as the ratio of the expected frequency that X occurs without Y (that is to say, the frequency that the rule makes an incorrect prediction) if X and Y were independent divided by the observed frequency of incorrect predictions. In this example, the conviction value of 1.2 shows that the rule {milk, bread} ⇒ {butter} would be incorrect 20% more often (1.2 times as often) if the association between X and Y was purely random chance.

Process

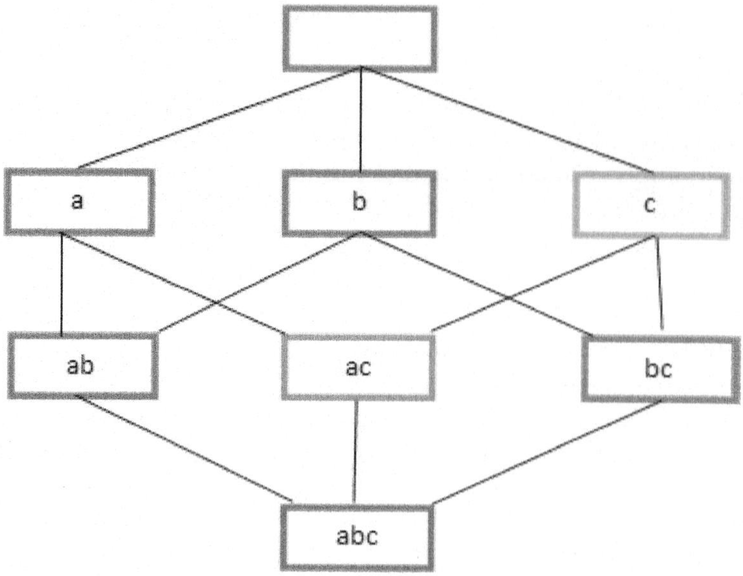

Fig.

Frequent item-set lattice, where the colour of the box indicates how many transactions contain the combination of items. Note that lower levels of the lattice can contain at most the minimum number of their parents' items; *e.g.* {ac} can have only at most min (a, c) items. This is called the *downward-closure property*.

Association rules are usually required to satisfy a user-specified minimum support and a user-specified minimum confidence at the same time. Association rule generation is usually split up into two separate steps :

1. First, minimum support is applied to find all *frequent item-sets* in a database.
2. Second, these frequent item-sets and the minimum confidence constraint are used to form rules.

While the second step is straightforward, the first step needs more attention.

Finding all frequent item-sets in a database is difficult since it involves searching all possible item-sets (item combinations). The set of possible item-sets is the power set over I and has size $2^n - 1$ (excluding the empty set which is not a valid item-set). Although the size of the powerset grows exponentially in the number of items n in I, efficient search is possible using the ***downward-closure property*** of support (also called *anti-monotonicity*) which guarantees that for a frequent item-set, all its subsets are also frequent and thus for an infrequent item-set, all its supersets must also be infrequent. Exploiting this property, efficient algorithms (*e.g.*, Apriori and Eclat) can find all frequent item-sets.

History

The concept of association rules was popularised particularly due to the 1993 article of Agrawal *et. al.*, which has acquired more than 6000 citations according to Google Scholar, as of March 2008, and is thus one of the most cited papers in the Data Mining field. However, it is possible that what is now called "association rules" is similar to what appears in the 1966 paper on GUHA, a general data mining method developed by Petr Hájek *et. al.*

Alternative Measures of Interestingness

In addition to confidence, other measures of *interestingness* for rules have been proposed. Some popular measures are :

- All-confidence
- Collective strength
- Conviction
- Leverage
- Lift (originally called interest).

A definition of these measures can be found here. Several more measures are presented and compared by Tan *et. al.* Looking for techniques that can model what the user has known (and using this models as interestingness measures) is currently an active research trend under the name of "Subjective Interestingness"

Statistically Sound Associations

One limitation of the standard approach to discovering associations is that by searching massive numbers of possible associations to look for collections of items that appear to be associated, there is a large risk of finding many spurious associations. These are collections of items that co-occur with unexpected frequency in the data, but only do so by chance. For example, suppose we are considering a collection of 10,000 items and looking for rules containing two items in the left-hand-side and 1 item in the right-hand-side. There are approximately 1,000,000,000,000 such rules. If we apply a statistical test for independence with a significance level of 0.05 it means there is only a 5% chance of accepting a rule if there is no association. If we assume there are no associations, we should nonetheless expect to find 50,000,000,000 rules. Statistically sound association discovery controls this risk, in most cases reducing the risk of finding *any* spurious associations to a user-specified significance level.

Algorithms

Many algorithms for generating association rules were presented over time.

Some well known algorithms are Apriori, Eclat and FP-Growth, but they only do half the job, since they are algorithms for mining frequent item-sets. Another step needs to be done after to generate rules from frequent item-sets found in a database.

Apriori Algorithm

Apriori is the best-known algorithm to mine association rules. It uses a breadth-first search strategy to count the support of item-sets and uses a candidate generation function which exploits the downward closure property of support.

Eclat Algorithm

Eclat (alt. ECLAT, stands for Equivalence Class Transformation) is a depth-first search algorithm using set intersection.

FP-growth Algorithm

FP stands for frequent pattern.

In the first pass, the algorithm counts occurrence of items (attribute-value pairs) in the data-set, and stores them to 'header table'. In the second pass, it builds the FP-tree structure by inserting instances. Items in each instance have to be sorted by descending order of their frequency in the data-set, so that the tree can be processed quickly. Items in each instance that do not meet minimum coverage threshold are discarded. If many instances share most frequent items, FP-tree provides high compression close to tree root.

Recursive processing of this compressed version of main data-set grows large item-sets directly, instead of generating candidate items and testing them against the entire database. Growth starts from the bottom of the header table (having longest branches), by finding all instances matching given condition. New tree is created, with counts projected from the original tree corresponding to the set of instances that are conditional on the attribute, with each node getting sum of its children counts. Recursive growth ends when no individual items conditional on the attribute meet minimum support threshold, and processing continues on the remaining header items of the original FP-tree.

Once the recursive process has completed, all large item-sets with minimum coverage have been found, and association rule creation begins.

Others

Node-set-based Algorithms

FIN, PrePost and PPV are three algorithms based on node sets. They use nodes in a coding FP-tree to represent item-sets, and employ a depth-first search strategy to discovery frequent item-sets using "intersection" of node sets.

GUHA Procedure ASSOC

GUHA is a general method for exploratory data analysis that has theoretical foundations in observational calculi.

The ASSOC procedure is a GUHA method which mines for generalized association rules using fast bitstrings operations. The association rules mined by this method are more general than those output by apriori, for example "items" can be connected both with conjunction and disjunctions and the relation between antecedent and consequent of the rule is not restricted to setting minimum support and confidence as in apriori : an arbitrary combination of supported interest measures can be used.

OPUS Search

OPUS is an efficient algorithm for rule discovery that, in contrast to most alternatives, does not require either monotone or anti-monotone constraints such as minimum support. Initially used to find rules for a fixed consequent it has subsequently been extended to find rules with any item as a consequent. OPUS search is the core technology in the popular Magnum Opus association discovery system.

Lore

A famous story about association rule mining is the "beer and diaper" story. A purported survey of behaviour of supermarket shoppers discovered that customers (presumably young men) who buy diapers tend also to buy beer. This anecdote became popular as an example of how unexpected association rules might be found from everyday data. There are varying opinions as to how much of the story is true. Daniel Powers says :

In 1992, Thomas Blischok, manager of a retail consulting group at Teradata, and his staff prepared an analysis of 1.2 million market baskets from about 25 Osco Drug stores. Database queries were developed to identify affinities. The analysis "did discover that between 5 :00 and 7 :00 p.m. that consumers bought beer and diapers". Osco managers did NOT exploit the beer and diapers relationship by moving the products closer together on the shelves.

Other Types of Association Mining

Contrast set learning is a form of associative learning. **Contrast set learners** use rules that differ meaningfully in their distribution across subsets.

Weighted class learning is another form of associative learning in which weight may be assigned to classes to give focus to a particular issue of concern for the consumer of the data mining results.

High-order pattern discovery facilitate the capture of high-order (polythetic) patterns or event associations that are intrinsic to complex real-world data.

K-optimal pattern discovery provides an alternative to the standard approach to association rule learning that requires that each pattern appear frequently in the data.

Generalized Association Rules hierarchical taxonomy (concept hierarchy)

Quantitative Association Rules categorical and quantitative data

Interval Data Association Rules *e.g.* partition the age into 5-year-increment ranged

Maximal Association Rules

Sequential pattern mining discovers subsequences that are common to more than minsup sequences in a sequence database, where minsup is set by the user. A sequence is an ordered list of transactions.

Sequential Rules discovering relationships between items while considering the time ordering. It is generally applied on a sequence database. For example, a sequential rule found in database of sequences of customer transactions can be that customers who bought a computer and CD-Roms, later bought a webcam, with a given confidence and support.

Warmr is shipped as part of the ACE data mining suite. It allows association rule learning for first order relational rules.

STATISTICAL CLASSIFICATION

In *machine learning* and *statistics*, **classification** is the problem of identifying to which of a set of *categories* (sub-populations) a new *observation* belongs, on the basis of a *training set* of data containing observations (or instances) whose category membership is known. The individual observations are analyzed into a set of quantifiable properties, known as various *explanatory variables*, *features*, etc. These properties may variously be *categorical* (*e.g.* "A", "B", "AB" or "O", for *blood type*), *ordinal* (*e.g.* "large", "medium" or "small"), *integer-valued* (*e.g.* the number of occurrences of a part word in an *email*) or *real-valued* (*e.g.* a measurement of *blood pressure*). Some *algorithms* work only in terms of discrete data and require that real-valued or integer-valued data be *discretized* into groups (*e.g.* less than 5, between 5 and 10, or greater than 10). An example would be assigning a given email into "spam" or "non-spam" classes or assigning a diagnosis to a given patient as described by observed characteristics of the patient (gender, blood pressure, presence or absence of certain symptoms, etc.).

An algorithm that implements classification, especially in a concrete implementation, is known as a *classifier*. The term "classifier" sometimes also refers to the mathematical *function*, implemented by a classification algorithm, that maps input data to a category.

In the terminology of machine learning, classification is considered an instance of *supervised learning*, *i.e.* learning where a training set of correctly identified observations is available. The corresponding *unsupervised* procedure is known as *clustering* or *cluster analysis*, and involves grouping data into categories based on some measure of inherent similarity (*e.g.* the *distance* between instances, considered as vectors in a multi-dimensional *vector space*).

Terminology across fields is quite varied. In *statistics*, where classification is often done with *logistic regression* or a similar procedure, the properties of observations are termed *explanatory variables* (or *independent variables*, regressors, etc.), and the categories to be predicted are known as outcomes, which are considered to be possible values of the *dependent variable*. In machine learning, the observations are often known as *instances*, the explanatory variables are termed *features* (grouped into a *feature vector*), and the possible categories to be predicted are *classes*. There is also some argument over whether classification methods that do not involve a *statistical model* can be considered "statistical".

Relation to Other Problems

Classification and clustering are examples of the more general problem of *pattern recognition*, which is the assignment of some sort of output value to a given input value. Other examples are *regression*, which assigns a real-valued output to each input; *sequence labelling*, which assigns a class to each member of a sequence of values (for example, *part of speech tagging*, which assigns a *part of speech* to each word in an input sentence); *parsing*, which assigns a *parse tree* to an input sentence, describing the *syntactic structure* of the sentence; etc.

A common sub-class of classification is **probabilistic classification**. Algorithms of this nature use *statistical inference* to find the best class for a given instance. Unlike other algorithms, which simply output a "best" class, probabilistic algorithms output a *probability* of the instance being a member of each of the possible classes. The best class is normally then selected as the one with the highest probability. However, such an algorithm has numerous advantages over non-probabilistic classifiers :

- It can output a confidence value associated with its choice (in general, a classifier that can do this is known as a *confidence-weighted classifier*).
- Correspondingly, it can *abstain* when its confidence of choosing any particular output is too low.
- Because of the probabilities which are generated, probabilistic classifiers can be more effectively incorporated into larger machine-learning tasks, in a way that partially or completely avoids the problem of *error propagation*.

Frequentist Procedures

Early work on statistical classification was undertaken by Fisher, in the context of two-group problems, leading to *Fisher's linear discriminant* function as the rule for assigning a group to a new observation. This early work assumed that data-values within each of the two groups had a *multivariate normal distribution*. The extension of this same context to more than two-groups has also been considered with a restriction imposed that the classification rule should be *linear*. Later work for the multivariate normal distribution allowed the classifier to be *non-linear* : several classification rules can be derived based on slight different adjustments of the *Mahalanobis distance*, with a new observation being assigned to the group whose centre has the lowest adjusted distance from the observation.

Bayesian Procedures

Unlike frequentist procedures, Bayesian classification procedures provide a natural way of taking into account any available information about the relative sizes of the sub-populations associated with the different groups within the overall population. Bayesian procedures tend to be computationally expensive and, in the days before *Markov chain Monte Carlo* computations were developed, approximations for Bayesian clustering rules were devised.

Some Bayesian procedures involve the calculation of *group membership probabilities* : these can be viewed as providing a more informative outcome of a data analysis than a simple attribution of a single group-label to each new observation.

Binary and Multiclass Classification

Classification can be thought of as two separate problems-*binary classification* and *multiclass classification*. In binary classification, a better understood task, only two classes are involved, whereas multiclass classification involves assigning an object to one of several classes. Since many classification methods have been developed specifically for binary classification, multiclass classification often requires the combined use of multiple binary classifiers.

Feature Vectors

Most algorithms describe an individual instance whose category is to be predicted using a *feature vector* of individual, measurable properties of the instance. Each property is termed a *feature*, also known in statistics as an *explanatory variable* (or *independent variable*, although in general different features may or may not be *statistically independent*). Features may variously be *binary* ("male" or "female"); *categorical* (*e.g.* "A", "B", "AB" or "O", for *blood type*); *ordinal* (*e.g.* "large", "medium" or "small"); *integer-valued* (*e.g.* the number of occurrences of a particular word in an email); or *real-valued* (*e.g.* a measurement of blood pressure). If the instance is an image, the feature values might correspond to the pixels of an image; if the instance is a piece of text, the feature values might be occurrence frequencies of different words. Some algorithms work only in terms of discrete data and require that real-valued or integer-valued data be *discretized* into groups (*e.g.* less than 5, between 5 and 10, or greater than 10).

The *vector space* associated with these vectors is often called the *feature space*. In order to reduce the dimensionality of the feature space, a number of *dimensionality reduction* techniques can be employed.

Linear Classifiers

A large number of *algorithms* for classification can be phrased in terms of a *linear function* that assigns a score to each possible category k by *combining* the feature vector of an instance with a vector of weights, using a *dot product*. The predicted category is the one with the highest score. This type of score function is known as a *linear predictor function* and has the following general form :

$$\text{score}(X_i, k) = \beta_k \cdot X_i,$$

where X_i is the feature vector for instance i, β_k is the vector of weights corresponding to category k, and score (X_i, k) is the score associated with assigning instance i to category k. In *discrete choice* theory, where instances represent people and categories represent choices, the score is considered the *utility* associated with person i choosing category k.

Algorithms with this basic setup are known as *linear classifiers*. What distinguishes them is the procedure for determining (training) the optimal weights/coefficients and the way that the score is interpreted.

Examples of such algorithms are :

- *Logistic regression* and *multinomial logit*
- *Probit regression*
- The *perceptron* algorithm
- *Support vector machines*
- *Linear discriminant analysis*.

Algorithms

The most widely used classifiers are the *neural network* (multi-layer perceptron), *support vector machines, k-nearest neighbours,* Gaussian mixture model, Gaussian, *naive Bayes, decision tree* and *RBF* classifiers.

Examples of classification algorithms include :

- Linear classifiers
  - Fisher's linear discriminant
  - Logistic regression
  - Naive Bayes classifier
  - Perceptron
- Support vector machines
  - Least squares support vector machines
- Quadratic classifiers
- Kernel estimation
  - k-nearest neighbour
- Boosting (meta-algorithm)
- Decision trees
  - Random forests
- Neural networks
- *Gene Expression Programming*

- *Bayesian networks*
- *Hidden Markov models*
- *Learning vector quantization.*

Evaluation

Classifier performance depends greatly on the characteristics of the data to be classified. There is no single classifier that works best on all given problems (a phenomenon that may be explained by the *no-free-lunch theorem*). Various empirical tests have been performed to compare classifier performance and to find the characteristics of data that determine classifier performance. Determining a suitable classifier for a given problem is however still more an art than a science.

The measures *precision and recall* are popular metrics used to evaluate the quality of a classification system. More recently, *receiver operating characteristic* (ROC) curves have been used to evaluate the tradeoff between true-and false-positive rates of classification algorithms.

As a performance metric, the *uncertainty coefficient* has the advantage over simple *accuracy* in that it is not affected by the relative sizes of the different classes. Further, it will not penalize an algorithm for simply *rearranging* the classes.

REGRESSION ANALYSIS

In *statistics*, **regression analysis** is a statistical process for estimating the relationships among variables. It includes many techniques for modelling and analyzing several variables, when the focus is on the relationship between a *dependent variable* and one or more *independent variables*. More specifically, regression analysis helps one understand how the typical value of the dependent variable (or 'criterion variable') changes when any one of the independent variables is varied, while the other independent variables are held fixed. Most commonly, regression analysis estimates the *conditional expectation* of the dependent variable given the independent variables—that is, the *average value* of the dependent variable when the independent variables are fixed. Less commonly, the focus is on a *quantile*, or other *location parameter* of the conditional distribution of the dependent variable given the independent variables. In all cases, the estimation target is a *function* of the independent variables called the **regression function**. In regression analysis, it is also of interest to characterize the variation of the dependent variable around the regression function which can be described by a *probability distribution*.

Regression analysis is widely used for *prediction* and *forecasting*, where its use has substantial overlap with the field of *machine learning*. Regression analysis is also used to understand which among the independent variables are related to the dependent variable, and to explore the forms of these relationships. In restricted circumstances, regression analysis can be used to infer *causal relationships* between the independent and dependent variables. However this can lead to illusions or false relationships, so caution is advisable; for example, *correlation does not imply causation*.

Many techniques for carrying out regression analysis have been developed. Familiar methods such as *linear regression* and *ordinary least squares* regression are *parametric*, in that the regression function is defined in terms of a finite number of unknown *parameters* that are estimated from the *data*. *Non-parametric regression* refers to techniques that allow the regression function to lie in a specified set of *functions*, which may be *infinite-dimensional*.

The performance of regression analysis methods in practice depends on the form of the *data generating process*, and how it relates to the regression approach being used. Since the true form of the data-generating process is generally not known, regression analysis often depends to some extent on making assumptions about this process. These assumptions are sometimes testable if a sufficient quantity of data is available. Regression models for prediction are often useful even when the assumptions are moderately violated, although they may not perform optimally. However, in many applications, especially with small *effects* or questions of *causality* based on *observational data*, regression methods can give misleading results.

History

The earliest form of regression was the *method of least squares*, which was published by *Legendre* in 1805, and by *Gauss* in 1809. Legendre and Gauss both applied the method to the problem of determining, from astronomical observations, the orbits of bodies about the Sun (mostly comets, but also later the then newly discovered minor planets). Gauss published a further development of the theory of least squares in 1821, including a version of the *Gauss–Markov theorem*.

The term "regression" was coined by *Francis Galton* in the nineteenth century to describe a biological phenomenon. The phenomenon was that the heights of descendants of tall ancestors tend to regress down towards a normal average (a phenomenon also known as *regression toward the mean*). For Galton, regression had only this biological meaning, but his work was later extended by *Udny Yule* and *Karl Pearson* to a more general statistical context. In the work of Yule and Pearson, the joint distribution of the response and explanatory variables is assumed to be *Gaussian*. This assumption was weakened by *R.A. Fisher* in his works of 1922 and 1925. Fisher assumed that the conditional distribution of the response variable is Gaussian, but the joint distribution need not be. In this respect, Fisher's assumption is closer to Gauss's formulation of 1821.

In the 1950s and 1960s, economists used electromechanical desk calculators to calculate regressions. Before 1970, it sometimes took up to 24 hours to receive the result from one regression.

Regression methods continue to be an area of active research. In recent decades, new methods have been developed for *robust regression*, regression involving correlated responses such as *time series* and *growth curves*, regression in which the predictor or response variables are curves, images, graphs, or other complex data objects, regression methods accommodating various types of missing data, *non-parametric regression*, *Bayesian* methods for regression, regression in which the predictor variables are measured with error, regression with more predictor variables than observations, and causal inference with regression.

Regression Models

Regression models involve the following variables :

- The **unknown parameters**, denoted as β, which may represent a *scalar* or a *vector*.
- The **independent variables, X.**
- The **dependent variable**, Y.

In various *fields of application*, different terminologies are used in place of *dependent and independent variables*.

A regression model relates Y to a function of $\mathbf{X}$ and β.

$$Y \approx f(X, \beta)$$

The approximation is usually formalized as $E(Y \mid \mathbf{X}) = f(\mathbf{X}, \beta)$. To carry out regression analysis, the form of the function f must be specified. Sometimes the form of this function is based on knowledge about the relationship between Y and $\mathbf{X}$ that does not rely on the data. If no such knowledge is available, a flexible or convenient form for f is chosen.

Assume now that the vector of unknown parameters β is of length k. In order to perform a regression analysis the user must provide information about the dependent variable Y :

- If N data points of the form $(Y, \mathbf{X})$ are observed, where $N<k$, most classical approaches to regression analysis cannot be performed : since the system of equations defining the regression model is underdetermined, there are not enough data to recover β.
- If exactly $N = k$ data points are observed, and the function f is linear, the equations $Y = f(\mathbf{X}, \beta)$ can be solved exactly rather than approximately. This reduces to solving a set of N equations with N unknowns (the elements of β), which has a unique solution as long as the $\mathbf{X}$ are linearly independent. If f is non-linear, a solution may not exist, or many solutions may exist.
- The most common situation is where $N>k$ data points are observed. In this case, there is enough information in the data to estimate a unique value for β that best fits the data in some sense, and the regression model when applied to the data can be viewed as an *overdetermined system* in β.

In the last case, the regression analysis provides the tools for :

1. Finding a solution for unknown parameters β that will, for example, minimize the distance between the measured and predicted values of the dependent variable Y (also known as method of *least squares*).

2. Under certain statistical assumptions, the regression analysis uses the surplus of information to provide statistical information about the unknown parameters β and predicted values of the dependent variable Y.

Necessary Number of Independent Measurements

Consider a regression model which has three unknown parameters, β_0, β_1, and β_2. Suppose an experimenter performs 10 measurements all at exactly the same value of independent variable vector **X** (which contains the independent variables X_1, X_2, and X_3). In this case, regression analysis fails to give a unique set of estimated values for the three unknown parameters; the experimenter did not provide enough information. The best one can do is to estimate the average value and the standard deviation of the dependent variable Y. Similarly, measuring at two different values of **X** would give enough data for a regression with two unknowns, but not for three or more unknowns.

If the experimenter had performed measurements at three different values of the independent variable vector **X**, then regression analysis would provide a unique set of estimates for the three unknown parameters in β.

In the case of *general linear regression*, the above statement is equivalent to the requirement that the matrix $\mathbf{X}^T\mathbf{X}$ is *invertible*.

Statistical Assumptions

When the number of measurements, N, is larger than the number of unknown parameters, k, and the measurement errors ε_i are normally distributed then *the excess of information* contained in $(N - k)$ measurements is used to make statistical predictions about the unknown parameters. This excess of information is referred to as the *degrees of freedom* of the regression.

Underlying Assumptions

Classical assumptions for regression analysis include :

- The sample is representative of the population for the inference prediction.
- The error is a *random variable* with a mean of zero conditional on the explanatory variables.
- The independent variables are measured with no error. (Note : If this is not so, modelling may be done instead using *errors-in-variables model* techniques).
- The predictors are *linearly independent, i.e.* it is not possible to express any predictor as a linear combination of the others.
- The errors are *uncorrelated,* that is, the *variance–covariance matrix* of the errors is *diagonal* and each non-zero element is the variance of the error.
- The variance of the error is constant across observations (*homoscedasticity*). If not, *weighted least squares* or other methods might instead be used.

These are sufficient conditions for the least-squares estimator to possess desirable properties; in particular, these assumptions imply that the parameter estimates will be *unbiased, consistent,* and *efficient* in the class of linear unbiased estimators. It is important to note that actual data rarely satisfies the assumptions. That is, the method is used even though the assumptions are not true. Variation from the assumptions can sometimes be used as a measure of how far the model is from being useful. Many of these assumptions may be relaxed in more advanced treatments. Reports of statistical analyses usually include analyses of tests on the sample data and methodology for the fit and usefulness of the model.

Assumptions include the geometrical support of the variables. Independent and dependent variables often refer to values measured at point locations. There may be spatial trends and spatial autocorrelation in the variables that violate statistical assumptions of regression. Geographic weighted regression is one technique to deal with such data. Also, variables may include values aggregated by areas. With aggregated data the *modifiable areal unit problem* can cause extreme variation in regression parameters. When analyzing data aggregated by political boundaries, postal codes or census areas results may be very distinct with a different choice of units.

Linear Regression

In linear regression, the model specification is that the dependent variable, y_i is a *linear combination* of the *parameters* (but need not be linear in the *independent variables*). For example, in *simple linear regression* for modelling n data points there is one independent variable : x_i, and two parameters, β_0 and β_1 :

straight line : $y_i = \beta_0 + \beta_1 x_1 + \varepsilon_i, \quad i = 1,\ldots, n.$

In multiple linear regression, there are several independent variables or functions of independent variables.

Adding a term in x_i^2 to the preceding regression gives :

parabola : $y_i = \beta_0 + \beta_1 x_i + \beta_2 x_i^2 + \varepsilon_i, i = 1,\ldots, n.$

This is still linear regression; although the expression on the right hand side is quadratic in the independent variable x_i, it is linear in the parameters β_0, β_1 and β_2.

In both cases, ε_i is an error term and the subscript i indexes a particular observation.

Given a random sample from the population, we estimate the population parameters and obtain the sample linear regression model :

$$\hat{y}_i = \hat{\beta}_0 + \hat{\beta}_1 x_i.$$

The *residual,* $e_i = y_i - \hat{y}_i$, is the difference between the value of the dependent variable predicted by the model, $\hat{y}_i$, and the true value of the dependent variable, y_i. One method of estimation is *ordinary least squares*. This method obtains parameter estimates that minimize the sum of squared *residuals,* SSE, also sometimes denoted *RSS* :

$$SSE = \sum_{i=1}^{n} e_i^2.$$

Minimization of this function results in a set of *normal equations*, a set of simultaneous linear equations in the parameters, which are solved to yield the parameter estimators, $\hat{\beta}_0, \hat{\beta}_1$.

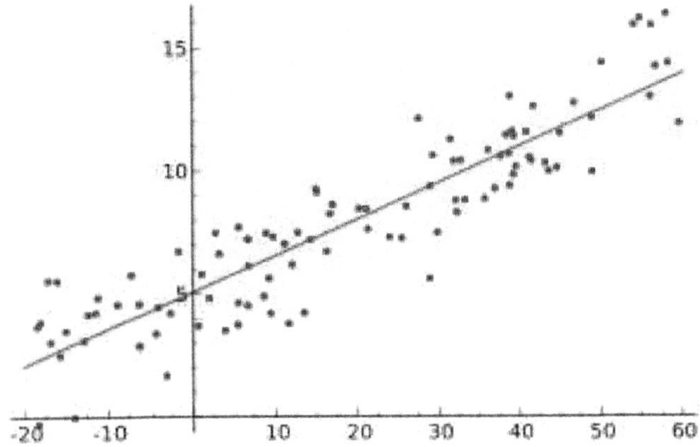

Fig. : Illustration of linear regression on a data set.

In the case of simple regression, the formulas for the least squares estimates are

$$\hat{\beta}_1 = \frac{\sum (x_i - \bar{x})(y_i - \bar{y})}{\sum (x_i - \bar{x})^2} \text{ and } \hat{\beta}_0 = \bar{y} - \hat{\beta}_1 \bar{x}$$

where $\bar{x}$ is the *mean* (average) of the x values and $\bar{y}$ is the mean of the y values.

Under the assumption that the population error term has a constant variance, the estimate of that variance is given by :

$$\hat{\sigma}_\varepsilon^2 = \frac{SSE}{n-2}.$$

This is called the *mean square error* (MSE) of the regression. The denominator is the sample size reduced by the number of model parameters estimated from the same data, (n-p) for *pregressors* or (n-p-1) if an intercept is used. In this case, p=1 so the denominator is n-2.

The *standard errors* of the parameter estimates are given by

$$\hat{\sigma}_{\beta_0} = \hat{\sigma}_\varepsilon \sqrt{\frac{1}{n} + \frac{\bar{x}^2}{\sum (x_i - \bar{x})^2}}$$

$$\hat{\sigma}_{\beta_1} = \hat{\sigma}_\varepsilon \sqrt{\frac{1}{\sum (x_i - \bar{x})^2}}.$$

Under the further assumption that the population error term is normally distributed, the researcher can use these estimated standard errors to create *confidence intervals* and conduct *hypothesis tests* about the *population parameters*.

General Linear Model

In the more general multiple regression model, there are p independent variables :

$$y_i = \beta_1 x_{i1} + \beta_2 x_{i2} + \cdots + \beta_p x_{ip} + \varepsilon_i,$$

where x_{ij} is the ith observation on the jth independent variable, and where the first independent variable takes the value 1 for all i (so β_1 is the regression intercept).

The least squares parameter estimates are obtained from p normal equations. The residual can be written as :

$$\varepsilon_i = y_i - \hat{\beta}_1 x_{i1} - \cdots - \hat{\beta}_p x_{ip}.$$

The **normal equations** are

$$\sum_{i=1}^{n}\sum_{k=1}^{p} X_{ij}X_{ik}\hat{\beta}_k = \sum_{i=1}^{n} X_{ij}y_i, j = 1,\ldots,p.$$

In matrix notation, the normal equations are written as :

$$(X^TX)\hat{\beta} = X^TY,$$

where the ij element of X is x_{ij}, the i element of the column vector Y is y_i, and the j element of $\hat{\beta}$ is $\hat{\beta}_j$. Thus X is $n\times p$, Y is $n\times 1$, and $\hat{\beta}$ is $p\times 1$. The solution is

$$\hat{\beta} = (X^TX)^{-1} X^TY.$$

Diagnostics

Once a regression model has been constructed, it may be important to confirm the *goodness of fit* of the model and the *statistical significance* of the estimated parameters. Commonly used checks of goodness of fit include the *R-squared,* analyses of the pattern of *residuals* and hypothesis testing. Statistical significance can be checked by an *F-test* of the overall fit, followed by *t-tests* of individual parameters.

Interpretations of these diagnostic tests rest heavily on the model assumptions. Although examination of the residuals can be used to invalidate a model, the results of a *t-test* or *F-test* are sometimes more difficult to interpret if the model's assumptions are violated. For example, if the error term does not have a normal distribution, in small samples the estimated parameters will not follow normal distributions and complicate inference. With relatively large samples, however, a *central limit theorem* can be invoked such that hypothesis testing may proceed using asymptotic approximations.

"Limited Dependent" Variables

The phrase "limited dependent" is used in *econometric* statistics for categorical and constrained variables.

The response variable may be non-continuous ("limited" to lie on some subset of the real line). For binary (zero or one) variables, if analysis proceeds with least-squares linear regression, the model is called the *linear probability model*. Non-linear models for binary dependent variables include the *probit* and *logit model*. The *multivariate probit* model is a standard method of estimating a joint relationship between several binary dependent variables and some independent variables. For *categorical variables* with more than two values there is the *multinomial logit*. For *ordinal variables* with more than two values, there are the *ordered logit* and *ordered probit* models. *Censored regression models* may be used when the dependent variable is only sometimes observed, and *Heckman correction* type models may be used when the sample is not randomly selected from the population of interest. An alternative to such procedures is linear regression based on *polychoric correlation* (or polyserial correlations) between the categorical variables. Such procedures differ in the assumptions made about the distribution of the variables in the population. If the variable is positive with low values and represents the repetition of the occurrence of an event, then count models like the *Poisson regression* or the *negative binomial* model may be used instead.

Interpolation and Extrapolation

Regression models predict a value of the *Y* variable given known values of the *X* variables. Prediction *within* the range of values in the data-set used for model-fitting is known informally as *interpolation*. Prediction *outside* this range of the data is known as *extrapolation*. Performing extrapolation relies strongly on the regression assumptions. The further the extrapolation goes outside the data, the more room there is for the model to fail due to differences between the assumptions and the sample data or the true values.

It is generally advised that when performing extrapolation, one should accompany the estimated value of the dependent variable with a prediction interval that represents the uncertainty. Such intervals tend to expand rapidly as the values of the independent variable(s) moved outside the range covered by the observed data.

For such reasons and others, some tend to say that it might be unwise to undertake extrapolation.

However, this does not cover the full set of modelling errors that may be being made : in particular, the assumption of a particular form for the relation between *Y* and X. A properly conducted regression analysis will include an assessment of how well the assumed form is matched by the observed data, but it can only do so within the range of values of the independent variables actually available. This means that any extrapolation is particularly reliant on the assumptions being made about the structural form of the regression relationship. Best-practice advice here is that a linear-in-variables and linear-in-parameters relationship should not be chosen simply for computational convenience, but that all available knowledge should be deployed in constructing a regression model. If this knowledge includes the fact that the dependent variable cannot go outside a certain range of values, this can be made use of in selecting the model-even if the observed data-set has no values particularly near such bounds. The implications of this step of choosing an appropriate functional form for the regression can be great when extrapolation is considered. At a minimum, it can ensure that any extrapolation arising from a fitted model is "realistic" (or in accord with what is known).

Non-linear Regression

When the model function is not linear in the parameters, the sum of squares must be minimized by an iterative procedure. This introduces many complications which are summarized in *Differences between linear and non-linear least squares*

Power and Sample Size Calculations

There are no generally agreed methods for relating the number of observations versus the number of independent variables in the model. One rule of thumb suggested by Good and Hardin is $N = m^n$, where N is the sample size, n is the number of independent variables and m is the number of observations needed to reach the desired precision if the model had only one independent variable. For example, a researcher is building a linear regression model using a data-set that contains 1000 patients (N). If he decides that five observations are needed to precisely define a straight line (m), then the maximum number of independent variables his model can support is 4, because

$$\frac{\log 1000}{\log 5} = 4.29.$$

Other Methods

Although the parameters of a regression model are usually estimated using the method of least squares, other methods which have been used include :

- *Bayesian methods, e.g. Bayesian linear regression*
- *Percentage regression,* for situations where reducing *percentage* errors is deemed more appropriate.
- *Least absolute deviations,* which is more robust in the presence of outliers, leading to *quantile regression*
- *Non-parametric regression,* requires a large number of observations and is computationally intensive
- *Distance metric learning,* which is learned by the search of a meaningful distance metric in a given input space.

Software

All major statistical software packages perform *least squares* regression analysis and inference. *Simple linear regression* and multiple regression using least squares can be done in some *spreadsheet* applications and on some calculators. While many statistical software packages can perform various types of non-parametric and robust regression, these methods are less standardized; different software packages implement different methods, and a method with a given name may be implemented differently in different packages. Specialized regression software has been developed for use in fields such as survey analysis and neuroimaging.

AUTOMATIC SUMMARIZATION

Automatic summarization is the process of reducing a text document with a computer program in order to create a summary that retains the most important points of the original document. As the problem of information overload has grown, and as the quantity of data has increased, so has interest in automatic summarization. Technologies that can make a coherent summary take into account variables such as length, writing style and syntax. An example of the use of summarization technology is search engines such as Google. Document summarization is another.

Generally, there are two approaches to automatic summarization : *extraction* and *abstraction*. Extractive methods work by selecting a subset of existing words, phrases, or sentences in the original text to form the summary. In contrast, abstractive methods build an internal semantic representation and then use natural language generation techniques to create a summary that is closer to what a human might generate. Such a summary might contain words not explicitly present in the original. The state-of-the-art abstractive methods are still quite weak, so most research has focused on extractive methods.

Methods

Methods of automatic summarization include extraction-based, abstraction-based, maximum entropy-based, and aided summarization.

Extraction-based Summarization

Two particular types of summarization often addressed in the literature are keyphrase extraction, where the goal is to select individual words or phrases to "tag" a document, and document summarization, where the goal is to select whole sentences to create a short paragraph summary.

Abstraction-based Summarization

Extraction techniques merely copy the information deemed most important by the system to the summary (for example, key clauses, sentences or paragraphs), while abstraction involves paraphrasing sections of the source document. In general, abstraction can condense a text more strongly than extraction, but the programs that can do this are harder to develop as they require the use of natural language generation technology, which itself is a growing field.

While some work has been done in abstractive summarization (creating an abstract synopsis like that of a human), the majority of summarization systems are extractive (selecting a subset of sentences to place in a summary).

Maximum Entropy-based Summarization

Even though automating abstractive summarization is the goal of summarization research, most practical systems are based on some form of extractive summarization. Extracted sentences can form a valid summary in itself or form a basis for further condensation operations. Furthermore, evaluation of extracted summaries can be automated, since it is essentially a classification task. During the DUC 2001 and 2002 evaluation workshops, TNO developed a sentence extraction system for multi-document summarization in the news domain. The system was based on a hybrid system using a naive Bayes classifier and statistical language models for modelling salience. Although the system exhibited good results, the researchers wanted to explore the effectiveness of a maximum entropy (ME) classifier for the meeting summarization task, as ME is known to be robust against feature dependencies. Maximum entropy has also been applied successfully for summarization in the broadcast news domain.

Aided Summarization

Machine learning techniques from closely related fields such as information retrieval or text mining have been successfully adapted to help automatic summarization.

Apart from Fully Automated Summarizers (FAS), there are systems that aid users with the task of summarization (MAHS = Machine Aided Human Summarization), for example by highlighting candidate passages to be included in the summary, and there are systems that depend on post-processing by a human (HAMS = Human Aided Machine Summarization).

Applications

There are different types of summaries depending what the summarization program focuses on to make the summary of the text, for example *generic summaries* or *query relevant summaries* (sometimes called *query-based summaries*).

Summarization systems are able to create both query relevant text summaries and generic machine-generated summaries depending on what the user needs. Summarization of multimedia documents, *e.g.* pictures or movies, is also possible.

Some systems will generate a summary based on a single source document, while others can use multiple source documents (for example, a cluster of news stories on the same topic). These systems are known as multi-document summarization systems.

Image collection summarization is other application example of automatic summarization. It consists in selecting a representative set of images from a larger set of images. A summary in this context is useful to show the most representative images of results in an image collection exploration system.

Keyphrase Extraction

Task Description and Example

The task is the following. You are given a piece of text, such as a journal article, and you must produce a list of keywords or keyphrases that capture the primary topics discussed in the text. In the case of research articles, many authors provide manually assigned keywords, but most text lacks pre-existing keyphrases. For example, news articles rarely have keyphrases attached, but it would be useful to be able to automatically do so for a number of applications discussed below. Consider the example text from a recent news article :

"The Army Corps of Engineers, rushing to meet President Bush's promise to protect New Orleans by the start of the 2006 hurricane season, installed defective flood-control pumps last year despite warnings from its own expert that the equipment would fail during a storm, according to documents obtained by The Associated Press".

An extractive keyphrase extractor might select "Army Corps of Engineers", "President Bush", "New Orleans", and "defective flood-control pumps" as keyphrases. These are pulled directly from the text. In contrast, an abstractive keyphrase system would somehow internalize the content and generate keyphrases that might be more descriptive and more like what a human would produce, such as "political negligence" or "inadequate protection from floods". Note that these terms do not appear in the text and require a deep understanding, which makes it difficult for a computer to produce such keyphrases. Keyphrases have many applications, such as to improve document browsing by providing a short summary. Also, keyphrases can improve information retrieval — if documents have keyphrases assigned, a user could search by keyphrase to produce more reliable hits than a full-text search. Also, automatic keyphrase extraction can be useful in generating index entries for a large text corpus.

Keyphrase Extraction as Supervised Learning

Beginning with the Turney paper, many researchers have approached keyphrase extraction as a supervised machine learning problem. Given a document, we construct an example for each unigram, bigram, and trigram found in the text (though other text units are also possible, as discussed below). We then compute various features describing each example (*e.g.*, does the phrase begin with an upper-case letter?). We assume there are known keyphrases available for a set of training documents. Using the known keyphrases, we can assign positive or negative labels to the examples. Then we learn a classifier that can discriminate between positive and negative examples as a function of the features. Some classifiers make a binary classification for a test example, while others assign a probability of being a keyphrase. For instance, in the above text, we might learn a rule that says phrases with initial capital letters are likely to be keyphrases. After training a learner, we can select keyphrases for test documents in the following manner. We apply the same example-generation strategy to the test documents, then run each example through the learner. We can determine the keyphrases by looking at binary classification decisions or probabilities returned from our learned model. If probabilities are given, a threshold is used to select the keyphrases. Keyphrase extractors are generally evaluated using precision and recall. Precision measures how many of the proposed keyphrases are actually correct. Recall measures how many of the true keyphrases your system proposed. The two measures can be combined in an F-score, which is the harmonic mean of the two ($F = 2PR/(P + R)$). Matches between the proposed keyphrases and the known keyphrases can be checked after stemming or applying some other text normalization.

Design choices

Designing a supervised keyphrase extraction system involves deciding on several choices (some of these apply to unsupervised, too) :

What are the examples?

The first choice is exactly how to generate examples. Turney and others have used all possible unigrams, bigrams, and trigrams without intervening punctuation and after removing stopwords. Hulth showed that you can get some improvement by selecting examples to be sequences of tokens that match certain patterns of part-of-speech tags. Ideally, the mechanism for generating examples produces all the known labelled keyphrases as candidates, though this is often not the case. For example, if we use only unigrams, bigrams, and trigrams, then we will never be able to extract a known keyphrase containing four words. Thus, recall may suffer. However, generating too many examples can also lead to low precision.

What are the features?

We also need to create features that describe the examples and are informative enough to allow a learning algorithm to discriminate keyphrases from non-keyphrases. Typically features involve various term frequencies (how many times a phrase appears in the current text or in a larger corpus), the length of the example, relative position of the first occurrence, various boolean syntactic features (*e.g.*, contains all caps), etc. The Turney paper used about 12 such features. Hulth uses a reduced set of features, which were found most successful in the KEA (Keyphrase Extraction Algorithm) work derived from Turney's seminal paper.

How many keyphrases to return?

In the end, the system will need to return a list of keyphrases for a test document, so we need to have a way to limit the number. Ensemble methods (*i.e.*, using votes from several classifiers) have been used to produce numeric scores that can be thresholded to provide a user-provided number of keyphrases. This is the technique used by Turney with C4.5 decision trees. Hulth used a single binary classifier so the learning algorithm implicitly determines the appropriate number.

What learning algorithm?

Once examples and features are created, we need a way to learn to predict keyphrases. Virtually any supervised learning algorithm could be used, such as decision trees, Naive Bayes, and rule induction. In the case of Turney's GenEx algorithm, a genetic algorithm is used to learn parameters for a domain-specific keyphrase extraction algorithm. The extractor follows a series of heuristics to identify keyphrases. The genetic algorithm optimizes parameters for these heuristics with respect to performance on training documents with known key phrases.

Unsupervised Keyphrase Extraction : Textrank

While supervised methods have some nice properties, like being able to produce interpretable rules for what features characterize a keyphrase, they also require a large amount of training data. Many documents with known keyphrases are needed. Furthermore, training on a specific domain tends to customize the extraction process to that domain, so the resulting classifier is not necessarily portable, as some of Turney's results demonstrate. Unsupervised keyphrase extraction removes the need for training data. It approaches the problem from a different angle. Instead of trying to learn explicit features that characterize keyphrases, the TextRank algorithm exploits the structure of the text itself to determine keyphrases that appear "central" to the text in the same way that PageRank selects important Web pages. Recall this is based on the notion of "prestige" or "recommendation" from social networks. In this way, TextRank does not rely on any previous training data at all, but rather can be run on any arbitrary piece of text, and it can produce output simply based on the text's intrinsic properties. Thus the algorithm is easily portable to new domains and languages.

TextRank is a general purpose graph-based ranking algorithm for NLP. Essentially, it runs PageRank on a graph specially designed for a particular NLP task. For keyphrase extraction, it builds a graph using some set of text units as vertices. Edges are based on some measure of semantic or lexical similarity between the text unit vertices. Unlike PageRank, the edges are typically undirected and can be weighted to reflect a degree of similarity. Once the graph is constructed, it is used to form a stochastic matrix, combined with a damping factor (as in the "random surfer model"), and the ranking over vertices is obtained by finding the eigenvector corresponding to eigenvalue 1 (*i.e.*, the stationary distribution of the random walk on the graph).

Design choices

What should vertices be?

The vertices should correspond to what we want to rank. Potentially, we could do something similar to the supervised methods and create a vertex for each unigram, bigram, trigram, etc. However, to keep the graph small, the authors decide to rank individual unigrams in a first step, and then include a second step that merges highly ranked adjacent unigrams to form multi-word phrases. This has a nice side effect of allowing us to produce keyphrases of arbitrary length. For example, if we rank unigrams and find that "advanced", "natural", "language", and "processing" all get high ranks, then we would look at the original text and see that these words appear consecutively and create a final keyphrase using all four together. Note that the unigrams placed in the graph can be filtered by part of speech. The authors found that adjectives and nouns were the best to include. Thus, some linguistic knowledge comes into play in this step.

How should we create edges?

Edges are created based on word co-occurrence in this application of TextRank. Two vertices are connected by an edge if the unigrams appear within a window of size N in the original text. N is typically around 2–10. Thus, "natural" and "language" might be linked in a text about NLP. "Natural" and "processing" would also be linked because they would both appear in the same string of N words. These edges build on the notion of "text cohesion" and the idea that words that appear near each other are likely related in a meaningful way and "recommend" each other to the reader.

How are the final keyphrases formed?

Since this method simply ranks the individual vertices, we need a way to threshold or produce a limited number of keyphrases. The technique chosen is to set a count T to be a user-specified fraction of the total number of vertices in the graph. Then the top T vertices/unigrams are selected based on their stationary probabilities. A post-processing step is then applied to merge adjacent instances of these T unigrams. As a result, potentially more or less than T final keyphrases will be produced, but the number should be roughly proportional to the length of the original text.

Why it works

It is not initially clear why applying PageRank to a co-occurrence graph would produce useful keyphrases. One way to think about it is the following. A word that appears multiple times throughout a text may have many different co-occurring neighbours. For example, in a text about machine learning, the unigram "learning" might co-occur with "machine", "supervised", "un-supervised", and "semi-supervised" in four different sentences. Thus, the "learning" vertex would be a central "hub" that connects to these other modifying words. Running PageRank/TextRank on the graph is likely to rank "learning" highly. Similarly, if the text contains the phrase "supervised classification", then there would be an edge between "supervised" and "classification". If "classification" appears several other places and thus has many neighbours, its importance would contribute to the importance of "supervised". If it ends up with a high rank, it will be selected as one of the top T unigrams, along with "learning" and probably "classification". In the final post-processing step, we would then end up with keyphrases "supervised learning" and "supervised classification".

In short, the co-occurrence graph will contain densely connected regions for terms that appear often and in different contexts. A random walk on this graph will have a stationary distribution that assigns large probabilities to the terms in the centers of the clusters. This is similar to densely connected Web pages getting ranked highly by PageRank.

Document Summarization

Like keyphrase extraction, document summarization hopes to identify the essence of a text. The only real difference is that now we are dealing with larger text units—whole sentences instead of words and phrases.

Before getting into the details of some summarization methods, we will mention how summarization systems are typically evaluated. The most common way is using the so-called ROUGE (Recall-Oriented Understudy for Gisting Evaluation) measure. This is a recall-based measure that determines how well a system-generated summary covers the content present in one or more human-generated model summaries known as references. It is recall-based to encourage systems to include all the important topics in the text. Recall can be computed with respect to unigram, bigram, trigram, or 4-gram matching, though ROUGE-1 (unigram matching) has been shown to correlate best with human assessments of system-generated summaries (*i.e.*, the summaries with highest ROUGE-1 values correlate with the summaries humans deemed the best). ROUGE-1 is computed as division of count of unigrams in reference that appear in system and count of unigrams in reference summary.

If there are multiple references, the ROUGE-1 scores are averaged. Because ROUGE is based only on content overlap, it can determine if the same general concepts are discussed between an automatic summary and a reference summary, but it cannot determine if the result is coherent or the sentences flow together in a sensible manner. High-order n-gram ROUGE measures try to judge fluency to some degree. Note that ROUGE is similar to the BLEU measure for machine translation, but BLEU is precision-based, because translation systems favour accuracy.

A promising line in document summarization is adaptive document/text summarization. The idea of adaptive summarization involves preliminary recognition of document/text genre and subsequent application of summarization algorithms optimized for this genre. First summarizes that perform adaptive summarization have been created.

Overview of Supervised Learning Approaches

Supervised text summarization is very much like supervised keyphrase extraction. Basically, if you have a collection of documents and human-generated summaries for them, you can learn features of sentences that make them good candidates for inclusion in the summary. Features might include the position in the document (*i.e.*, the first few sentences are probably important), the number of words in the sentence, etc. The main difficulty in supervised extractive summarization is that the known summaries must be manually created by extracting sentences so the sentences in an original training document can be labelled as "in summary" or "not in summary". This is not typically how people create summaries, so simply using journal abstracts or existing summaries is usually not sufficient. The sentences in these summaries do not necessarily match up with sentences in the original text, so it would be difficult to assign labels to examples for training. Note, however, that these natural summaries can still be used for evaluation purposes, since ROUGE-1 only cares about unigrams.

Unsupervised Approaches : Textrank and LexRank

The unsupervised approach to summarization is also quite similar in spirit to unsupervised keyphrase extraction and gets around the issue of costly training data. Some unsupervised summarization approaches are based on finding a "centroid" sentence, which is the mean word vector of all the sentences in the document. Then the sentences can be ranked with regard to their similarity to this centroid sentence.

A more principled way to estimate sentence importance is using random walks and eigenvector centrality. LexRank is an algorithm essentially identical to TextRank, and both use this approach for document summarization. The two methods were developed by different groups at the same time, and LexRank simply focused on summarization, but could just as easily be used for keyphrase extraction or any other NLP ranking task.

Design Choices

What are the vertices?

In both LexRank and TextRank, a graph is constructed by creating a vertex for each sentence in the document.

What are the edges?

The edges between sentences are based on some form of semantic similarity or content overlap. While LexRank uses cosine similarity of TF-IDF vectors, TextRank uses a very similar measure based on the number of words two sentences have in common (normalized by the sentences' lengths). The LexRank paper explored using unweighted edges after applying a threshold to the cosine values, but also experimented with using edges with weights equal to the similarity score. TextRank uses continuous similarity scores as weights.

How are summaries formed?

In both algorithms, the sentences are ranked by applying PageRank to the resulting graph. A summary is formed by combining the top ranking sentences, using a threshold or length cutoff to limit the size of the summary.

TextRank and LexRank differences

It is worth noting that TextRank was applied to summarization exactly as described here, while LexRank was used as part of a larger summarization system (MEAD) that combines the LexRank score (stationary probability) with other features like sentence position and length using a linear combination with either user-specified or automatically tuned weights. In this case, some training documents might be needed, though the TextRank results show the additional features are not absolutely necessary.

Another important distinction is that TextRank was used for single document summarization, while LexRank has been applied to multi-document summarization. The task remains the same in both cases—only the number of sentences to choose from has grown. However, when summarizing multiple documents, there is a greater risk of selecting duplicate or highly redundant sentences to place in the same summary. Imagine you have a cluster of news articles on a particular event, and you want to produce one summary. Each article is likely to have many similar sentences, and you would only want to include distinct ideas in the summary. To address this issue, LexRank applies a heuristic post-processing step that builds up a summary by adding sentences in rank order, but discards any sentences that are too similar to ones already placed in the summary. The method used is called Cross-Sentence Information Subsumption (CSIS).

Why Unsupervised Summarization Works

These methods work based on the idea that sentences "recommend" other similar sentences to the reader. Thus, if one sentence is very similar to many others, it will likely be a sentence of great importance. The importance of this sentence also stems from the importance of the sentences "recommending" it. Thus, to get ranked highly and placed in a summary, a sentence must be similar to many sentences that are in turn also similar to many other sentences. This makes intuitive sense and allows the algorithms to be applied to any arbitrary new text. The methods are domain-independent and easily portable. One could imagine the features indicating important sentences in the news domain might vary considerably from the biomedical domain. However, the unsupervised "recommendation"-based approach applies to any domain.

Multi-document Summarization

Multi-document summarization is an automatic procedure aimed at extraction of information from multiple texts written about the same topic. Resulting summary report allows individual users, such as professional information consumers, to quickly familiarize themselves with information contained in a large cluster of documents. In such a way, multi-document summarization systems are complementing the news aggregators performing the next step down the road of coping with information overload.

Multi-document summarization creates information reports that are both concise and comprehensive. With different opinions being put together & outlined, every topic is described from multiple perspectives within a single document. While the goal of a brief summary is to simplify information search and cut the time by pointing to the most relevant source documents, comprehensive multi-document summary should itself contain the required information, hence limiting the need for accessing original files to cases when refinement is required. Automatic summaries present information extracted from multiple sources algorithmically, without any editorial touch or subjective human intervention, thus making it completely unbiased.

Incorporating Diversity : GRASSHOPPER Algorithm

Multi-document extractive summarization faces a problem of potential redundancy. Ideally, we would like to extract sentences that are both "central" (*i.e.*, contain the main ideas) and "diverse" (*i.e.*, they differ from one another). LexRank deals with diversity as a heuristic final stage using CSIS, and other systems have used similar methods, such as Maximal Marginal Relevance (MMR), in trying to eliminate redundancy in information retrieval results.

There is a general purpose graph-based ranking algorithm like Page/Lex/TextRank that handles both "centrality" and "diversity" in a unified mathematical framework based on absorbing Markov chain random walks. (An absorbing random walk is like a standard random walk, except some states are now absorbing states that act as "black holes" that cause the walk to end abruptly at that state.) The algorithm is called GRASSHOPPER. In addition to explicitly promoting diversity during the ranking process, GRASSHOPPER incorporates a prior ranking (based on sentence position in the case of summarization).

Evaluation Techniques

The most common way to evaluate the informativeness of automatic summaries is to compare them with human-made model summaries.

Evaluation techniques fall into intrinsic and extrinsic, inter-texual and intra-texual.

Intrinsic and Extrinsic Evaluation

An intrinsic evaluation tests the summarization system in of itself while an extrinsic evaluation tests the summarization based on how it affects the completion of some other task. Intrinsic evaluations have assessed mainly the coherence and informativeness of summaries. Extrinsic evaluations, on the other hand, have tested the impact of summarization on tasks like relevance assessment, reading comprehension, etc.

Inter-textual and Intra-textual

Intra-texual methods assess the output of a specific summarization system, and the inter-texual ones focus on contrastive analysis of outputs of several summarization systems.

Human judgement often has wide variance on what is considered a "good" summary, which means that making the evaluation process automatic is particularly difficult. Manual evaluation can be used, but this is both time and labour intensive as it requires humans to read not only the summaries but also the source documents. Other issues are those concerning coherence and coverage.

One of the metrics used in NIST's annual Document Understanding Conferences, in which research groups submit their systems for both summarization and translation tasks, is the ROUGE metric (Recall-Oriented Understudy for Gisting Evaluation [3]). It essentially calculates n-gram overlaps between automatically generated summaries and previously-written human summaries. A high level of overlap should indicate a high level of shared concepts between the two summaries. Note that overlap metrics like this are unable to provide any feedback on a summary's coherence. Anaphor resolution remains another problem yet to be fully solved.

Current Difficulties in Evaluating Summaries Automatically

Evaluating summaries, either manually or automatically, is a hard task. The main difficulty in evaluation comes from the impossibility of building a fair gold-standard against which the results of the systems can be compared. Furthermore, it is also very hard to determine what a correct summary is, because there is always the possibility of a system to generate a good summary that is quite different from any human summary used as an approximation to the correct output.

Content selection is not a deterministic problem. People are subjective, and different authors would choose different sentences. And individuals may not be consistent. A particular person may chose different sentences at different times. Two distinct sentences expressed in different words can express the same meaning. This phenomenon is known as paraphrasing. We can find an approach to automatically evaluating summaries using paraphrases (ParaEval).

Most summarization systems perform an extractive approach, selecting and copying important sentences from the source documents. Although humans can also cut and paste relevant information of a text, most of the times they rephrase sentences when necessary, or they join different related information into one sentence.

Evaluating Summaries Qualitatively

The main drawback of the evaluation systems existing so far is that we need at least one reference summary, and for some methods more than one, to be able to compare automatic summaries with models. This is a hard and expensive task. Much effort has to be done in order to have corpus of texts and their corresponding summaries. Furthermore, for some methods, not only do we need to have human-made summaries available for comparison, but also manual annotation has to be performed in some of them (*e.g.* SCU in the Pyramid Method). In any case, what the evaluation methods need as an input, is a set of summaries to serve as gold standards and a set of automatic summaries. Moreover, they all perform a quantitative evaluation with regard to different similarity metrics. To overcome these problems, we think that the quantitative evaluation might not be the only way to evaluate summaries, and a qualitative automatic evaluation would be also important.

WHAT ARE DATA MINING TOOLS

Data mining tools are software components and theories that allow users to extract information from data. The tools provide individuals and companies with the ability to gather large amounts of data and use it to make determinations about a particular user or groups of users. Some of the most common uses of data mining tools are in the fields of marketing, fraud protections and surveillance.

The manual extraction of data has existed for hundreds of years. However, the automation of data mining has been most prevalent since the dawn of the computer age. During the 20th century, various computer sciences emerged to help support the concept of developing data mining tools. The overall goal of the utilization of the tools is to uncover hidden patterns. For example, if a marketing company finds that a person takes a monthly trip from New York City to Los Angeles, it becomes beneficial for that company to advertise details of the destination to the individual.

Within the data *mining industry,* standards have been established to define the parameters of the use of data mining tools. Annually, the Association for Computing Machinery's Special Interest Group on Knowledge Discovery and Data Mining (SIGKDD) holds a meeting to determine what processes are used. The same group is also responsible for assessing the ethical implications of the analysis of data from individuals and companies. A biannual journal is published by the group entitled *SIGKDD Explorations.*

The most prevalent tool used in data mining is the process called Knowledge Discovery in Databases (KDD). KDD was developed in 1989 by Gregory Piatetsky-Shapiro. Using this data mining tool, users are able to process raw data, mine the data for information and interpret the various results in the form of information management.

One of the most important forms of data mining tools is used for combating terrorism in the 21st century. In the United States, the National Research Council uses the concepts of pattern mining and subject-based data mining to identify terrorist activity in the large pool of information around the world. Pattern mining is defined by the process of locating patterns within a large volume of data. Subject-based data mining attempts to identify relationships between individuals. Both techniques can also be utilized in general business practice by defining the mind-set of a customer base and the interactive relationship between customers.

DATA CLASSIFICATION

In business intelligence, **data classification** has close ties to data clustering, but where data clustering is *descriptive,* data classification is *predictive.* In essence data classification consists of using variables with known values to predict the unknown or future values of other variables. It can be used in *e.g.* direct marketing, insurance fraud detection or medical diagnosis.

The first step in doing a data classification is to cluster the data-set used for category training, to create the wanted number of categories. An algorithm, called the *classifier,* is then used on the categories, creating a descriptive model for each. These models can then be used to categorize new items in the created classification system.

According to Golfarelli and Rizzi, these are the measures of effectiveness of the classifier :

- *Predictive accuracy* : How well does it predict the categories for new observations?
- *Speed* : What is the computational cost of using the classifier?
- *Robustness* : How well do the models created perform if data quality is low?
- *Scalability* : Does the classifier function efficiently with large amounts of data?
- *Interpretability* : Are the results understandable to users?

Typical examples of input for data classification could be variables such as demographics, lifestyle information, or economical behaviour.

Challenges for Data Classification

There are several challenges in working with data classification. One in particular is that it is necessary for all using categories on *e.g.* customers or clients, to do the modelling in an iterative process. This is to make sure that change in the characteristics of customer groups does not go unnoticed, making the existing categories outdated and obsolete, without anyone noticing.

This could be of special importance to insurance or banking companies, where fraud detection is extremely relevant. New fraud patterns may come unnoticed, if the methods to surveil these changes and alert when categories are changing, disappearing or new ones emerge, are not developed and implemented.

DECISION TREE LEARNING

Decision tree learning uses a *decision tree* as a *predictive model* which maps observations about an item to conclusions about the item's target value. It is one of the predictive modelling approaches used in *statistics, data mining* and *machine learning*. More descriptive names for such tree models are **classification trees** or **regression trees.** In these tree structures, *leaves* represent class labels and branches represent *conjunctions* of features that lead to those class labels.

In decision analysis, a decision tree can be used to visually and explicitly represent decisions and *decision-making*. In *data mining*, a decision tree describes data but not decisions; rather the resulting classification tree can be an input for *decision-making*. This page deals with decision trees in *data mining*.

General

A tree showing survival of passengers on the *Titanic* ("sibsp" is the number of spouses or siblings aboard). The figures under the leaves show the probability of survival and the percentage of observations in the leaf.

Decision tree learning is a method commonly used in data mining. The goal is to create a model that predicts the value of a target variable based on several input variables. An example is shown on the right. Each *interior node* corresponds to one of the input variables; there are edges to children for each of the possible values of that input variable. Each leaf represents a value of the target variable given the values of the input variables represented by the path from the root to the leaf.

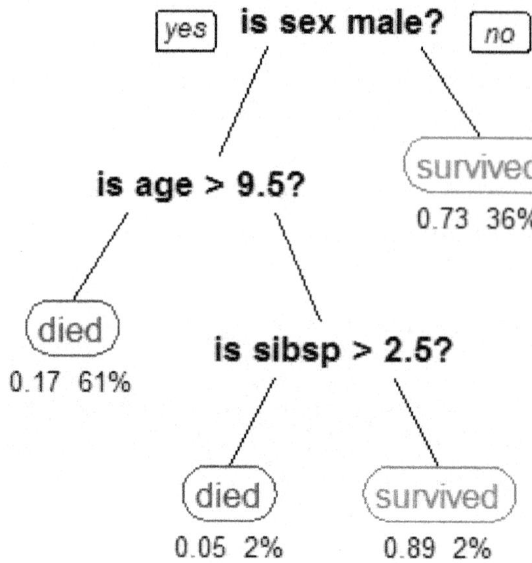

A decision tree is a simple representation for classifying examples. Decision tree learning is one of the most successful techniques for supervised classification learning. For this section, assume that all of the features have finite discrete domains, and there is a single target feature called the classification. Each element of the domain of the classification is called a class. A decision tree or a classification tree is a tree in which each internal (non-leaf) node is labelled with an input feature. The arcs coming from a node labelled with a feature are labelled with each of the possible values of the feature. Each leaf of the tree is labelled with a class or a probability distribution over the classes.

A tree can be "learned" by splitting the source *set* into subsets based on an attribute value test. This process is repeated on each derived subset in a recursive manner called *recursive partitioning*. The *recursion* is completed when the subset at a node has all the same value of the target variable, or when splitting no longer adds value to the predictions. This process of *top-down induction of decision trees* (TDIDT) is an example of a *greedy algorithm*, and it is by far the most common strategy for learning decision trees from data.

In *data mining*, decision trees can be described also as the combination of mathematical and computational techniques to aid the description, categorisation and generalisation of a given set of data.

Data comes in records of the form :

$$(x, Y) = (x_1, x_2, x_3, \ldots, x_k, Y)$$

The dependent variable, Y, is the target variable that we are trying to understand, classify or generalize. The vector **x** is composed of the input variables, x_1, x_2, x_3 etc., that are used for that task.

Types

Decision trees used in data mining are of two main types :

- **Classification tree** analysis is when the predicted outcome is the class to which the data belongs.
- **Regression tree** analysis is when the predicted outcome can be considered a real number (*e.g.* the price of a house, or a patient's length of stay in a hospital).

The term **Classification And Regression Tree (CART)** analysis is an *umbrella term* used to refer to both of the above procedures, first introduced by *Breiman et. al.* Trees used for regression and trees used for classification have some similarities-but also some differences, such as the procedure used to determine where to split.

Some techniques, often called ensemble methods, construct more than one decision tree :

- *Bagging* decision trees, an early ensemble method, builds multiple decision trees by repeatedly resampling training data with replacement, and voting the trees for a consensus prediction.
- A *Random Forest* classifier uses a number of decision trees, in order to improve the classification rate.
- *Boosted Trees* can be used for regression-type and classification-type problems.
- *Rotation forest*—in which every decision tree is trained by first applying *principal component analysis* (PCA) on a random subset of the input features.

Decision tree learning is the construction of a decision tree from class-labelled training tuples. A decision tree is a flow-chart-like structure, where each internal (non-leaf) node denotes a test on an attribute, each branch represents the outcome of a test, and each leaf (or terminal) node holds a class label. The topmost node in a tree is the root node.

There are many specific decision-tree algorithms. Notable ones include :

- *ID3* (Iterative Dichotomiser 3)
- *C4.5* (successor of ID3)
- *CART* (Classification And Regression Tree)
- *CHAID* (CHi-squared Automatic Interaction Detector). Performs multi-level splits when computing classification trees.
- *MARS* : extends decision trees to better handle numerical data.

ID3 and CART were invented independently at around same time, yet follow a similar approach for learning decision tree from training tuples.

Formula

Algorithms for constructing decision trees usually work top-down, by choosing a variable at each step that best splits the set of items. Different algorithms use different metrics for measuring "best". These generally measure the homogeneity of the target variable within the subsets. Some examples are given below.

These metrics are applied to each candidate subset, and the resulting values are combined (*e.g.*, averaged) to provide a measure of the quality of the split.

Gini Impurity

Used by the CART (classification and regression tree) algorithm, Gini impurity is a measure of how often a randomly chosen element from the set would be incorrectly labelled if it were randomly labelled according to the distribution of labels in the subset. Gini impurity can be computed by summing the probability of each item being chosen times the probability of a mistake in categorizing that item. It reaches its minimum (zero) when all cases in the node fall into a single target category.

To compute Gini impurity for a set of items, suppose i takes on values in {1, 2,..., m}, and let f_i be the fraction of items labelled with value i in the set.

$$I_G(f) = \sum_{i=1}^{m} f_i(1-f_i) = \sum_{i=1}^{m}(f_i - f_i^2) = \sum_{i=1}^{m} f_i - \sum_{i=1}^{m} f_i^2 = 1 - \sum_{i=1}^{m} f_i^2$$

Information Gain

Used by the *ID3, C4.5* and C5.0 tree-generation algorithms. *Information gain* is based on the concept of *entropy* from *information theory.*

$$I_E(f) = -\sum_{i=1}^{m} f_i \log_2 f_i$$

Decision Tree Advantages

Amongst other data mining methods, decision trees have various advantages :

- **Simple to understand and interpret.** People are able to understand decision tree models after a brief explanation.
- **Requires little data preparation.** Other techniques often require data normalisation, *dummy variables* need to be created and blank values to be removed.
- **Able to handle both numerical and** *categorical* **data.** Other techniques are usually specialised in analysing data-sets that have only one type of variable. (For example, relation rules can be used only with nominal variables while neural networks can be used only with numerical variables.)
- **Uses a** *white box* **model.** If a given situation is observable in a model the explanation for the condition is easily explained by boolean logic. (An example of a black box model is an *artificial neural network* since the explanation for the results is difficult to understand.)
- **Possible to validate a model using statistical tests.** That makes it possible to account for the reliability of the model.
- *Robust.* Performs well even if its assumptions are somewhat violated by the true model from which the data were generated.
- **Performs well with large data-sets.** Large amounts of data can be analysed using standard computing resources in reasonable time.

Limitations

- The problem of learning an optimal decision tree is known to be *NP-complete* under several aspects of optimality and even for simple concepts. Consequently, practical decision-tree learning algorithms are based on heuristics such as the *greedy algorithm* where locally-optimal decisions are made at each node. Such algorithms cannot guarantee to return the globally-optimal decision tree. To reduce the greedy effect of local-optimality some methods such as the dual information distance (DID) tree were proposed.
- Decision-tree learners can create over-complex trees that do not generalise well from the training data. (This is known as *overfitting*.) Mechanisms such as *pruning* are necessary to avoid this problem.
- There are concepts that are hard to learn because decision trees do not express them easily, such as *XOR*, *parity* or *multiplexer* problems. In such cases, the decision tree becomes prohibitively large. Approaches to solve the problem involve either changing the representation of the problem domain (known as propositionalisation) or using learning algorithms based on more expressive representations (such as *statistical relational learning* or *inductive logic programming*).
- For data including categorical variables with different numbers of levels, *information gain in decision trees* is biased in favour of those attributes with more levels.

Extensions

Decision Graphs

In a decision tree, all paths from the root node to the leaf node proceed by way of conjunction, or *AND*. In a decision graph, it is possible to use disjunctions (ORs) to join two more paths together using *Minimum message length* (MML). Decision graphs have been further extended to allow for previously unstated new attributes to be learnt dynamically and used at different places within the graph. The more general coding scheme results in better predictive accuracy and log-loss probabilistic scoring. In general, decision graphs infer models with fewer leaves than decision trees.

Alternative Search Methods

Evolutionary algorithms have been used to avoid local optimal decisions and search the decision tree space with little *a priori* bias.

It is also possible for a tree to be sampled using MCMC in a bayesian paradigm.

The tree can be searched for in a bottom-up fashion.

Chapter 2

PREDICTING QUALITY OUTCOMES THROUGH DATA MINING

The market for new analytical methods for statistical process control (SPC) and optimization is driven by a number of factors. In particular, companies that have invested in Six Sigma leadership, training and cultural development have emphasized quantitative assessment and problem solving. Using traditional Six Sigma tools and processes, they've realized significant improvements and cost savings. They've also recognized that a number of significant manufacturing problems are still unsolved.

Manufacturers are look-ing for ways to apply quantitative analysis to help their employees create better products. These companies are clear about their business objectives and have the historical data to characterize their problem domains. What they lack are methods to gain the insights necessary to approach these problems.

Traditional methods such as design of experiments (DOE), linear regression and correlation, and quality control charts aren't suitable in many circumstances for a variety of reasons :

- The number of relevant factors is large.
- The underlying relationship between the factors and critical outcomes is complex.
- The important factors interact with each other.
- The nature of the data violates the assumptions.

Data-mining methods can offer effective solutions to manufacturers facing these problems. Although data mining has been widely adopted in other industries, applying it specifically to manufacturing has been hindered by a lack of required expertise, the need to constrain solutions to practical implementations, the tendency to optimize outcomes to target specifications, the creation of the necessary tools to allow engineering professionals to perform what-if scenarios, and the absence of other tools to deploy predictive models to the factory floor for ongoing monitoring and decision making.

The return on investment from applying data mining in manufacturing has resulted in improved product performance, less scrap and rework, decreased testing, improved product performance, and fewer field service and warranty claims. Data-mining methods can provide significant value to all the data collected, stored and managed by manufacturers during the last several years.

CATERPILLAR INC. : A CASE STUDY

During the 24 years that principle manufacturing engineer Bill Matthews has worked at Caterpillar Inc., he's seen products fail quality tests for many reasons. "We found that some of what contributed to product failures was tribal knowledge," says Matthews. "So now we use software to look at empirical data from our manufacturing processes equipment in relation to product performance, which gives us a much more accurate idea of what contributes to product compliance."

Like many large companies, Peoria, Illinois-based Caterpillar has developed sophisticated data-collection techniques. A single manufacturing plant can use thousands of gages, sensors and other automated devices to collect data from machines and other manufacturing equipment. But much of the massive amounts of data the company collected for monitoring hundreds of process parameters went unused.

Beyond the data was the dilemma of what to do to improve quality. What knobs get turned and in which direction? What design feature gets modified and by how much? What exactly in the manufacturing stream gets adjusted, altered, tuned or tweaked to positively affect quality and compliance outcomes?

Using predictive data mining on their mountains of data, Caterpillar's quality managers uncovered the manageable process parameters most important to the quality outcomes of finished products. Tribal knowledge, which contributed to process variation, was now consolidated and focused by providing a line of sight between the manufacturing processes and the finished product. By distilling the relevant relationships from a large candidate pool of data, the manufacturing process was fine-tuned to consistently produce a finished product that met engineering and customer requirements without added costs.

Building on years of experience with SPC and Six Sigma quality initiatives, engineers at Caterpillar are now deploying data-mining software developed with StatSoft Inc.

When the software was implemented on a Caterpillar rotating assembly found in industrial machine equipment, the results were improved product quality and streamlined manufacturing processes that produced cost savings in many categories. When engineers set out to solve an intermittent vibration problem found in some engines, they applied the define, measure, analyze, improve and control (DMAIC) methodology to the problem.

During the "measure" phase, engineers examined what data were useful to consider in relation to the vibration problem, which drove unnecessary cost by interrupting the finished product testing to "trim balance" the rotating assembly. Here, the abundance of data included 113 different assembly features measured during the manufacturing process. During the "analyze" phase, Caterpillar distilled these raw data down to a subset of predictor variables—such as clearances and fits in the rotating assembly that affect trim balance in an engine.

The data were categorized as input, output or constraints data, and the software then predicted a subset of variables that caused trim balance outcomes during finished product testing. To validate predictions, the software simulates the probability of trim balance problems. If the simulation results meet a minimum criteria for accuracy and uncertainty for validation, the software then optimizes the model to minimize trim balance.

Moving into the "improve" phase, engineers used the information rendered in the software's Actionable Decision Environment to explore what-if alternatives. In Caterpillar's case, the model revealed six assembly features that could affect trim balance. Engineers explored scenarios for implementing cost-effective changes using the actionable decision environment and determined that making two of the changes would have a disproportionately high effect on reducing the frequency of trim balance problems.

The actionable decision environment indicated what actions would produce the desired outcome. "The model allowed the Caterpillar team to determine that a reduction in run-out of two interacting features on the assembly would reduce [the occurrence of] trim balance problems by approximately 50 per cent," says Matthews. The necessary manufacturing process changes were implemented to reduce the run-out in these features. "It was especially beneficial that the software model could provide us with an empirical line of sight between what we do during the assembly process and its effect on finished product," notes Matthews. "The tribal knowledge that previously contributed to process variation now seamlessly shifted to minimize the variation of features known to contribute to trim balance. The process changes were implemented with no additional cost of resources."

It's important to point out that although a reduction in trim balance problems was the single requirement in this example, many problems tend to be more complicated because they have competing requirements. The software's simultaneous optimization module effectively addresses these situations. For example, industrial machinery typically has power, fuel efficiency and emissions requirements. The software finds the optimum balance among competing goals by giving ranges of inputs relative to the desired outcomes.

The results of using purposeful data mining for this particular Caterpillar rotating assembly included a 45-per cent reduction in trim balance problems. Beside the typical reduction costs associated with finished-product discrepancies, such as reduced rework and scrap, and less visible effects such as increased throughput, a decrease in resources required to support testing also materialized. Additional return on investment (ROI) is derived from the fact that useful and actionable data were distilled. The investment in metrology equipment to collect data, storage and management applications for the data, and the added product cycle time to collect the data are just a few of the costs that are easily recovered when the data are converted to knowledge and useful actions.

Caterpillar reduces post-manufacturing product adjustments and effectively speeds time to market because it "predicts" the manufacturing process parameters driving the need for expensive and time-consuming product adjustments. More ROI is achieved through cost savings from the reduction in personnel and time previously used to perform the adjustments.

WHAT IS DATA MINING?

The term "data-mining methods" refers to a category of analytical methods geared toward determining useful relationships in huge, complex sets of data. The term arose partially to distinguish these methods from traditional statistical ones.

The interest in data-mining methods originally began in non-manufacturing domains through a set of interrelated developments :

- *Data storage :* Relational database management systems have become commonplace during the last few decades. Storage hardware continues to be more scalable and less expensive. As a result, companies can collect, store and manage more data effectively. Whereas analyses and decisions in the past were limited to small data sets dealing with an immediate issue, opportunities now abound to build predictive models holistically using the available historical data.
- *Computing power :* At the same time that databases have matured and storage is becoming less expensive, the power available on standard workstation computers continues to improve. Knowledge workers today have a comparatively large amount of computing power available to them in terms of processor speed, working memory and hard-disk space.

These factors have contributed to organizations› readiness to adopt a new set of analytic methods to address data-mining opportunities. Originally, applications for data mining of business problems weren't found in manufacturing or quality-related disciplines but rather with credit-risk scoring in financial services companies, customer up-selling and cross-selling, and customer-retention applications in marketing domains. Quality practitioners in manufacturing are now realizing the applicability and advantages of these methods to their databases. The wealth of existing data in SPC software databases provides a great opportunity to derive valuable insights about a company's processes and how they contribute to product quality outcomes.

A popular category of data-mining methods is recursive partitioning, often called "tree" methods because the graphical outputs from these methods resemble a tree. Tree methods come in different algorithms, including classification and regression trees (C&RT) and chi-squared automatic interaction detectors (CHAID). This means that data-mining practitioners should apply multiple methods to the same problem to determine which provide the best predictive accuracy.

Tree methods are popular for predictive data mining because :

- They're non-parametric and will model both simple (*e.g.*, linear) and complex (*e.g.*, non-linear) relationships, including interactions between factors and data sets with variables that aren't normally distributed.
- They train fast, even on large data sets.
- They're easy to interpret. Reviewing a tree graph entails simply studying a series of if/then statements.

THE DATA-MINING PROCESS

Data mining is an analytic process designed to explore large amounts of data in search of consistent patterns or systematic relationships between variables, and then validating the findings by applying the detected patterns to new data subsets. The ultimate goal of data mining is prediction. Predictive data mining is the most common type and the one with the most direct business applications.

The process of data mining consists of three stages : initial exploration, model building or pattern identification with validation and/or verification, and deployment (*i.e.*, applying the model to new data to generate predictions).

- *Stage 1 : Exploration :* This stage usually starts with data preparation, which may involve cleaning data, data transformations, selecting subsets of records and—in the case of data sets with large numbers of variables (or "fields")—performing preliminary feature selection operations to bring the number of variables to a manageable range, depending on the statistical methods being considered.
- *Stage 2 : Model building and validation :* This stage involves considering various models and choosing the best one based on its predictive performance (*i.e.*, explaining the variability in question and producing stable results across samples). This might sound like a simple operation, but it can involve a very elaborate process. To achieve this goal, a variety of techniques have been developed, many of which are based on so-called "competitive evaluation of models," that is, applying different models to the same data set and then comparing their performance to choose the best.
- *Stage 3 : deployment :* This final stage involves using the model selected in the previous stage and applying it to new data to generate predictions of the expected outcome.

The data-mining process aligns very well with the DMAIC methodology. In fact, many predictive modelling projects in manufacturing follow the DMAIC project process and include the data-mining steps of modelling (analyze), validation (analyze), and deployment (improve and control).

Comparisons to Traditional SPC

Typical analytical tools in the Six Sigma practitioner's toolkit for SPC are quality control charting, design of experiments and multiple linear regression. Data-mining methods can augment these methods in circumstances where historical data are abundant and a predictive model between input factors and quality outcomes would derive business benefits. In fact, a predictive data-mining model is often instrumental in determining the important factors driving product quality that should be monitored through quality control charting.

Predictive data-mining methods augment traditional SPC for a variety of reasons. First, they're robust enough for complex, non-linear and non-normally distributed data scenarios. With traditional methods, it's important that the analyst checks assumptions to ensure that the method can be used and the interpretation of results is justified. With predictive data mining, those assumptions (often not met) are no longer relevant.

Second, predictive data-mining methods are holistic and multivariate. When reviewing a standard quality control chart, for example, the data under scrutiny are univariate (*i.e.*, for one process parameter or one quality dimension in isolation). In contrast, a predictive data-mining model will take into consideration multi-dimensional relationships between the predictors and one or more quality outcomes, so the analyst is provided with a holistic understanding of how the process, from raw materials through process steps and in-process tests, contributes to the product attributes important to customers.

Third, predictive data-mining methods make use of historical data. Many companies have invested heavily in data capture, collection, storage and management. Predictive data-mining methods are used to derive a ROI for these data. In contrast, design of experiments requires that a project team defines and conducts experimental runs to derive its data.

MANUFACTURING-SPECIFIC DATA MINING

A breakthrough in Caterpillar's approach overcame a critical barrier in attempting to deploy data mining in manufacturing : making the recommendations from the specific predictive models relevant to practical implementation.

Quality process managers, manufacturing engineers, design engineers and other decision makers at Caterpillar manipulated process settings on virtual models to observe the effect on product quality outcomes. For example, setting a particular process parameter at its theoretical optimum might require a costly upgrade to the production process not justified by the benefits of the change. Or, it might be beyond the capabilities of the materials supplier. In either case, the quality decision-makers could see the likely outcomes that affect product quality. Interacting with the model helps participants see opportunities to add their collective expertise to the software predictions and derive reality-based settings for optimal quality results.

Included with the Actionable Decision Environment is a set of graphical tools to help users interact with the virtual process models. The interface is a mechanism that allows users to poke the model with what-if scenarios so that they can see the implications of their actions.

Using this approach, Caterpillar engineers were able to simultaneously improve multiple, competitive outcomes (*e.g.*, power vs. fuel efficiency) driven by multiple upstream processes.

At the same time and equally important, the software helped Caterpillar identify processes where additional variation could be allowed with no penalty in product quality. The upside of the findings is simple : Why spend money and time to control a material or process variable when it doesn't contribute to product quality?

WHERE AND WHEN TO USE IT

The application of data-mining software is relevant in complex manufacturing processes that involve many steps in different and diverse industries. The software can model relationships between specific manufacturing process parameters and outcomes even when data are sparse and relationships complex.

Manufacturing operations that have spent the last several years investing in measurement and data-storage systems are candidates for adopting data-mining and predictive modelling software. The software builds on existing data-collection methodologies and makes full use of the data already collected and stored.

Manufacturing industries adopting data mining for predictive modelling and optimization of their processes include heavy equipment, automotive, aerospace, machine tool, packaging, pharmaceuticals, robotics, semi-conductor, medical and others with complex products.

Chapter 3

Decision Analysis and Cluster Analysis

DECISION TREE

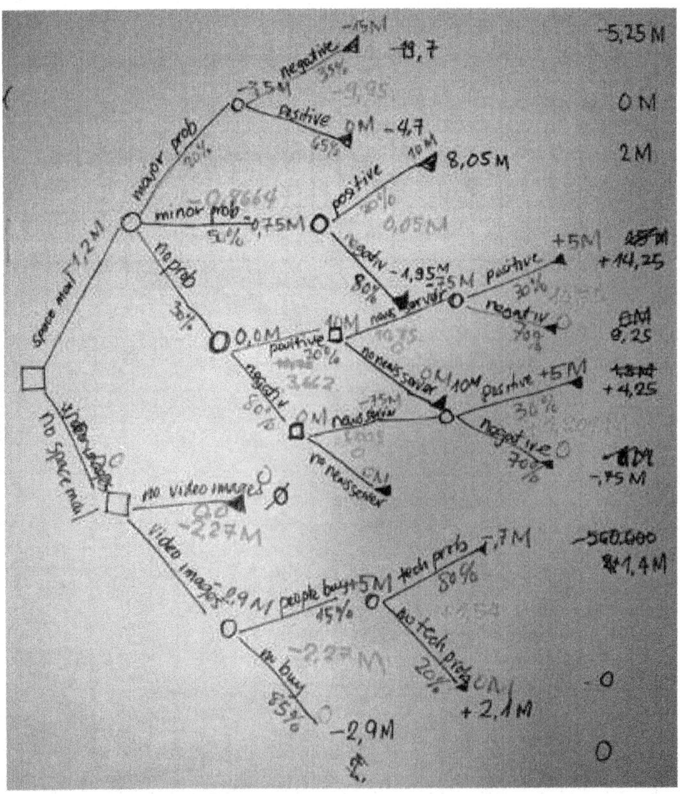

Fig. : Traditionally, decision trees have been created manually.

A **decision tree** is a decision support tool that uses a tree-like graph or model of decisions and their possible consequences, including chance event outcomes, resource costs, and utility. It is one way to display an algorithm.

Decision trees are commonly used in operations research, specifically in decision analysis, to help identify a strategy most likely to reach a goal.

Overview

A decision tree is a flowchart-like structure in which internal node represents test on an attribute, each branch represents outcome of test and each leaf node represents class label (decision taken after computing all attributes). A path from root to leaf represents classification rules.

In decision analysis a decision tree and the closely related influence diagram is used as a visual and analytical decision support tool, where the expected values (or expected utility) of competing alternatives are calculated.

A decision tree consists of 3 types of nodes :

1. Decision nodes—commonly represented by squares
2. Chance nodes—represented by circles
3. End nodes—represented by triangles

Decision trees are commonly used in operations research, specifically in decision analysis, to help identify a strategy most likely to reach a goal. If in practice decisions have to be taken online with no recall under incomplete knowledge, a decision tree should be paralleled by a probability model as a best choice model or online selection model algorithm. Another use of decision trees is as a descriptive means for calculating conditional probabilities.

Decision trees, influence diagrams, utility functions, and other decision analysis tools and methods are taught to undergraduate students in schools of business, health economics, and public health, and are examples of operations research or management science methods.

Decision Tree Building Blocks

Decision Tree Elements

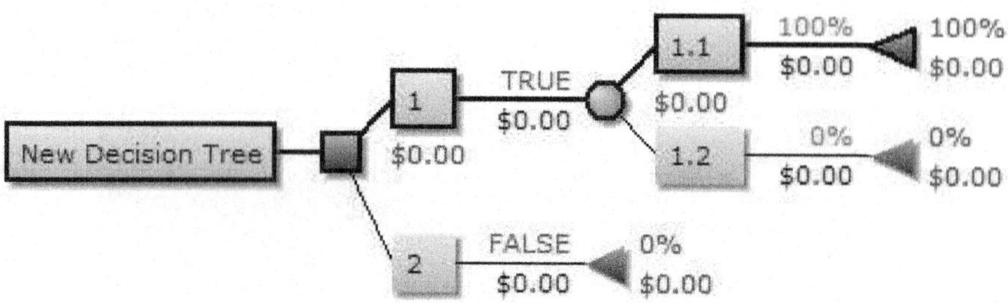

Drawn from left to right, a decision tree has only burst nodes (splitting paths) but no sink nodes (converging paths). Therefore, used manually, they can grow very big and are then often hard to draw fully by hand. Traditionally, decision trees have been created manually-as the aside example shows-although increasingly, specialized software is employed.

Decision Tree Using Flow Chart Symbols

Commonly a decision tree is drawn using flow chart symbols as it is easier for many to read and understand.

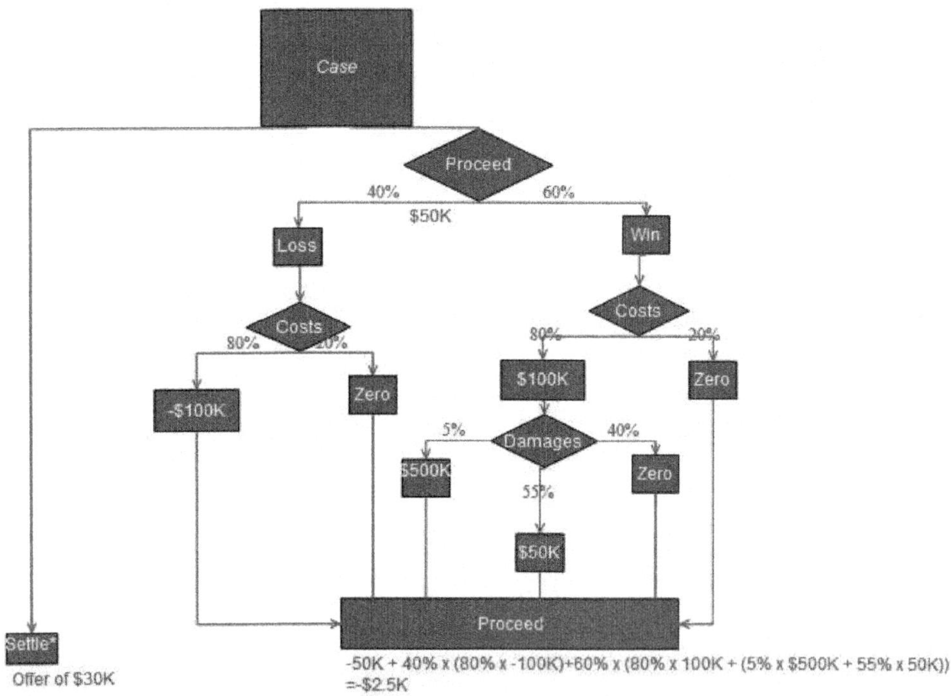

Analysis Example

Analysis can take into account the decision maker's (*e.g.*, the company's) preference or utility function, for example :

The basic interpretation in this situation is that the company prefers B's risk and payoffs under realistic risk preference coefficients (greater than $400K—in that range of risk aversion, the company would need to model a third strategy, "Neither A nor B").

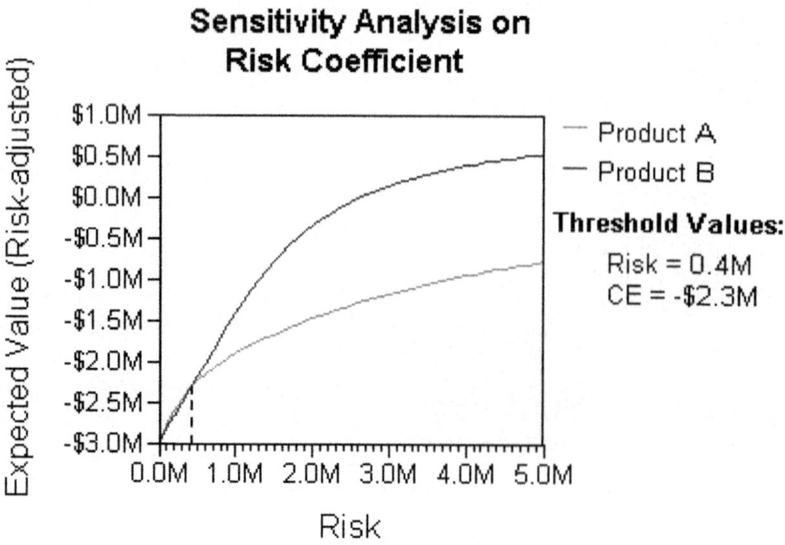

Another Example

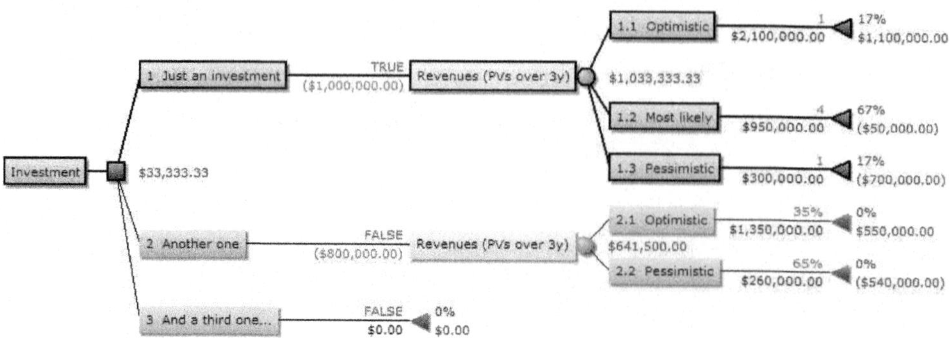

Decision trees can be used to optimize an investment portfolio. The following example shows a portfolio of 7 investment options (projects). The organization has $10,000,000 available for the total investment. Bold lines mark the best selection 1, 3, 5, 6, and 7, which will cost $9,750,000 and create a payoff of 16,175,000. All other combinations would either exceed the budget or yield a lower payoff.

Influence Diagram

A decision tree can be represented more compactly as an influence diagram, focusing attention on the issues and relationships between events.

The squares represent decisions, the ovals represent action, and the diamond represents results.

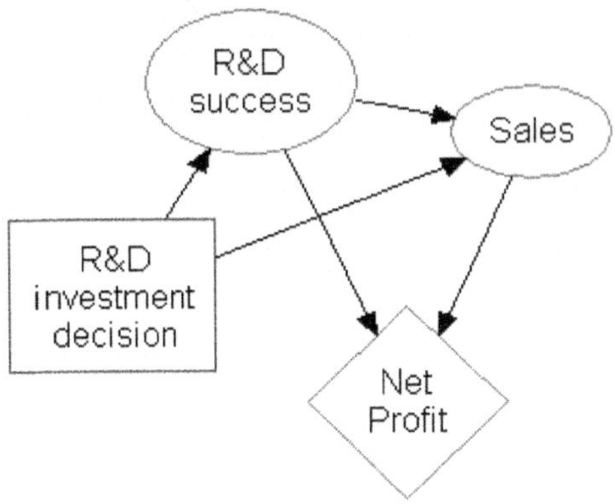

Advantages and Disadvantages

Amongst decision support tools, decision trees (and influence diagrams) have several advantages. Decision trees :

- Are simple to understand and interpret. People are able to understand decision tree models after a brief explanation.
- Have value even with little hard data. Important insights can be generated based on experts describing a situation (its alternatives, probabilities, and costs) and their preferences for outcomes.
- Possible scenarios can be added.
- Worst, best and expected values can be determined for different scenarios.
- Use a white box model. If a given result is provided by a model.
- Can be combined with other decision techniques. The following example uses Net Present Value calculations, PERT 3-point estimations and a linear distribution of expected outcomes :

Disadvantages of decision trees :

- For data including categorical variables with different number of levels, information gain in decision trees are biased in favour of those attributes with more levels.
- Calculations can get very complex particularly if many values are uncertain and/or if many outcomes are linked.

CLUSTER ANALYSIS

Cluster analysis or **clustering** is the task of grouping a set of objects in such a way that objects in the same group (called a **cluster**) are more similar (in some sense or another) to each other than to those in other groups (clusters). It is a main task of exploratory data mining, and a common technique for statistical data analysis, used in many fields, including machine learning, pattern recognition, image analysis, information retrieval, and bio-informatics.

Cluster analysis itself is not one specific algorithm, but the general task to be solved. It can be achieved by various algorithms that differ significantly in their notion of what constitutes a cluster and how to efficiently find them. Popular notions of clusters include groups with small distances among the cluster members, dense areas of the data space, intervals or particular statistical distributions. Clustering can therefore be formulated as a multi-objective optimization problem. The appropriate clustering algorithm and parameter settings (including values such as the distance function to use, a density threshold or the number of expected clusters) depend on the individual data set and intended use of the results. Cluster analysis as such is not an automatic task, but an iterative process of knowledge discovery or interactive multi-objective optimization that involves trial and failure. It will often be necessary to modify data preprocessing and model parameters until the result achieves the desired properties.

Besides the term *clustering*, there are a number of terms with similar meanings, including *automatic classification*, *numerical taxonomy*, *botryology* and *typological analysis*. The subtle differences are often in the usage of the results : while in data mining, the resulting groups are the matter of interest, in automatic classification the resulting discriminative power is of interest. This often leads to misunderstandings between researchers coming from the fields of data mining and machine learning, since they use the same terms and often the same algorithms, but have different goals.

Cluster analysis was originated in anthropology by Driver and Kroeber in 1932 and introduced to psychology by Zubin in 1938 and Robert Tryon in 1939 and famously used by Cattell beginning in 1943 for trait theory classification in personality psychology.

Clusters and Clusterings

According to Vladimir Estivill-Castro, the notion of a "cluster" cannot be precisely defined, which is one of the reasons why there are so many clustering algorithms. There is a common denominator : a group of data objects. However, different researchers employ different cluster models, and for each of these cluster models again different algorithms can be given. The notion of a cluster, as found by different algorithms, varies significantly in its properties. Understanding these "cluster models" is key to understanding the differences between the various algorithms. Typical cluster models include :

- *Connectivity models :* For example hierarchical clustering builds models based on distance connectivity.
- *Centroid models :* For example the k-means algorithm represents each cluster by a single mean vector.
- *Distribution models :* Clusters are modelled using statistical distributions, such as multivariate normal distributions used by the Expectation-maximization algorithm.

- *Density models :* For example DBSCAN and OPTICS defines clusters as connected dense regions in the data space.
- *Sub-space models :* In Biclustering (also known as Co-clustering or two-mode-clustering), clusters are modelled with both cluster members and relevant attributes.
- *Group models :* Some algorithms do not provide a refined model for their results and just provide the grouping information.
- *Graph-based models :* A clique, *i.e.*, a subset of nodes in a graph such that every two nodes in the subset are connected by an edge can be considered as a prototypical form of cluster. Relaxations of the complete connectivity requirement (a fraction of the edges can be missing) are known as quasi-cliques.

A "clustering" is essentially a set of such clusters, usually containing all objects in the data set. Additionally, it may specify the relationship of the clusters to each other, for example a hierarchy of clusters embedded in each other. Clusterings can be roughly distinguished as :

- Hard clustering : each object belongs to a cluster or not
- Soft clustering (also : fuzzy clustering) : each object belongs to each cluster to a certain degree (*e.g.* a likelihood of belonging to the cluster)

There are also finer distinctions possible, for example :

- *Strict partitioning clustering :* Here each object belongs to exactly one cluster.
- *Strict partitioning clustering with outliers :* Objects can also belong to no cluster, and are considered outliers.
- *Overlapping clustering (also : alternative clustering, multi-view clustering) :* While usually a hard clustering, objects may belong to more than one cluster.
- *Hierarchical clustering :* Objects that belong to a child cluster also belong to the parent cluster.
- *Sub-space clustering :* While an overlapping clustering, within a uniquely defined sub-space, clusters are not expected to overlap.

Clustering Algorithms

Clustering algorithms can be categorized based on their cluster model, as listed above. The following overview will only list the most prominent examples of clustering algorithms, as there are possibly over 100 published clustering algorithms. Not all provide models for their clusters and can thus not easily be categorized. An overview of algorithms explained in Wikipedia can be found in the list of statistics algorithms.

There is no objectively "correct" clustering algorithm, but as it was noted, "clustering is in the eye of the beholder." The most appropriate clustering algorithm for a particular problem often needs to be chosen experimentally, unless there is a mathematical reason to prefer one cluster model over another. It should be noted that an algorithm that is designed for one kind of model has no chance on a data set that contains a radically different kind of model. For example, k-means cannot find non-convex clusters.

Connectivity Based Clustering (Hierarchical Clustering)

Connectivity based clustering, also known as *hierarchical clustering*, is based on the core idea of objects being more related to nearby objects than to objects farther away. These algorithms connect "objects" to form "clusters" based on their distance. A cluster can be described largely by the maximum distance needed to connect parts of the cluster. At different distances, different clusters will form, which can be represented using a dendrogram, which explains where the common name "hierarchical clustering" comes from : these algorithms do not provide a single partitioning of the data set, but instead provide an extensive hierarchy of clusters that merge with each other at certain distances. In a dendrogram, the y-axis marks the distance at which the clusters merge, while the objects are placed along the x-axis such that the clusters don't mix.

Connectivity based clustering is a whole family of methods that differ by the way distances are computed. Apart from the usual choice of distance functions, the user also needs to decide on the linkage criterion (since a cluster consists of multiple objects, there are multiple candidates to compute the distance to) to use. Popular choices are known as single-linkage clustering (the minimum of object distances), complete linkage clustering (the maximum of object distances) or UPGMA ("Unweighted Pair Group Method with Arithmetic Mean", also known as average linkage clustering). Furthermore, hierarchical clustering can be agglomerative (starting with single elements and aggregating them into clusters) or divisive (starting with the complete data set and dividing it into partitions).

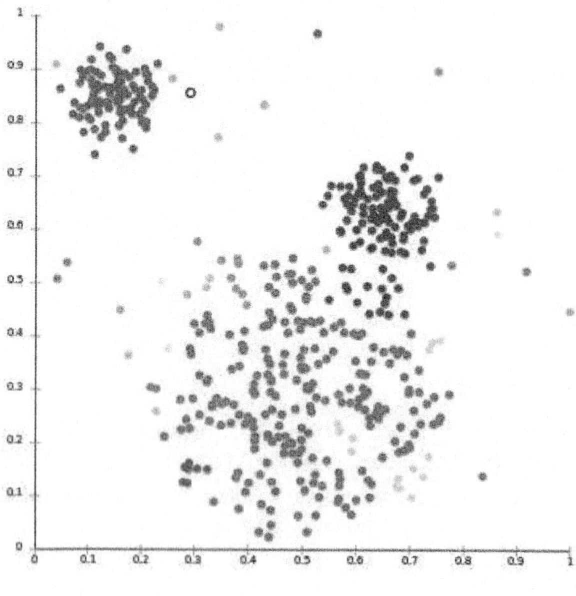

These methods will not produce a unique partitioning of the data set, but a hierarchy from which the user still needs to choose appropriate clusters. They are not very robust towards outliers, which will either show up as additional clusters or even cause other clusters to merge (known as "chaining phenomenon", in particular with single-linkage clustering). In the general case, the complexity is $O(n^3)$, which makes them too slow for large data sets. For some special cases, optimal efficient methods (of complexity $O(n^2)$ are known : SLINK for single-linkage and CLINK for complete-linkage clustering. In the data mining community these methods are recognized as a theoretical foundation of cluster analysis, but often considered obsolete. They did however provide inspiration for many later methods such as density based clustering.

- *Linkage clustering examples* : Single-linkage on Gaussian data. At 35 clusters, the biggest cluster starts fragmenting into smaller parts, while before it was still connected to the second largest due to the single-link effect.

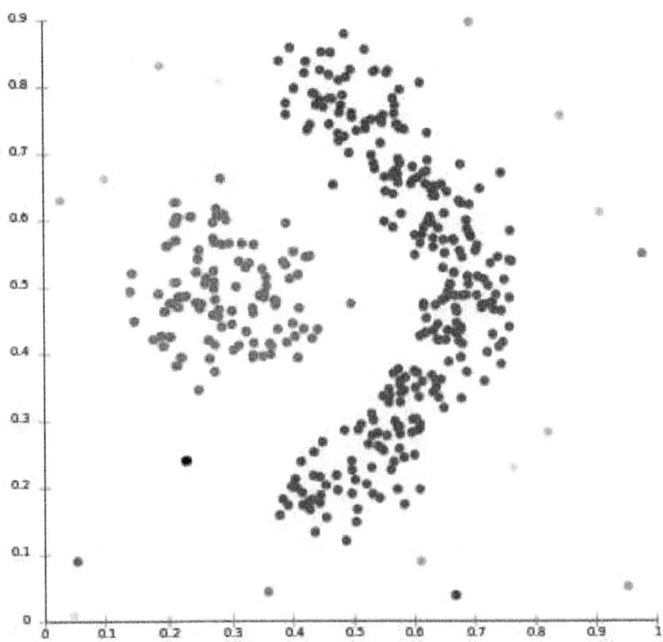

Single-linkage on density-based clusters. 20 clusters extracted, most of which contain single elements, since linkage clustering does not have a notion of "noise".

Centroid-based Clustering

In centroid-based clustering, clusters are represented by a central vector, which may not necessarily be a member of the data set. When the number of clusters is fixed to *k*, *k*-means clustering gives a formal definition as an optimization problem : find the *k* cluster centers and assign the objects to the nearest cluster center, such that the squared distances from the cluster are minimized.

The optimization problem itself is known to be NP-hard, and thus the common approach is to search only for approximate solutions. A particularly well known approximative method is Lloyd's algorithm, often actually referred to as "*k-means algorithm*". It does however only find a local optimum, and is commonly run multiple times with different random initializations. Variations of k-means often include such optimizations as choosing the best of multiple runs, but also restricting the centroids to members of the data set (k-medoids), choosing medians (k-medians clustering), choosing the initial centers less randomly (K-means++) or allowing a fuzzy cluster assignment (Fuzzy c-means).

Most k-means-type algorithms require the number of clusters-*k*-to be specified in advance, which is considered to be one of the biggest drawbacks of these algorithms. Furthermore, the algorithms prefer clusters of approximately similar size, as they will always assign an object to the nearest centroid. This often leads to incorrectly cut borders in between of clusters (which is not surprising, as the algorithm optimized cluster centers, not cluster borders).

K-means has a number of interesting theoretical properties. On the one hand, it partitions the data space into a structure known as a Voronoi diagram. On the other hand, it is conceptually close to nearest neighbour classification, and as such is popular in machine learning. Third, it can be seen as a variation of model based classification, and Lloyd's algorithm as a variation of the Expectation-maximization algorithm for this model discussed below :

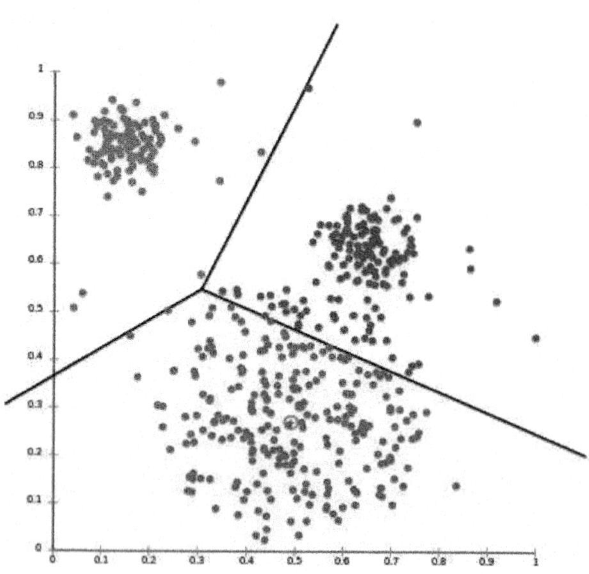

Fig. : *k-Means clustering examples* : K-means separates data into Voronoi-cells, which assumes equal-sized clusters (not adequate here)

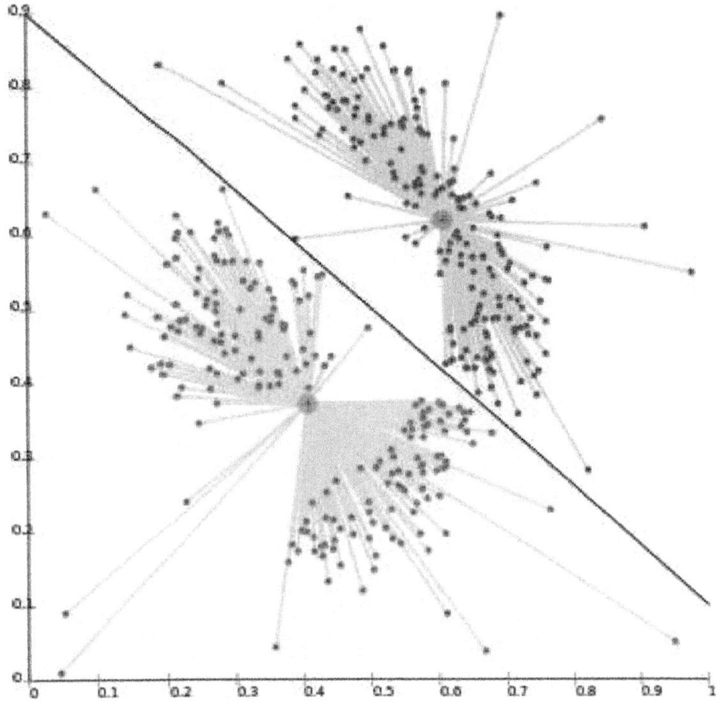

Fig. : K-means cannot represent density-based clusters.

Distribution-based Clustering

The clustering model most closely related to statistics is based on distribution models. Clusters can then easily be defined as objects belonging most likely to the same distribution. A nice property of this approach is that this closely resembles the way artificial data sets are generated : by sampling random objects from a distribution.

While the theoretical foundation of these methods is excellent, they suffer from one key problem known as overfitting, unless constraints are put on the model complexity. A more complex model will usually always be able to explain the data better, which makes choosing the appropriate model complexity inherently difficult.

One prominent method is known as Gaussian mixture models (using the expectation-maximization algorithm). Here, the data set is usually modelled with a fixed (to avoid overfitting) number of Gaussian distributions that are initialized randomly and whose parameters are iteratively optimized to fit better to the data set. This will converge to a local optimum, so multiple runs may produce different results. In order to obtain a hard clustering, objects are often then assigned to the Gaussian distribution they most likely belong to; for soft clusterings, this is not necessary.

Distribution-based clustering is a semantically strong method, as it not only provides you with clusters, but also produces complex models for the clusters that can also capture correlation and dependence of attributes. However, using these algorithms puts an extra burden on the user : to choose appropriate data models to optimize, and for many real data sets, there may be no mathematical model available the algorithm is able to optimize (*e.g.* assuming Gaussian distributions is a rather strong assumption on the data).

- Expectation-Maximization (EM) clustering examples

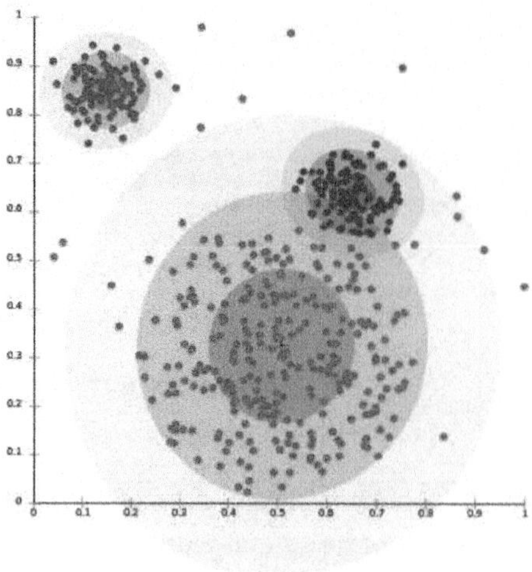

On Gaussian-distributed data, EM works well, since it uses Gaussians for modelling clusters.

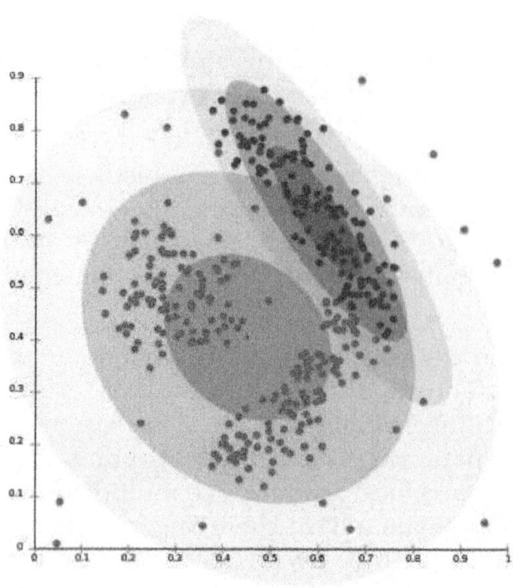

Density-based clusters cannot be modelled using Gaussian distributions.

Density-based Clustering

In density-based clustering, clusters are defined as areas of higher density than the remainder of the data set. Objects in these sparse areas—that are required to separate clusters—are usually considered to be noise and border points.

The most popular density based clustering method is DBSCAN. In contrast to many newer methods, it features a well-defined cluster model called "density-reachability". Similar to linkage based clustering, it is based on connecting points within certain distance thresholds. However, it only connects points that satisfy a density criterion, in the original variant defined as a minimum number of other objects within this radius. A cluster consists of all density-connected objects (which can form a cluster of an arbitrary shape, in contrast to many other methods) plus all objects that are within these objects' range. Another interesting property of DBSCAN is that its complexity is fairly low—it requires a linear number of range queries on the database—and that it will discover essentially the same results (it is deterministic for core and noise points, but not for border points) in each run, therefore there is no need to run it multiple times. OPTICS is a generalization of DBSCAN that removes the need to choose an appropriate value for the range parameter ε, and produces a hierarchical result related to that of linkage clustering. DeLi-Clu, Density-Link-Clustering combines ideas from single-linkage clustering and OPTICS, eliminating the ε parameter entirely and offering performance improvements over OPTICS by using an R-tree index.

The key drawback of DBSCAN and OPTICS is that they expect some kind of density drop to detect cluster borders. Moreover, they cannot detect intrinsic cluster structures which are prevalent in the majority of real life data. A variation of DBSCAN, EnDBSCAN, efficiently detects such kinds of structures. On data sets with, for example, overlapping Gaussian distributions-a common use case in artificial data-the cluster borders produced by these algorithms will often look arbitrary, because the cluster density decreases continuously. On a data set consisting of mixtures of Gaussians, these algorithms are nearly always outperformed by methods such as EM clustering that are able to precisely model this kind of data.

- Density-based clustering examples

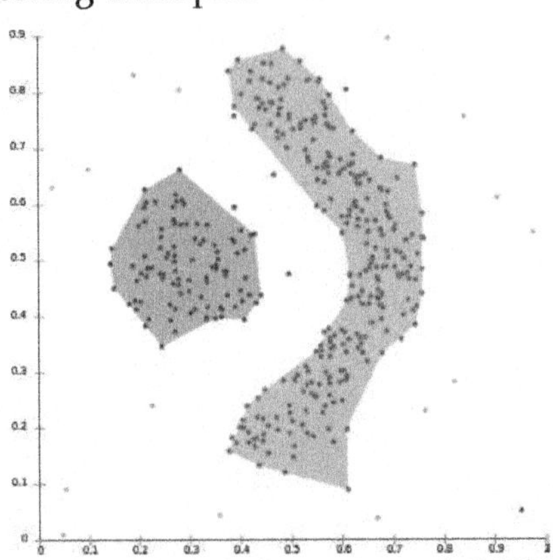

Fig. : Density-based clustering with DBSCAN.

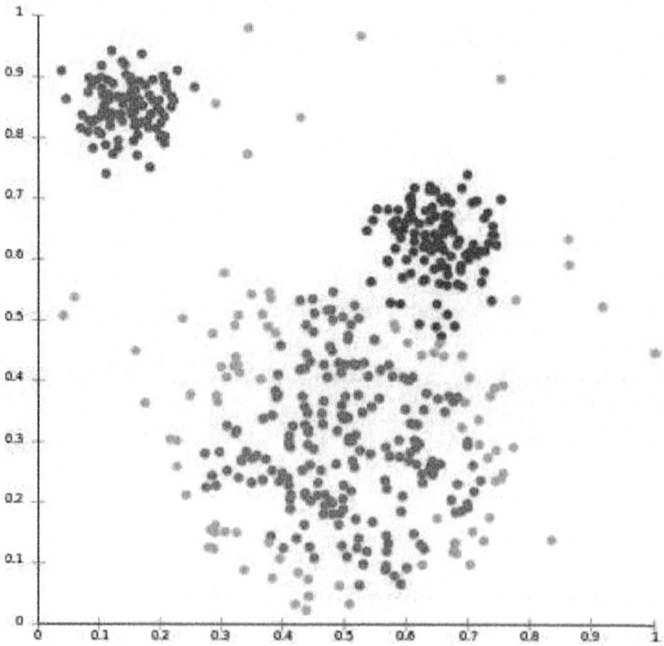

DBSCAN assumes clusters of similar density, and may have problems separating nearby clusters.

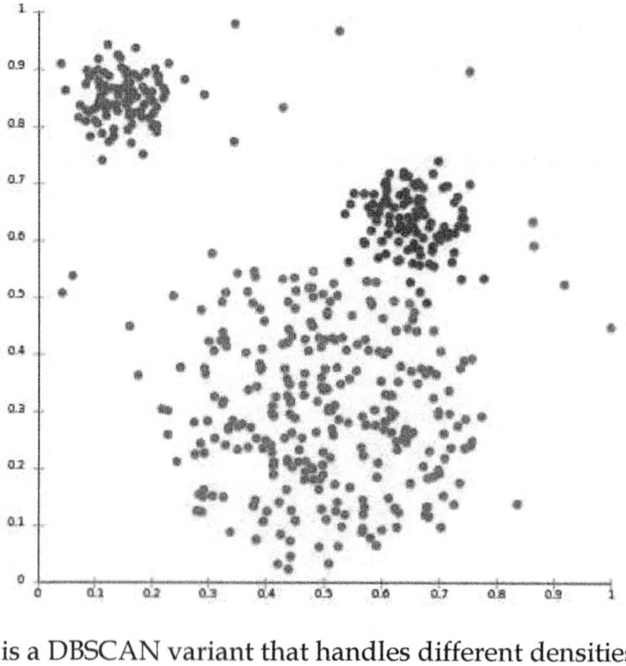

OPTICS is a DBSCAN variant that handles different densities much better.

Recent Developments

In recent years considerable effort has been put into improving algorithm performance of the existing algorithms. Among them are *CLARANS,* and *BIRCH.* With the recent need to process larger and larger data sets (also known as big data), the willingness to trade semantic meaning of the generated clusters for performance has been increasing. This led to the development of pre-clustering methods such as canopy clustering, which can process huge data sets efficiently, but the resulting "clusters" are merely a rough pre-partitioning of the data set to then analyze the partitions with existing slower methods such as k-means clustering. Various other approaches to clustering have been tried such as seed based clustering.

For high-dimensional data, many of the existing methods fail due to the curse of dimensionality, which renders particular distance functions problematic in high-dimensional spaces. This led to new clustering algorithms for high-dimensional data that focus on sub-space clustering (where only some attributes are used, and cluster models include the relevant attributes for the cluster) and correlation clustering that also looks for arbitrary rotated ("correlated") sub-space clusters that can be modelled by giving a correlation of their attributes. Examples for such clustering algorithms are CLIQUE and SUBCLU.

Ideas from density-based clustering methods (in particular the DBSCAN/OPTICS family of algorithms) have been adopted to sub-space clustering (HiSC, hierarchical sub-space clustering and DiSH) and correlation clustering (HiCO, hierarchical correlation clustering, 4C using "correlation connectivity" and ERiC exploring hierarchical density-based correlation clusters).

Several different clustering systems based on mutual information have been proposed. One is Marina Meilă's *variation of information* metric; another provides hierarchical clustering. Using genetic algorithms, a wide range of different fit-functions can be optimized, including mutual information. Also message passing algorithms, a recent development in Computer Science and Statistical Physics, has led to the creation of new types of clustering algorithms.

Evaluation of Clustering Results

Evaluation of clustering results sometimes is referred to as cluster validation.

There have been several suggestions for a measure of similarity between two clusterings. Such a measure can be used to compare how well different data clustering algorithms perform on a set of data. These measures are usually tied to the type of criterion being considered in assessing the quality of a clustering method.

Internal Evaluation

When a clustering result is evaluated based on the data that was clustered itself, this is called internal evaluation. These methods usually assign the best score to the algorithm that produces clusters with high similarity within a cluster and low similarity between clusters. One drawback of using internal criteria in cluster evaluation is that high scores on an internal measure do not necessarily result in effective information retrieval applications. Additionally, this evaluation is biased towards algorithms that use the same cluster model. For example, k-Means clustering naturally optimizes object distances, and a distance-based internal criterion will likely overrate the resulting clustering.

Therefore, the internal evaluation measures are best suited to get some insight into situations where one algorithm performs better than another, but this shall not imply that one algorithm produces more valid results than another. Validity as measured by such an index depends on the claim that this kind of structure exists in the data set. An algorithm designed for some kind of models has no chance if the data set contains a radically different set of models, or if the evaluation measures a radically different criterion. For example, k-means clustering can only find convex clusters, and many evaluation indexes assume convex clusters. On a data set with non-convex clusters neither the use of k-means, nor of an evaluation criterion that assumes convexity, is sound.

The following methods can be used to assess the quality of clustering algorithms based on internal criterion :

- *Davies–Bouldin index* : The Davies–Bouldin index can be calculated by the following formula :

$$DB = \frac{1}{n}\sum_{n}^{n} \max_{i=1}\left(\frac{\sigma_i + \sigma_j}{d(c_i, c_j)}\right)$$

where n is the number of clusters, c_x is the centroid of cluster x, σ_x is the average distance of all elements in cluster x to centroid c_x, and $d(c_i, c_j)$ is the distance between centroids c_i and c_j. Since algorithms that produce clusters with low intra-cluster distances (high intra-cluster similarity) and high inter-cluster distances (low inter-cluster similarity) will have a low Davies–Bouldin index, the clustering algorithm that produces a collection of clusters with the smallest Davies–Bouldin index is considered the best algorithm based on this criterion.

- *Dunn index* : The Dunn index aims to identify dense and well-separated clusters. It is defined as the ratio between the minimal inter-cluster distance to maximal intra-cluster distance. For each cluster partition, the Dunn index can be calculated by the following formula :

$$D = \min_{1\le i\le n}\left\{\min_{1\le j\le n, i\ne j}\left\{\frac{d(i,j)}{\max_{1\le k\le n} d'(k)}\right\}\right\}$$

where d(i, j) represents the distance between clusters i and j, and $d'(k)$ measures the intra-cluster distance of cluster k. The inter-cluster distance $d(i, j)$ between two clusters may be any number of distance measures, such as the distance between the centroids of the clusters. Similarly, the intra-cluster distance $d'(k)$ may be measured in a variety ways, such as the maximal distance between any pair of elements in cluster k. Since internal criterion seek clusters with high intra-cluster similarity and low inter-cluster similarity, algorithms that produce clusters with high Dunn index are more desirable.

External Evaluation

In external evaluation, clustering results are evaluated based on data that was not used for clustering, such as known class labels and external benchmarks. Such benchmarks consist of a set of pre-classified items, and these sets are often created by human (experts). Thus, the benchmark sets can be thought of as a gold standard for evaluation. These types of evaluation methods measure how close the clustering is to the predetermined benchmark classes. However, it has recently been discussed whether this is adequate for real data, or only on synthetic data sets with a factual ground truth, since classes can contain internal structure, the attributes present may not allow separation of clusters or the classes may contain anomalies. Additionally, from a knowledge discovery point of view, the reproduction of known knowledge may not necessarily be the intended result.

Some of the measures of quality of a cluster algorithm using external criterion include :

- Rand measure : The Rand index computes how similar the clusters (returned by the clustering algorithm) are to the benchmark classifications. One can also view the Rand index as a measure of the percentage of correct decisions made by the algorithm. It can be computed using the following formula :

$$RI = \frac{TP + TN}{TP + FP + FN + TN}$$

where TP is the number of true positives, TN is the number of true negatives, FP is the number of false positives, and FN is the number of false negatives. One issue with the Rand index is that false positives and false negatives are equally weighted. This may be an undesirable characteristic for some clustering applications. The F-measure addresses this concern.

- *F-measure* : The F-measure can be used to balance the contribution of false negatives by weighting recall through a parameter $\beta \geq 0$. Let precision and recall be defined as follows :

$$P = \frac{TP}{TP + FP}$$

$$R = \frac{TP}{TP + FN}$$

where P is the precision rate and R is the recall rate. We can calculate the F-measure by using the following formula :

$$F_\beta = \frac{(\beta^2 + 1) \cdot P \cdot R}{\beta^2 \cdot P + R}$$

Notice that when $\beta = 0$, $F_0 = P$. In other words, recall has no impact on the F-measure when $\beta = 0$, and increasing β allocates an increasing amount of weight to recall in the final F-measure.

- *Pair-counting F-Measure* : Is the F-Measure applied to the set of object pairs, where objects are paired with each other when they are part of the same cluster. This measure is able to compare clusterings with different numbers of clusters.
- *Jaccard index* : The Jaccard index is used to quantify the similarity between two data-sets. The Jaccard index takes on a value between 0 and 1. An index of 1 means that the two data-set are identical, and an index of 0 indicates that the data-sets have no common elements. The Jaccard index is defined by the following formula :

$$J(A, B) = \frac{|A \cap B|}{|A \cup B|} = \frac{TP}{TP + FP + FN}$$

This is simply the number of unique elements common to both sets divided by the total number of unique elements in both sets.

- ***Fowlkes–Mallows index*** *(E. B. Fowlkes & C. L. Mallows 1983)* : The Fowlkes-Mallows index computes the similarity between the clusters returned by the clustering algorithm and the benchmark classifications. The higher the value of the Fowlkes-Mallows index the more similar the clusters and the benchmark classifications are. It can be computed using the following formula :

$$FM = \sqrt{\frac{TP}{TP + FP} \cdot \frac{TP}{TP + FN}}$$

where TP is the number of true positives, FP is the number of false positives, and FN is the number of false negatives. The FM index is the geometric mean of the precision and recall P and R, while the F-measure is their harmonic mean. Moreover, precision and recall are also known as Wallace's indices B^I and B^{II}.

- *Confusion matrix* : A confusion matrix can be used to quickly visualize the results of a classification (or clustering) algorithm. It shows how different a cluster is from the gold standard cluster.
- The **Mutual Information** is an information theoretic measure of how much information is shared between a clustering and a ground-truth classification that can detect a non-linear similarity between two clusterings. Adjusted mutual information is the corrected-for-chance variant of this that has a reduced bias for varying cluster numbers.

Clustering Axioms

Given that there is a myriad of clustering algorithms and objectives, it is helpful to reason about clustering independently of any particular algorithm, objective function, or generative data model. This can be achieved by defining a clustering function as one that satisfies a set of properties. This is often termed as an Axiomatic System. Functions that satisfy the basic axioms are called clustering functions.

Formal Preliminaries

A partitioning function acts on a set S of $n \geq 2$ points along with an integer $k < 0$, and pairwise distances among the points in S. The points are not assumed to belong to any specific larger set or space; the pairwise distances are the only data the partitioning function has about them. We may label the points in S using the numbers $\{1, 2,..., n\}$. The pairwise distances define a distance function $d : S \times S \rightarrow R$ which should have the properties of a semimetric : for any $i, j \in S$, we must have $d(i, j) \geq 0$, $d(i, j) = d(j, i)$, and $d(i, j) = 0$ if and only if $i = j$. In other words, the distances must be non-negative, symmetric, and two points have distance zero if and only if they are the same point.

A partitioning function F takes a distance function d on $S \times S$ and an integer $k \geq 1$and returns a k-partition of S, a collection of non-empty disjoint sub-sets of S whose union is S. The sets making up the k-partition are the clusters. Two clustering functions are equivalent if and only if they output the same partitioning on all values of d and k.

Axioms

Now in an effort to distinguish clustering functions from partitioning functions, we lay down some properties that one may like a clustering function to satisfy. Here is the first one. If d is a distance function, then we define $\alpha \cdot d$ to be the same function with all distances multiplied by α.

Scale-invariance : For any distance function d, number of clusters k, and scalar $\alpha > 0$, we have $F(d, k) = F(\alpha \cdot d, k)$

This property simply requires the function to be immune to stretching or shrinking the data points linearly. It effectively disallows clustering functions to be sensitive to changes in units of measurement—which is desirable. We would like clustering functions to not have any predefined hard-coded distance values in their decision process.

The next property ensures that the clustering function is "rich" in types of partitioning it could output. For a fixed S and k, Let Range $(F(\bullet, k))$ be the set of all possible outputs while varying d.

k-richess : For any number of clusters k, Range $(F(\bullet, k))$ is equal to the set of all k-partitions of S.

In other words, if we are given a set of points such that all we know about the points are pairwise distances, then for any partitioning Γ, there should exist a d such that $F(d, k) = \Gamma$. By varying distances amongst points, we should be able to obtain all possible k-partitionings.

The next property is more subtle. We call a partitioning function "consistent" if it satisfies the following : when we shrink distances between points in the same cluster and expand distances between points in different clusters, we get the same result. Formally, we say that d' is a Γ-*transformation* of d if (a) for all $i, j \in S$ belonging to the same cluster of Γ, we have $d'(i, j) \leq d(i, j)$; and (b) for all

$i, j \in S$ belonging to different clusters of Γ, we have $d'(i, j) \geq d(i, j)$. In other words, d' is a transformation of d such that points inside the same cluster are brought closer together and points not inside the same cluster are moved further away from one another.

Consistency : Fix k. Let d be a distance function, and d' be a $F(d, k)$-transformation of d. Then $F(d, k) = F(d' k)$.

In other words, suppose that we run the partitioning function F on d to get back a particular partitioning Γ. Now, with respect to Γ, if we shrink in-cluster distances or expand between-cluster distances and run F again, we should still get back the same result-namely Γ.

The partitioning function F is forced to return a fixed number of clusters : k. If this were not the case, then the above three properties could never be satisfied by any function. In many popular clustering algorithms such as k-means, Single-Linkage, and spectral clustering, the number of clusters to be returned is determined beforehand—by the human user or other methods—and passed into the clustering function as a parameter.

Applications

Biology, computational biology and bioinformatics

Plant and animalecology : Cluster analysis is used to describe and to make spatial and temporal comparisons of communities (assemblages) of organisms in heterogeneous environments; it is also used in plant systematics to generate artificial phylogenies or clusters of organisms (individuals) at the species, genus or higher level that share a number of attributes

Transcriptomics : Clustering is used to build groups of genes with related expression patterns (also known as coexpressed genes). Often such groups contain functionally related proteins, such as enzymes for a specific pathway, or genes that are co-regulated. High throughput experiments using expressed sequence tags (ESTs) or DNA microarrays can be a powerful tool for genome annotation, a general aspect of genomics.

Sequence analysis : Clustering is used to group homologous sequences into gene families. This is a very important concept in bioinformatics, and evolutionary biology in general.

High-throughput genotyping platforms : Clustering algorithms are used to automatically assign genotypes.

Human genetic clustering : The similarity of genetic data is used in clustering to infer population structures.

Medicine

Medical imaging : On PET scans, cluster analysis can be used to differentiate between different types of tissue and blood in a three-dimensional image. In this application, actual position does not matter, but the voxel intensity is considered as a vector, with a dimension for each image that was taken over time. This technique allows, for example, accurate measurement of the rate a radio-active tracer is delivered to the area of interest, without a separate sampling of arterial blood, an intrusive technique that is most common today.

IMRT segmentation : Clustering can be used to divide a fluence map into distinct regions for conversion into deliverable fields in MLC-based Radiation Therapy.

Business and Marketing

Market research : Cluster analysis is widely used in market research when working with multivariate data from surveys and test panels. Market researchers use cluster analysis to partition the general population of consumers into market segments and to better understand the relationships between different groups of consumers/potential customers, and for use in market segmentation, Product positioning, New product development and Selecting test markets.

Grouping of shopping items : Clustering can be used to group all the shopping items available on the web into a set of unique products. For example, all the items on eBay can be grouped into unique products. (eBay doesn't have the concept of a SKU)

World Wide Web

Social network analysis : In the study of social networks, clustering may be used to recognize communities within large groups of people.

Search result grouping : In the process of intelligent grouping of the files and websites, clustering may be used to create a more relevant set of search results compared to normal search engines like Google. There are currently a number of web based clustering tools such as Clusty.

Slippy map optimization : Flickr's map of photos and other map sites use clustering to reduce the number of markers on a map. This makes it both faster and reduces the amount of visual clutter.

Computer Science

Software evolution : Clustering is useful in software evolution as it helps to reduce legacy properties in code by reforming functionality that has become dispersed. It is a form of restructuring and hence is a way of directly preventative maintenance.

Image segmentation : Clustering can be used to divide a digital image into distinct regions for border detection or object recognition.

Evolutionary algorithms : Clustering may be used to identify different niches within the population of an evolutionary algorithm so that reproductive opportunity can be distributed more evenly amongst the evolving species or subspecies.

Recommender systems : Recommender systems are designed to recommend new items based on a user's tastes. They sometimes use clustering algorithms to predict a user's preferences based on the preferences of other users in the user's cluster.

Markov chain Monte Carlo methods : Clustering is often utilized to locate and characterize extrema in the target distribution.

Social Science

Crime analysis : Cluster analysis can be used to identify areas where there are greater incidences of particular types of crime. By identifying these distinct areas or "hot spots" where a similar crime has happened over a period of time, it is possible to manage law enforcement resources more effectively.

Educational data mining : Cluster analysis is for example used to identify groups of schools or students with similar properties.

Typologies : From poll data, projects such as those undertaken by the Pew Research Center use cluster analysis to discern typologies of opinions, habits, and demographics that may be useful in politics and marketing.

Others

Field robotics : Clustering algorithms are used for robotic situational awareness to track objects and detect outliers in sensor data.

Mathematical chemistry : To find structural similarity, etc., for example, 3000 chemical compounds were clustered in the space of 90 topological indices.

Climatology : To find weather regimes or preferred sea level pressure atmospheric patterns.

Petroleum geology : Cluster analysis is used to reconstruct missing bottom hole core data or missing log curves in order to evaluate reservoir properties.

Physical geography : The clustering of chemical properties in different sample locations.

Chapter 4

ASSOCIATION RULES MINING IN INVENTORY DATABASE

APRIORI ALGORITHM

Apriori is an algorithm for frequent item set mining and association rule learning over transactional databases. It proceeds by identifying the frequent individual items in the database and extending them to larger and larger item-sets as long as those item-sets appear sufficiently often in the database. The frequent item-sets determined by Apriori can be used to determine association rules which highlight general trends in the database : this has applications in domains such as market basket analysis.

Setting

Apriori is designed to operate on databases containing transactions (for example, collections of items bought by customers, or details of a website frequentation). Other algorithms are designed for finding association rules in data having no transactions (Winepi and Minepi), or having no time-stamps (DNA sequencing). Each transaction is seen as a set of items (an *item-set*). Given a threshold C, the Apriori algorithm identifies the item-sets which are sub-sets of at least C transactions in the database.

Apriori uses a "bottom up" approach, where frequent sub-sets are extended one item at a time (a step known as *candidate generation*), and groups of candidates are tested against the data. The algorithm terminates when no further successful extensions are found.

Apriori uses breadth-first search and a Hash tree structure to count candidate item-sets efficiently. It generates candidate item-sets of length k from item-sets of length $k - 1$. Then it prunes the candidates which have an infrequent sub-pattern. According to the downward closure lemma, the candidate set contains all frequent

k-length item-sets. After that, it scans the transaction database to determine frequent item-sets among the candidates.

The pseudo code for the algorithm is given below for a transaction database T, and a support threshold of $\in$. Usual set theoretic notation is employed, though note that T is a multiset. C_k is the candidate set for level k. Generate() algorithm is assumed to generate the candidate sets from the large item-sets of the preceding level, heeding the downward closure lemma. count $[c]$ accesses a field of the data structure that represents candidate set c, which is initially assumed to be zero. Many details are omitted below, usually the most important part of the implementation is the data structure used for storing the candidate sets, and counting their frequencies.

Apriori (T, $\in$)

$L_1 \leftarrow$ {large 1 – item-sets}

$k \leftarrow 2$

while $L_{k-1} \neq 0$

$C_k \leftarrow \{a \cup \{b\} \mid a \in L_{k-1} \wedge b \in \cup L_{k-1} \wedge b \notin a\}$

for transactions $t \in T$

$C_t \leftarrow \{c \mid c \in C_k \wedge c \subseteq t\}$

for candidates $c \in C_t$

count$[c] \leftarrow$ count $[c] + 1$

$L_k \leftarrow \{c \mid c \in C_k \wedge \text{count}[c] \geq \in\}$

$k \leftarrow k + 1$

return $\bigcup_k L_k$

Examples

Example 1

Consider the following database, where each row is a transaction and each cell is an individual item of the transaction :

alpha	beta	epsilon
alpha	beta	Theta
alpha	beta	Epsilon
alpha	beta	Theta

The association rules that can be determined from this database are the following :

1. 100% of sets with alpha also contain beta
2. 50% of sets with alpha, beta also have epsilon

3. 50% of sets with alpha, beta also have theta
   we can also illustrate this through variety of examples.

Example 2

Assume that a large supermarket tracks sales data by stock-keeping unit (SKU) for each item : each item, such as "butter" or "bread", is identified by a numerical SKU. The supermarket has a database of transactions where each transaction is a set of SKUs that were bought together.

Let the database of transactions consist of following item-sets :

Item-sets
{1,2,3,4}
{1,2,4}
{1,2}
{2,3,4}
{2,3}
{3,4}
{2,4}

We will use Apriori to determine the frequent item-sets of this database. To do so, we will say that an item set is frequent if it appears in at least 3 transactions of the database : the value 3 is the *support threshold.*

The first step of Apriori is to count up the number of occurrences, called the support, of each member item separately, by scanning the database a first time. We obtain the following result :

Item	*Support*
{1}	3
{2}	6
{3}	4
{4}	5

All the item-sets of size 1 have a support of at least 3, so they are all frequent.

The next step is to generate a list of all pairs of the frequent items :

Item	*Support*
{1,2}	3
{1,3}	1
{1,4}	2
{2,3}	3
{2,4}	4
{3,4}	3

The pairs {1,2}, {2,3}, {2,4}, and {3,4} all meet or exceed the minimum support of 3, so they are frequent. The pairs {1,3} and {1,4} are not. Now, because {1,3} and {1,4} are not frequent, any larger set which contains {1,3} or {1,4} cannot be frequent. In this way, we can *prune* sets : we will now look for frequent triples in the database, but we can already exclude all the triples that contain one of these two pairs :

Item	*Support*
{2,3,4}	2

In the example, there are no frequent triplets—{2,3,4} is below the minimal threshold, and the other triplets were excluded because they were super sets of pairs that were already below the threshold.

We have thus determined the frequent sets of items in the database, and illustrated how some items were not counted because one of their sub-sets was already known to be below the threshold.

Limitations

Apriori, while historically significant, suffers from a number of inefficiencies or trade-offs, which have spawned other algorithms. Candidate generation generates large numbers of sub-sets (the algorithm attempts to load up the candidate set with as many as possible before each scan). Bottom-up sub-set exploration (essentially a breadth-first traversal of the sub-set lattice) finds any maximal sub-set S only after all $2^{|S|} - 1$ of its proper sub-sets.

Later algorithms such as Max-Miner try to identify the maximal frequent item-sets without enumerating their sub-sets, and perform "jumps" in the search space rather than a purely bottom-up approach.

SEQUENTIAL PATTERN MINING

Sequential Pattern mining is a topic of data mining concerned with finding statistically relevant patterns between data examples where the values are delivered in a sequence. It is usually presumed that the values are discrete, and thus time series mining is closely related, but usually considered a different activity. Sequential pattern mining is a special case of structured data mining.

There are several key traditional computational problems addressed within this field. These include building efficient databases and indexes for sequence information, extracting the frequently occurring patterns, comparing sequences for similarity, and recovering missing sequence members. In general, sequence mining problems can be classified as *string mining* which is typically based on string processing algorithms and *item-set mining* which is typically based on association rule learning.

String Mining

String mining typically deals with a limited alphabet for items that appear in a sequence, but the sequence itself may be typically very long. Examples of an alphabet can be those in the ASCII character set used in natural language text, nucleotide bases 'A', 'G', 'C' and 'T' in DNA sequences, or amino acids for protein sequences. In biology applications analysis of the arrangement of the alphabet in strings can be used to examine gene and protein sequences to determine their properties. Knowing the sequence of letters of a DNA a protein is not an ultimate goal in itself. Rather, the major task is to understand the sequence, in terms of its structure and biological function. This is typically achieved first by identifying individual regions or structural units within each sequence and then assigning a function to each structural unit. In many cases this requires comparing a given sequence with previously studied ones. The comparison between the strings becomes complicated when insertions, deletions and mutations occur in a string.

A survey and taxonomy of the key algorithms for sequence comparison for bioinformatics is presented by Abouelhoda & Ghanem, which include :

- *Repeat-related problems :* That deal with operations on single sequences and can be based on exact string matching or approximate string matching methods for finding dispersed fixed length and maximal length repeats, finding tandem repeats, and finding unique subsequences and missing (un-spelled) subsequences.
- *Alignment problems :* That deal with comparison between strings by first aligning one or more sequences; examples of popular methods include BLAST for comparing a single sequence with multiple sequences in a database, and ClustalW for multiple alignments. Alignment algorithms can be based on either exact or approximate methods, and can also be classified as global alignments, semi-global alignments and local alignment.

Item-set Mining

Some problems in sequence mining lend themselves discovering frequent itemsets and the order they appear, for example, one is seeking rules of the form "if a {customer buys a car}, he or she is likely to {buy insurance} within 1 week", or in the context of stock prices, "if {Nokia up and Ericsson Up}, it is likely that {Motorola up and Samsung up} within 2 days". Traditionally, item-set mining is used in marketing applications for discovering regularities between frequently co-occurring items in large transactions. For example, by analysing transactions of customer shopping baskets in a supermarket, one can produce a rule which reads "if a customer buys onions and potatoes together, he or she is likely to also buy hamburger meat in the same transaction".

A survey and taxonomy of the key algorithms for item set mining is presented by Han *et. al.*.

The two common techniques that are applied to sequence databases for frequent item-set mining are the influential apriori algorithm and the more-recent FP-Growth technique.

Application

With a great variation of products and user buying behaviours, shelf on which products are being displayed is one of the most important resources in retail environment. Retailers can not only increase their profit but, also decrease cost by proper management of shelf space allocation and products display. To solve this problem, George and Binu have proposed an approach to mine user buying patterns using PrefixSpan algorithm and place the products on shelves based on the order of mined purchasing patterns.

Chapter 5

Fuzzy Modelling and Optimization : Theory and Methods

FUZZY LOGIC

Fuzzy logic is a form of many-valued logic; it deals with reasoning that is approximate rather than fixed and exact. Compared to traditional binary sets (where variables may take on true or false values) fuzzy logic variables may have a truth value that ranges in degree between 0 and 1. Fuzzy logic has been extended to handle the concept of partial truth, where the truth value may range between completely true and completely false. Furthermore, when linguistic variables are used, these degrees may be managed by specific functions. Irrationality can be described in terms of what is known as the fuzzjective.

The term "fuzzy logic" was introduced with the 1965 proposal of fuzzy set theory by Lotfi A. Zadeh. Fuzzy logic has been applied to many fields, from control theory to artificial intelligence. Fuzzy logics however had been studied since the 1920s as infinite-valued logics notably by Łukasiewicz and Tarski.

Overview

Classical logic only permits propositions having a value of truth or falsity. The notion of whether 1+1=2 is absolute, immutable, mathematical truth. However, there exist certain propositions with variable answers, such as asking various people to identify a colour. The notion of truth doesn't fall by the wayside, but rather a means of representing and reasoning over partial knowledge is afforded, by aggregating all possible outcomes into a dimensional spectrum.

Both degrees of truth and probabilities range between 0 and 1 and hence may seem similar at first. For example, let a 100 ml glass contain 30 ml of water. Then we may consider two concepts : Empty and Full. The meaning of each of them can be represented by a certain fuzzy set. Then one might define the glass as being

0.7 empty and 0.3 full. Note that the concept of emptiness would be subjective and thus would depend on the observer or designer. Another designer might equally well design a set membership function where the glass would be considered full for all values down to 50 ml. It is essential to realize that fuzzy logic uses truth degrees as a mathematical model of the vagueness phenomenon while probability is a mathematical model of ignorance.

Applying Truth Values

A basic application might characterize sub-ranges of a continuous variable. For instance, a temperature measurement for anti-lock brakes might have several separate membership functions defining particular temperature ranges needed to control the brakes properly. Each function maps the same temperature value to a truth value in the 0 to 1 range. These truth values can then be used to determine how the brakes should be controlled.

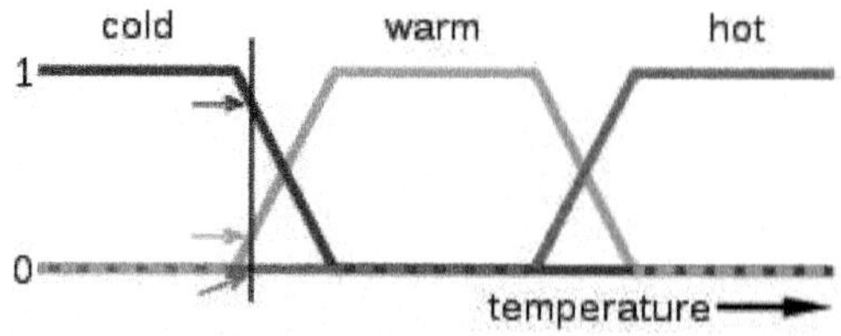

Fig. : Fuzzy logic temperature.

In this image, the meanings of the expressions *cold, warm,* and *hot* are represented by functions mapping a temperature scale. A point on that scale has three "truth values" – one for each of the three functions. The vertical line in the image represents a particular temperature that the three arrows (truth values) gauge. Since the red arrow points to zero, this temperature may be interpreted as "not hot". The orange arrow (pointing at 0.2) may describe it as "slightly warm" and the blue arrow (pointing at 0.8) "fairly cold".

Linguistic Variables

While variables in mathematics usually take numerical values, in fuzzy logic applications, the non-numeric are often used to facilitate the expression of rules and facts.

A linguistic variable such as *age* may have a value such as *young* or its antonym *old*. However, the great utility of linguistic variables is that they can be modified via linguistic hedges applied to primary terms. The linguistic hedges can be associated with certain functions.

Early Applications

The Japanese were the first to utilize fuzzy logic for practical applications. The first notable application was on the high-speed train in Sendai, in which fuzzy logic was able to improve the economy, comfort, and precision of the ride. It has also been used in recognition of hand written symbols in Sony pocket computers, Canon auto-focus technology, Omron auto-aiming cameras, earthquake prediction and modelling at the Institute of Seismology Bureau of Metrology in Japan, etc.

Example

Hard Science with IF-THEN Rules

Fuzzy set theory defines fuzzy operators on fuzzy sets. The problem in applying this is that the appropriate fuzzy operator may not be known. For this reason, fuzzy logic usually uses IF-THEN rules, or constructs that are equivalent, such as fuzzy associative matrices.

Rules are usually expressed in the form : IF *variable* IS *property* THEN *action*

For example, a simple temperature regulator that uses a fan might look like this :

IF temperature IS very cold THEN stop fan

IF temperature IS cold THEN turn down fan

IF temperature IS normal THEN maintain level

IF temperature IS hot THEN speed up fan.

There is no "ELSE"-all of the rules are evaluated, because the temperature might be "cold" and "normal" at the same time to different degrees.

The AND, OR, and NOT operators of boolean logic exist in fuzzy logic, usually defined as the minimum, maximum, and complement; when they are defined this way, they are called the *Zadeh operators*. So for the fuzzy variables x and y :

NOT x = (1-truth(x))

x AND y = minimum(truth(x), truth(y))

x OR y = maximum(truth(x), truth(y))

There are also other operators, more linguistic in nature, called *hedges* that can be applied. These are generally adverbs such as "very", or "somewhat", which modify the meaning of a set using a mathematical formula.

Logical Analysis

In mathematical logic, there are several formal systems of "fuzzy logic"; most of them belong among so-called t-norm fuzzy logics.

Propositional Fuzzy Logics

The most important propositional fuzzy logics are :

- Monoidal t-norm-based propositional fuzzy logic MTL is an axiomatization of logic where conjunction is defined by a left continuous t-norm, and implication is defined as the residuum of the t-norm. Its models correspond to MTL-algebras that are pre-linear commutative bounded integral residuated lattices.

- Basic propositional fuzzy logic BL is an extension of MTL logic where conjunction is defined by a continuous t-norm, and implication is also defined as the residuum of the t-norm. Its models correspond to BL-algebras.
- Łukasiewicz fuzzy logic is the extension of basic fuzzy logic BL where standard conjunction is the Łukasiewicz t-norm. It has the axioms of basic fuzzy logic plus an axiom of double negation, and its models correspond to MV-algebras.
- Gödel fuzzy logic is the extension of basic fuzzy logic BL where conjunction is Gödel t-norm. It has the axioms of BL plus an axiom of idempotence of conjunction, and its models are called G-algebras.
- Product fuzzy logic is the extension of basic fuzzy logic BL where conjunction is product t-norm. It has the axioms of BL plus another axiom for cancellativity of conjunction, and its models are called product algebras.
- Fuzzy logic with evaluated syntax (sometimes also called Pavelka's logic), denoted by EVŁ, is a further generalization of mathematical fuzzy logic. While the above kinds of fuzzy logic have traditional syntax and many-valued semantics, in EVŁ is evaluated also syntax. This means that each formula has an evaluation. Axiomatization of EVŁ stems from Łukasziewicz fuzzy logic. A generalization of classical Gödel completeness theorem is provable in EVŁ.

Predicate Fuzzy Logics

These extend the above-mentioned fuzzy logics by adding universal and existential quantifiers in a manner similar to the way that predicate logic is created from propositional logic. The semantics of the universal (resp. existential) quantifier in t-norm fuzzy logics is the infimum (resp. supremum) of the truth degrees of the instances of the quantified subformula.

Decidability issues for fuzzy logic : The notions of a "decidable sub-set" and "recursively enumerable sub-set" are basic ones for classical mathematics and classical logic. Thus the question of a suitable extension of these concepts to fuzzy set theory arises. A first proposal in such a direction was made by E.S. Santos by the notions of *fuzzy Turing machine, Markov normal fuzzy algorithm* and *fuzzy program*. Successively, L. Biacino and G. Gerla argued that the proposed definitions are rather questionable and therefore they proposed the following ones. Denote by $\ddot{U}$ the set of rational numbers in [0,1]. Then a fuzzy sub-set $s : S \rightarrow [0,1]$ of a set S is *recursively enumerable* if a recursive map $h : S{\times}N \rightarrow \ddot{U}$ exists such that, for every x in S, the function $h(x,n)$ is increasing with respect to n and $s(x) = \lim h(x,n)$. We say that s is *decidable* if both s and its complement $-s$are recursively enumerable. An extension of such a theory to the general case of the L-sub-sets is possible. The proposed definitions are well related with fuzzy logic. Indeed, the following theorem holds true (provided that the deduction apparatus of the considered fuzzy logic satisfies some obvious effectiveness property).

Theorem : Any axiomatizable fuzzy theory is recursively enumerable. In particular, the fuzzy set of logically true formulas is recursively enumerable in spite of the fact that the crisp set of valid formulas is not recursively enumerable, in general. Moreover, any axiomatizable and complete theory is decidable.

It is an open question to give supports for a *Church thesis* for fuzzy mathematics the proposed notion of recursive enumerability for fuzzy sub-sets is the adequate one. To this aim, an extension of the notions of fuzzy grammar and fuzzy Turing machine should be necessary. Another open question is to start from this notion to find an extension of Gödel's theorems to fuzzy logic.

Synthesis of Fuzzy Logic Functions Given in Tabular Form

It is known that any boolean logic function could be represented using a truth table mapping each set of variable values into set of values {0,1}. The task of synthesis of boolean logic function given in tabular form is one of basic tasks in traditional logic that is solved via disjunctive (conjunctive) perfect normal form.

Each fuzzy (continuous) logic function could be represented by a choice table containing all possible variants of comparing arguments and their negations. A choice table maps each variant into value of an argument or a negation of an argument. For instance, for two arguments a row of choice table contains a variant of comparing values x_1, $\neg x_1$, x_2, $\neg x_2$ and the corresponding function value

$$f(x_2 \leq \neg x_1 \leq x_1 \leq \neg x_2) = \neg x_1$$

The task of synthesis of fuzzy logic function given in tabular form was solved in. New concepts of constituents of minimum and maximum were introduced. The sufficient and necessary conditions that a choice table defines a fuzzy logic function were derived.

Fuzzy Databases

Once fuzzy relations are defined, it is possible to develop fuzzy relational databases. The first fuzzy relational database, FRDB, appeared in Maria Zemankova's dissertation. Later, some other models arose like the Buckles-Petry model, the Prade-Testemale Model, the Umano-Fukami model or the GEFRED model by J.M. Medina, M.A. Vila *et. al.* In the context of fuzzy databases, some fuzzy querying languages have been defined, highlighting the SQLf by P. Bosc *et. al.* and the FSQL by J. Galindo *et. al.* These languages define some structures in order to include fuzzy aspects in the SQL statements, like fuzzy conditions, fuzzy comparators, fuzzy constants, fuzzy constraints, fuzzy thresholds, linguistic labels and so on.

Much progress has been made to take fuzzy logic database applications to the web and let the world easily use them, for example : http://sullivansoftwaresystems.com/cgi-bin/fuzzy-logic-match-algorithm.cgi?SearchString=gariaThis enables fuzzy logic matching to be incorporated into a database system or application.

Comparison to Probability

Fuzzy logic and probability are different ways of expressing uncertainty. While both fuzzy logic and probability theory can be used to represent subjective belief, fuzzy set theory uses the concept of fuzzy set membership (*i.e., how much* a variable is in a set), and probability theory uses the concept of subjective probability (*i.e., how probable* do I think that a variable is in a set). While this distinction is mostly philosophical, the fuzzy-logic-derived possibility measure is inherently different from the probability measure, hence they are not *directly* equivalent. However, many statisticians are persuaded by the work of Bruno de Finetti that only one kind of mathematical uncertainty is needed and thus fuzzy logic is unnecessary. On the other hand, Bart Koskoargues that probability is a subtheory of fuzzy logic, as probability only handles one kind of uncertainty. He also claims to have proven a derivation of Bayes' theorem from the concept of fuzzy sub-sethood. Lotfi A. Zadeh argues that fuzzy logic is different in character from probability, and is not a replacement for it. He fuzzified probability to fuzzy probability and also generalized it to what is called possibility theory. (cf.) More generally, fuzzy logic is one of many different proposed extensions to classical logic, known as probabilistic logics, intended to deal with issues of uncertainty in classical logic, the inapplicability of probability theory in many domains, and the paradoxes of Dempster-Shafer theory.

Relation to Ecorithms

Harvard's Dr. Leslie Valiant, co-author of the Valiant-Vazirani theorem, uses the term "ecorithms" to describe how many less exact systems and techniques like fuzzy logic (and "less robust" logic) can be applied to learning algorithms. Valiant essentially redefines machine learning as evolutionary. Ecorithms and fuzzy logic also have the common property of dealing with possibilities more than probabilities, although feedback and feedforward, basically stochastic "weights," are a feature of both when dealing with, for example, dynamical systems.

In general use, ecorithms are algorithms that learn from their more complex environments (hence eco) to generalize, approximate and simplify solution logic. Like fuzzy logic, they are methods used to overcome continuous variables or systems too complex to completely enumerate or understand discretely or exactly.

FUZZY SET

In mathematics, **Fuzzy sets** are sets whose elements have degrees of membership. Fuzzy sets were introduced by Lotfi A. Zadeh and Dieter Klaua in 1965 as an extension of the classical notion of set. At the same time, Salii defined a more general kind of structures called *L*-relations, which were studied by him in an abstract algebraic context. Fuzzy relations, which are used now in different areas, such as linguistics, decision-making and clustering, are special cases of *L*-relations when *L* is the unit interval.

In classical set theory, the membership of elements in a set is assessed in binary terms according to a bivalent condition – an element either belongs or does not belong to the set. By contrast, fuzzy set theory permits the gradual assessment of the membership of elements in a set; this is described with the aid of a membership function valued in the real unit interval [0, 1]. Fuzzy sets generalize classical sets, since the indicator functions of classical sets are special cases of the membership functions of fuzzy sets, if the latter only take values 0 or 1. In fuzzy set theory, classical bivalent sets are usually called *crisp* sets. The fuzzy set theory can be used in a wide range of domains in which information is incomplete or imprecise, such as bioinformatics.

It has been suggested by Thayer Watkins that Zadeh's ethnicity is an example of a fuzzy set because "His father was Turkish-Iranian (Azerbaijani) and his mother was Russian. His father was a journalist working in Baku, Azerbaijan in the Soviet Union...Lotfi was born in Baku in 1921 and lived there until his family moved to Tehran in 1931."

Definition(h)

A fuzzy set is a pair (U, m) where U is a set and $m : U = [0, 1]$.

For each $x \in U$, the value $m(x)$ is called the **grade** of membership of x in (U, m). For a finite set $U = \{x_1, \ldots, x_n\}$, the fuzzy set (U, m) is often denoted by $\{m(x_1)/x_1, \ldots, m(x_n)/x_n\}$.

Let $x \in U$. Then x is called **not included** in the fuzzy set (U, m) if $m(x) = 0$, x is called **fully included** if $m(x) = 1$, and x is called a **fuzzy member** if $0 < m(x) < 1$. The set $\{x \in U \mid m(x) > 0\}$ is called the **support** of (U, m) and the set $\{x \in U \mid m(x) = 1\}$ is called its **kernel**. The function m is called the **membership function** of the fuzzy set (U, m).

Sometimes, more general variants of the notion of fuzzy set are used, with membership functions taking values in a (fixed or variable) algebra or structure L of a given kind; usually it is required that L be at least a poset or lattice. These are usually called ***L*-fuzzy sets**, to distinguish them from those valued over the unit interval. The usual membership functions with values in [0, 1] are then called [0, 1]-valued membership functions. These kinds of generalizations were first considered in 1967 by Joseph Goguen, who was a student of Zadeh.

Fuzzy Logic

As an extension of the case of multi-valued logic, valuations ($\mu : V_o \rightarrow W$) of propositional variables (V_o) into a set of membership degrees (W) can be thought of as membership functions mapping predicates into fuzzy sets (or more formally, into an ordered set of fuzzy pairs, called a fuzzy relation). With these valuations, many-valued logic can be extended to allow for fuzzy premises from which graded conclusions may be drawn.

This extension is sometimes called "fuzzy logic in the narrow sense" as opposed to "fuzzy logic in the wider sense," which originated in the engineering fields of automated control and knowledge engineering, and which encompasses many topics involving fuzzy sets and "approximated reasoning."

Industrial applications of fuzzy sets in the context of "fuzzy logic in the wider sense" can be found at fuzzy logic.

Fuzzy Number

A **fuzzy number** is a convex, normalized fuzzy set $\tilde{A} \subseteq R$ whose membership function is at least segmentally continuous and has the functional value $\mu_A(x) = 1$ at precisely one element.

This can be likened to the funfair game "guess your weight," where someone guesses the contestant's weight, with closer guesses being more correct, and where the guesser "wins" if he or she guesses near enough to the contestant's weight, with the actual weight being completely correct (mapping to 1 by the membership function).

Fuzzy Interval

A **fuzzy interval** is an uncertain set $\tilde{A} \subseteq R$ with a mean interval whose elements possess the membership function value $\mu_A(x) = 1$. As in fuzzy numbers, the membership function must be convex, normalized, at least segmentally continuous.

Fuzzy Relation Equation

The fuzzy relation equation is an equation of the form $A \cdot R = B$, where A and B are fuzzy sets, R is a fuzzy relation, and $A \cdot R$ stands for the composition of A with R.

Axiomatic Definition of Credibility

Let A be a non-empty set and P(A) be the power set of A. The set function is known as credibility measure if it satisfies following condition

- *Axiom 1* : $Cr\{A\} = 1$
- *Axiom 2* : If B is sub-set of C, then, $Cr\{B\} \leq Cr\{C\}$
- *Axiom 3* : $Cr\{B\} + Cr\{B^c\} = 1$
- *Axiom 4* : $Cr\{\cup A_i\} = sup_i(A_i)$, for any event A_i with $sup_i Cr\{A_i\} < 0.5$

  Cr{B} indicates how frequently event B will occur.

Credibility Inversion Theorem

Let A be a fuzzy variable with membership function u. Then for any set B of real numbers, we have

$$Cr\{A \in B\} = \frac{1}{2}(Sup_{t \in B} u(t) + 1 - Sup_{t \in B^c} u(t))$$

Expected Value

Let A be a fuzzy variable. Then the expected value is

$$E[A] = \int_0^\infty Cr\{A \geq t\}dt - \int_{-\infty}^0 Cr\{A \leq t\}dt.$$

Entropy

Let A be a fuzzy variable with a continuous membership function. Then its entropy is

$$H[A] = \int_{-\infty}^\infty S(Cr\{A \geq t\})dt.$$

Where

$S(y) = -ylny - (1 - y)$ In $(1-y)$.

FUZZY ENGINEERING

There is more to fuzzy logic than some interesting math, it has some impressive applications in engineering. The main application of fuzzy logic in engineering is in the area of control systems. The definition of a control system, given by Richard Dorf in *Modern Control Systems* is : "An interconnection of components forming a system configuration that will provide a desired response." This means that a control system needs to know the desired response (input) and it needs to process this input and attempt to acheive it. The general control system can then be summarized with the following diagram :

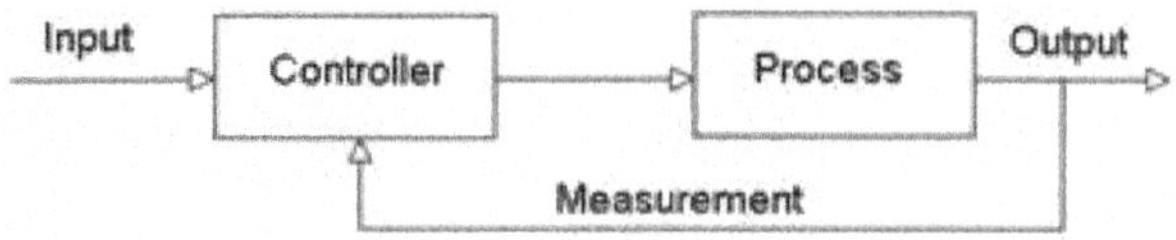

The process is the system that is being controlled and cannot typically be changed. The controller then, must take the input and also take measurements from the process and use this information to generate the appropriate input to the process. A basic example of a controller would be a summing point that will provide the difference between input and output to the process, whereas a more advanced controller would be a PID controller. A fuzzy logic based controller will use fuzzy membership functions and inference rules to determine the appropriate process input. Designing a fuzzy controller is a more intuitive approach to controller design since it uses a comprehendable linguistic rule base.

A fuzzy controller can be broken down into three main processes. The first of these is the fuzzification, this uses defined membership functions to process the inputs and to fuzzify them. These fuzzified inputs are then used in the second part, the rule-based inference system. This system uses previously defined linguistic rules to generate a fuzzy response. The fuzzy response is then defuzzified in the final process : defuzzification. This process will provide a real number as an output.

Chapter 6

DATA-MINING THEORY

VAPNIK–CHERVONENKIS THEORY

Introduction

Vapnik–Chervonenkis theory (also known as **VC theory**) was developed during 1960–1990 by Vladimir Vapnik and Alexey Chervonenkis. The theory is a form of computational learning theory, which attempts to explain the learning process from a statistical point of view.

VC theory is related to **statistical learning theory** and to empirical processes. Richard M. Dudley and Vladimir Vapnik himself, among others, apply VC-theory to empirical processes.

VC theory covers at least four parts (as explained in *The Nature of Statistical Learning Theory*) :

- Theory of consistency of learning processes
  - What are (necessary and sufficient) conditions for consistency of a learning process based on the empirical risk minimization principle?
- Non-asymptotic theory of the rate of convergence of learning processes
  - How fast is the rate of convergence of the learning process?
- Theory of controlling the generalization ability of learning processes
  - How can one control the rate of convergence (the generalization ability) of the learning process?
- Theory of constructing learning machines
  - How can one construct algorithms that can control the generalization ability?

VC Theory is a major sub-branch of statistical learning theory. One of its main applications in statistical learning theory is to provide generalization conditions for learning algorithms. From this point of view, VC theory is related to **stability**, which is an alternative approach for characterizing generalization.

In addition, VC theory and VC dimension are instrumental in the theory of empirical processes, in the case of processes indexed by VC classes. Arguably these are the most important applications of the VC theory, and are employed in proving generalization. Several techniques will be introduced that are widely used in the empirical process and VC theory. The discussion is mainly based on the book "Weak Convergence and Empirical Processes : With Applications to Statistics".

Overview of VC theory in Empirical Processes

Background on Empirical Processes

Let $X_1, X_2..., X_n$ are random elements defined on a measurable space (X, A). Define the empirical measure $P_n = n^{-1}\sum_{i=1}^{n} \delta X_i$, where δ here stands for the dirac measure.

Denote with $Qf = \int fdQ$ for a measure Q. Measurability issues, will be ignored here. For more technical detail consult.

Let F be a class of measurable functions $f : X \to R$. The empirical measure induces a map from F to R given by :

$f \to P_n f$

Let $||Q||_F = \sup\{|Qf| : f \in F\}$. Empirical Processes theory aims at identifying classes F for which statements like :

- $||P_n - P||_F \to 0$, aka uniform law of large numbers
- $G_n = \sqrt{n}(P_n - P) \rightsquigarrow G$,in $\ell^{\infty}(F)$, aka uniform central limit theorem hold. Here

  P is the underlying true distribution of the data, which is unknown in practice. In the former case the class F is called *Glivenko-Cantelli*, and in the latter case (under the assumption $\sup_{f\in F} |f(x) - Pf| < \infty.\forall x$) the class F is called *Donsker* or P-Donsker. Obviously, a Donsker class is Glivenko-Cantelli in probability by an application of Slutsky's theorem.

These statements are true for a single f, by standard LLN, CLT arguments under regularity conditions, and the difficulty in the Empirical Processes comes in because joint statements are being made for all $f \in F$. Intuitively then, the set F cannot be too large, and as it turns out that the geometry of F plays a very important role.

One way of measuring how big the function set F is to use the so-called covering numbers. The covering number $N(\in, F, ||\cdot||)$ is the minimal number of balls $\{g : ||g - f|| < \varepsilon\}$ needed to cover the set F (here it is obviously assumed that there is an underlying norm on F). The entropy is the logarithm of the covering number.

Two sufficient conditions are provided below, under which it can be proved that the set F is Glivenko-Cantelli or Donsker.

A class F is P-Glivenko-Cantelli if it is P-measurable with envelope F such that $P^*F < \infty$ and satisfies :

$$\sup_Q N(\varepsilon \| F \|_Q, F, L_1(Q)) < \infty \text{ for every } \varepsilon > 0.$$

The next condition is a version of the celebrated Dudley's theorem. If F is a class of functions such that

$$\int_0^\infty \sup_Q \sqrt{\log N(\varepsilon \| F \|_{Q,2}, F, L_2(Q))}\, d\varepsilon < \infty$$

then F is P-Donsker for every probability measure P such that $P^*F^2 < \infty$. In the last integral, the notation means $\| f \|_{Q,2} = \left(\int | f |^2 \, dQ\right)^{1/2}$.

Symmetrization

The majority of the arguments of how to bound the empirical process, rely on symmetrization, maximal and concentration inequalities and chaining. Symmetrization is usually the first step of the proofs, and since it is used in many machine learning proofs on bounding empirical loss functions (including the proof of the VC inequality which is discussed in the next section) it is presented here.

Consider the empirical process :

$$f \to (P_n - P)f = \frac{1}{n}\sum_{i=1}^{n}(f(X_i) - Pf)$$

Turns out that there is a connection between the empirical and the following symmetrized process :

$$f \to P_n^0 = \frac{1}{n}\sum_{i=1}^{n}\varepsilon_i f(X_i)$$

The symmetrized process is a Rademacher process, conditionally on the data X_i. Therefore it is a sub-Gaussian process by Hoeffding's inequality.

Lemma (Symmetrization) : For every non-decreasing, convex $\Phi : R \to R$ and class of measurable functions F,

$$E\Phi(\| P_n - P \|_F) < E\Phi(2 \| P^0_n \|_F)$$

The proof of the Symmetrization lemma relies on introducing independent copies of the original variables X_i (sometimes referred to as a *ghost sample*) and replacing the inner expectation of the LHS by these copies. After an application of Jensen's inequality different signs could be introduced (hence the name symmetrization) without changing the expectation. The proof can be found below because of its instructive nature.

Introduce the "ghost sample" $Y_1,\ldots, Y_n$ to be independent copies of $X_1,\ldots, X_n$. For fixed values of $X_1,\ldots, X_n$ one has :

$$\| P_n - P \|_F = \sup_{f\in F}\frac{1}{n}\left|\sum_{i=1}^{n}[f(X_i) - Ef(Y_i)]\right| \le E_Y \sup_{f\in F}\frac{1}{n}\left|\sum_{i=1}^{n}[f(X_i) - f(Y_i)]\right|$$

Therefore by Jensen's inequality :

$$f(\|P_n - P\|_F) \le E_Y \Phi\left(\left\|\frac{1}{n}\sum_{i=1}^{n}[f(X_i) - f(Y_i)]\right\|_F\right)$$

Taking expectation with respect to X gives :

$$E\Phi(\|P_n - P\|_F \le E_X E_Y \Phi\left(\left\|\frac{1}{n}\sum_{i=1}^{n}[f(X_i) - f(Y_i)]\right\|_F\right)$$

Note that adding a minus sign in front of a term $[f(X_i) - f(Y_i)]$ doesn't change the RHS, because it's a symmetric function of

X and Y. Therefore the RHS remains the same under "sign perturbation" :

$$E\Phi\left(\left\|\frac{1}{n}\sum_{i=1}^{n}e_i[f(x_i) - f(Y_i)]\right\|_F\right)$$

for any $(e_1, e_2,..., e_n) \in \{-1, 1\}^n$. Therefore :

$$E\Phi(\|P_n - P\|_F) \le E_\epsilon E\Phi\left(\left\|\frac{1}{n}\sum_{i=1}^{n}\varepsilon_i[f(X_i) - f(Y_i)]\right\|_F\right)$$

Finally using first triangle inequality and then convexity of Φ gives :

$$E\Phi(\|P_n - P\|_F) \le \frac{1}{2}E_\epsilon E\Phi\left(2\left\|\frac{1}{n}\sum_{i=1}^{n}\varepsilon_i f(X_i)\right\|_F\right) + \frac{1}{2}E_\varepsilon E\Phi\left(2\left\|\frac{1}{n}\sum_{i=1}^{n}\varepsilon_i f(Y_i)\right\|_F\right)$$

Where the last two expressions on the RHS are the same, which concludes the proof.

A typical way of proving empirical CLTs, first uses symmetrization to pass the empirical process to P^0_n and then argue conditionally on the data, using the fact that Rademacher processes are simple processes with nice properties.

VC Connection

It turns out that there is a fascinating connection between certain combinatorial properties of the set F and the entropy numbers. Uniform covering numbers can be controlled by the notion of *Vapnik-Cervonenkis classes of sets*-or shortly *VC sets*.

Take a collection of sub-sets of the sample space X - C. A collection of sets C is said to *pick out* a certain sub-set of the finite set $S = \{x_1,..., x_n\} \subset X$ if $S = S \cap C$ for some $C \,\varepsilon\, C$. C is said to *shatter* S if it picks out each of its 2^n sub-sets. The *VC-index* (similar to VC dimension + 1 for an appropriately chosen classifier set) $V(C)$ of C is the smallest n for which no set of size n is shattered by C.

Sauer's lemma then states that the number $\Delta_n(C, x_1,..., x_n)$ of sub-sets picked out by a VC-class C satisfies :

$$\max_{x_1,...,x_n} \Delta_n(C, x_1,...,x_n) \le \sum_{j=0}^{V(C)-1}\binom{n}{j} \le \left(\frac{ne}{V(C)-1}\right)^{V(C)-1}$$

Which is a polynomial number $O(n^{V(C)-1})$ of sub-sets rather than an exponential number. Intuitively this means that a finite VC-index implies that C has an apparent simplistic structure.

A similar bound can be shown (with a different constant, same rate) for the so called *VC sub-graph classes*. For a function $f: X \rightarrow R$ the *sub-graph* is a sub-set of $X \times R$ such that : $\{(x, t) \ : t < f(x)\}$. A collection of F is called a VC sub-graph class if all sub-graphs form a VC-class.

Consider a set of indicator functions $I_C = \{1_C \ : C \in \mathcal{C}\}$ in $L_1(Q)$ for discrete empirical type of measure Q (or equivalently for any probability measure Q). It can then be shown that quite remarkably, for $r \geq 1$:

$$N(\varepsilon, I_C, L_r(Q)) \leq KV(C)(4e)^{V(C)}\left(\frac{1}{\varepsilon}\right)^{r(V(C)-1)}$$

Further consider the *symmetric convex hull* of a set F : sconv F being the collection of functions of the form $\sum_{i=1}^{m} \alpha_i f_i$ with $\sum_{i=1}^{m} |\alpha_i| \leq 1$. Then if

$$N(\varepsilon \, | \, |F| \, |_{Q,2}, F, L_2(Q)) \leq C\left(\frac{1}{\varepsilon}\right)^{V}$$

the following is valid for the convex hull of F :

$$\log N(\varepsilon \, | \, |F| \, |_{Q,2}, \text{sconv } F, L_2(Q)) \leq K\left(\frac{1}{\varepsilon}\right)^{\frac{2V}{V+2}}$$

The important consequence of this fact is that the power of $1/\varepsilon - 2V/(V+2)$ is strictly less than 2, which is just enough so that the entropy integral is going to converge, and therefore the class sconv F is going to be P-Donsker.

Finally an example of a VC-sub-graph class is considered. Any finite-dimensional vector space F of measurable functions $f : X \rightarrow R$ is VC-sub-graph of index smaller than or equal to $\dim(F) + 2$.

Take $n = \dim(F) + 2$ points $(x_1, t_1), \ldots, (x_n, t_n)$. The vectors :

$(f(x_1), \ldots, f(x_n)) - (t_1, \ldots, t_n)$

are in a $n - 1$ dimensional sub-space of R^n. Take $a \neq 0$, a non-zero vector that is orthogonal to this sub-space. Therefore :

$$\sum_{a_i > 0} a_i (f(x_i) - t_i) = \sum_{a_i < 0} (-a_i)(f(x_i) - t_i), \ \forall f \in F$$

Consider the set $S = \{(x_i, t_i) \ : a_i > 0\}$. This set cannot be picked out since if there is some f such that $S = \{(x_i, t_i) \ : f(x_i, t_i) \ : f(x_i) > t_i\}$ that would imply that the LHS is strictly positive but the RHS is non-negative.

There are generalizations of the notion VC sub-graph class, *e.g.* there is the notion of pseudo-dimension. The interested reader can look into.

VC Inequality

A similar setting is considered, which is more common to machine learning. Let X is a feature space and $y = \{0, 1\}$. A function $f : X = y$ is called a classifier. Let F be a set of classifiers. Similarly to the previous section, define the *shattering coefficient* $S(F,n) = \max_{x_1,\dots,x_n} |\{(f(x_1),\dots,f(x_n)), f \in F\}|$. The shattering coefficient is also known as growth function. Note here that there is a 1-1 mapping between each of the functions in F and the set on which the function is 1. Therefore in terms of the previous section the shattering coefficient is precisely $\max_{x_1,\dots x_n} \Delta_n(C, x_1,\dots,x_n)$ for C being the collection of all sets described above. Now for the same reasoning as before, namely using Sauer's Lemma it can be shown that $S(F, n)$ is going to be polynomial in n provided that the class F has a finite VC-dimension or equivalently the collection C has finite VC-index.

Let $D_n = \{(X_1, Y_1),\dots,(X_n, Y_m)\}$ is an observed data-set. Assume that the data is generated by an unknown probability distribution P_{XY}. Define $R(f) = P(f(X) \neq Y)$ to be the expected 0/1 loss. Of course since P_{XY} is unknown in general, one has no access to $R(f)$. However the *empirical risk*, given by :

$$\hat{R}_n(f) = \frac{1}{n}\sum_{i=1}^{n} I(f(X_n) \neq Y_n)$$

can certainly be evaluated. Then one has the following Theorem :

Theorem (VC Inequality) For binary classification and the 0/1 loss function we have the following generalization bounds :

$$P\left(\sup_{f\in F} |\hat{R}_n(f) - R(f)| > \varepsilon\right) \leq 8S(F,n)e^{-n\varepsilon^2/32}$$

and

$$E\left[\sup_{f\in F} \hat{R}_n(f) - R(f)|\right] \leq 2\sqrt{\frac{\log S(F,n) + \log 2}{n}}$$

In words the VC inequality is saying that as the sample increases, provided that F has a finite VC dimension, the empirical 0/1 risk becomes a good proxy for the expecred 0/1 risk. Note that both RHS of the two inequalities will converge to 0, provided that $S(F, n)$ grows polynomially in n.

The connection between this framework and the Empirical Process framework is evident. Here one is dealing with a modified empirical process $\left|\hat{R}_n - R\right|_F$ but not surprisingly the ideas are the same. The proof of the (first part of) VC inequality, relies on symmetrization, and then argue conditionally on the data using concentration inequalities (in particular Hoeffding's inequality). The interested reader can check the book Theorems 12.4 and 12.5.

BIAS-VARIANCE DILEMMA

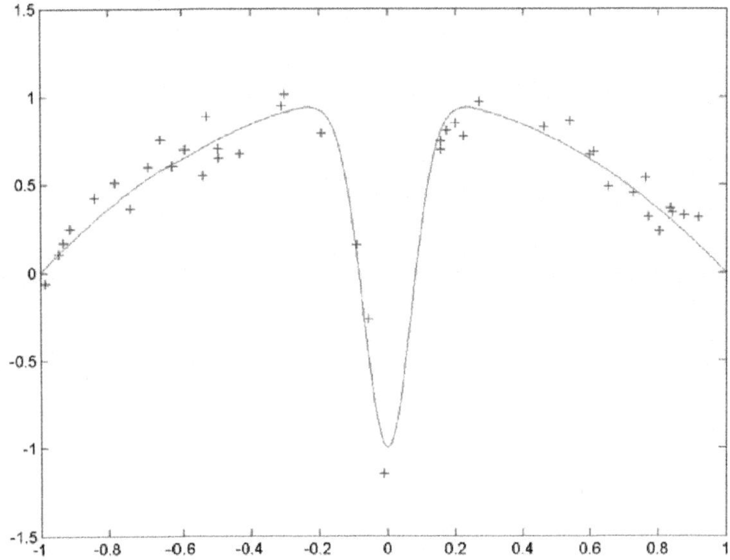

Fig. : Function and noisy data.

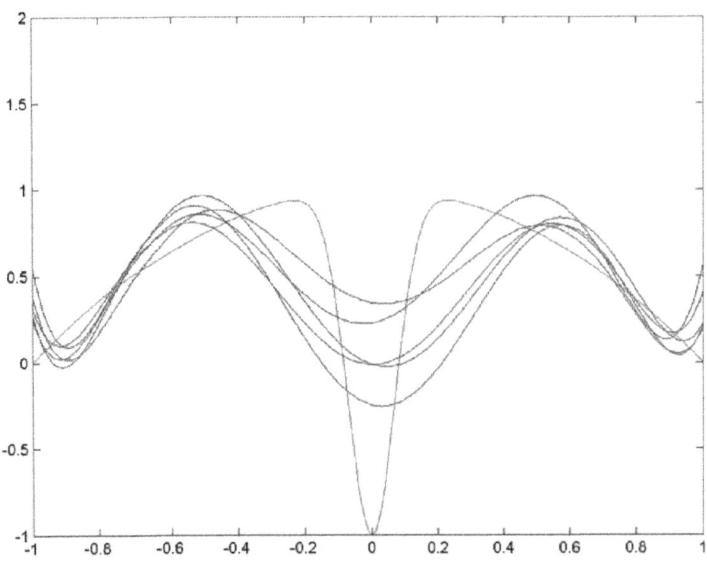

spread=5

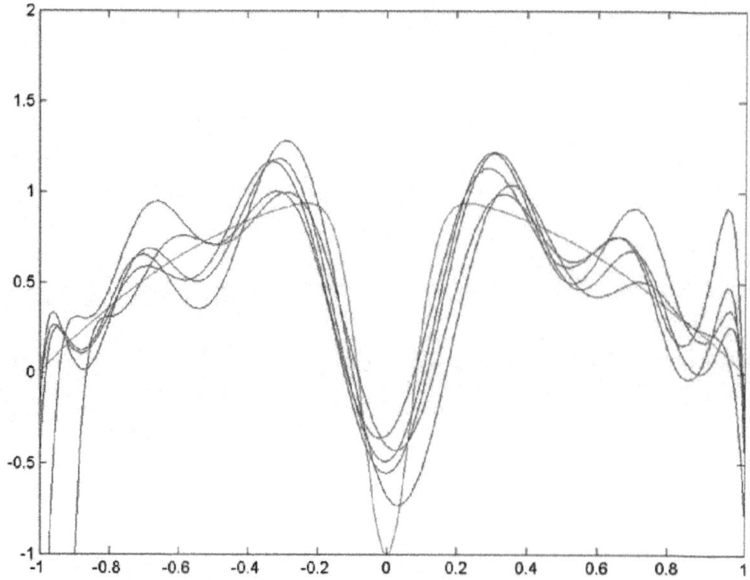

spread=1

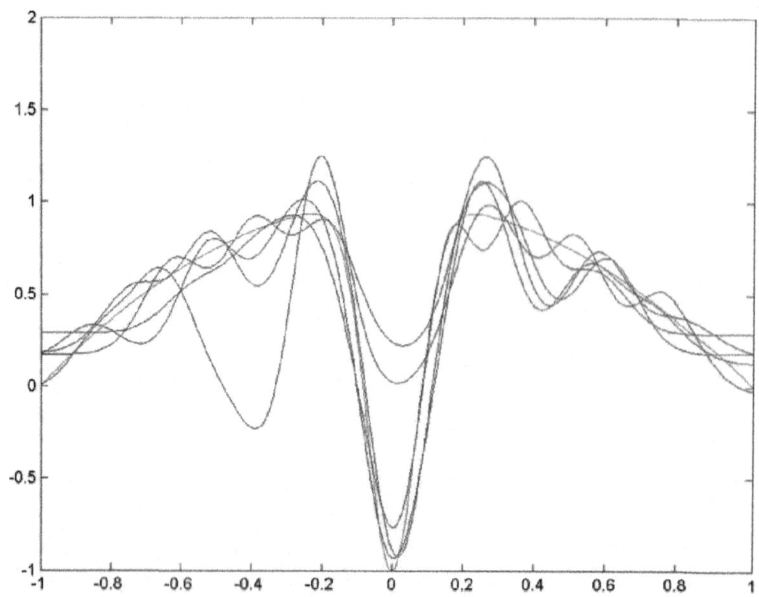

spread=0.1

A function (red) is approximated using radial basis functions (blue). Several trials are shown in each graph. For each trial, a few noisy data points are provided as training set (top). For a wide spread the bias is high : the RBFs cannot fully approximate the function (especially the central dip), but the variance between different trials is low. As spread decreases the bias decreases : the blue curves more closely approximate the red. However, depending on the noise in different trials the variance between trials increases. In the lowermost image the approximated values for x=0 varies wildly depending on where the data points were located.

In machine learning, the **bias–variance dilemma** or **bias–variance tradeoff** is the problem of simultaneously minimizing the bias (how accurate a model is across different training sets) and variance of the model error (how sensitive the model is to small changes in training set). This tradeoff applies to all forms of supervised learning : classification, function fitting, and structured output learning.

The bias–variance tradeoff is a central problem in supervised learning. Ideally, one wants to choose a model that at the same time captures the regularities in its training data, but also generalizes well to unseen data. Models with high bias are intuitively simple models : they impose restrictions on the kind of regularities that can be learned (examples include linear classifiers). The problem with these models is that they underfit, *i.e.* not learn the relationship between predicted (target) variables and features. Models with high variance are those that can learn many kinds of complex regularities—but that includes the possibility to learn noise in the training data, *i.e.* overfitting. To achieve good performance on data outside the training set, a tradeoff must be made.

Relevance

This trade-off is related to issues of overfitting and underfitting. Models that reliably have small deviations from the training set are typically bad predictors for non-training set inputs and will change significantly if the set changes (making them sensitive to outliers and noise). Meanwhile models that systematically deviate *can* resist noise and generalise well, although too strong deviation decreases performance.

Typically the goal is to find an optimal trade-off between bias and variance. A common model selection criterion is that the decrease of bias with increasing model complexity becomes equal to the increase of variance. There are however other possible considerations, such as error losses or complexity costs, that may lead to other trade-offs. The choice of model may also introduce biases that correspond to useful prior information, for example that the output must be within a given interval.

Bias–variance Decomposition

(After)

Suppose we have data $D = \{(x_1, y_1), (x_2, y_2),..., (x_N, y_N)\}$ derived from the true model $y = f(x) + \in$, where $\in$ is a random variable with $E[\in] = 0$.

Given the data, we train a model $g(x \mid D)$ to approximate f. For brevity and clarity $g(x_i \mid D)$ will be written as I_i below. Similarly $f_i = f(x_i)$.

The mean-squared erroris $MSE = (1/N)\sum_{i=1}^{N}(y_i - g(x_i \mid D))^2$.

The value of interest is the expectation of the MSE across different realisations of data :

$$E[MSE] = (1/N)\sum_{i=1}^{N} E[(y_i - g_i)^2]$$

This formula can be split apart :

$$\begin{aligned} E[y_i - g_i)^2] &= E[(y_i - f_i + f_i - g_i)^2] \\ &= E[y_i - f_i)^2] + E[(f_i - g_i)^2] + 2E[(f_i - g_i)(y_i - f_i)] \\ &= E[\in^2\} + E[(f_i - g_i)^2] + 2(E[f_i y_i] - E[f_i^2] - E[g_i y_i] + E[g_i f_i]) \end{aligned}$$

Now, $E[f_i y_i] = f_i^2$ since f is deterministic and $E[y_i] = f_i$, $E[f_i^2] = f_i^2$, and $E[g_i y_i] = E[g_i (f_i + \in)] = E[g_i f_i + g_i \in] = E[g_i f_i] + 0$. Hence the last term above sums to zero :

$$E[(y_i - g_i)^2] = E[\in^2] + E[(f_i - g_i)^2]$$

The second term can be decomposed in the same way using the same trick :

$$E[f_i - g_i)^2] = E[(f_i - E[g_i])^2] + E[(E[g_i] - g_i)^2] + 0$$

Hence

$$\begin{aligned} E[y_i - g_i)^2] &= E[\in^2] + E[(f_i - E[g_i])^2] + E[(E[g_i] - g_i)^2] \\ &= \mathrm{Var}(\in) + (\mathrm{bias})^2 + \mathrm{Var}(g_i) \end{aligned}$$

The expected MSE is hence composed of an irreducible error due to the noise (the first term), a bias due to the model systematically deviating from f (the second term), and a variance due to the model not perfectly predicting the particular realization of the data (the third term).

In the above the data was written as scalar values, but the decomposition works for vectorial data too.

Approaches

Cross-validation is widely used to check model error by testing on data not part of the training set. Multiple rounds with randomly selected test sets are averaged together to reduce variability of the cross-validation; high variability of the model will produce high average errors on the test set.

Dimensionality reduction and feature selection can decrease variance by simplifying models. Similarly, a larger training set tends to decrease variance. Adding features (predictors) tends to decrease bias, at the expense of introducing additional variance. Learning algorithms typically have some tunable parameters that control bias and variance, *e.g.* :

- (Generalized) linear models can be regularized to increase their bias.

- In neural nets, deeper models will have stronger variance. Like in GLMs, regularization is typically applied.
- In k-nearest neighbour models, a high value of *k* leads to high variance.
- In Instance-based learning, regularization can be achived varying the mixture of prototypes and exemplars.
- In decision trees, the depth of the tree determines the variance. Decision trees are commonly pruned to control variance.

One way of resolving the trade-off is to use mixture models and ensemble learning. For example, boosting combines many "weak" (high bias) models in an ensemble that has greater variance than the individual models, while bagging combines "strong" learners in a way that reduces their variance.

COMPUTATIONAL LEARNING THEORY

In theoretical computer science, **computational learning theory** is a mathematical field related to the analysis of machine learning algorithms.

Overview

Theoretical results in machine learning mainly deal with a type of inductive learning called supervised learning. In supervised learning, an algorithm is given samples that are labelled in some useful way. For example, the samples might be descriptions of mushrooms, and the labels could be whether or not the mushrooms are edible. The algorithm takes these previously labelled samples and uses them to induce a classifier. This classifier is a function that assigns labels to samples including the samples that have never been previously seen by the algorithm. The goal of the supervised learning algorithm is to optimize some measure of performance such as minimizing the number of mistakes made on new samples.

In addition to performance bounds, computational learning theorists study the time complexity and feasibility of learning. In computational learning theory, a computation is considered feasible if it can be done in polynomial time. There are two kinds of time complexity results :

- *Positive results* : Showing that a certain class of functions is learnable in polynomial time.
- *Negative results* : Showing that certain classes cannot be learned in polynomial time.

Negative results often rely on commonly believed, but yet unproven assumptions, such as :

- Computational complexity—$P \neq NP$
- Cryptographic—One-way functions exist.

There are several different approaches to computational learning theory. These differences are based on making assumptions about the inference principles used to generalize from limited data. This includes different definitions of probability and different assumptions on the generation of samples. The different approaches include :

- Probably approximately correct learning (PAC learning), proposed by Leslie Valiant;
- VC theory, proposed by Vladimir Vapnik;
- Bayesian inference
- Algorithmic learning theory, from the work of E. M. Gold.
- Online machine learning, from the work of Nick Littlestone.

Computational learning theory has led to several practical algorithms. For example, PAC theory inspired boosting, VC theory led to support vector machines, and Bayesian inference led to belief networks (by Judea Pearl).

EMPIRICAL RISK MINIMIZATION

Empirical risk minimization (ERM) is a principle in statistical learning theory which defines a family of learning algorithms and is used to give theoretical bounds on the performance of learning algorithms.

Background

Consider the following situation, which is a general setting of many supervised learning problems. We have two spaces of objects X and Y and would like to learn a function $h : X \to Y$ (often called *hypothesis*) which outputs an object $y \in Y$, given $x \in X$. To do so, we have at our disposal a *training set* of a few examples $(x_1, y_1),\ldots,(x_m, y_m)$ where $x_i \in X$ is an input and $y_i \in Y$ is the corresponding response that we wish to get from $h(x_i)$.

To put it more formally, we assume that there is a joint probability distribution $P(x, y)$ over X and Y, and that the training set consists of m instances $(x_1, y_1),\ldots,(x_m, y_m)$ drawn i.i.d. from $P(x, y)$. Note that the assumption of a joint probability distribution allows us to model uncertainty in predictions (*e.g.* from noise in data) because y is not a deterministic function of x, but rather a random variable with conditional distribution $P(y \mid x)$ for a fixed x.

We also assume that we are given a non-negative real-valued loss function $L(\hat{y}, y)$ which measures how different the prediction $\hat{y}$ of a hypothesis is from the true outcome y. The risk associated with hypothesis $h(x)$ is then defined as the expectation of the loss function :

$$R(h(= E[L(h(x), y)] = \int L(h(x), y)dP(x, y).$$

A loss function commonly used in theory is the 0–1 loss function : $L(\hat{y}, y) = I(\hat{y} \neq y)$, where $I(\ldots)$ is the indicator notation.

The ultimate goal of a learning algorithm is to find a hypothesis h^* among a fixed class of functions H for which the risk $R(h)$ is minimal :

$$h^* = \arg\min_{h \in H} R(h).$$

Empirical Risk Minimization

In general, the risk $R(h)$ cannot be computed because the distribution $P(x, y)$ is unknown to the learning algorithm (this situation is referred to as agnostic learning). However, we can compute an approximation, called *empirical risk*, by averaging the loss function on the training set :

$$R_{\text{emp}}(h) = \frac{1}{m}\sum_{i=1}^{m} L(h(x_i), y_i).$$

Empirical risk minimization principle states that the learning algorithm should choose a hypothesis $\hat{h}$ which minimizes the empirical risk :

$$\hat{h} = \arg\min_{h \in H} \mathrm{R}_{\text{emp}}(h).$$

Thus the learning algorithm defined by the ERM principle consists in solving the above optimization problem.

Properties

Computational Complexity

Empirical risk minimization for a classification problem with 0-1 loss function is known to be an NP-hard problem even for such relatively simple class of functions as linear classifiers. Though, it can be solved efficiently when minimal empirical risk is zero, *i.e.* data is linearly separable.

In practice, machine learning algorithms cope with that either by employing a convex approximation to 0-1 loss function (like hinge loss for SVM), which is easier to optimize, or by posing assumptions on the distribution $P(x, y)$ (and thus stop being agnostic learning algorithms to which the above result applies,)

PROBABLY APPROXIMATELY CORRECT LEARNING

In computational learning theory, **probably approximately correct learning (PAC learning)** is a framework for mathematical analysis of machine learning. It was proposed in 1984 by Leslie Valiant.

In this framework, the learner receives samples and must select a generalization function (called the *hypothesis*) from a certain class of possible functions. The goal is that, with high probability (the "probably" part), the selected function will have low generalization error (the "approximately correct" part). The learner must be able to learn the concept given any arbitrary approximation ratio, probability of success, or distribution of the samples.

The model was later extended to treat noise (misclassified samples).

An important innovation of the PAC framework is the introduction of computational complexity theory concepts to machine learning. In particular, the learner is expected to find efficient functions (time and space requirements bounded to a polynomial of the example size), and the learner itself must implement an efficient procedure (requiring an example count bounded to a polynomial of the concept size, modified by the approximation and likelihood bounds).

Definitions and Terminology

In order to give the definition for something that is PAC-learnable, we first have to introduce some terminology.

For the following definitions, two examples will be used. The first is the problem of character recognition given an array of n bits encoding a binary-valued image. The other example is the problem of finding an interval that will correctly classify points within the interval as positive and the points outside of the range as negative.

Let X be a set called the *instance space* or the encoding of all the samples, and each *instance* have length assigned. In the character recognition problem, the instance space is $X = \{0, 1\}^n$. In the interval problem the instance space is $X = R$, where R denotes the set of all real numbers.

A *concept* is a sub-set $c \subset X$. One concept is the set of all patterns of bits in $X = \{0, 1\}^n$ that encode a picture of the letter "*P*". An example concept from the second example is the set of all of the numbers between $\pi/2$ and $\sqrt{10}$. A *concept class* C is a set of concepts over X. This could be the set of all sub-sets of the array of bits that are skeletonized 4-connected (width of the font is 1).

Let $EX(c, D)$ be a procedure that draws an example, x, using a probability distribution D and gives the correct label $c(x)$, that is 1 if $x \in c$ and 0 otherwise.

Say that there is an algorithm A that given access to $EX(c, D)$ and inputs ϵ and δ that, with probability of at least $1 - \delta$, A outputs a hypothesis $h \in C$ that has error less than or equal to ϵ with examples drawn from X with the distribution D. If there is such an algorithm for every concept $c \in C$, for every distribution D over X, and for all $0 < \epsilon < 1/2$ and $0 < \delta < 1/2$ then C is **PAC learnable** (or *distribution-free PAC learnable*). We can also say that A is a **PAC learning algorithm** for C.

An algorithm runs in time t if it draws at most t examples and requires at most t time steps. A concept class is **efficiently PAC learnable** if it is *PAC learnable* by an algorithm that runs in time polynomial in $1/\epsilon, 1/\delta$ and *instance* length.

Equivalence

Under some regularity conditions these three conditions are equivalent :

1. The concept class C is **PAC learnable**.
2. The VC dimension of C is finite.
3. C is a uniform Glivenko-Cantelli class.

STATISTICAL LEARNING THEORY

Statistical learning theory is a framework for machine learning drawing from the fields of statistics and functional analysis. Statistical learning theory deals with the problem of finding a predictive function based on data. Statistical learning theory has led to successful applications in fields such as computer vision, speech recognition, bioinformatics, and baseball. It is the theoretical framework underlying support vector machines.

Introduction

The goal of learning is prediction. Learning falls into many categories, including supervised learning, unsupervised learning, online learning, and reinforcement learning. From the perspective of statistical learning theory, supervised learning is best understood. Supervised learning involves learning from a training set of data. Every point in the training is an input-output pair, where the input maps to an output. The learning problem consists of inferring the function that maps between the input and the output in a predictive fashion, such that the learned function can be used to predict output from future input.

Depending of the type of output, supervised learning problems are either problems of regression or problems of classification. If the output takes a continuous range of values, it is a regression problem. Using Ohm's Law as an example, a regression could be performed with voltage as input and current as output. The regression would find the functional relationship between voltage and current to be $\frac{1}{R}$, such that

$$I = \frac{1}{R}V$$

Classification problems are those for which the output will be an element from a discrete set of labels. Classification is very common for machine learning applications. In facial recognition, for instance, a picture of a person's face would be the input, and the output label would be that person's name. The input would be represented by a large multi-dimensional vector, in which each dimension represents the value of one of the pixels.

After learning a function based on the training set data, that function is validated on a test set of data, data that did not appear in the training set. Classification functions can use the percentage of inputs that are correctly classified as a metric for how predictive the learned function is, while regression functions must use some distance metric, called a loss function, for how accurate the predicted value is. A familiar example of a loss function is the square of the difference between the actual value and the predicted value; this is the loss function used in ordinary least squares regression.

Formal Description

Take X to be the vector space of all possible inputs, and Y to be the vector space of all possible outputs. Statistical learning theory takes the perspective that there is some unknown probability distribution over the product space $Z = X \otimes Y$, *i.e.* there exists some unknown $p(z) = p(\vec{x}, y)$. The training set is made up of n samples from this probability distribution, and is notated

$$S = \{\vec{x}_1, y_1), \ldots, (\vec{x}_n, y_n)\} = \{\vec{z}_1, \ldots, \vec{z}_n\}$$

Every $\vec{x}_i$ is an input vector from the training data, and y_i is the output that corresponds to it.

In this formalism, the inference problem consists of finding a function $f : X \to Y$ such that $f(\vec{x}) \sim y$. Let H be a space of functions $f : X \to Y$ called the hypothesis space. The hypothesis space is the space of functions the algorithm will search through. Let $V(f(\vec{x}), y)$ be the loss functional, a metric for the difference between the predicted value $f(\vec{x})$ and the actual value y. The expected risk is defined to be

$$I[f] = \int_{X \otimes Y} V(f(\vec{x}), y) p(\vec{x}, y) d\vec{x} dy$$

The target function, the best possible function f that can be chosen, is given by the f that satisfies

$$\inf_{f \in H} I[f]$$

Because the probability distribution $p(\vec{x}, y)$ is unknown, a proxy measure for the expected risk must be used. This measure is based on the training set, a sample from this unknown probability distribution. It is called the empirical risk

$$I_S[f] = \frac{1}{n} \sum_{i=1}^{n} V(f(\vec{x}_i), y_i)$$

A learning algorithm that chooses the function f_S that minimizes the empirical risk is called empirical risk minimization.

Loss Functions

The choice of loss function is a determining factor on the function f_S that will be chosen by the learning algorithm. The loss function also affects the convergence rate for an algorithm. It is important for the loss function to be convex.

Different loss functions are used depending on whether the problem is one of regression or one of classification.

Regression

The most common loss function for regression is the square loss function. This familiar loss function is used in ordinary least squares regression. The form is :

$$V(f(\vec{x}),y)=(y-f(\vec{x}))^2$$

The absolute value loss is also sometimes used :

$$V(f(\vec{x}),y)=|y-f(\vec{x})|$$

Classification

In some sense the 0–1 loss is the most natural loss function for classification. It takes the value 0 if the predicted output is the same as the actual output, and it takes the value 1 if the predicted output is different from the actual output. For binary classification, this is :

$$V(f(\vec{x},y))=\theta\,(-yf(\vec{x}))$$

where θ is the Heaviside step function.

The 0-1 loss function, however, is not convex. The hinge loss is thus often used :

$$V(f(\vec{x},y))=(-yf(\vec{x}))+$$

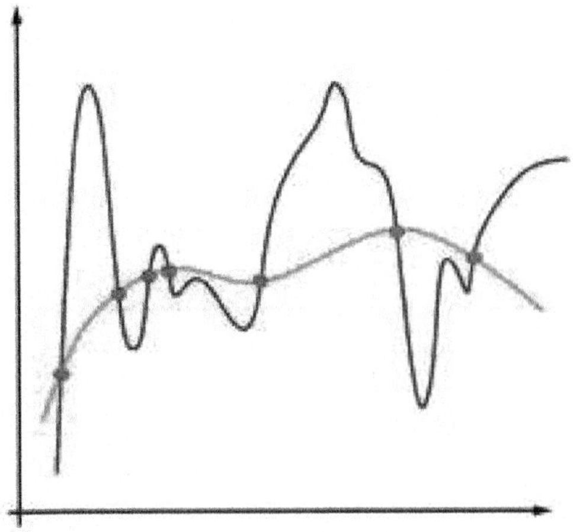

Fig. : This image represents an example of overfitting in machine learning. The red dots represent training set data. The green line represents the true functional relationship, while the blue line shows the learned function, which has fallen victim to overfitting.

Regularization

In machine learning problems, a major problem that arises is that of overfitting. Because learning is a prediction problem, the goal is not to find a function that most closely fits the data, but to find one that will most accurately will predict output from future input. Empirical risk minimization runs this risk of overfitting : finding a function that matches the data exactly but does not predict future output well.

Overfitting is symptomatic of unstable solutions; a small perturbation in the training set data would cause a large variation in the learned function. It can be shown that if the stability for the solution can be guaranteed, generalization and consistency are guaranteed as well. Regularization can solve the overfitting problem and give the problem stability.

Regularization can be accomplished by restricting the hypothesis space *H*. A common example would be restricting *H* to linear functions : this can be seen as a reduction to the standard problem of linear regression. *H* could also be restricted to polynomial of degree *p*, exponentials, or bounded functions on L1. Restriction of the hypothesis space avoids overfitting because the form of the potential functions are limited, and so does not allow for the choice of a function that gives empirical risk arbitrarily close to zero.

Regularization can also be accomplished through Tikhonov regularization. This consists of minimizing

$$\frac{1}{n}\sum_{i=1}^{n} V(f(\vec{x}_i, y_i)) + \gamma \| f \|_H^2$$

where γ is a fixed and positive parameter, the regularization parameter. Tikhonov regularization ensures existence, uniqueness, and stability of the solution.

METROLOGY/TEST : THE EMERGING ROLE FOR DATA-MINING

The complexity of semi-conductor manufacturing and its wealth of data provide an ideal environment for the application of data-mining to discover patterns of behaviour leading to knowledge. The demands are, however, such that data-mining technology needs to be extended further to achieve its full potential. Companies successful in developing and harnessing this technology will be rewarded with powerful new capabilities for diagnosing process control, yield, and equipment problems.

Each generation of semi-conductor process technology demands more detailed data to support process monitoring and improvement. This is being driven by additional process steps and by new forms of metrology and flexibility in sampling plans. In turn, the wealth of data drives storage, integration, analysis, and automation trends in wafer processing.

Data storage : Islands of engineering data, generated and stored at the tool or cell level, are beginning to be integrated into repositories for analysis packages. These repositories make possible new types of queries that request data from across regions of the process. Thus, new sources of process variation can be traced to their root cause, leading to quicker yield ramps.

Data integration : To access data repositories, integration must allow analysis tools to pull data from across the network. This can be done via a configurable data integration language or forced consolidation of all data into a proprietary in-house or third party database. Both techniques will remain in use until specific standards are developed to allow plug-and-play among commercial data storage, access, and analysis components.

Data analysis : Analysis tools range from those provided on process tools to domain-specific applications in statistical process control (SPC), yield improvement, and equipment maintenance to *ad hoc* applications with third-party software packages (*e.g.*, MS Excel, SAS JMP, and SPLUS). Many large IC manufacturers have built in-house domain-specific applications because comprehensive commercial products were limited.

Data analysis automation : Data volume and the number of relationships being monitored dictate a need for improved automation of line and yield monitoring. It is no longer sufficient to generate volumes of trend charts; charts must be analyzed automatically for exceptions and action must be taken if anomalies are found.

Users and suppliers of manufacturing execution systems (MES) tend to look at these problems from the top down; equipment suppliers from the bottom up. MES suppliers must create and maintain such capabilities at the factory level. Tool suppliers see market potential in increasing the communications capabilities of tools and packaging localized automation solutions within a work cell. Automated cells will slowly become more prevalent, similar to the evolution of multi-chamber tools. This will drive the need for cell management systems that treat the cell as a virtual tool. SPC used within the cell will need to be comprehensive, and the cell will need to be capable of tracing local process and defect related SPC failures to root causes. Eventually, the cell will need to communicate with neighbouring cells to elicit information needed to maintain local process targets and anticipate tool wear and breakdown before they have an impact on the process.

Such automated systems will require infrastructure and behavioural models for their various components. The infrastructure is emerging through Sematech's work on a CIM Framework [1]. Behavioural models are already in use at the cell-level for such applications as run-to-run process control of overlay, critical dimensions (CDs), and chemical mechanical planarization (CMP) uniformity.

Developing control models involves tedious manual steps : gathering data, cleaning it, and repeatedly exploring it until an analytical model can be established. Techniques—specifically involving data-mining technologies—are evolving, however, that allow some such modelling to be derived, or at least revised, automatically. The promise of such methods is that automated systems guiding tools, cells, and eventually perhaps even factories, will be able to learn and adapt their own control schemes to best achieve high-level directives. We believe that data-mining is a key component of the technologies needed to achieve quicker yield ramps, fewer yield busts, higher capacities, and higher productivity metrics.

Current Uses of Data-mining

Data-mining is being used extensively in the retail, finance, security, medicine, and insurance industries. Many of these applications were built originally as expert systems. Today, a sub-set of such systems can be built automatically using data-mining technology to derive rules from contextual historical data. This has opened up new vistas of modelling heretofore deemed impractical because of the volumes of data involved and the transient nature of the models, especially in retail.

Some applications in these industries have analogies in manufacturing. As in the medical industry, for example, use of data-mining in manufacturing requires that the derived models have physical understanding associated with their components. It is not enough to predict system behaviour, since the underlying root cause of the phenomena usually needs to be identified to drive corrective actions. Also, as with medicine, most applications in manufacturing are for diagnostic systems. Manufacturing groups would like to know the circumstances under which they will encounter certain types of process control excursions, equipment events, and final quality levels. Having a system that can crunch historical data and establish related models improves response time when such events occur and follow-up is needed.

Caveats for Manufacturing

Statistical significance does not always imply causality. Given enough variables, one or more will show up as significant regardless of the question asked. In other cases, the answer to a problem may be in the data, but the analytical methods are too narrow to uncover them. Both types of errors are more easily tolerated in off-line than in real-time systems.

Avoidance of false signals requires ongoing work to improve models by better "cleaning" of the data, linking to previous knowledge, or including certain variables only if certain conditions are true. Today, statisticians or engineers do this weeding out process through the benefit of considerable domain knowledge beyond that contained in the data-set being analyzed.

When data-mining misses the correct model, it is usually the result of the narrowness of the algorithms used. For instance, the data-mining algorithm may not be robust to "outliers" and therefore be thrown off-track by "dirty" data. Or, in the decision tree method, models at each node may not be complete enough to catch a behaviour (*e.g.*, cyclic vs. linear behaviours).

Data-mining can also encounter problems when there are correlations among the variables used in the modelling. Such is often the case when the model involves variables related to the same region of the process. In these cases, the data-mining routine might choose to use one of the correlated variables based on its analytical criteria, even though a statistician or engineer would deem it to have no causal relationship and wonder why the proper variable was not selected. Data-mining tools that offer features for engineers to interact with the analysis and override variable selections help alleviate such black-box limitations.

Given the variety of caveats, the use of data-mining in manufacturing is just beginning. The current interactive use will need to reach a level of maturity before being followed by embedding in real-time applications. To be viable for real-time use, data-mining engines must be applied to well-understood phenomena so that output is meaningful and trusted. To do so may require new forms of boundary conditions to define the limits for the solution space. Such rules define what is required of the process outside of the immediate context and what knowledge has been gained outside of the immediate data-set. Today such knowledge is referenced and used through participation of real engineers in the engineering change process.

Applications to Wafer Processing

In semi-conductor manufacturing, possible uses for data-mining include process and tool control, yield improvement, and equipment maintenance—analyzing historical data for patterns to derive physically meaningful models for predicting future behaviour.

Process control is ripe for use of data-mining because of the complexities of problem-solving processes and the richness of the data sources that need to be brought together. SPC is usually supplemented with troubleshooting guides—out of control action plans (OCAPs). An OCAP is derived from protocol and engineering experience in solving previous problems. Data-mining can assist in solving new problems by looking for commonalties in the processing history of previous occurrences—for example, diagnosing excursions of in-line defect metrology and wafer electrical test (WET) data.

In-line defect metrology data is fraught with analytical difficulties. Sophisticated kill-ratio schemes have been derived by correlating with sort data to assist in prioritizing the many out-of-control signals obtained from application of standard SPC methods to random defect data. Once the decision is made to work on a given defect type, analytical information from defect metrology, context, and history of the affected product, and other data, are assimilated by knowledgeable engineers to suggest and eliminate possible causes. This proceeds until the defect goes away or is traced to a root cause.

Given access to this same data, it should be possible to develop a data-mining application to assist in this complex task. Previous results would be embodied in an expert system against which new results would be compared. With a new failure, the system would interrogate other automation components and metrology tools to gather the data needed to suggest a likely cause.

Another possible application is diagnosis of excursion failures at WET. When lots fail at WET, the engineer responsible for disposition must assimilate the pattern of failing tests and either recognize the problem from previous experience or begin an analysis of process history to find the root cause. In the latter case, knowledge of the nature of the failure narrows the process steps to be investigated. Here, a data-mining application could encapsulate relationships between failing tests and the process flow, together with a methodology for gathering data on failing lots and analyzing it for commonalties.

Advanced process control applications close the loop. Process drifts are not only detected, but corrected by feeding process corrections either forward or backward in the process flow. Run-to-run feedback control applications have become critical to achieving production-worthy processes for CD, overlay, and CMP uniformity. Such model-based process control applications require well-founded models for factors that cause drift in the response, tight integration with the factory control system, and an appropriate control algorithm.

The Sematech APC Framework is the foundation for an infrastructure to build such applications. Once this infrastructure is in place, the process model is crucial for achieving success. Modelling needs to take into account the primary factors influencing the response and secondary factors that determine under what sets of conditions the model shall remain valid. For the most part, statisticians and highly knowledgeable engineers have derived these factors through exploratory data analysis. Data-mining offers the opportunity to increase the productivity of the tasks leading to the initial model as well as the means to improve it once in place. For instance, the residual errors from a run-to-run overlay control system could be analyzed to look for patterns that would allow the model to be improved further.

Equipment Predictions

On-board tool control systems manage the tool as a self-contained electromechanical system. With the challenge of each succeeding generation of wafer processing technology, these systems have become increasingly complex. This complexity demands sophisticated control systems that monitor and react to problems. During the course of development, extensive data is collected on the failure modes of each subsystem as well as that of the integrated tool.

Data-mining has potential application in building models based on historical failure data to detect the precursors of failures. This would permit systems to increase their capabilities for self-repair or graceful shutdown before compromising the process.

Yield Management

Yield management covers a diverse set of tasks in both front-end and back-end processing. Yield ramps on new technologies and yield busts on existing technologies demand powerful routines to profile yield loss and trace it to root causes. These types of correlation activities can be tedious and time-consuming because of the time it takes to consolidate data for analysis. Sometimes, it is still a matter of trial and error before a signal is found.

Although various companies have developed automated search routines based on traditional methods like ANOVA and regression, these are of limited use when an interaction exists between two tools or when time is a factor. In such cases, data-mining has potential to supplement existing techniques. This is especially true when the data-mining tool is embedded within an exploratory data-analysis environment and can be integrated with existing automation methods.

Macros could be developed to search for certain types of phenomena on a daily or weekly basis and if found, to run data-mining to pinpoint the source of the process variation. Similar to an in-line process control application, an example of this might be a monitor that looks at the level of a certain failing bin. When the level exceeds a threshold, the monitor would trigger a data-mining analysis of the last 50 lots to determine the cause of the high counts. This type of system could be extended to trigger data-mining analysis when certain spatial patterns are detected on the wafers.

Fault Detection, Classification

Fault-detection systems supplement those built within process tools to monitor trace data from tool sensors and look for anomalies. They allow IC manufacturers to customize patterns being monitored and associated action plans.

Data-mining is useful in correlating the presence and severity of such faults to downstream quality metrics. This provides an ability to prioritize the response to detected faults. As with excursions detected in defect metrology data, some signals from fault detection on trace data are cosmetic, without a detrimental impact on the wafer; others are indicators of trouble.

Equipment Maintenance

Equipment maintenance involves diagnosis and repair of failures, and preventative maintenance (PM). Diagnosis of a tool failure can be a difficult task requiring several pieces of information to be assimilated as clues until a consistent picture can be drawn and the faults isolated. Here again, data-mining can make use of historical data to derive expert system-like diagnostic rules to suggest next steps and allow faults to be isolated more efficiently.

PM anticipates the effects of wear and tear to replace parts before they fail and compromise the process. PM frequencies and special PMs are often dependent on processes and recipe changes being performed. For instance, there may be interactions between recipe changes and the process performance immediately after a change. These types of effects are difficult for a semi-conductor equipment supplier to anticipate and test. They can have major effects on a manufacturer's success with a tool, however.

Data-mining would provide the opportunity to look for such interactions between recipes, as well as establish correlation between the number of lots processed since the last PM and downstream quality measures. Once such models are established they can be used to customize PM procedures and frequencies to improve process performance under given load conditions.

Conclusion and Future Trends

Although the need for data-mining tools is great, and their value has been proven in other industries, data-mining capabilities needed for semi-conductor manufacturing are just beginning to be developed. As with applications in medicine, potential manufacturing applications are hard to picture without involving humans. Current practice necessarily involves an engineer or statistician to guide, interpret, and evaluate the results of data-mining.

Today, data-mining remains a complementary tool to more traditional statistical and graphical methods for exploratory data-analysis and model building. Automated results still need to be reviewed and interpreted prior to direct application in the form of a process change or a new control algorithm for a tool.

Increasingly, however, methods will be developed to reduce these limitations by increasing the breadth of models that can be developed and by reducing the frequency of false signals. Eventually it will be possible to couple data-mining to repositories of knowledge as well as data. This will lead to self-optimizing systems of increasing complexity as witnessed by the emerging trend of new offerings for work cell management from the major tool suppliers. This trend will continue and lead to larger centers of automation that will rely on data-mining to anticipate and recover from potential problems.

The rate at which this vision is achieved will hinge on the development and adoption of standards for data storage and access between and within tools. Such standards will also need to be extended to the more difficult topic of knowledge storage. Scripting languages will need to be developed or enhanced to manage the complex interactions among tool, cell, and factory control systems and data-mining engines.

Data-mining technology presents an opportunity to increase significantly the rate at which the volumes of data generated on the manufacturing process can be turned into information. Truly, its time has come !

Chapter 7

GENETIC ALGORITHM

In the computer science field of artificial intelligence, **genetic algorithm (GA)** is a searchheuristic that mimics the process of natural selection. This heuristic (also sometimes called a metaheuristic) is routinely used to generate useful solutions to optimization and search problems. Genetic algorithms belong to the larger class of evolutionary algorithms (EA), which generate solutions to optimization problems using techniques inspired by natural evolution, such as inheritance, mutation, selection, and cross-over.

Genetic algorithms find application in bioinformatics, phylogenetics, computational science, engineering, economics, chemistry, manufacturing, mathematics, physics, pharmacometrics and other fields.

METHODOLOGY

In a genetic algorithm, a population of candidate solutions (called individuals, creatures, or phenotypes) to an optimization problem is evolved toward better solutions. Each candidate solution has a set of properties (its chromosomes or genotype) which can be mutated and altered; traditionally, solutions are represented in binary as strings of 0s and 1s, but other encodings are also possible.

The evolution usually starts from a population of randomly generated individuals, and is an iterative process, with the population in each iteration called a *generation*. In each generation, the fitness of every individual in the population is evaluated; the fitness is usually the value of the objective function in the optimization problem being solved. The more fit individuals are stochastically selected from the current population, and each individual's genome is modified (recombined and possibly randomly mutated) to form a new generation. The new generation of candidate solutions is then used in the next iteration of the algorithm. Commonly, the algorithm terminates when either a maximum number of generations has been produced, or a satisfactory fitness level has been reached for the population.

A typical genetic algorithm requires :

1. A genetic representation of the solution domain,
2. A fitness function to evaluate the solution domain.

A standard representation of each candidate solution is as an array of bits. Arrays of other types and structures can be used in essentially the same way. The main property that makes these genetic representations convenient is that their parts are easily aligned due to their fixed size, which facilitates simple cross-over operations. Variable length representations may also be used, but cross-over implementation is more complex in this case. Tree-like representations are explored in genetic programming and graph-form representations are explored in evolutionary programming; a mix of both linear chromosomes and trees is explored in gene expression programming.

Once the genetic representation and the fitness function are defined, a GA proceeds to initialize a population of solutions and then to improve it through repetitive application of the mutation, cross-over, inversion and selection operators.

Initialization of Genetic Algorithm

Initially many individual solutions are (usually) randomly generated to form an initial population. The population size depends on the nature of the problem, but typically contains several hundreds or thousands of possible solutions. Traditionally, the population is generated randomly, allowing the entire range of possible solutions (the *search space*). Occasionally, the solutions may be "seeded" in areas where optimal solutions are likely to be found.

Selection

During each successive generation, a proportion of the existing population is selected to breed a new generation. Individual solutions are selected through a *fitness-based* process, where fitter solutions (as measured by a fitness function) are typically more likely to be selected. Certain selection methods rate the fitness of each solution and preferentially select the best solutions. Other methods rate only a random sample of the population, as the former process may be very time-consuming.

The fitness function is defined over the genetic representation and measures the *quality* of the represented solution. The fitness function is always problem dependent. For instance, in the knapsack problem one wants to maximize the total value of objects that can be put in a knapsack of some fixed capacity. A representation of a solution might be an array of bits, where each bit represents a different object, and the value of the bit (0 or 1) represents whether or not the object is in the knapsack. Not every such representation is valid, as the size of objects may exceed the capacity of the knapsack. The *fitness* of the solution is the sum of values of all objects in the knapsack if the representation is valid, or 0 otherwise.

In some problems, it is hard or even impossible to define the fitness expression; in these cases, a simulation may be used to determine the fitness function value of a phenotype (*e.g.* computational fluid dynamics is used to determine the air resistance of a vehicle whose shape is encoded as the phenotype), or even interactive genetic algorithms are used.

Genetic Operators

The next step is to generate a second generation population of solutions from those selected through a combination of genetic operators : cross-over (also called recombination), and mutation.

For each new solution to be produced, a pair of "parent" solutions is selected for breeding from the pool selected previously. By producing a "child" solution using the above methods of cross-over and mutation, a new solution is created which typically shares many of the characteristics of its "parents". New parents are selected for each new child, and the process continues until a new population of solutions of appropriate size is generated. Although reproduction methods that are based on the use of two parents are more "biology inspired", some research suggests that more than two "parents" generate higher quality chromosomes.

These processes ultimately result in the next generation population of chromosomes that is different from the initial generation. Generally the average fitness will have increased by this procedure for the population, since only the best organisms from the first generation are selected for breeding, along with a small proportion of less fit solutions. These less fit solutions ensure genetic diversity within the genetic pool of the parents and therefore ensure the genetic diversity of the subsequent generation of children.

Opinion is divided over the importance of cross-over *versus* mutation. There are many references in Fogel that support the importance of mutation-based search.

Although cross-over and mutation are known as the main genetic operators, it is possible to use other operators such as regrouping, colonization-extinction, or migration in genetic algorithms.

It is worth tuning parameters such as the mutation probability, cross-over probability and population size to find reasonable settings for the problem class being worked on. A very small mutation rate may lead to genetic drift (which is non-ergodic in nature). A recombination rate that is too high may lead to premature convergence of the genetic algorithm. A mutation rate that is too high may lead to loss of good solutions unless there is elitist selection.

Termination

This generational process is repeated until a termination condition has been reached. Common terminating conditions are :

- A solution is found that satisfies minimum criteria
- Fixed number of generations reached

- Allocated budget (computation time/money) reached
- The highest ranking solution's fitness is reaching or has reached a plateau such that successive iterations no longer produce better results
- Manual inspection
- Combinations of the above.

THE BUILDING BLOCK HYPOTHESIS

Genetic algorithms are simple to implement, but their behaviour is difficult to understand. In particular it is difficult to understand why these algorithms frequently succeed at generating solutions of high fitness when applied to practical problems. The building block hypothesis (BBH) consists of :

1. A description of a heuristic that performs adaptation by identifying and recombining "building blocks", *i.e.* low order, low defining-length schemata with above average fitness.
2. A hypothesis that a genetic algorithm performs adaptation by implicitly and efficiently implementing this heuristic.

Goldberg describes the heuristic as follows :

"Short, low order, and highly fit schemata are sampled, recombined [crossed over], and resampled to form strings of potentially higher fitness. In a way, by working with these particular schemata [the building blocks], we have reduced the complexity of our problem; instead of building high-performance strings by trying every conceivable combination, we construct better and better strings from the best partial solutions of past samplings.

"Because highly fit schemata of low defining length and low order play such an important role in the action of genetic algorithms, we have already given them a special name : building blocks. Just as a child creates magnificent fortresses through the arrangement of simple blocks of wood, so does a genetic algorithm seek near optimal performance through the juxtaposition of short, low-order, high-performance schemata, or building blocks."

LIMITATIONS

There are limitations of the use of a genetic algorithm compared to alternative optimization algorithms :

- Repeated fitness function evaluation for complex problems is often the most prohibitive and limiting segment of artificial evolutionary algorithms. Finding the optimal solution to complex high-dimensional, multi-modal problems often requires very expensive fitness function evaluations. In real world problems such as structural optimization problems, a single function evaluation may require several hours to several days of complete simulation. Typical optimization methods can not deal with such types of problem. In this case, it may be necessary to forgo an exact evaluation and use an approximated fitness that is computationally efficient. It is apparent that amalgamation of approximate models may be one of the most promising approaches to convincingly use GA to solve complex real life problems.

- Genetic algorithms do not scale well with complexity. That is, where the number of elements which are exposed to mutation is large there is often an exponential increase in search space size. This makes it extremely difficult to use the technique on problems such as designing an engine, a house or plane. In order to make such problems tractable to evolutionary search, they must be broken down into the simplest representation possible. Hence we typically see evolutionary algorithms encoding designs for fan blades instead of engines, building shapes instead of detailed construction plans, airfoils instead of whole aircraft designs. The second problem of complexity is the issue of how to protect parts that have evolved to represent good solutions from further destructive mutation, particularly when their fitness assessment requires them to combine well with other parts.
- The "better" solution is only in comparison to other solutions. As a result, the stop criterion is not clear in every problem.
- In many problems, GAs may have a tendency to converge towards local optima or even arbitrary points rather than the global optimum of the problem. This means that it does not "know how" to sacrifice short-term fitness to gain longer-term fitness. The likelihood of this occurring depends on the shape of the fitness landscape : certain problems may provide an easy ascent towards a global optimum, others may make it easier for the function to find the local optima. This problem may be alleviated by using a different fitness function, increasing the rate of mutation, or by using selection techniques that maintain a diverse population of solutions, although the No Free Lunch theorem proves that there is no general solution to this problem. A common technique to maintain diversity is to impose a "niche penalty", wherein, any group of individuals of sufficient similarity (niche radius) have a penalty added, which will reduce the representation of that group in subsequent generations, permitting other (less similar) individuals to be maintained in the population. This trick, however, may not be effective, depending on the landscape of the problem. Another possible technique would be to simply replace part of the population with randomly generated individuals, when most of the population is too similar to each other. Diversity is important in genetic algorithms (and genetic programming) because crossing over a homogeneous population does not yield new solutions. In evolution strategies and evolutionary programming, diversity is not essential because of a greater reliance on mutation.
- Operating on dynamic data sets is difficult, as genomes begin to converge early on towards solutions which may no longer be valid for later data. Several methods have been proposed to remedy this by increasing genetic diversity somehow and preventing early convergence, either by increasing the probability of mutation when the solution quality drops (called *triggered hypermutation*), or by occasionally introducing entirely new, randomly

generated elements into the gene pool (called *random immigrants*). Again, evolution strategies and evolutionary programming can be implemented with a so-called "comma strategy" in which parents are not maintained and new parents are selected only from offspring. This can be more effective on dynamic problems.

- GAs cannot effectively solve problems in which the only fitness measure is a single right/wrong measure (like decision problems), as there is no way to converge on the solution (no hill to climb). In these cases, a random search may find a solution as quickly as a GA. However, if the situation allows the success/failure trial to be repeated giving (possibly) different results, then the ratio of successes to failures provides a suitable fitness measure.
- For specific optimization problems and problem instances, other optimization algorithms may be more efficient than genetic algorithms in terms of speed of convergence. Alternative and complementary algorithms include evolution strategies, evolutionary programming, simulated annealing, Gaussian adaptation, hill climbing, and swarm intelligence (*e.g.* : ant colony optimization, particle swarm optimization) and methods based on integer linear programming. The suitability of genetic algorithms is dependent on the amount of knowledge of the problem; well known problems often have better, more specialized approaches.

VARIANTS

Chromosome Representation

The simplest algorithm represents each chromosome as a bit string. Typically, numeric parameters can be represented by integers, though it is possible to use floating point representations. The floating point representation is natural to evolution strategies and evolutionary programming. The notion of real-valued genetic algorithms has been offered but is really a misnomer because it does not really represent the building block theory that was proposed by John Henry Holland in the 1970s. This theory is not without support though, based on theoretical and experimental results. The basic algorithm performs cross-over and mutation at the bit level. Other variants treat the chromosome as a list of numbers which are indexes into an instruction table, nodes in a linked list, hashes, objects, or any other imaginable data structure. Cross-over and mutation are performed so as to respect data element boundaries. For most data types, specific variation operators can be designed. Different chromosomal data types seem to work better or worse for different specific problem domains.

When bit-string representations of integers are used, Gray coding is often employed. In this way, small changes in the integer can be readily effected through mutations or cross-overs. This has been found to help prevent premature convergence at so called *Hamming walls*, in which too many simultaneous mutations (or cross-over events) must occur in order to change the chromosome to a better solution.

Other approaches involve using arrays of real-valued numbers instead of bit strings to represent chromosomes. Results from the theory of schemata suggest that in general the smaller the alphabet, the better the performance, but it was initially surprising to researchers that good results were obtained from using real-valued chromosomes. This was explained as the set of real values in a finite population of chromosomes as forming a *virtual alphabet* (when selection and recombination are dominant) with a much lower cardinality than would be expected from a floating point representation.

Elitism

A very successful (slight) variant of the general process of constructing a new population is to allow some of the better organisms from the current generation to carry over to the next, unaltered. This strategy is known as *elitist selection*.

Parallel Implementations

Parallel implementations of genetic algorithms come in two flavours. Coarse-grained parallel genetic algorithms assume a population on each of the computer nodes and migration of individuals among the nodes. Fine-grained parallel genetic algorithms assume an individual on each processor node which acts with neighbouring individuals for selection and reproduction. Other variants, like genetic algorithms for online optimization problems, introduce time-dependence or noise in the fitness function.

Adaptive GAs

Genetic algorithms with adaptive parameters (adaptive genetic algorithms, AGAs) is another significant and promising variant of genetic algorithms. The probabilities of cross-over (pc) and mutation (pm) greatly determine the degree of solution accuracy and the convergence speed that genetic algorithms can obtain. Instead of using fixed values of pc and pm, AGAs utilize the population information in each generation and adaptively adjust the pc and pm in order to maintain the population diversity as well as to sustain the convergence capacity. In AGA (adaptive genetic algorithm), the adjustment of pc and pm depends on the fitness values of the solutions. In *CAGA* (clustering-based adaptive genetic algorithm), through the use of clustering analysis to judge the optimization states of the population, the adjustment of pc and pm depends on these optimization states. It can be quite effective to combine GA with other optimization methods. GA tends to be quite good at finding generally good global solutions, but quite inefficient at finding the last few mutations to find the absolute optimum. Other techniques (such as simple hill climbing) are quite efficient at finding absolute optimum in a limited region. Alternating GA and hill climbing can improve the efficiency of GA while overcoming the lack of robustness of hill climbing.

This means that the rules of genetic variation may have a different meaning in the natural case. For instance-provided that steps are stored in consecutive order-crossing over may sum a number of steps from maternal DNA adding a number of steps from paternal DNA and so on. This is like adding vectors that more probably may follow a ridge in the phenotypic landscape. Thus, the efficiency of the process may be increased by many orders of magnitude. Moreover, the inversion operator has the opportunity to place steps in consecutive order or any other suitable order in favour of survival or efficiency.

A variation, where the population as a whole is evolved rather than its individual members, is known as gene pool recombination.

A number of variations have been developed to attempt to improve performance of GAs on problems with a high degree of fitness epistasis, *i.e.* where the fitness of a solution consists of interacting sub-sets of its variables. Such algorithms aim to learn (before exploiting) these beneficial phenotypic interactions. As such, they are aligned with the Building Block Hypothesis in adaptively reducing disruptive recombination. Prominent examples of this approach include the mGA, GEMGA and LLGA.

PROBLEM DOMAINS

Problems which appear to be particularly appropriate for solution by genetic algorithms include time tabling and scheduling problems, and many scheduling software packages are based on GAs. GAs have also been applied to engineering. Genetic algorithms are often applied as an approach to solve global optimization problems.

As a general rule of thumb genetic algorithms might be useful in problem domains that have a complex fitness landscape as mixing, *i.e.*, mutation in combination with cross-over, is designed to move the population away from local optima that a traditional hill climbing algorithm might get stuck in. Observe that commonly used cross-over operators cannot change any uniform population. Mutation alone can provide ergodicity of the overall genetic algorithm process (seen as a Markov chain).

Examples of problems solved by genetic algorithms include : mirrors designed to funnel sunlight to a solar collector, antennae designed to pick up radio signals in space, and walking methods for computer figures. Many of their solutions have been highly effective, unlike anything a human engineer would have produced, and inscrutable as to how they arrived at that solution.

HISTORY

Computer simulations of evolution started as early as in 1954 with the work of Nils Aall Barricelli, who was using the computer at the Institute for Advanced Study in Princeton, New Jersey. His 1954 publication was not widely noticed. Starting in 1957, the Australian quantitative geneticist Alex Fraser published a series of papers on simulation of artificial selection of organisms with multiple loci controlling a measurable trait. From these beginnings, computer simulation of evolution by biologists became more common in the early 1960s, and the methods were described in books by Fraser and Burnell and Crosby. Fraser's simulations included all of the essential elements of modern genetic algorithms. In addition, Hans-Joachim Bremermann published a series of papers in the 1960s that also adopted a population of solution to optimization problems, undergoing recombination, mutation, and selection. Bremermann's research also included the elements of modern genetic algorithms. Other noteworthy early pioneers include Richard Friedberg, George Friedman, and Michael Conrad. Many early papers are reprinted by Fogel.

Although Barricelli, in work he reported in 1963, had simulated the evolution of ability to play a simple game, artificial evolution became a widely recognized optimization method as a result of the work of Ingo Rechenberg and Hans-Paul Schwefel in the 1960s and early 1970s-Rechenberg's group was able to solve complex engineering problems through evolution strategies. Another approach was the evolutionary programming technique of Lawrence J. Fogel, which was proposed for generating artificial intelligence. Evolutionary programming originally used finite state machines for predicting environments, and used variation and selection to optimize the predictive logics. Genetic algorithms in particular became popular through the work of John Holland in the early 1970s, and particularly his book *Adaptation in Natural and Artificial Systems*. His work originated with studies of cellular automata, conducted by Holland and his students at the University of Michigan. Holland introduced a formalized framework for predicting the quality of the next generation, known as Holland's Schema Theorem. Research in GAs remained largely theoretical until the mid-1980s, when The First International Conference on Genetic Algorithms was held in Pittsburgh, Pennsylvania.

As academic interest grew, the dramatic increase in desktop computational power allowed for practical application of the new technique. In the late 1980s, General Electric started selling the world's first genetic algorithm product, a mainframe-based toolkit designed for industrial processes. In 1989, Axcelis, Inc. released Evolver, the world's first commercial GA product for desktop computers. The New York Times technology writer John Markoffwrote about Evolver in 1990, and it remained the only interactive commercial genetic algorithm until 1995. Evolver was sold to Palisade in 1997, translated into several languages, and is currently in its 6th version. Genetic Algorithm for Rule Set Production

Genetic Algorithm for Rule Set Production (GARP) is a computer program based on genetic algorithm that creates ecological niche models for species. The generated models describe environmental conditions (precipitation, temperatures, elevation, etc.) under which the species should be able to maintain populations. As input, local observations of species and related environmental parameters are used which describe potential limits of the species' capabilities to survive. Such environmental parameters are commonly stored in geographical information systems. A GARP model is a random set of mathematical rules which can be read as limiting environmental conditions. Each rule is considered as a gene; the set of genes is combined in random ways to further generate many possible models describing the potential of the species to occur.

GENETIC ALGORITHM SCHEDULING

To be competitive, corporations must minimize inefficiencies and maximize productivity. In manufacturing, productivity is inherently linked to how well you can optimize the resources you have, reduce waste and increase efficiency. Finding the best way to maximize efficiency in a manufacturing process can be extremely complex. Even on simple projects, there are multiple inputs, multiple steps, many constraints and limited resources. In general a resource constrained scheduling problem consists of :

- A set of jobs that must be executed
- A finite set of resources that can be used to complete each job
- A set of constraints that must be satisfied
    - Temporal Constraints–the time window to complete the task
    - Procedural Constraints–the order each task must be completed
    - Resource Constraints-is the resource available
- A set of objectives to evaluate the scheduling performance

A typical factory floor setting is a good example of this where scheduling which jobs need to be completed on which machines, by which employees in what order and at what time. In very complex problems such as scheduling there is no known way to get to a final answer, so we resort to searching for it trying to find a "good" answer. Scheduling problems most often use heuristic algorithms to search for the optimal solution. Heuristic search methods suffer as the inputs become more complex and varied. This type of problem is known in computer science as an NP-Hard problem. This means that there are no known algorithms for finding an optimal solution in polynomial time.

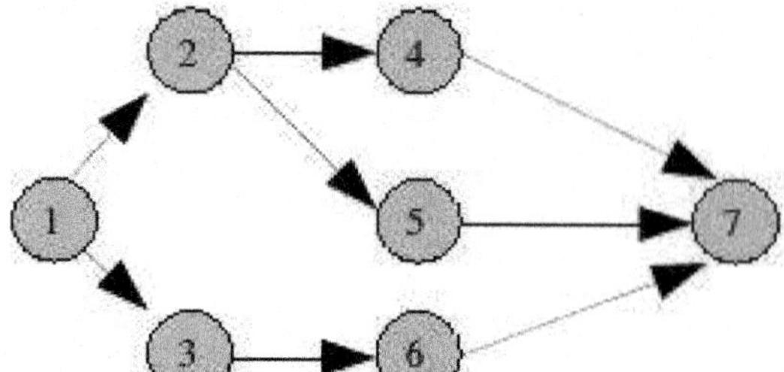

Fig. : Precedence in scheduling.

Genetic algorithms are well suited to solving production scheduling problems, because unlike heuristic methods genetic algorithms operate on a population of solutions rather than a single solution. In production scheduling this population of solutions consists of many answers that may have different sometimes conflicting objectives. For example, in one solution we may be optimizing a production process to be completed in a minimal amount of time. In another solution we may be optimizing for a minimal amount of defects. By cranking up the speed at which we produce we may run into an increase in defects in our final product.

As we increase the number of objectives we are trying to achieve we also increase the number of constraints on the problem and similarly increase the complexity. Genetic algorithms are ideal for these types of problems where the search space is large and the number of feasible solutions is small.

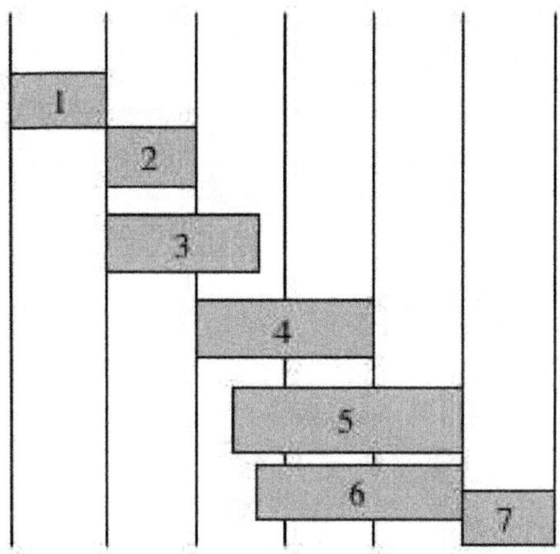

Fig. A. Example Schedule genome.

To apply a genetic algorithm to a scheduling problem we must first represent it as a genome. One way to represent a scheduling genome is to define a sequence of tasks and the start times of those tasks relative to one another. Each task and its corresponding start time represents a gene.

A specific sequence of tasks and start times (genes) represents one genome in our population. To make sure that our genome is a feasible solution we must take care that it obeys our precedence constraints. We generate an initial population using random start times within the precedence constraints. With genetic algorithms we then take this initial population and cross it, combining genomes along with a small amount of randomness (mutation). The offspring of this combination is selected based on a fitness function that includes one or many of our constraints, such as minimizing time and minimizing defects. We let this process continue either for a pre-allotted time or until we find a solution that fits our minimum criteria. Overall each successive generation will have a greater average fitness *i.e.* taking less time with higher quality than the proceeding generations. In scheduling problems, as with other genetic algorithm solutions, we must make sure that we do not select offspring that are infeasible, such as offspring that violate our precedence constraint. We of course may have to add further fitness values such as minimizing costs however each constraint that we add greatly increases the search space and lowers the number of solutions that are good matches.

GENETIC ALGORITHMS IN ECONOMICS

Genetic algorithms have increasingly been applied to economics since the pioneering work by John H. Miller in 1986. It has been used to characterize a variety of models including the cobweb model, the overlapping generations model, game theory, schedule optimization and asset pricing. Specifically, it has been used as a model to represent learning, rather than as a means for fitting a model.

Genetic Algorithm in the Cobweb Model

The cobweb model is a simple supply and demand model for a good over *t* periods. Firms (agents) make a production quantity decision in a given period, however their output is not produced until the following period. Thus, the firms are going to have to use some sort of method to forecast what the future price will be. The GA is used as a sort of learning behaviour for the firms. Initially their quantity production decisions are random, however each period they learn a little more. The result is the agents converge within the area of the rational expectations (RATEX) equilibrium for the stable and unstable case. If the election operator is used, the GA converges exactly to the RATEX equilibrium.

There are two types of learning methods these agents can be deployed with: social learning and individual learning. In social learning, each firm is endowed with a single string which is used as its quantity production decision. It then compares this string against other firms' strings. In the individual learning case, agents are endowed with a pool of strings. These strings are then compared against other strings within the agent's population pool. This can be thought of as mutual competing ideas within a firm whereas in the social case, it can be thought of as a firm learning from more successful firms. Note that in the social case and in the individual learning case with identical cost functions, that this is a homogeneous solution, that is all agents' production decisions are identical. However, if the cost functions are not identical, this will result in a heterogeneous solution, where firms produce different quantities (note that they are still locally homogeneous, that is within the firm's own pool all the strings are identical).

After all agents have made a quantity production decision, the quantities are aggregated and plugged into a demand function to get a price. Each firm's profit is then calculated. Fitness values are then calculated as a function of profits. After the offspring pool is generated, hypothetical fitness values are calculated. These hypothetical values are based on some sort of estimation of the price level, often just by taking the previous price level.

QUALITY CONTROL AND GENETIC ALGORITHMS

The combination of **quality control and genetic algorithms** led to novel solutions of complex quality control design and optimization problems. Quality control is a process by which entities review the quality of all factors involved in production. Quality is the degree to which a set of inherent characteristics fulfils a need or expectation that is stated, general implied or obligatory. Genetic algorithms are search algorithms, based on the mechanics of natural selection and natural genetics.

Quality Control

Alternative quality control (QC) procedures can be applied on a process to test statistically the null hypothesis, that the process conforms to the quality requirements, therefore that the process is in control, against the alternative, that the process is out of control. When a true null hypothesis is rejected, a statistical type I error is committed. We have then a false rejection of a run of the process. The probability of a type I error is called probability of false rejection. When a false null hypothesis is accepted, a statistical type II error is committed. We fail then to detect a significant change in the process. The probability of rejection of a false null hypothesis equals the probability of detection of the non-conformity of the process to the quality requirements.

The QC procedure to be designed or optimized can be formulated as :

$$Q_1(n_1, \mathbf{X}_1)\# \; Q_2(n_2, \mathbf{X}_2) \; \#...\# \; Q_q(n_q, \mathbf{X}_q) \qquad ...1$$

where $Q_i(n_i, \mathbf{X}_i)$ denotes a statistical decision rule, n_i denotes the size of the sample $\mathbf{S}_i$, that is the number of the samples the rule is applied upon, and $\mathbf{X}_i$ denotes the vector of the rule specific parameters, including the decision limits. Each symbol # denotes either the Boolean operator AND or the operator OR. Obviously, for # denoting AND, and for $n_1 < n_2 < ... < n_q$, that is for $\mathbf{S}_1 \subset \mathbf{S}_2 \subset .... \subset \mathbf{S}_q$, the (1) denotes a q-sampling QC procedure.

Each statistical decision rule is evaluated by calculating the respective statistic of a monitored variable of samples taken from the process. Then, if the statistic is out of the interval between the decision limits, the decision rule is considered to be true. Many statistics can be used, including the following : a single value of the variable of a sample, the range, the mean, and the standard deviation of the values of the variable of the samples, the cumulative sum, the smoothed mean, and the smoothed standard deviation. Finally, the QC procedure is evaluated as a Boolean proposition. If it is true, then the null hypothesis is considered to be false, the process is considered to be out of control, and the run is rejected.

A quality control procedure is considered to be optimum when it minimizes (or maximizes) a context specific objective function. The objective function depends on the probabilities of detection of the non-conformity of the process and of false rejection. These probabilities depend on the parameters of the quality control procedure (1) and on the probability density functions.

Genetic Algorithms

Genetic algorithms are robust search algorithms, that do not require knowledge of the objective function to be optimized and search through large spaces quickly. Genetic algorithms have been derived from the processes of the molecular biology of the gene and the evolution of life. Their operators, cross-over, mutation, and reproduction, are isomorphic with the synonymous biological processes. Genetic algorithms have been used to solve a variety of complex optimization problems. Additionally the classifier systems and the genetic programmingparadigm have shown us that genetic algorithms can be used for tasks as complex as the program induction.

Quality Control and Genetic Algorithms

In general, we can not use algebraic methods to optimize the quality control procedures. Usage of enumerative methods would be very tedious, especially with multi-rule procedures, as the number of the points of the parameter space to be searched grows exponentially with the number of the parameters to be optimized. Optimization methods based on the genetic algorithms offer an appealing alternative.

Furthermore, the complexity of the design process of novel quality control procedures is obviously greater than the complexity of the optimization of pre-defined ones.

In fact, since 1993, genetic algorithms have been used successfully to optimize and to design novel quality control procedures.

COMPUTER-AUTOMATED DESIGN

Design Automation usually refers to electronic design automation. Extending Computer-Aided Design (CAD), automated design and **Computer-Automated Design (CAutoD)** are more concerned with a broader range of applications, such as automotive engineering, civil engineering, composite material design, control engineering, dynamic system identification, financial systems, industrial equipment, mechatronic systems, steel construction, structural optimisation, and the invention of novel systems.

The concept of CAutoD perhaps first appeared in 1963, in the IBM Journal of Research and Development, where a computer program was written (1) to search for logic circuits having certain constraints on hardware design and (2) to evaluate these logics in terms of their discriminating ability over samples of the character set they are expected to recognize. More recently, traditional CAD simulation is seen to be transformed to CAutoD by biologically-inspired machine learning or search techniques such as evolutionary computation, including swarm intelligence algorithms.

Guiding Designs by Performance Improvements

To meet the ever growing demand of quality and competitiveness, iterative physical prototyping is now often replaced by 'digital prototyping' of a 'good design', which aims to meet multiple objectives such as maximised output, energy efficiency, highest speed and cost-effectiveness. The design problem concerns both finding the best design within a known range (*i.e.*, through 'learning' or 'optimisation') and finding a new and better design beyond the existing ones (*i.e.*, through creation and invention). This is equivalent to a search problem in an, almost certainly, multi-dimensional (multi-variate), multi-modal space with a single (or weighted) objective or multiple objectives.

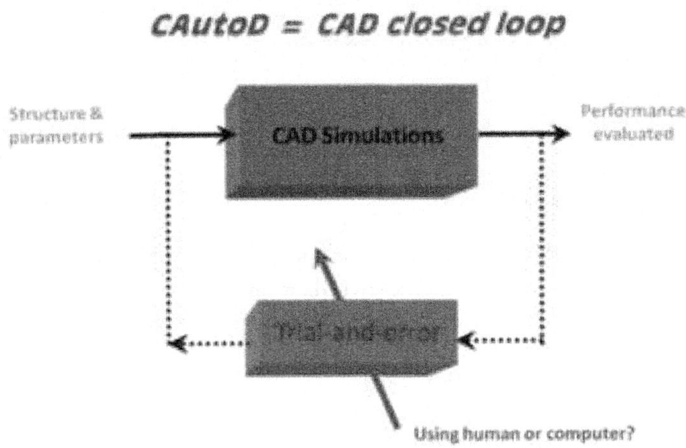

Fig. : Interaction in computer-automated design.

Normalized Objective Function : Cost vs. Fitness

Using single-objective C Auto D as an example, if the objective function, either as a cost function J $\in [0, \infty]$, or inversely, as a fitness function $f \in (0, 1]$, where

$$f = \frac{J}{1+J},$$

is differentiable under practical constraints in the multi-dimensional space, the design problem may be solved analytically. Finding the parameter sets that result in a zero first-order derivative and that satisfy the second-order derivative conditions would reveal all local optima. Then comparing the values of the performance index of all the local optima, together with those of all boundary parameter sets, would lead to the global optimum, whose corresponding 'parameter' set will thus represent the best design. However, in practice, the optimization usually involves multiple object gives and the matters involving derivatives are lot more complex.

Dealing with Practical Objectives

In practice, the objective value may be noisy or even non-numerical, and hence its gradient information may be unreliable or unavailable. This is particularly true when the problem is multi-objective. At present, many designs and refinements are mainly made through a manual trial-and-error process with the help of a CAD simulation package. Usually, such *a posteriori* learning or adjustments need to be repeated many times until a 'satisfactory' or 'optimal' design emerges.

Exhaustive Search

In theory, this adjustment process can be automated by computerised search, such as exhaustive search. As this is an exponential algorithm, it may not deliver solutions in practice within a limited period of time.

Search in Polynomial Time

One approach to virtual engineering and automated design is evolutionary computation such as evolutionary algorithms.

Evolutionary Algorithms

To reduce the search time, the biologically-inspired evolutionary algorithm (EA) can be used instead, which is a (non-deterministic) polynomial algorithm. The EA based multi-objective "search team" can be interfaced with an existing CAD simulation package in a batch mode. The EA encodes the design parameters (encoding being necessary if some parameters are non-numerical) to refine multiple candidates through parallel and interactive search. In the search process, 'selection' is performed using 'survival of the fittest' *a posteriori* learning. To obtain the next 'generation' of possible solutions, some parameter values are exchanged between two candidates (by an operation called 'cross-over') and new values introduced (by an operation called 'mutation'). This way, the evolutionary technique makes use of past trial information in a similarly intelligent manner to the human designer.

The EA based optimal designs can start from the designer's existing design database or from an initial generation of candidate designs obtained randomly. A number of finally evolved top-performing candidates will represent several automatically optimized digital prototypes.

There are websites that demonstrate interactive evolutionary algorithms for design. EndlessForms.com allows you to evolve 3D objects online and have them 3D printed. PicBreeder.org allows you to do the same for 2D images.

DECISION SUPPORT SYSTEM

A **Decision Support System (DSS)** is a computer-based information system that supports business or organizational decision-making activities. DSSs serve the management, operations, and planning levels of an organization (usually mid and higher management) and help to make decisions, which may be rapidly changing and not easily specified in advance (Unstructured and Semi-Structured decision problems). Decision support systems can be either fully computerized, human or a combination of both.

While academics have perceived DSS as a tool to support decision-making process, DSS users see DSS as a tool to facilitate organizational processes. Some authors have extended the definition of DSS to include any system that might support decision-making. Sprague defines DSS by its characteristics :

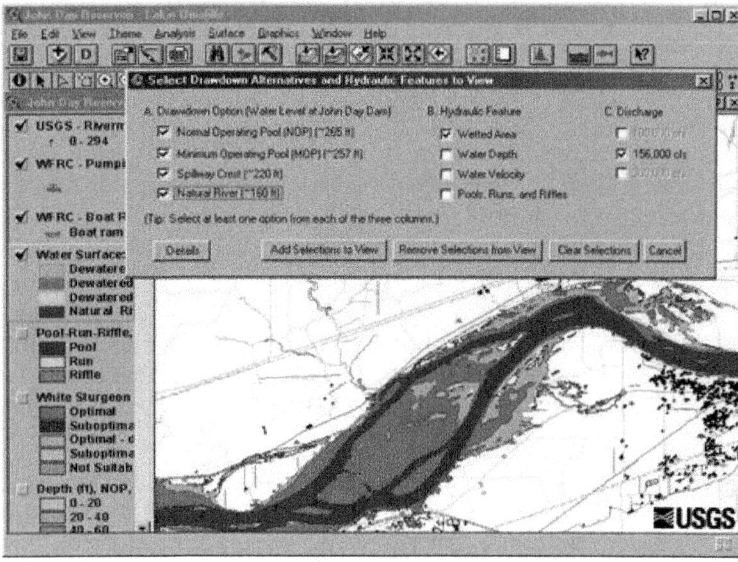

Fig. : Example of a Decision Support System for John Day Reservoir.

1. DSS tends to be aimed at the less well structured, underspecified problem that upper level managers typically face;
2. DSS attempts to combine the use of models or analytic techniques with traditional data access and retrieval functions;
3. DSS specifically focuses on features which make them easy to use by non-computer people in an interactive mode; and
4. DSS emphasizes flexibility and adaptability to accommodate changes in the environment and the decision-making approach of the user.

DSSs include knowledge-based systems. A properly designed DSS is an interactive software-based system intended to help decision makers compile useful information from a combination of raw data, documents, and personal knowledge, or business models to identify and solve problems and make decisions.

Typical information that a decision support application might gather and present includes :

- Inventories of information assets (including legacy and relational data sources, cubes, data warehouses, and data marts),
- Comparative sales figures between one period and the next,
- Projected revenue figures based on product sales assumptions.

History

The concept of decision support has evolved from two main areas of research : The theoretical studies of organizational decision-making done at the Carnegie Institute of Technology during the late 1950s and early 1960s, and the technical work on Technology in the 1960s. DSS became an area of research of its own in the middle of the 1970s, before gaining in intensity during the 1980s. In the middle and late 1980s, executive information systems (EIS), group decision support systems (GDSS), and organizational decision support systems (ODSS) evolved from the single user and model-oriented DSS.

According to Sol the definition and scope of DSS has been migrating over the years. In the 1970s DSS was described as "a computer-based system to aid decision-making". In the late 1970s the DSS movement started focusing on "interactive computer-based systems which help decision-makers utilize data bases and models to solve ill-structured problems". In the 1980s DSS should provide systems "using suitable and available technology to improve effectiveness of managerial and professional activities", and towards the end of 1980s DSS faced a new challenge towards the design of intelligent workstations.

In 1987, Texas Instruments completed development of the Gate Assignment Display System (GADS) for United Airlines. This decision support system is credited with significantly reducing travel delays by aiding the management of ground operations at various airports, beginning with O'Hare International Airport in Chicago and Stapleton Airport in Denver Colorado.

Beginning in about 1990, data warehousing and on-line analytical processing (OLAP) began broadening the realm of DSS. As the turn of the millennium approached, new Web-based analytical applications were introduced.

The advent of better and better reporting technologies has seen DSS start to emerge as a critical component of management design. Examples of this can be seen in the intense amount of discussion of DSS in the education environment.

DSS also have a weak connection to the user interface paradigm of hypertext. Both the University of Vermont PROMIS system (for medical decision-making) and the Carnegie Mellon ZOG/KMS system (for military and business decision-making) were decision support systems which also were major breakthroughs in user interface research. Furthermore, although hypertext researchers have generally been concerned with information overload, certain researchers, notably Douglas Engelbart, have been focused on decision-makers in particular.

Taxonomies

Using the relationship with the user as the criterion, Haettenschwiler differentiates *passive*, *active*, and *co-operative DSS*. A *passive DSS* is a system that aids the process of decision-making, but that cannot bring out explicit decision suggestions or solutions. An *active DSS* can bring out such decision suggestions or solutions. A *cooperative DSS* allows the decision-maker (or its advisor) to modify, complete, or refine the decision suggestions provided by the system, before sending them back to the system for validation. The system again improves, completes, and refines the suggestions of the decision-maker and sends them back to them for validation. The whole process then starts again, until a consolidated solution is generated.

Another taxonomy for DSS has been created by Daniel Power. Using the mode of assistance as the criterion, Power differentiates *communication-driven DSS*, *data-driven DSS*, *document-driven DSS*, *knowledge-driven DSS*, and *model-driven DSS*.

- A **communication-driven DSS** supports more than one person working on a shared task; examples include integrated tools like Google Docs or Groove
- A **data-driven DSS** or data-oriented DSS emphasizes access to and manipulation of a time series of internal company data and, sometimes, external data.
- A **document-driven DSS** manages, retrieves, and manipulates unstructured information in a variety of electronic formats.
- A **knowledge-driven DSS** provides specialized problem-solving expertise stored as facts, rules, procedures, or in similar structures.
- A **model-driven DSS** emphasizes access to and manipulation of a statistical, financial, optimization, or simulation model. Model-driven DSS use data and parameters provided by users to assist decision makers in analyzing a situation; they are not necessarily data-intensive. Dicodess is an example of an open source model-driven DSS generator.

Using scope as the criterion, Power differentiates *enterprise-wide DSS* and *desktop DSS*. An *enterprise-wide DSS* is linked to large data warehouses and serves many managers in the company. A *desktop, single-user DSS* is a small system that runs on an individual manager's PC.

Components

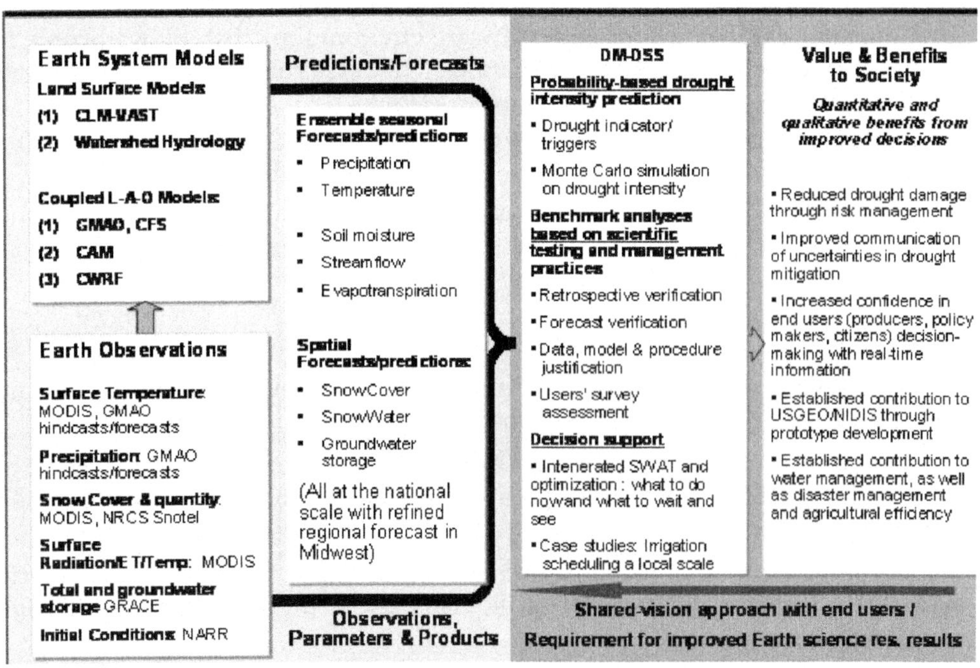

Fig. : Design of a drought mitigation decision support system.

Three fundamental components of a DSS architecture are :

1. The database (or knowledge base),

2. The model (*i.e.*, the decision context and user criteria), and
3. Theuser interface.

The users themselves are also important components of the architecture.

Development Frameworks

DSS systems are not entirely different from other systems and require a structured approach. Such a framework includes people, technology, and the development approach.

The Early Framework of Decision Support System consists of four phases :

Intelligence Searching for conditions that call for decision.

Design Inventing, developing and analyzing possible alternative actions of solution.

Choice Selecting a course of action among those.

Implementation Adopting the selected course of action in decision situation.

DSS technology levels (of hardware and software) may include :

1. The actual application that will be used by the user. This is the part of the application that allows the decision-maker to make decisions in a particular problem area. The user can act upon that particular problem.
2. Generator contains Hardware/software environment that allows people to easily develop specific DSS applications. This level makes use of case tools or systems such as Crystal, Analytica and iThink.
3. Tools include lower level hardware/software. DSS generators including special languages, function libraries and linking modules.

An iterative developmental approach allows for the DSS to be changed and redesigned at various intervals. Once the system is designed, it will need to be tested and revised where necessary for the desired outcome.

Classification

There are several ways to classify DSS applications. Not every DSS fits neatly into one of the categories, but may be a mix of two or more architectures.

Holsapple and Whinston classify DSS into the following six frameworks : text-oriented DSS, database-oriented DSS, spreadsheet-oriented DSS, solver-oriented DSS, rule-oriented DSS, and compound DSS.

A compound DSS is the most popular classification for a DSS. It is a hybrid system that includes two or more of the five basic structures described by Holsapple and Whinston.

The support given by DSS can be separated into three distinct, interrelated categories : Personal Support, Group Support, and Organizational Support.

DSS components may be classified as :

1. *Inputs :* Factors, numbers, and characteristics to analyze

2. *User Knowledge and Expertise :* Inputs requiring manual analysis by the user
3. *Outputs :* Transformed data from which DSS "decisions" are generated
4. *Decisions :* Results generated by the DSS based on user criteria.

DSSs which perform selected cognitive decision-making functions and are based on artificial intelligence or intelligent agents technologies are called Intelligent Decision Support Systems (IDSS).

The nascent field of Decision engineering treats the decision itself as an engineered object, and applies engineering principles such as Design and Quality assurance to an explicit representation of the elements that make up a decision.

Applications

As mentioned above, there are theoretical possibilities of building such systems in any knowledge domain.

One is the clinical decision support system for medical diagnosis. Other examples include a bank loan officer verifying the credit of a loan applicant or an engineering firm that has bids on several projects and wants to know if they can be competitive with their costs.

DSS is extensively used in business and management. Executive dashboard and other business performance software allow faster decision-making, identification of negative trends, and better allocation of business resources. Due to DSS all the information from any organization is represented in the form of charts, graphs *i.e.* in a summarized way, which helps the management to take strategic decision.

A growing area of DSS application, concepts, principles, and techniques is in agricultural production, marketing for sustainable development. For example, the DSSAT4 package, developed through financial support of USAID during the 80s and 90s, has allowed rapid assessment of several agricultural production systems around the world to facilitate decision-making at the farm and policy levels. There are, however, many constraints to the successful adoption on DSS in agriculture.

DSS are also prevalent in forest management where the long planning time frame demands specific requirements. All aspects of Forest management, from log transportation, harvest scheduling to sustainability and ecosystem protection have been addressed by modern DSSs.

A specific example concerns the Canadian National Railway system, which tests its equipment on a regular basis using a decision support system. A problem faced by any railroad is worn-out or defective rails, which can result in hundreds of derailments per year. Under a DSS, CN managed to decrease the incidence of derailments at the same time other companies were experiencing an increase.

Benefits

1. Improves personal efficiency.
2. Speed up the process of decision-making.

3. Increases organizational control.
4. Encourages exploration and discovery on the part of the decision-maker.
5. Speeds up problem solving in an organization.
6. Facilitates interpersonal communication.
7. Promotes learning or training.
8. Generates new evidence in support of a decision.
9. Creates a competitive advantage over competition.
10. Reveals new approaches to thinking about the problem space.
11. Helps automate managerial processes.
12. Create Innovative ideas to speed up the performance.

DSS Characteristics and Capabilities

1. Solve semi-structured and unstructured problems.
2. Support managers at all levels.
3. Support individuals and groups.
4. Interdependence and sequence of decisions.
5. Support Intelligence, Design, Choice.
6. Adaptable and flexible.
7. Interactive and ease of use.
8. Interactive and efficiency.
9. Human control of the process.
10. Ease of development by end user.
11. Modeling and analysis.
12. Data access.
13. Standalone and web-based integration.
14. Support varieties of decision processes.
15. Support varieties of decision trees.
16. Quick response.

MORPHOLOGICAL ANALYSIS (PROBLEM-SOLVING)

Morphological Analysis or **General Morphological Analysis** is a method developed by Fritz Zwicky for exploring all the possible solutions to a multi-dimensional, non-quantified complex problem.

Overview

General Morphology was developed by Fritz Zwicky, the Bulgarian-born, Swiss-national astrophysicist based at the California Institute of Technology. Among others, Zwicky applied Morphological Analysis (MA) to astronomical studies and the development of jet and rocket propulsion systems. As a problem-structuring and problem-solving technique, MA was designed for multi-dimensional, non-quantifiable problems where causal modelling and simulation do not function well, or at all. Zwicky developed this approach to address seemingly non-reducible complexity : using the technique of cross-consistency assessment (CCA), the system allows for reduction by identifying the possible solutions that actually exist, eliminating the illogical solution combinations in a grid box rather than reducing the number of variables involved. A detailed introduction to morphological modelling is given in Ritchey.

Morphological Analysis of Real-world Problems

Consider a complex, real-world problem, like those of marketing or making policies for a nation, where there are many governing factors, and most of them cannot be expressed as numerical time series data, as one would like to have for building mathematical models.

The conventional approach here would be to break the system down into parts, isolate the vital parts (dropping the 'trivial' components) for their contributions to the output and solve the simplified system for creating desired models or scenarios. The disadvantage of this method is that real-world scenarios do not behave rationally : more often than not, a simplified model will break down when the contribution of the 'trivial' components becomes significant. Also, importantly, the behaviour of many components will be governed by the states of, and their relations with, other components-ones that may be seen to be minor before the analysis.

Morphological Analysis, on the other hand, does not drop any of the components from the system itself, but works backwards from the output towards the system internals. Again, the interactions and relations get to play their parts in MA and their effects are accounted for in the analysis.

Chapter 8

DECISION ENGINEERING

Decision engineering is a framework that unifies a number of best practices for organizational decision-making. It is based on the recognition that, in many organizations, decision-making could be improved if a more structured approach were used. Decision engineering seeks to overcome a decision-making "complexity ceiling", which is characterized by a mismatch between the sophistication of organizational decision-making practices and the complexity of situations in which those decisions must be made. As such, it seeks to solve some of the issues

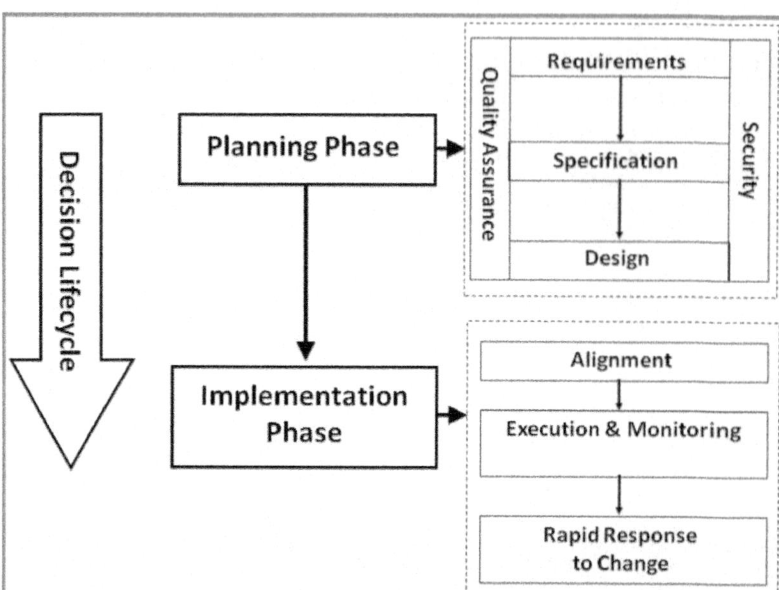

Fig. : Decision engineering framework.

identified around complexity theory and organizations. In this sense, decision engineering represents a practical application of the field of complex systems, which helps organizations to navigate the complex systems in which they find themselves. Decision engineering can also be thought of as a framework that brings advanced analytics techniques to the desktop of the non-expert decision-maker, as well as incorporating, and then extending, inductive reasoning and machine learning techniques to overcome the problems articulated in Black swan theory.

Decision engineering proponents believe that many organizations continue to make poor decisions. In response, decision engineering seeks to unify a number of decision-making best practices, described in more detail below.

Decision engineering builds on the insight that it is possible to *design* the decision itself, using principles previously used for designing more tangible objects like bridges and buildings.

The use of a *visual* design language representing decisions is an important element of decision engineering, since it provides an intuitive common language readily understood by all decision participants. A visual metaphor improves the ability to reason about complex systems as well as to enhance collaboration.

In addition to visual decision design, there are other two aspects of engineering disciplines that aid mass adoption. These are : (1) the creation of a shared language of design elements and (2) the use of a common methodology or process, as illustrated in the diagram above.

MOTIVATION

The need for a unified methodology of decision-making is driven by a number of factors that organizations face as they make difficult decisions in a complex internal and external environment.

Recognition of the broad-based inability of current methods to solve decision-making issues in practice comes from several sources, including government sources and industries such as telecommunications, media, the automotive industry, and pharmaceuticals.

Examples

- The *outcomes* of decisions are becoming more complex, going well beyond next quarter's revenues or other tangible outcomes to multiple goals that must be satisfied together, some of which are often intangible :

The car is becoming an expression of identity, values, and personal control in ways that move far beyond traditional segmentation and branding. For example, fuel efficiency will be only one consideration for a socially responsible vehicle (SRV). What percent of the parts are recyclable? What is the vehicle's total carbon footprint? Are there child labour inputs? Toxic paints, glues, or plastics? How transparent is the supply chain? Is the seller accountable for recycling? What methods are used? Are fair labour practices employed?

- Shoshana Zuboff, *The GM Solution : Life Boats, Not Life Support. Business Week, November 18, 2008*
- Global increase in complexity :

We live in a dynamic world in which the pace, scope, and complexity of change are increasing. The continued march of globalization, the growing number of independent actors, and advancing technology have increased global connectivity, interdependence and complexity, creating greater uncertainties, systemic risk and a less predictable future.

TRANSFERRING ENGINEERING PRINCIPLES

Unlike other decision-making tools and methodologies, decision engineering seeks to bring to bear a number of engineering practices to the process of creating a decision. These include requirements analysis, specification, scenario planning, quality assurance, security, and the use of design principles as described above. During the decision execution phase, outputs produced during the design phase can be used in a number of ways; monitoring approaches like business dashboards and assumption based planning are used to track the outcome of a decision and to trigger replanning as appropriate. One view of how some of these elements combine is shown in the diagram at the start of this article.

Decision engineering has the potential to improve the quality of decisions made, the ability to make them more quickly, the ability to align organizational resources more effectively around a change in decisions, and lowers the risks associated with decisions. Furthermore, a designed decision can be reused and modified as new information is obtained.

BRINGING NUMERICAL METHODS TO THE DESKTOP

Although many elements of decision engineering, such as Sensitivity analysis and analytics are mature disciplines, they are not in wide use by decision-makers. Decision engineering seeks to create a visual language that serves to facilitate communication between them and quantitative experts, allowing broader utilization of these and other numerical and technical approaches.

In particular, dependency links in a decision model represent cause-and-effect (as in a causal loop diagram), data flow (as in a data flow diagram), or other relationships. As an example, one link might represent the connection between "mean time to repair a problem with telephone service" and "customer satisfaction", where a short repair time would presumably raise customer satisfaction. The functional form of these dependencies can be determined by a number of approaches. Numerical approaches, which analyze data to determine these functions, include machine learning and analytics algorithms (including artificial neural networks), as well as more traditional regression analysis. Results from operations research and many other quantitative approaches have a similar role to play.

When data is not available (or is too noisy, uncertain, or incomplete), these dependency links can take on the form of rules as might be found in an expert system or rule-based system, and so can be obtained through knowledge engineering.

In this way, a decision model represents a mechanism for combining multiple relationships, as well as symbolic and subsymbolic reasoning, into a complete solution to determining the outcome of a practical decision.

ORIGINS

It is interesting to note that, despite decades of development of decision support system and methodologies (like decision analysis), these are still less popular than spreadsheets as primary tools for decision-making. Decision engineering seeks to bridge this gap, creating a critical mass of users of a common methodology and language for the core entities included in a decision, such as assumptions, external values, facts, data, and conclusions. If a pattern from previous industries holds, such a methodology will also facilitate technology adoption, by clarifying common maturity models and road maps that can be shared from one organization to another.

The decision engineering approach is multi-disciplinary, unifying findings on cognitive bias and decision-making, situational awareness, critical and creative thinking, collaboration and organizational design, with engineering technologies.

Decision engineering is considered an improvement upon current organizational decision-making practices, which include the use of spreadsheets (difficult to QA, hard to collaborate and discuss), text (sequential in nature, so is not a good fit for how information flows through a decision structure), and verbal argument. The movement from these largely informal structures to one in which a decision is documented in a well understood, visual language, echoes the creation of common blueprint methodologies in construction, with promise of similar benefits.

Decision engineering is both a very new and also a very old discipline. Many of its elements—such as the language of assessing assumptions, using logic to support an argument, the necessity of critical thinking to evaluate a decision, and understanding the impacts of bias—are ancient. Yet the realization that these elements can form a coherent whole that provides significant benefits to organizations by focusing on a common methodology is relatively new.

VISUAL DECISION DESIGN

Because it makes visible the otherwise invisible reasoning structures used in complex decisions, the design aspect of decision engineering draws from other conceptual representation technologies like mind mapping, conceptual graphs, and semantic networks.

The basic idea is that a visual metaphor enhances intuitive thinking, inductive reasoning, and pattern recognition—important cognitive skills usually less accessible in a verbal or text discussion. A business decision map can be seen as one approach to a formal decision language to support decision engineering.

Explicit Representation of Intangibles

Decision engineering recognizes that many aspects of decision-making are based on intangible elements, including opportunity costs, employee morale, intellectual capital, brand recognition and other forms of business value that are not captured in traditional quantitative or financial models. Value network analysis—most notably *Value network maps*—are therefore relevant here.

NOTES

- Enterprise Decision Management (EDM) is a closely related discipline that focuses on automating decisions across an enterprise. Decision engineering is from this point of view a superset of EDM, since it encompasses both manual and automated decision-making processes, unifying them into a common methodology that, when effective, breaks down barriers between quantitative analysis/analytics tools and departments and those with a more qualitative/strategic/management focus.
- The term *"decision engineering"* is used in several industries with more specific meaning than the framework described here. For instance, the *Australian Software Research Centre* has an IT evaluation approach called *Decision Engineering; Idea* focuses on emergency management under the heading of *"Decision Engineering Analysis"*; and the National University of Singapore includes an organization called the Biomedical Decision Engineering Group. Each of these has a meaning that is distinct from what is discussed in the present article.
- In behavioural economics, *"decision engineering"* can mean the deliberate manipulation of consumer choices, as in this Journal of Consumer Research study : *People choose healthy meals, if given more choice : Study*. In this use of the term, decision engineering is roughly analogous to soft paternalism-a quite different meaning than is covered in the present article, referring as it does to the engineering of decisions made by consumers, rather than the use of engineering principles to aid in complex decision-making. Although distinctly different, this practice draws on much of the same decision-making research as does decision engineering (such as, for the example, the work of Richard Thaler as described in *this article about Barack Obama's University of Chicago connections to this school of thought*).
- Cost engineering measures the costs of engineering projects. Cost engineering is sometimes grouped into product engineering and design optimization as *"decision engineering"*. This can be distinguished from the broader framework of this article, which goes beyond the arena of engineering decisions to all decisions faced by organizations.
- Operations research is a largely quantitative approach to decision-making that attempts to identify optimal or near-optimal solutions to decision-making problems.

PROCESS OPTIMIZATION

Process optimization is the discipline of adjusting a process so as to optimize some specified set of parameters without violating some constraint. The most common goals are minimizing cost, maximizing throughput, and/or efficiency. This is one of the major quantitative tools in industrial decision-making.

When optimizing a process, the goal is to maximize one or more of the process specifications, while keeping all others within their constraints.

Areas

Fundamentally, there are three parameters that can be adjusted to affect optimal performance. They are :

- *Equipment optimization :* The first step is to verify that the existing equipment is being used to its fullest advantage by examining operating data to identify equipment bottlenecks.
- *Operating procedures :* Operating procedures may vary widely from person-to-person or from shift-to-shift. Automation of the plant can help significantly. But automation will be of no help if the operators take control and run the plant in manual.
- *Control optimization :* In a typical processing plant, such as a chemical plant or oil refinery, there are hundreds or even thousands of control loops. Each control loop is responsible for controlling one part of the process, such as maintaining a temperature, level, or flow.

If the control loop is not properly designed and tuned, the process runs below its optimum. The process will be more expensive to operate, and equipment will wear out prematurely. For each control loop to run optimally, identification of sensor, valve, and tuning problems is important. It has been well documented that over 35% of control loops typically have problems.

The process of continuously monitoring and optimizing the entire plant is sometimes called performance supervision.

Calculation of Glass Properties

The **calculation of glass properties (glass modelling)** is used to predict glass properties of interest or glass behaviour under certain conditions (*e.g.*, during production) without experimental investigation, based on past data and experience, with the intention to save time, material, financial, and environmental resources, or to gain scientific insight. It was first practised at the end of the 19th century by A. Winkelmann and O. Schott. The combination of several glass models together with other relevant functions can be used for optimization and six sigma procedures. In the form of statistical analysis glass modelling can aid with accreditation of new data, experimental procedures, and measurement institutions (glass laboratories).

History

Historically, the calculation of glass properties is directly related to the founding of glass science. At the end of the 19th century the physicist Ernst Abbe developed equations that allow calculating the design of optimized optical microscopes in Jena, Germany, stimulated by co-operation with the optical workshop of Carl Zeiss. Before Ernst Abbe's time the building of microscopes was mainly a work of art and experienced craftsmanship, resulting in very expensive optical microscopes with variable quality. Now Ernst Abbe knew exactly how to construct an excellent microscope, but unfortunately, the required lenses and prisms with specific ratios of refractive index and dispersion did not exist. Ernst Abbe was not able to find answers to his needs from glass artists and engineers; glass making was not based on science at this time.

In 1879 the young glass engineer Otto Schott sent Abbe glass samples with a special composition (lithium silicate glass) that he had prepared himself and that he hoped to show special optical properties. Following measurements by Ernst Abbe, Schott's glass samples did not have the desired properties, and they were also not as homogeneous as desired. Nevertheless, Ernst Abbe invited Otto Schott to work on the problem further and to evaluate all possible glass components systematically. Finally, Schott succeeded in producing homogeneous glass samples, and he invented borosilicate glass with the optical properties Abbe needed. These inventions gave rise to the well-known companies Zeiss and Schott Glass. Systematic glass research was born. In 1908, Eugene Sullivan founded glass research also in the United States.

At the beginning of glass research it was most important to know the relation between the glass composition and its properties. For this purpose Otto Schott introduced the **additivity principle** in several publications for calculation of glass properties. This principle implies that the relation between the glass composition and a specific property is linear to all glass component concentrations, assuming an ideal mixture, with C_i and b_i representing specific glass component concentrations and related coefficients respectively in the equation below. The additivity principle is a simplification and only valid within narrow composition ranges as seen in the displayed diagrams for the refractive index and the viscosity. Nevertheless, the application of the additivity principle lead the way to many of Schott's inventions, including optical glasses, glasses with low thermal expansion for cooking and laboratory ware (Duran), and glasses with reduced freezing point depression for mercury thermometers. Subsequently, English and Gehlhoff *et. al.*published similar additive glass property calculation models. Schott's additivity principle is still widely in use today in glass research and technology.

$$\textit{Additivity Principle}: \text{Glass Property} = b_0 + \sum_{i=1}^{n} b_i C_i$$

Global Models

Schott and many scientists and engineers afterwards applied the additivity principle to experimental data measured in their own laboratory within sufficiently narrow composition ranges (**local glass models**). This is most convenient because disagreements between laboratories and non-linear glass component interactions do not need to be considered. In the course of several decades of systematic glass research thousands of glass compositions were studied, resulting in millions of published glass properties, collected in glass databases. This huge pool of experimental data was not investigated as a whole, until Bottinga, Kucuk, Priven, Choudhary, Mazurin, and Fluegel published their **global glass models**, using various approaches. In contrast to the models by Schott the global models consider many independent data sources, making the model estimates more reliable. In addition, global models can reveal and quantify *non-additive* influences of certain glass component combinations on the properties, such as the *mixed-alkali effect* as seen in the diagram on the right, or the *boron anomaly*. Global models also reflect interesting developments of glass property measurement accuracy, *e.g.*, a decreasing accuracy of experimental data in modern scientific literature for some glass properties, shown in the diagram. They can be used for accreditation of new data, experimental procedures, and measurement institutions (glass laboratories). In the following sections (except melting enthalpy) *empirical* modelling techniques are presented, which seem to be a successful way for handling huge amounts of experimental data. The resulting models are applied in contemporary engineering and research for the calculation of glass properties.

Non-empirical (*deductive*) glass models exist. They are often not created to obtain reliable glass property predictions in the first place (except melting enthalpy), but to establish relations among several properties (*e.g.* atomic radius, atomic mass, chemical bond strength and angles, chemical valency, heat capacity) to gain scientific insight. In future, the investigation of property relations in deductive models may ultimately lead to reliable predictions for all desired properties, provided the property relations are well understood and all required experimental data are available.

Methods

Glass properties and glass behaviour during production can be calculated through statistical analysis of glass databases such as GE-SYSTEM SciGlass and Interglad, sometimes combined with the finite element method. For estimating the melting enthalpy thermodynamic databases are used.

Linear Regression

If the desired glass property is not related to crystallization (*e.g.*, liquidus temperature) or phase separation, linear regression can be applied using common polynomial functions up to the third degree. Below is an example equation of the second degree. The C-values are the glass component concentrations like Na_2O or CaO in per cent or other fractions, the *b*-values are coefficients, and *n* is the total number of glass components. The glass main component silica (SiO_2) is excluded in the equation below because of over-parametrization due to the constraint that all components sum up to 100%. Many terms in the equation below can be neglected based on correlation and significance analysis. Systematic errors such as seen in the picture are quantified by dummy variables. Further details and examples are available in an online tutorial by Fluegel.

$$\text{Glass Property} = b_0 + \sum_{i=1}^{n}\left(b_i C_i + \sum_{k=i}^{n} b_{ik} C_i C_k\right)$$

Non-linear Regression

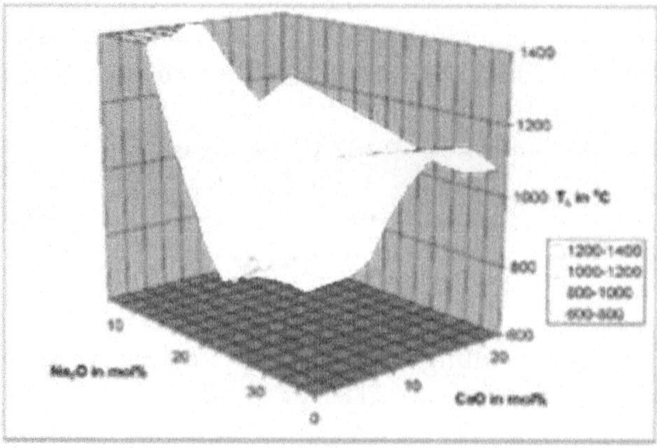

Fig. : Liquidus surface in the system SiO_2-Na_2O-CaO using disconnected peak functions based on 237 experimental data from 28 investigators. Error = 15°C.

The liquidus temperature has been modelled by non-linear regression using neural networks and disconnected peak functions. The disconnected peak functions approach is based on the observation that within one primary crystalline phase field linear regression can be applied and at eutectic points sudden changes occur.

Glass Melting Enthalpy

The glass melting enthalpy reflects the amount of energy needed to convert the mix of raw materials (batch) to a melt glass. It depends on the batch and glass compositions, on the efficiency of the furnace and heat regeneration systems, the average residence time of the glass in the furnace, and many other factors. A pioneering article about the subject was written by Carl Kröger in 1953.

Finite Element Method

For modelling of the glass flow in a glass melting furnace the finite element method is applied commercially, based on data or models for viscosity, density, thermal conductivity, heat capacity, absorption spectra, and other relevant properties of the glass melt. The finite element method may also be applied to glass forming processes.

Optimization

It is often required to optimize several glass properties simultaneously, including production costs. GE-SYSTEM. This can be performed, *e.g.*, by simplex search, or in a spreadsheet as follows :

1. Listing of the desired properties;
2. Entering of models for the reliable calculation of properties based on the glass composition, including a formula for estimating the production costs;
3. Calculation of the squares of the differences (errors) between desired and calculated properties;
4. Reduction of the sum of square errors using the Solver option in Microsoft Excel with the glass components as variables. Other software (*e.g.* Microcal Origin) can also be used to perform these optimizations.

It is possible to weight the desired properties differently. Basic information about the principle can be found in an article by Huff *et. al.* The combination of several glass models together with further relevant technological and financial functions can be used in six sigma optimization.

GLASS BATCH CALCULATION

Glass batch calculation or **glass batching** is used to determine the correct mix of raw materials (batch) for a glass melt.

Principle

The raw materials mixture for glass melting is termed "batch". The batch must be measured properly to achieve a given, desired glass formulation. This batch calculation is based on the common linear regression equation :

$$N_B = (B^T \cdot B)^{-1} \cdot B^T \cdot N_G$$

with N_B and N_G being the molarities 1-column matrices of the batch and glass components respectively, and B being the batching matrix. The symbol "T" stands for the matrix transpose operation, "-1" indicates matrix inversion, and the sign "·" means the scalar product. From the molarities matrices N, percentages by weight (wt%) can easily be derived using the appropriate molar masses.

Example Calculation

An example batch calculation may be demonstrated here. The desired glass composition in wt% is : 67 SiO_2, 12 Na_2O, 10 CaO, 5 Al_2O_3, 1 K_2O, 2 MgO, 3 B_2O_3, and as raw materials are used sand, trona, lime, albite, orthoclase, dolomite, and borax. The formulas and molar masses of the glass and batch components are listed in the following table :

Formula of glass component	*Desired concentration of glass component, wt%*	*Molar mass of glass component, g/mol*	*Batch component*	*Formula of batch component*	*Molar mass of batch component, g/mol*
SiO_2	67	60.0843	Sand	SiO_2	60.0843
Na_2O	12	61.9789	Trona	$Na_3H(CO_3)_2*2H_2O$	226.0262
CaO	10	56.0774	Lime	$CaCO_3$	100.0872
Al_2O_3	5	101.9613	Albite	$Na_2O*Al_2O_3*6SiO_2$	524.4460
K_2O	1	94.1960	Orthoclase	$K_2O*Al_2O_3*6SiO_2$	556.6631
MgO	2	40.3044	Dolomite	$MgCa(CO_3)_2$	184.4014
B_2O_3	3	69.6202	Borax	$Na_2B_4O_7*10H_2O$	381.3721

The batching matrix B indicates the relation of the molarity in the batch (columns) and in the glass (rows). For example, the batch component SiO_2 adds 1 mol SiO_2 to the glass, therefore, the intersection of the first column and row shows "1". Trona adds 1.5 mol Na_2O to the glass; albite adds 6 mol SiO_2, 1 mol Na_2O, and 1 mol Al_2O_3, and so on. For the example given above, the complete batching matrix is listed below. The molarity matrix N_G of the glass is simply determined by dividing the desired wt% concentrations by the appropriate molar masses, *e.g.*, for SiO_2 67/60.0843 = 1.1151.

$$B = \begin{bmatrix} 1 & 0 & 0 & 6 & 6 & 0 & 0 \\ 0 & 1.5 & 0 & 1 & 0 & 0 & 1 \\ 0 & 0 & 1 & 0 & 0 & 1 & 0 \\ 0 & 0 & 0 & 1 & 1 & 0 & 0 \\ 0 & 0 & 0 & 0 & 1 & 0 & 0 \\ 0 & 0 & 0 & 0 & 0 & 1 & 0 \\ 0 & 0 & 0 & 0 & 0 & 0 & 2 \end{bmatrix} \quad N_G = \begin{bmatrix} 1.1151 \\ 0.1936 \\ 0.1783 \\ 0.0490 \\ 0.0106 \\ 0.0496 \\ 0.0431 \end{bmatrix}$$

The last row of the B matrix has a mistake : it must be 0 1 0 0 0 0 2

The resulting molarity matrix of the batch, N_B, is given here. After multiplication with the appropriate molar masses of the batch ingredients one obtains the batch mass fraction matrix M_B :

$$N_B = \begin{bmatrix} 0.82087 \\ 0.08910 \\ 0.12870 \\ 0.03842 \\ 0.01062 \\ 0.04962 \\ 0.02155 \end{bmatrix} \quad M_B = \begin{bmatrix} 49.321 \\ 20.138 \\ 12.881 \\ 20.150 \\ 5.910 \\ 9.150 \\ 8.217 \end{bmatrix}$$

or

$$M_B(100\% \text{ normalized}) = \begin{bmatrix} 39.216 \\ 16.012 \\ 10.242 \\ 16.022 \\ 4.699 \\ 7.276 \\ 6.533 \end{bmatrix}$$

The matrix M_B, normalized to sum up to 100% as seen above, contains the final batch composition in wt% : 39.216 sand, 16.012 trona, 10.242 lime, 16.022 albite, 4.699 orthoclase, 7.276 dolomite, 6.533 borax. If this batch is melted to a glass, the desired composition given above is obtained. During glass melting, carbon dioxide (from trona, lime, dolomite) and water (from trona, borax) evaporate.

Another simple glass batch calculation can be found at the website of the University of Washington.

Advanced Batch Calculation by Optimization

If the number of glass and batch components is not equal, if it is impossible to exactly obtain the desired glass composition using the selected batch ingredients, or if the matrix equation is not soluble for other reasons (*e.g.*, correlation), the batch composition must be determined by optimization techniques.

Chapter 9

Dimensionality Reduction

FACTOR ANALYSIS

Factor analysis is a statistical method used to describe variability among observed, correlated variables in terms of a potentially lower number of unobserved variables called **factors**. For example, it is possible that variations in four observed variables mainly reflect the variations in two unobserved variables. Factor analysis searches for such joint variations in response to unobserved latent variables. The observed variables are modelled as linear combinations of the potential factors, plus "error" terms. The information gained about the interdependencies between observed variables can be used later to reduce the set of variables in a data-set. Computationally this technique is equivalent to low rank approximation of the matrix of observed variables. Factor analysis originated in psychometrics, and is used in behavioral sciences, social sciences, marketing, product management, operations research, and other applied sciences that deal with large quantities of data.

Factor analysis is related to principal component analysis (PCA), but the two are not identical. Latent variable models, including factor analysis, use regression modelling techniques to test hypotheses producing error terms, while PCA is a descriptive statistical technique. There has been significant controversy in the field over the equivalence or otherwise of the two techniques.

Statistical Model

Definition

Suppose we have a set of p observable random variables, $x_1,\ldots, x_p$ with means $\mu_1,\ldots, \mu_p$.

Suppose for some unknown constants l_{ij} and k unobserved random variables F_j, where $i \in 1,\ldots, p$ and $j \in 1,\ldots, k$, where $k < p$, we have

$$x_i - \mu_i = l_{i1}F_1 + \cdots + l_{ik}F_k + \varepsilon_i.$$

Here, the ε_i are independently distributed error terms with zero mean and finite variance, which may not be the same for all i. Let $\text{Var}(\varepsilon_i) = \psi_i$, so that we have

$\text{Cov}(\varepsilon) = \text{Diag}(\psi_1,\ldots, \psi_p) = \Psi$ and $E(\varepsilon) = 0$.

In matrix terms, we have

$$x - \mu = LF + \varepsilon.$$

If we have n observations, then we will have the dimensions $x_p \times n$, $L_p \times k$, and $F_k \times n$. Each column of x and F denote values for one particular observation, and matrix L does not vary across observations.

Also we will impose the following assumptions on F.

1. F and ε are independent.
2. $E(F) = 0$
3. $\text{Cov}(F) = I$ (to make sure that the factors are uncorrelated)

Any solution of the above set of equations following the constraints for F is defined as the *factors*, and L as the *loading matrix*.

Suppose $\text{Cov}(x - \mu) = \Sigma$. Then note that from the conditions just imposed on F, we have

$$\text{Cov}(x - \mu) = \text{Cov}(LF + \varepsilon),$$

or

$$\Sigma = L\text{Cov}(F)L^T + \text{Cov}(\varepsilon),$$

or

$$\Sigma = LL^T + \Psi,$$

Note that for any orthogonal matrix Q if we set $L = LQ$ and $F = Q^TF$, the criteria for being factors and factor loadings still hold. Hence a set of factors and factor loadings is identical only up to orthogonal transformation.

Example

The following example is for expository purposes, and should not be taken as being realistic. Suppose a psychologist proposes a theory that there are two kinds of intelligence, "verbal intelligence" and "mathematical intelligence", neither of which is directly observed. Evidence for the theory is sought in the examination scores from each of 10 different academic fields of 1000 students. If each student is chosen randomly from a large population, then each student's 10 scores are random variables. The psychologist's theory may say that for each of the 10 academic fields, the score averaged over the group of all students who share some common pair of values for verbal and mathematical "intelligences" is some constant times their level of verbal intelligence plus another constant times their level of mathematical intelligence, *i.e.*, it is a combination of those two "factors". The numbers for a particular subject, by which the two kinds of intelligence are multiplied to obtain the expected score, are posited by the theory to be the same for all intelligence level pairs, and are called **"factor loadings"** for this subject.

For example, the theory may hold that the average student's aptitude in the field of taxonomy is

{10 × the student's verbal intelligence} + {6 × the student's mathematical intelligence}.

The numbers 10 and 6 are the factor loadings associated with taxonomy. Other academic subjects may have different factor loadings.

Two students having identical degrees of verbal intelligence and identical degrees of mathematical intelligence may have different aptitudes in taxonomy because individual aptitudes differ from average aptitudes. That difference is called the "error" – a statistical term that means the amount by which an individual differs from what is average for his or her levels of intelligence.

The observable data that go into factor analysis would be 10 scores of each of the 1000 students, a total of 10,000 numbers. The factor loadings and levels of the two kinds of intelligence of each student must be inferred from the data.

Mathematical Model of the Same Example

In the following, matrices will be indicated by indexed variables. "Subject" indices will be indicated using letters a,b and c, with values running from 1 to N_a which is equal to 10 in the above example. "Factor" indices will be indicated using letters p, q and r, with values running from 1 to N_p which is equal to 2 in the above example. "Instance" or "sample" indices will be indicated using letters i,j and k, with values running from 1 to N_i. In the example above, if a sample of $N_i = 1000$ students responded to the $N_a = 10$ questions, the ith student's score for the ath question are given by x_{ai}. The purpose of factor analysis is to characterize the correlations between the variables x_a of which the x_{ai} are a particular instance, or set of observations. In order that the variables be on equal footing, they are standardized :

$$z_{ai} = \frac{x_{ai} - \mu_a}{\sigma_a}$$

where the sample mean is :

$$\mu_a = \frac{1}{N}\sum_i x_{ai}$$

and the sample variance is given by :

$$\sigma_a^2 = \frac{1}{N_a}\sum_i (x_{ai} - \mu_a)^2$$

The factor analysis model for this particular sample is then :

$$\begin{array}{ccccccc} z_{1,i} & = & \ell_{1,1}F_{1,i} & + & \ell_{1,2}F_{2,i} & + & \varepsilon_{1,i} \\ \vdots & & \vdots & & \vdots & & \vdots \\ z_{10,i} & = & \ell_{10,1}F_{1,i} & + & \ell_{10,2}F_{2,i} & + & \varepsilon_{10,2} \end{array}$$

or, more succintly :

$$z_{ai} = \sum_p \ell_{ap} F_{pi} + \varepsilon_{ai}$$

where :

- $F_{1,i}$ is the ith student's "verbal intelligence",
- $F_{2,i}$ is the ith student's "mathematical intelligence",
- ℓ_{ap} are the factor loadings for the ath subject, for $p = 1, 2$.

In matrix notation, we have

$$Z = LF + \in$$

Observe that by doubling the scale on which "verbal intelligence" – the first component in each column of F – is measured, and simultaneously halving the factor loadings for verbal intelligence makes no difference to the model. Thus, no generality is lost by assuming that the standard deviation of verbal intelligence is 1. Likewise for mathematical intelligence. Moreover, for similar reasons, no generality is lost by assuming the two factors are uncorrelated with each other. In other words :

$$\sum_i F_{pi} F_{qi} = \delta_{pd}$$

where δ_{pq} is the Kronecker delta (0 when $p \neq q$ and 1 when $p = q$).The errors are assumed to be independent of the factors :

$$\sum_i F_{pi} \varepsilon_{ai} = 0$$

Note that, since any rotation of a solution is also a solution, this makes interpreting the factors difficult. In this particular example, if we do not know beforehand that the two types of intelligence are uncorrelated, then we cannot interpret the two factors as the two different types of intelligence. Even if they are uncorrelated, we cannot tell which factor corresponds to verbal intelligence and which corresponds to mathematical intelligence without an outside argument.

The values of the loadings L, the averages μ, and the variances of the "errors" ε must be estimated given the observed data X and F (the assumption about the levels of the factors is fixed for a given F). The "fundamental theorem" may be derived from the above conditions :

$$\sum_i z_{ai} z_{bi} = \sum_P \ell_{ap} \ell_{bp} + \sum_i \varepsilon_{ai} \varepsilon_{bi}$$

The term on the left is just the correlation matrix of the observed data, and its N_a diagonal elements will be 1's. The last term on the right will be a diagonal matrix with terms less than unity. The first term on the right is the "reduced correlation matrix" and will be equal to the correlation matrix except for its diagonal values which will be less than unity. These diagonal elements of the reduced correlation matrix are called "communalities" :

$$h_a^2 = 1 - \psi_a = \sum_p \ell_{ap} \ell_{ap}$$

The sample data z_{ai} will not, of course, exactly obey the fundamental equation given above due to sampling errors, inadequacy of the model, etc. The goal of any analysis of the above model is to find the factors F_{pi} and loadings ℓ_{ap} which, in some sense, give a "best fit" to the data. In factor analysis, the best fit is defined as the minimum of the mean square error in the off-diagonal residuals of the correlation matrix :

$$\varepsilon^2 \sum_{ab,a\neq b}\left[\sum_i z_{ai}z_{bi} - \sum_p \ell_{ap}\ell_{bp}\right]^2$$

This is equivalent to minimizing the off-diagonal components of the error covariance which, in the model equations have expected values of zero. This is to be contrasted with principal component analysis which seeks to minimize the mean square error of all residuals. Before the advent of high speed computers, considerable effort was devoted to finding approximate solutions to the problem, particularly in estimating the communalities by other means, which then simplifies the problem considerably by yielding a known reduced correlation matrix. This was then used to estimate the factors and the loadings. With the advent of high-speed computers, the minimization problem can be solved quickly and directly, and the communalities are calculated in the process, rather than being needed beforehand. The MinRes algorithm is particularly suited to this problem, but is hardly the only means of finding an exact solution.

Geometric Interpretation

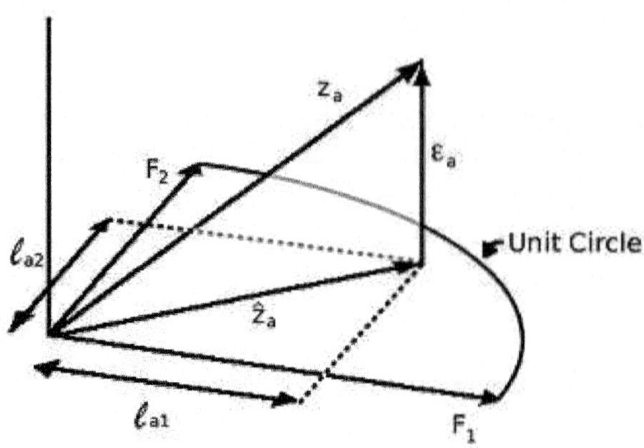

Geometric interpretation of Factor Analysis parameters for 3 respondents to question "a". The "answer" is represented by the unit vector z_a, which is projected onto a plane defined by two orthonormal vectors F_1 and F_2. The projection vector is $\hat{z}_a$ and the error ε_a is perpendicular to the plane, so that $z_a = \hat{z}_a + \varepsilon_a$. The projection vector $\hat{z}_a$ may be represented in terms of the factor vectors as $\hat{z}_a = \ell_{a1}F_1 + \ell_{a2}F_2$. The square of the length of the projection vector is the communality : $|\hat{z}_a|^2 = h^2_a$.

If another data vector z_b were plotted, the cosine of the angle between z_a and z_b would be r_{ab}: the (a, b) entry in the correlation matrix. (Adapted from Harman Fig.)

The parameters and variables of factor analysis can be given a geometrical interpretation. The data (z_{ai}), the factors (F_{pi}) and the errors (ε_{ai}) can be viewed as vectors in an N_i-dimensional Euclidean space (sample space), represented as z_a, F_p and ε_a respectively. Since the data is standardized, the data vectors are of unit length $(z_a \cdot z_a = 1)$. The factor vectors define an N_p-dimensional linear sub-space (*i.e.* a hyperplane) in this space, upon which the data vectors are projected orthogonally. This follows from the model equation

$$z_a = \sum_p \ell_{ap} F_p + \varepsilon_a$$

and the independence of the factors and the errors : $F_p \cdot \varepsilon_a = 0$. In the above example, the hyperplane is just a 2-dimensional plane defined by the two factor vectors. The projection of the data vectors onto the hyperplane is given by

$$\hat{z}_a = \sum_p \ell_{ap} F_p$$

and the errors are vectors from that projected point to the data point and are perpendicular to the hyperplane. The goal of factor analysis is to find a hyperplane which is a "best fit" to the data in some sense, so it doesn't matter how the factor vectors which define this hyperplane are chosen, as long as they are independent and lie in the hyperplane. We are free to specify them as both orthogonal and normal $(F_p \cdot F_q = \delta_{pq})$ with no loss of generality. After a suitable set of factors are found, they may also be arbitrarily rotated within the hyperplane, so that any rotation of the factor vectors will define the same hyperplane, and also be a solution. As a result, in the above example, in which the fitting hyperplane is two dimensional, if we do not know beforehand that the two types of intelligence are uncorrelated, then we cannot interpret the two factors as the two different types of intelligence. Even if they are uncorrelated, we cannot tell which factor corresponds to verbal intelligence and which corresponds to mathematical intelligence, or whether the factors are linear combinations of both, without an outside argument.

The data vectors z_a have unit length. The correlation matrix for the data is given by $r_{ab} = z_a \cdot z_b$. The correlation matrix can be geometrically interpreted as the cosine of the angle between the two data vectors z_a and z_b. The diagonal elements will clearly be 1's and the off diagonal elements will have absolute values less than or equal to unity. The "reduced correlation matrix" is defined as

$$\hat{r}_{ab} = \hat{z}_a \cdot \hat{z}_b.$$

The goal of factor analysis is to choose the fitting hyperplane such that the reduced correlation matrix reproduces the correlation matrix as nearly as possible, except for the diagonal elements of the correlation matrix which are known to have unit value. In other words, the goal is to reproduce as accurately as possible the cross-correlations in the data. Specifically, for the fitting hyperplane, the mean square error in the off-diagonal components

$$\varepsilon^2 = \sum_{a,b\neq a} (r_{ab} - \hat{r}_{ab})^2$$

is to be minimized, and this is accomplished by minimizing it with respect to a set of orthonormal factor vectors. It can be seen that

$$r_{ab} - \hat{r}_{ab} = \varepsilon_a \cdot \varepsilon_b$$

The term on the right is just the covariance of the errors. In the model, the error covariance is stated to be a diagonal matrix and so the above minimization problem will in fact yield a "best fit" to the model : It will yield a sample estimate of the error covariance which has its off-diagonal components minimized in the mean square sense. It can be seen that since the $\hat{z}_a$ are orthogonal projections of the data vectors, their length will be less than or equal to the length of the projected data vector, which is unity. The square of these lengths are just the diagonal elements of the reduced correlation matrix. These diagonal elements of the reduced correlation matrix are known as "communalities" :

$$h_a^2 = \hat{z}_a \cdot \hat{z}_a = \sum_p \ell_{ap} \ell_{ap}$$

Large values of the commmunalities will indicate that the fitting hyperplane is rather accurately reproducing the correlation matrix. It should be noted that the mean values of the factors must also be constrained to be zero, from which it follows that the mean values of the errors will also be zero.

Practical Implementation

Type of Factor Analysis

Exploratory factor analysis (EFA) is used to identify complex interrelationships among items and group items that are part of unified concepts. The researcher makes no "a priori" assumptions about relationships among factors.

Confirmatory factor analysis (CFA) is a more complex approach that tests the hypothesis that the items are associated with specific factors. CFA uses structural equation modelling to test a measurement model whereby loading on the factors allows for evaluation of relationships between observed variables and unobserved variables. Structural equation modelling approaches can accommodate measurement error, and are less restrictive than least-squares estimation. Hypothesized models are tested against actual data, and the analysis would demonstrate loadings of observed variables on the latent variables (factors), as well as the correlation between the latent variables.

Types of Factoring

Principal component analysis (PCA) : PCA is a widely used method for factor extraction, which is the first phase of EFA. Factor weights are computed in order to extract the maximum possible variance, with successive factoring continuing until there is no further meaningful variance left. The factor model must then be rotated for analysis.

Canonical factor analysis, also called Rao's canonical factoring, is a different method of computing the same model as PCA, which uses the principal axis method. Canonical factor analysis seeks factors which have the highest canonical correlation with the observed variables. Canonical factor analysis is unaffected by arbitrary rescaling of the data.

Common factor analysis, also called principal factor analysis (PFA) or principal axis factoring (PAF), seeks the least number of factors which can account for the common variance (correlation) of a set of variables.

Image factoring : based on the correlation matrix of predicted variables rather than actual variables, where each variable is predicted from the others using multiple regression.

Alpha factoring : based on maximizing the reliability of factors, assuming variables are randomly sampled from a universe of variables. All other methods assume cases to be sampled and variables fixed.

Factor regression model : a combinatorial model of factor model and regression model; or alternatively, it can be viewed as the hybrid factor model, whose factors are partially known.

Terminology

Factor loadings : The factor loadings, also called component loadings in PCA, are the correlation coefficients between the cases (rows) and factors (columns). Analogous to Pearson's r, the squared factor loading is the per cent of variance in that indicator variable explained by the factor. To get the per cent of variance in all the variables accounted for by each factor, add the sum of the squared factor loadings for that factor (column) and divide by the number of variables. (Note the number of variables equals the sum of their variances as the variance of a standardized variable is 1.) This is the same as dividing the factor's eigenvalue by the number of variables.

Interpreting factor loadings : By one rule of thumb in confirmatory factor analysis, loadings should be.7 or higher to confirm that independent variables identified a priori are represented by a particular factor, on the rationale that the.7 level corresponds to about half of the variance in the indicator being explained by the factor. However, the.7 standard is a high one and real-life data may well not meet this criterion, which is why some researchers, particularly for exploratory purposes, will use a lower level such as.4 for the central factor and.25 for other factors. In any event, factor loadings must be interpreted in the light of theory, not by arbitrary cutoff levels.

In oblique rotation, one gets both a pattern matrix and a structure matrix. The structure matrix is simply the factor loading matrix as in orthogonal rotation, representing the variance in a measured variable explained by a factor on both a unique and common contributions basis. The pattern matrix, in contrast, contains coefficients which just represent unique contributions. The more factors, the lower the pattern coefficients as a rule since there will be more common contributions to variance explained. For oblique rotation, the researcher looks at both the structure and pattern coefficients when attributing a label to a factor. Principles of oblique rotation can be derived from both cross entropy and its dual entropy.

Communality : The sum of the squared factor loadings for all factors for a given variable (row) is the variance in that variable accounted for by all the factors, and this is called the communality. The communality measures the per cent of variance in a given variable explained by all the factors jointly and may be interpreted as the reliability of the indicator.

Spurious solutions : If the communality exceeds 1.0, there is a spurious solution, which may reflect too small a sample or the researcher has too many or too few factors.

Uniqueness of a variable : That is, uniqueness is the variability of a variable minus its communality.

Eigenvalues :/Characteristic roots : The eigenvalue for a given factor measures the variance in all the variables which is accounted for by that factor. The ratio of eigenvalues is the ratio of explanatory importance of the factors with respect to the variables. If a factor has a low eigenvalue, then it is contributing little to the explanation of variances in the variables and may be ignored as redundant with more important factors. Eigenvalues measure the amount of variation in the total sample accounted for by each factor.

Extraction sums of squared loadings : Initial eigenvalues and eigenvalues after extraction (listed by SPSS as "Extraction Sums of Squared Loadings") are the same for PCA extraction, but for other extraction methods, eigenvalues after extraction will be lower than their initial counterparts. SPSS also prints "Rotation Sums of Squared Loadings" and even for PCA, these eigenvalues will differ from initial and extraction eigenvalues, though their total will be the same.

Factor scores (also called component scores in PCA) : are the scores of each case (row) on each factor (column). To compute the factor score for a given case for a given factor, one takes the case's standardized score on each variable, multiplies by the corresponding loadings of the variable for the given factor, and sums these products. Computing factor scores allows one to look for factor outliers. Also, factor scores may be used as variables in subsequent modelling.

Criteria for Determining the Number of Factors

Using one or more of the methods below, the researcher determines an appropriate range of solutions to investigate. Methods may not agree. For instance, the Kaiser criterion may suggest five factors and the scree test may suggest two, so the researcher may request 3-, 4-, and 5-factor solutions discuss each in terms of their relation to external data and theory.

Comprehensibility : A purely subjective criterion would be to retain those factors whose meaning is comprehensible to the researcher. This is not recommended.

Kaiser criterion : The Kaiser rule is to drop all components with eigenvalues under 1.0-this being the eigenvalue equal to the information accounted for by an average single item. The Kaiser criterion is the default in SPSS and most statistical software but is not recommended when used as the sole cut-off criterion for estimating the number of factors as it tends to overextract factors.

Variance explained criteria : Some researchers simply use the rule of keeping enough factors to account for 90% (sometimes 80%) of the variation. Where the researcher's goal emphasizes parsimony (explaining variance with as few factors as possible), the criterion could be as low as 50%.

Scree plot : The Cattell scree test plots the components as the X axis and the corresponding eigenvalues as the Y-axis. As one moves to the right, toward later components, the eigenvalues drop. When the drop ceases and the curve makes an elbow toward less steep decline, Cattell's scree test says to drop all further components after the one starting the elbow. This rule is sometimes criticised for being amenable to researcher-controlled "fudging". That is, as picking the "elbow" can be subjective because the curve has multiple elbows or is a smooth curve, the researcher may be tempted to set the cut-off at the number of factors desired by his or her research agenda.

Horn's Parallel Analysis (PA) : A Monte-Carlo based simulation method that compares the observed eigenvalues with those obtained from uncorrelated normal variables. A factor or component is retained if the associated eigenvalue is bigger than the 95th of the distribution of eigenvalues derived from the random data. PA is one of the most recommendable rules for determining the number of components to retain, but only few programs include this option.

However, before dropping a factor below one's cutoff, the analyst(s) should create scores for a data-set based on the factor loadings and check the scores' correlation with any given dependent variable(s) of interest. Scores based on a factor with a very small eigenvalue can correlate strongly with dependent variables, in which case dropping such a factor from a theoretical model may reduce its predictive validity.

Rotation Methods

The unrotated output maximises variance accounted for by the first and subsequent factors, and forcing the factors to be orthogonal. This data-compression comes at the cost of having most items load on the early factors, and usually, of having many items load substantially on more than one factor. Rotation serves to make the output more understandable, by seeking so-called "Simple Structure" : A pattern of loadings where items load most strongly on one factor, and much more weakly on the other factors. Rotations can be orthogonal or oblique (allowing the factors to correlate).

Varimax rotation is an orthogonal rotation of the factor axes to maximize the variance of the squared loadings of a factor (column) on all the variables (rows) in a factor matrix, which has the effect of differentiating the original variables by extracted factor. Each factor will tend to have either large or small loadings of any particular variable. A varimax solution yields results which make it as easy as possible to identify each variable with a single factor. This is the most common rotation option. However, the orthogonality (*i.e.*, independence) of factors is often an unrealistic assumption. Oblique rotations are inclusive of orthogonal rotation, and for that reason, oblique rotations are a preferred method.

Quartimax rotation is an orthogonal alternative which minimizes the number of factors needed to explain each variable. This type of rotation often generates a general factor on which most variables are loaded to a high or medium degree. Such a factor structure is usually not helpful to the research purpose.

Equimax rotation is a compromise between Varimax and Quartimax criteria.

Direct oblimin rotation is the standard method when one wishes a non-orthogonal (oblique) solution-that is, one in which the factors are allowed to be correlated. This will result in higher eigenvalues but diminished interpretability of the factors.

Promax rotation is an alternative non-orthogonal (oblique) rotation method which is computationally faster than the direct oblimin method and therefore is sometimes used for very large data-sets.

Factor Analysis in Psychometrics

History

Charles Spearman pioneered the use of factor analysis in the field of psychology and is sometimes credited with the invention of factor analysis. He discovered that school children's scores on a wide variety of seemingly unrelated subjects were positively correlated, which led him to postulate that a general mental ability, or *g*, underlies and shapes human cognitive performance. His postulate now enjoys broad support in the field of intelligence research, where it is known as the *g* theory.

Raymond Cattell expanded on Spearman's idea of a two-factor theory of intelligence after performing his own tests and factor analysis. He used a multi-factor theory to explain intelligence. Cattell's theory addressed alternate factors in intellectual development, including motivation and psychology. Cattell also developed several mathematical methods for adjusting psychometric graphs, such as his "scree" test and similarity coefficients. His research led to the development of his theory of fluid and crystallized intelligence, as well as his 16 Personality Factors theory of personality. Cattell was a strong advocate of factor analysis and psychometrics. He believed that all theory should be derived from research, which supports the continued use of empirical observation and objective testing to study human intelligence.

Applications in Psychology

Factor analysis is used to identify "factors" that explain a variety of results on different tests. For example, intelligence research found that people who get a high score on a test of verbal ability are also good on other tests that require verbal abilities. Researchers explained this by using factor analysis to isolate one factor, often called crystallized intelligence or verbal intelligence, which represents the degree to which someone is able to solve problems involving verbal skills.

Factor analysis in psychology is most often associated with intelligence research. However, it also has been used to find factors in a broad range of domains such as personality, attitudes, beliefs, etc. It is linked to psychometrics, as it can assess the validity of an instrument by finding if the instrument indeed measures the postulated factors.

Advantages

- Reduction of number of variables, by combining two or more variables into a single factor. For example, performance at running, ball throwing, batting, jumping and weight lifting could be combined into a single factor such as general athletic ability. Usually, in an item by people matrix, factors are selected by grouping related items. In the Q factor analysis technique, the matrix is transposed and factors are created by grouping related people : For example, liberals, libertarians, conservatives and socialists, could form separate groups.
- Identification of groups of inter-related variables, to see how they are related to each other. For example, Carroll used factor analysis to build his Three Stratum Theory. He found that a factor called "broad visual perception" relates to how good an individual is at visual tasks. He also found a "broad auditory perception" factor, relating to auditory task capability. Furthermore, he found a global factor, called "g" or general intelligence, that relates to both "broad visual perception" and "broad auditory perception". This means someone with a high "g" is likely to have both a high "visual perception" capability and a high "auditory perception" capability, and that "g" therefore explains a good part of why someone is good or bad in both of those domains.

Disadvantages

- "...each orientation is equally acceptable mathematically. But different factorial theories proved to differ as much in terms of the orientations of factorial axes for a given solution as in terms of anything else, so that model fitting did not prove to be useful in distinguishing among theories." (Sternberg, 1977). This means all rotations represent different underlying processes, but all rotations are equally valid outcomes of standard factor analysis optimization. Therefore, it is impossible to pick the proper rotation using factor analysis alone.

- Factor analysis can be only as good as the data allows. In psychology, where researchers often have to rely on less valid and reliable measures such as self-reports, this can be problematic.
- Interpreting factor analysis is based on using a "heuristic", which is a solution that is "convenient even if not absolutely true". More than one interpretation can be made of the same data factored the same way, and factor analysis cannot identify causality.

Exploratory Factor Analysis *versus* Principal Components Analysis

While exploratory factor analysis and principal component analysis are treated as synonymous techniques in some fields of statistics, this has been criticised (*e.g.* Fabrigar *et. al.*, 1999; Suhr, 2009). In factor analysis, the researcher makes the assumption that an underlying causal model exists, whereas PCA is simply a variable reduction technique. Researchers have argued that the distinctions between the two techniques may mean that there are objective benefits for preferring one over the other based on the analytic goal.

Arguments Contrasting PCA and EFA

Fabrigar *et. al.* address a number of reasons used to suggest that principal components analysis is equivalent to factor analysis :

1. It is sometimes suggested that principal components analysis is computationally quicker and requires fewer resources than factor analysis. Fabrigar *et. al.* suggest that the ready availability of computer resources have rendered this practical concern irrelevant.
2. PCA and factor analysis can produce similar results. This point is also addressed by Fabrigar *et. al.*; in certain cases, whereby the communalities are low (*e.g.*,.40), the two techniques produce divergent results. In fact, Fabrigar *et. al.* argue that in cases where the data correspond to assumptions of the common factor model, the results of PCA are inaccurate results.
3. There are certain cases where factor analysis leads to 'Heywood cases'. These encompass situations whereby 100% or more of the variance in a measured variable is estimated to be accounted for by the model. Fabrigar *et. al.* suggest that these cases are actually informative to the researcher, indicating a misspecified model or a violation of the common factor model. The lack of Heywood cases in the PCA approach may mean that such issues pass unnoticed.
4. Researchers gain extra information from a PCA approach, such as an individual's score on a certain component—such information is not yielded from factor analysis. However, as Fabrigar *et. al.* contend, the typical aim of factor analysis-*i.e.* to determine the factors accounting for the structure of the correlations between measured variables—does not require knowledge of factor scores and thus this advantage is negated. It is also possible to compute factor scores from a factor analysis.

Variance versus Covariance

Factor analysis takes into account the random error that is inherent in measurement, whereas PCA fails to do so. This point is exemplified by Brown, who indicated that, in respect to the correlation matrices involved in the calculations :

"In PCA, 1.00s are put in the diagonal meaning that all of the variance in the matrix is to be accounted for (including variance unique to each variable, variance common among variables, and error variance). That would, therefore, by definition, include all of the variance in the variables. In contrast, in EFA, the communalities are put in the diagonal meaning that only the variance shared with other variables is to be accounted for (excluding variance unique to each variable and error variance). That would, therefore, by definition, include only variance that is common among the variables."

— Brown, *Principal components analysis and exploratory factor analysis—Definitions, differences and choices*

For this reason, Brown recommends using factor analysis when theoretical ideas about relationships between variables exist, whereas PCA should be used if the goal of the researcher is to explore patterns in their data.

Differences in Procedure and Results

The differences between principal components analysis and factor analysis are further illustrated by Suhr :

- PCA results in principal components that account for a maximal amount of variance for observed variables; FA account for common variance in the data.
- PCA inserts ones on the diagonals of the correlation matrix; FA adjusts the diagonals of the correlation matrix with the unique factors.
- PCA minimizes the sum of squared perpendicular distance to the component axis; FA estimates factors which influence responses on observed variables.
- The component scores in PCA represent a linear combination of the observed variables weighted by eigenvectors; the observed variables in FA are linear combinations of the underlying and unique factors.
- In PCA, the components yielded are uninterpretable, *i.e.* they do not represent underlying 'constructs'; in FA, the underlying constructs can be labelled and readily interpreted, given an accurate model specification.

Factor Analysis in Marketing

The basic steps are :

- Identify the salient attributes consumers use to evaluate products in this category.
- Use quantitative marketing research techniques (such as surveys) to collect data from a sample of potential customers concerning their ratings of all the product attributes.

- Input the data into a statistical program and run the factor analysis procedure. The computer will yield a set of underlying attributes (or factors).
- Use these factors to construct perceptual maps and other product positioning devices.

Information Collection

The data collection stage is usually done by marketing research professionals. Survey questions ask the respondent to rate a product sample or descriptions of product concepts on a range of attributes. Anywhere from five to twenty attributes are chosen. They could include things like : ease of use, weight, accuracy, durability, colourfulness, price, or size. The attributes chosen will vary depending on the product being studied. The same question is asked about all the products in the study. The data for multiple products is coded and input into a statistical program such as R, SPSS, SAS, Stata, STATISTICA, JMP, and SYSTAT.

Analysis

The analysis will isolate the underlying factors that explain the data using a matrix of associations. Factor analysis is an interdependence technique. The complete set of interdependent relationships is examined. There is no specification of dependent variables, independent variables, or causality. Factor analysis assumes that all the rating data on different attributes can be reduced down to a few important dimensions. This reduction is possible because some attributes may be related to each other. The rating given to any one attribute is partially the result of the influence of other attributes. The statistical algorithm deconstructs the rating (called a raw score) into its various components, and reconstructs the partial scores into underlying factor scores. The degree of correlation between the initial raw score and the final factor score is called a *factor loading*.

Advantages

- Both objective and subjective attributes can be used provided the subjective attributes can be converted into scores.
- Factor analysis can identify latent dimensions or constructs that direct analysis may not.
- It is easy and inexpensive.

Disadvantages

- Usefulness depends on the researchers' ability to collect a sufficient set of product attributes. If important attributes are excluded or neglected, the value of the procedure is reduced.
- If sets of observed variables are highly similar to each other and distinct from other items, factor analysis will assign a single factor to them. This may obscure factors that represent more interesting relationships.

- Naming factors may require knowledge of theory because seemingly dissimilar attributes can correlate strongly for unknown reasons.

Factor Analysis in Physical Sciences

Factor analysis has also been widely used in physical sciences such as geochemistry, ecology, and hydrochemistry.

In groundwater quality management, it is important to relate the spatial distribution of different chemical parameters to different possible sources, which have different chemical signatures. For example, a sulfide mine is likely to be associated with high levels of acidity, dissolved sulfates and transition metals. These signatures can be identified as factors through R-mode factor analysis, and the location of possible sources can be suggested by contouring the factor scores.

In geochemistry, different factors can correspond to different mineral associations, and thus to mineralisation.

Factor Analysis in Microarray Analysis

Factor analysis can be used for summarizing high-density oligonucleotideDNA microarrays data at probe level for Affymetrix GeneChips. In this case, the latent variable corresponds to the RNA concentration in a sample.

Implementation

Factor analysis has been implemented in several statistical analysis programs since the 1980s : SAS, BMDP and SPSS. It is also implemented in the *R* programming language (with the *factanal* function), OpenOpt, and the statistical software package Stata. Rotations are implemented in the *GPArotation* R package.

CANONICAL CORRELATION

In statistics, **canonical-correlation analysis (CCA)** is a way of making sense of cross-covariance matrices. If we have two vectors $X = (X_1,..., X_n)$ and $Y = (Y_1,..., Y_m)$ of random variables, and there are correlations among the variables, then canonical-correlation analysis will find linear combinations of the X_i and Y_j which have maximum correlation with each other. T. R. Knapp notes "virtually all of the commonly encountered parametric tests of significance can be treated as special cases of canonical-correlation analysis, which is the general procedure for investigating the relationships between two sets of variables." The method was first introduced by Harold Hotelling in 1936.

Definition

Given two column vectors $X = (x_1,..., x_n)'$ and $Y = (y_1,..., y_m)'$ of random variables with finite second moments, one may define the cross-covariance $\Sigma_{XY} = \text{cov}(X, Y)$ to be the $n \times m$ matrix whose (i, j) entry is the covariance $\text{cov}(x_i, y_j)$. In practice, we would estimate the covariance matrix based on sampled data from X and Y (*i.e.* from a pair of data matrices).

Canonical-correlation analysis seeks vectors a and b such that the random variables $a'X$ and $b'Y$ maximize the correlation $\sigma = \mathrm{corr}(a'X, b'Y)$. The random variables $U = a'X$ and $V = b'Y$ are the ***first pair of canonical variables***. Then one seeks vectors maximizing the same correlation subject to the constraint that they are to be uncorrelated with the first pair of canonical variables; this gives the ***second pair of canonical variables***. This procedure may be continued up to min $\{m, n\}$ times.

Computation

Derivation

Let $\Sigma_{XX} = \mathrm{cov}(X, X)$ and $\Sigma_{YY} = \mathrm{cov}(Y, Y)$. The parameter to maximize is

$$\rho = \frac{a'\sum_{XY} b}{\sqrt{a'\sum_{XX} a}\sqrt{b'\sum_{YY} b}}.$$

The first step is to define a change of basis and define

$$c = \sum\nolimits_{XX}^{1/2} a,$$

$$d = \sum\nolimits_{YY}^{1/2} b.$$

And thus we have

$$\rho = \frac{c'\sum_{XX}^{-1/2}\sum_{XX}\sum_{YY}^{-1/2} d}{\sqrt{c'c}\sqrt{d'd}}.$$

By the Cauchy-Schwarz inequality, we have

$$\left(c'\sum\nolimits_{XX}^{-1/2}\sum\nolimits_{XY}\sum\nolimits_{YY}^{-1/2}\right)d \le \left(c'\sum\nolimits_{XX}^{-1/2}\sum\nolimits_{XY}\sum\nolimits_{YY}^{-1/2}\sum\nolimits_{YY}^{-1/2}\sum\nolimits_{YX}\sum\nolimits_{XX}^{-1/2} c\right)^{1/2} (d'd)^{1/2},$$

$$\rho \le \frac{\left(c'\sum_{XX}^{-1/2}\sum_{XY}\sum_{YY}^{-1}\sum_{YX}\sum_{XX}^{-1/2} c\right)^{1/2}}{(c'c)^{1/2}}.$$

There is equality if the vectors d and $\sum_{YY}^{-1/2}\sum_{YX}\sum_{XX}^{-1/2} c$ are collinear. In addition, the maximum of correlation is attained if c is the eigenvector with the maximum eigenvalue for the matrix $\sum_{XX}^{-1/2}\sum_{XY}\sum_{YY}^{-1}\sum_{YX}\sum_{XX}^{-1/2}$. The subsequent pairs are found by using eigenvalues of decreasing magnitudes. Orthogonality is guaranteed by the symmetry of the correlation matrices.

Solution

The solution is therefore :

- c is an eigenvector of

$$\Sigma_{XX}^{-1/2}\Sigma_{XY}\Sigma_{YY}^{-1}\Sigma_{YX}\Sigma_{XX}^{-1/2}$$

- d is proportional to $\Sigma_{YY}^{-1/2}\Sigma_{YX}\Sigma_{XX}^{-1/2}c$

Reciprocally, there is also :

- d is an eigenvector of $\Sigma_{YY}^{-1/2}\Sigma_{YX}\Sigma_{XX}^{-1}\Sigma_{XY}\Sigma_{YY}^{-1/2}$
- c is proportional to $\Sigma_{XX}^{-1/2}\Sigma_{XY}\Sigma_{YY}^{-1/2}d$

Reversing the change of co-ordinates, we have that

- a is an eigenvector of $\Sigma_{XX}^{-1}\Sigma_{XY}\Sigma_{YY}^{-1}\Sigma_{YX}$
- b is an eigenvector of $\Sigma_{YY}^{-1}\Sigma_{YX}\Sigma_{XX}^{-1}\Sigma_{XY}$
- a is proportional to $\Sigma_{XX}^{-1}\Sigma_{XY}b$
- b is proportional to $\Sigma_{YY}^{-1}\Sigma_{YX}a$

The canonical variables are defined by :

$$U = c'\Sigma_{XX}^{-1/2}X = a'X$$

$$V = d'\Sigma_{YY}^{-1/2}Y = b'Y$$

Implementation

CCA can be computed using singular value decomposition on a correlation matrix. It is available as a function in

- MATLAB as canoncorr
- R as cancor or in FactoMineR
- SAS as proc cancorr
- Scikit-Learn, Python as Cross decomposition.

Hypothesis Testing

Each row can be tested for significance with the following method. Since the correlations are sorted, saying that row i is zero implies all further correlations are also zero. If we have P independent observations in a sample and $\hat{\rho}_i$ is the estimated correlation for i = 1,..., min{m, n}. For the ith row, the test statistic is :

$$\chi^2 = -\left(p-1-\frac{1}{2}(m+n+1)\right)\ln\prod_{j=i}^{\min\{m,n\}}(1-\hat{\rho}_j^2),$$

which is asymptotically distributed as a chi-squared with $(m - i + 1)$ $(n - i + 1)$ degrees of freedom for large p. Since all the correlations from min {m, n} to p are logically zero (and estimated that way also) the product for the terms after this point is irrelevant.

Practical Uses

A typical use for canonical correlation in the experimental context is to take two sets of variables and see what is common amongst the two sets. For example in psychological testing, you could take two well established multi-dimensional personality tests such as the Minnesota Multiphasic Personality Inventory (MMPI-2) and the NEO. By seeing how the MMPI-2 factors relate to the NEO factors, you could gain insight into what dimensions were common between the tests and how much variance was shared. For example you might find that an extraversion or neuroticism dimension accounted for a substantial amount of shared variance between the two tests.

One can also use canonical-correlation analysis to produce a model equation which relates two sets of variables, for example a set of performance measures and a set of explanatory variables, or a set of outputs and set of inputs. Constraint restrictions can be imposed on such a model to ensure it reflects theoretical requirements or intuitively obvious conditions. This type of model is known as a maximum correlation model.

Visualization of the results of canonical correlation is usually through bar plots of the coefficients of the two sets of variables for the pairs of canonical variates showing significant correlation. Some authors suggest that they are best visualized by plotting them as heliographs, a circular format with ray like bars, with each half representing the two sets of variables.

Examples

Let $X = x_1$ with zero expected value, *i.e.*, $E(X) = 0$. If $Y = X$, *i.e.*, X and Y are perfectly correlated, then, *e.g.*, $a = 1$ and $b = 1$, so that the first (and only in this example) pair of canonical variables is $U = X$ and $V = Y = X$. If $Y = -X$, *i.e.*, X and Y are perfectly anti-correlated, then, *e.g.*, $a = 1$ and $b = -1$, so that the first (and only in this example) pair of canonical variables is $U = X$ and $V = -Y = X$. We notice that in both cases $U = V$, which illustrates that the canonical-correlation analysis treats correlated and anti-correlated variables similarly.

Connection to Principal Angles

Assuming that $X = (x_1,\ldots, x_n)'$ and $Y = (y_1,\ldots, y_m)'$ have zero expected values, *i.e.*, $E(X) = E(Y) = 0$, their covariance matrices $\Sigma_{XX} = \mathrm{Cov}(X, X) = E[XX']$ and $\Sigma_{YY} = \mathrm{Cov}(Y, Y) = E[YY']$ can be viewed as Gram matrices in an inner product for the entries of X and Y, correspondingly. In this interpretation, the random variables, entries x_i of X and y_i of Y are treated as elements of a vector space with an inner product given by the covariance $\mathrm{cov}(x_i, y_j)$.

The definition of the canonical variables U and V is then equivalent to the definition of principal vectors for the pair of sub-spaces spanned by the entries of X and Y with respect to this inner product. The canonical correlations $\mathrm{corr}(U, V)$ is equal to the cosine of principal angles.

INDEPENDENT COMPONENT ANALYSIS

Introduction

In signal processing, **independent component analysis (ICA)** is a computational method for separating a multi-variate signal into additive subcomponents by assuming that the sub-components are non-Gaussian signals and that they are all statistically independent from each other. ICA is a special case of blind source separation. As an example, sound is usually a signal that is composed of the numerical addition, at each time t, of signals from several sources. The question then is whether it is possible to separate these contributing sources from the observed total signal. When the statistical independence assumption is correct, blind ICA separation of a mixed signal gives very good results. It is also used for signals that are not supposed to be generated by a mixing for analysis purposes. A simple application of ICA is the "cocktail party problem", where the underlying speech signals are separated from a sample data consisting of people talking simultaneously in a room. Usually the problem is simplified by assuming no time delays or echoes. An important note to consider is that if N sources are present, at least N observations (*e.g.* microphones) are needed to recover the original signals. This constitutes the square case ($J = D$, where D is the input dimension of the data and J is the dimension of the model). Other cases of underdetermined ($J > D$) and overdetermined ($J < D$) have been investigated.

That the ICA separation of mixed signals gives very good results are based on two assumptions and three effects of mixing source signals. Two assumptions :

1. The source signals are independent of each other.
2. The distribution of the values in each source signals are non-gaussian.

Three effects of mixing source signals :

1. *Independence :* As what we assume, the source signals are independent, however, their signal mixtures are not. That is because the signal mixtures share the same source signals.
2. *Normality :* Based on the Central Limit Theorem, the distribution of a sum of independent random variables tends towards a gaussian distribution. Loosely speaking, a sum of two independent random variables usually has a distribution that is closer to gaussian than any of the two original variables. Here we consider the value of each signals as the random variable.
3. *Complexity :* The temporal complexity of any signal mixture is greater than that of its simplest constituent source signal.

Those principles contributes to the basic establishment of ICA. If the signals we happen to extract from a set of mixtures are independent like sources signals, or have non-gaussian histograms like source signals, or have low complexity like source signals, then they must be source signals.

Defining Component Independence

ICA finds the independent components (also called factors, latent variables or sources) by maximizing the statistical independence of the estimated components. We may choose one of many ways to define independence, and this choice governs the form of the ICA algorithm. The two broadest definitions of independence for ICA are :

1. Minimization of mutual information
2. Maximization of non-Gaussianity.

The Minimization-of-Mutual information (MMI) family of ICA algorithms uses measures like Kullback-Leibler Divergence and maximum entropy. The non-Gaussianity family of ICA algorithms, motivated by the central limit theorem, uses kurtosis and negentropy.

Typical algorithms for ICA use centering (subtract the mean to create a zero mean signal), whitening (usually with the eigenvalue decomposition), and dimensionality reduction as preprocessing steps in order to simplify and reduce the complexity of the problem for the actual iterative algorithm. Whitening and dimension reduction can be achieved with principal component analysis or singular value decomposition. Whitening ensures that all dimensions are treated equally *a priori* before the algorithm is run. Well-known algorithms for ICA include infomax, FastICA, and JADE, but there are many others.

In general, ICA cannot identify the actual number of source signals, a uniquely correct ordering of the source signals, nor the proper scaling (including sign) of the source signals.

ICA is important to blind signal separation and has many practical applications. It is closely related to (or even a special case of) the search for a factorial code of the data, *i.e.*, a new vector-valued representation of each data vector such that it gets uniquely encoded by the resulting code vector (loss-free coding), but the code components are statistically independent.

Mathematical Definitions

Linear independent component analysis can be divided into noiseless and noisy cases, where noiseless ICA is a special case of noisy ICA. Non-linear ICA should be considered as a separate case.

General Definition

The data are represented by the random vector $x = (x_1,..., x_m)^T$ and the components as the random vector $s = (s_1,..., s_n)^T$. The task is to transform the observed data x, using a linear static transformation W as $s = Wx$ into maximally independent components s measured by some function $F(s_1,..., s_n)$ of independence.

Generative Model

Linear Noiseless ICA

The components x_i of the observed random vector $x = (x_1,..., x_m)^T$ are generated as a sum of the independent components s_k, $k = 1,..., n$:

$$x_i = a_{i,1}\, s_1 + \cdots + a_{i,k} s_k + \cdots + a_{i,n}\, s_n$$

weighted by the mixing weights $a_{i,k}$.

The same generative model can be written in vectorial form as $x = \sum_{k=1}^{n} s_k a_k$,

where the observed random vector x is represented by the basis vectors $a_k = (a_{1,k},..., a_m,k)^T$. The basis vectors a_k form the columns of the mixing matrix $A = (a_1,..., a_n)$ and the generative formula can be written as $x = As$, where $s = (s_1,..., s_n)^T$.

Given the model and realizations (samples) $x_1,..., x_N$ of the random vector x, the task is to estimate both the mixing matrix A and the sources s. This is done by adaptively calculating the w vectors and setting up a cost function which either maximizes the non-gaussianity of the calculated $s_k = (w^T * x)$ or minimizes the mutual information. In some cases, a priori knowledge of the probability distributions of the sources can be used in the cost function.

The original sources s can be recovered by multiplying the observed signals x with the inverse of the mixing matrix $W = A^{-1}$, also known as the unmixing matrix. Here it is assumed that the mixing matrix is square ($n = m$). If the number of basis vectors is greater than the dimensionality of the observed vectors, $n > m$, the task is overcomplete but is still solvable with the pseudo inverse.

Linear Noisy ICA

With the added assumption of zero-mean and uncorrelated Gaussian noise $n \sim N(0, \mathrm{diag}(\Sigma))$, the ICA model takes the form $x = As + n$.

Non-linear ICA

The mixing of the sources does not need to be linear. Using a non-linear mixing function $f(\cdot \mid \theta)$ with parameters θ the non-linear ICA model is $x = f(s \mid \theta) + n$.

Identifiability

The independent components are identifiable up to a permutation and scaling of the sources. This identifiability requires that :

- At most one of the sources s_k is Gaussian,
- The number of observed mixtures, m, must be at least as large as the number of estimated components $n : m \geq n$. It is equivalent to say that the mixing matrix A must be of full rank for its inverse to exist.

Binary independent component analysis

A special variant of ICA is **Binary ICA** in which both signal sources and monitors are in binary form and observations from monitors are disjunctive mixtures of binary independent sources. The problem was shown to have applications in many domains including medical diagnosis, multi-cluster assignment, network tomography and internet resource management.

Let $x_1, x_2,..., x_m$ be the set of binary variables from m monitors and $y_1, y_2,..., y_n$ be the set of binary variables from n sources. Source-monitor connections are represented by the (unknown) mixing matrix G, where $g_{ij} = 1$ indicates that signal from the i-th source can be observed by the j-th monitor. The system works as follows : at any time, if a source i is active ($y_i = 1$) and it is connected to the monitor $j(g_{ij} = 1)$ then the monitor j will observe some activity ($x_j = 1$). Formally we have :

$$x_i = \bigvee_{j=1}^{n} (g_{ij} \wedge y_i), i = 1,2,...,m,$$

where $\wedge$ is Boolean AND and $\vee$ is Boolean OR. Note that noise is not explicitly modelled, rather, can be treated as independent sources.

The above problem can be heuristically solved by assuming variables are continuous and running FastICA on binary observation data to get the mixing matrix G(real values), then apply round number techniques on G to obtain the binary values. This approach has been shown to produce a highly inaccurate result.

Another method is to use dynamic programming : recursively breaking the observation matrix X into its sub-matrices and run the inference algorithm on these sub-matrices. The key observation which leads to this algorithm is the sub-matrix X^0 of X where $x_{ij} = 0, \forall j$ corresponds to the unbiased observation matrix of hidden components that do not have connection to the i-th monitor. Experimental results from show that this approach is accurate under moderate noise levels.

Methods for Blind Source Separation

Projection Pursuit

We know that signal mixtures tend to have Gaussian probability density functions, and that source signals have non-Gaussian probability density functions. We also know that each source signal can be extracted from a set of signal mixtures by taking the inner product of a weight vector and those signal mixtures where this inner product provides an orthogonal projection of the signal mixtures. But we do not yet know precisely how to find such a weight vector. One type of method for doing so is projection pursuit.

Projection pursuit seek one projection at a time such that the extracted signal is as non-Gaussian as possible. This contrasts with ICA, which typically extracts M signals simultaneously from M signal mixtures, which requires estimating a$M \times M$ unmixing matrix. One practical advantage of projection pursuit over ICA is that less than M signals can be extracted if required, where each source signal is extracted from M signal mixtures using an M-element weight vector.

We can use kurtosis to recover the multiple source signal by finding the correct weight vectors with the use of projection pursuit.

The kurtosis of the probability density function of a signal, for a finite sample, is computed as

$$K = \frac{E[y-\bar{y})^4]}{(E[(y-\bar{y})^2])^2} - 3$$

where $\bar{y}$ is the sample mean of y, the extracted signals. The constant 3 ensures that Gaussian signals have zero kurtosis, Super-Gaussian signals have positive kurtosis, and Sub-Gaussian signals have negative kurtosis. The denominator is the variance of y, and ensures that the measured kurtosis takes account of signal variance. The goal of projection pursuit is to maximize the kurtosis, and make the extracted signal as non-normality as possible.

Using kurtosis as a measure of non-normality, we can now examine how the kurtosis of a signal $y = w^T x$ extracted from a set of M mixtures $x = (x_1, x_2, \ldots, x_M)^T$ varies as the weight vector w is rotated around the origin. Given our assumption that each source signal s is super-gaussian we would expect :

1. The kurtosis of the extracted signal y to be maximal precisely when $y = s$.
2. The kurtosis of the extracted signal y to be maximal when w is orthogonal to the projected axes S_1 or S_2, because we know the optimal weight vector should be orthogonal to a transformed axis S_1 or S_2.

For multiple source mixture signals, we can use kurtosis and Gram-Schmidt Orthogonalizaton (GSO) to recover the signals. Given M signal mixtures in an M-dimensional space, GSO project these data points onto an (M-1)-dimensional space by using the weight vector. We can guarantee the independence of the extracted signals with the use of GSO.

In order to find the correct value of w, we can use gradient descent method. We first of all whitening the data, and transform x into a new mixture z, which has unit variance, and $z = (z_1, z_2, \ldots, z_m)^T$. This process can be achieved by applying Singular value decompositionto x,

$$x = UDV^T$$

Rescaling each vector $U_i = U_i / E(U_i^2)$, and let $z = U$. The signal extracted by a weighted vector w is $y = w^T z$. If the weight vector **w** has unit length, that is $E[(w^T z)^2] = 1$, then the kurtosis can be written as :

$$K = \frac{E[y^4]}{(E[y^2])^2} - 3 = E[w^T z)^4] - 3.$$

The updating process for w is :

$$w_{new} = w_{old} - \eta\, E[z(\mathbf{w}^T_{old}\, z)^3].$$

where η is a small constant to guarantee that w converge to the optimal solution. After each update, we normalized $w_{new} = \frac{w_{new}}{|w_{new}|}$, and set $w_{old} = w_{new}$, and repeat the updating process till it converges. We can also use another algorithm to update the weight vector w.

Another approach is using Negentropy instead of kurtosis. Negentropy is a robust method for kurtosis, as kurtosis is very sensitive to outliers. The negentropy method are based on an important property of gaussian distribution : a gaussian variable has the largest entropy among all random variables of equal variance. This is also the reason why we want to find the most non-gaussian variables. A simple proof can be found in wiki page Differential entropy.

$$J(x) = S(y) - S(x)$$

y is a Gaussian random variable of the same covariance matrix as x

$$S(x) = -\int p_x(u)\log p_x(u)du$$

A approximation for negentropy is

$$J(x) = \frac{1}{12}(E(x^3))^2 + \frac{1}{48}(\text{kurt}(x))^2$$

A prove can be found one the page 131 in the book Independent Components Analysis written by Aapo Hyv¨arinen, Juha Karhunen, and Erkki Oja (They contributes great works to ICA) This approximation also suffers the same problem as kurtosis (sensitive to outliers) Another approach were developed.

$$J(y) = k_1(E(G_1(y)))^2 + k_2(E(G_2(y)) - E(G_2(v))^2$$

A choice of G_1 and G_2 are

$$G_1 = \frac{1}{a_1}\log(\cosh(a_1 u)) \text{ and } G_2 = -\exp\left(-\frac{u^2}{2}\right)$$

Independent Component Analysis Based on Infomax

ICA is essentially a multi-variate, parallel version of projection pursuit. Whereas projection pursuit extracts a series of signals one at a time from a set of M signal mixtures, ICA extracts M signals in parallel. This tends to make ICA more robust than projection pursuit.

The projection pursuit method use Gram-Schmidt Orthogonalizaton to ensure the independence of the extracted signal, while ICA use infomax and maximum likelihood estimate to ensure the independence of the extracted signal. The Non-Normality of the extracted signal is achieved by assigning an appropriate model, or prior, for the signal.

The process of ICA based on infomax in short is : given a set of signal mixtures x and a set of identical independent model cumulative distribution functions(cdfs) g, we seek the unmixing matrix W which maximizes the joint entropy of the signals

$Y = g(y)$, where $y = Wx$ are the signals extracted by W. Given the optimal W, the signals Y have maximum entropy and are therefore independent, which ensures that the extracted signals $y = g^{-1}(Y)$ are also independent. I is an invertible function, and is the signal model. Note that if the source signal model probability density function p_s matches the probability density function of the extracted signal p_y, then maximizing the joint entropy of Y also maximizes the amount of mutual information between x and Y. For this reason, using entropy to extract independent signals is known as infomax.

Consider the entropy of the vector variable $Y = g(y)$, where $y = Wx$ is the set of signals extracted by the unmixing matrix W. For a finite set of values sampled from a distribution with pdf p_y, the entropy of Y can be estimated as :

$$H(Y) = -\frac{1}{N}\sum_{t=1}^{N} \text{In } pY(Y^t)$$

The joint pdf pY can be shown to be related to the joint pdf p_y of the extracted signals by the multi-variate form :

$$pY(Y) = \frac{p_y(y)}{\left|\frac{\partial Y}{\partial y}\right|}$$

where $J = \frac{\partial Y}{\partial y}$ is the Jacobian matrix. We have $|J| = g'(y)$, and g' is the pdf assumed for source signals $g' = p_s$, therefore,

$$p_Y(Y) = \frac{p_y(y)}{\left|\frac{\partial Y}{\partial y}\right|} = \frac{p_y(y)}{p_s(y)}$$

therefore,

$$H(Y) = -\frac{1}{N}\sum_{t=1}^{N} \text{In } \frac{p_y(y)}{p_s(y)}$$

We know that when $p_y = p_s$, p_Y is of uniform distribution, and $H(Y)$ is maximized. Since

$$p_y(y) = \frac{p_x(x)}{\left|\frac{\partial y}{\partial x}\right|} = \frac{p_x(x)}{|W|}$$

where $|W|$ is the absolute value of the determinant of the unmixing matix < math>\mathbf{W}</math>. Therefore,

$$H(Y) = -\frac{1}{N}\sum_{t=1}^{N} \text{In } \frac{p_x(x^t)}{|W|\, p_s(y^t)}$$

so,

$$H(Y) = -\frac{1}{N}\sum_{t=1}^{N} \text{In } p_s(y^t) + \text{In} \mid W \mid - H(x)$$

since $H(x) = -\frac{1}{N}\sum_{t=1}^{N} \text{In } p_x(x^t)$, and maximizing W does not affect H_x, so we can maximized function

$$h(Y) = -\frac{1}{N}\sum_{t=1}^{N} \text{In } p_s(y^t) + \text{In } \mid W \mid$$

to achieve the independence of extracted signal.

If there are M marginal pdfs of the model joint pdf p_s are independent and use the commonly super-gaussian model pdf for the source signals $p_s = (1 - \tanh(s)^2)$, then we have

$$h(Y) = \frac{1}{N}\sum_{i=1}^{M}\sum_{t=1}^{N} \text{In } (1 - \tanh(w_i^T x^t)^2) + \text{In} \mid W \mid$$

In sum, given an observed signal mixture x, corresponding set of extracted signals y and source signal model $p_s = g'$, we can find the optimal unmixing matrix W, and make the extracted signals independent and non-gaussian. Like the projection pursuit situation, we can use gradient descent method to find the optimal solution of the unmixing matrix.

Independent Component Analysis based on Maximum Likelihood Estimation

Maximum likelihood estimation (MLE) is a standard statistical tool for finding parameter values (*e.g.* the unmixing matrix W) that provide the best fit of some data (*e.g.*, the extracted signals y) to a given a model (*e.g.*, the assumed joint probability density function (pdf) p_s of source signals).

The **ML** "model" includes a specification of a pdf, which in this case is the pdf p_s of the unknown source signals *s*. Using **ML ICA**, the objective is to find an unmixing matrix that yields extracted signals $y = W_x$ with a joint pdf as similar as possible to the joint pdf p_s of the unknown source signals *s*.

MLE is thus based on the assumption that if the model pdf p_s and the model parameters A are correct then a high probability should be obtained for the data x that were actually observed. Conversely, if A is far from the correct parameter values then a low probability of the observed data would be expected.

Using **MLE**, we call the probability of the observed data for a given set of model parameter values (*e.g.*, a pdf p_s and a matrix A) the *likelihood* of the model parameter values given the observed data.

We define a *likelihood* function $L(W)$ of W:

$$L(W) = p_s(W_x) \mid W \mid .$$

Thus, if we wish to find a W that is most likely to have generated the observed mixtures x from the unknown source signals s with pdf p_s then we need only find that W which maximizes the *likelihood* $L(W)$. The unmixing matrix that maximizes equation is known as the **MLE** of the optimal unmixing matrix.

It is common practice to use the log *likelihood,* because this is easier to evaluate. As the logarithm is a monotonic function, the W that maximizes the function $L(W)$ also maximizes its logarithm In $L(W)$. This allows us to take the logarithm of equation above, which yields the log *likelihood* function

$$\text{In } L(W) = \sum_i \sum_t \text{In } p_s(w_i^T x_t) + N \text{ In} \mid W \mid$$

If we substitute a commonly used high-Kurtosis model pdf for the source signals $p_s = (1 - \tanh(s)^2)$ then we have

$$\text{In } L(W) = \frac{1}{N} \sum_i^M \sum_t^N \text{In } (1 - \tan h(w_i^T x_t)^2) + \text{In} \mid W \mid$$

This matrix W that maximizes this function is the ***maximum likelihood estimation***.

History and Background

The general framework for independent component analysis was introduced by Jeanny Herault and Christian Jutten in 1986 and was most clearly stated by Pierre Comon in 1994. In 1995, Tony Bell and Terry Sejnowski introduced a fast and efficient ICA algorithm based on infomax, a principle introduced by Ralph Linsker in 1987.

There are many algorithms available in the literature which do ICA. A largely used one, including in industrial applications, is the FastICA algorithm, developed by Aapo Hyvärinen and Erkki Oja, which uses the kurtosis as cost function. Other examples are rather related to blind source separation where a more general approach is used. For example, one can drop the independence assumption and separate mutually correlated signals, thus, statistically "dependent" signals. Sepp Hochreiter and Jürgen Schmidhuber showed how to obtain non-linear ICA or source separation as a by-product of regularization. Their method does not require a priori knowledge about the number of independent sources..

Applications

ICA can be extended to analyze non-physical signals. For instance, ICA has been applied to discover discussion topics on a bag of news list archives.....->

Some ICA applications are listed below :

- Optical Imaging of neurons
- Neuronal spike sorting
- Face recognition
- Modelling receptive fields of primary visual neurons
- Predicting stock market prices

- Mobile phone communications
- Colour based detection of the ripeness of tomatoes.

LINEAR DISCRIMINANT ANALYSIS

Linear discriminant analysis (LDA) and the related **Fisher's linear discriminant** are methods used in statistics, pattern recognition and machine learning to find a linear combination of features which characterizes or separates two or more classes of objects or events. The resulting combination may be used as a linear classifier, or, more commonly, for dimensionality reduction before later classification.

LDA is closely related to ANOVA (analysis of variance) and regression analysis, which also attempt to express one dependent variable as a linear combination of other features or measurements. However, ANOVA uses categorical independent variables and a continuous dependent variable, whereas discriminant analysis has continuous independent variables and a categorical dependent variable (*i.e.* the class label). Logistic regression and probit regression are more similar to LDA, as they also explain a categorical variable by the values of continuous independent variables. These other methods are preferable in applications where it is not reasonable to assume that the independent variables are normally distributed, which is a fundamental assumption of the LDA method.

LDA is also closely related to principal component analysis (PCA) and factor analysis in that they both look for linear combinations of variables which best explain the data. LDA explicitly attempts to model the difference between the classes of data. PCA on the other hand does not take into account any difference in class, and factor analysis builds the feature combinations based on differences rather than similarities. Discriminant analysis is also different from factor analysis in that it is not an interdependence technique : a distinction between independent variables and dependent variables (also called criterion variables) must be made.

LDA works when the measurements made on independent variables for each observation are continuous quantities. When dealing with categorical independent variables, the equivalent technique is discriminant correspondence analysis.

LDA for Two Classes

Consider a set of observations $\vec{x}$ (also called features, attributes, variables or measurements) for each sample of an object or event with known class y. This set of samples is called the training set. The classification problem is then to find a good predictor for the class y of any sample of the same distribution (not necessarily from the training set) given only an observation $\vec{x}$.

LDA approaches the problem by assuming that the conditional probability density functions $p(\vec{x} \mid y = 0)$ and $p(\vec{x} \mid y = 1)$ are both normally distributed with mean and covariance parameters $(\vec{\mu}_0, \Sigma_{y=0})$ and $(\vec{\mu}_1, \Sigma_{y=1})$, respectively. Under this assumption, the Bayes optimal solution is to predict points as being from the second class if the log of the likelihood ratios is below some threshold T, so that;

$$(\vec{x}-\vec{\mu}_0)^T \Sigma_{y=0}^{-1}(\vec{x}-\vec{\mu}_0)+\text{In}\,|\Sigma_{y=0}|-(\vec{x}-\vec{\mu}_1)^T \Sigma_{y=1}^{-1}(\vec{x}-\vec{\mu}_1)-\text{In}\,|\Sigma_{y=1}|<T$$

Without any further assumptions, the resulting classifier is referred to as QDA (quadratic discriminant analysis). LDA also makes the simplifying homoscedasticity assumption (*i.e.* that the class covariances are identical, so $\Sigma_{y=0} = \Sigma_{y=1} = \Sigma$) and that the covariances have full rank. In this case, several terms cancel and the above decision criterion becomes a threshold on the dot product

$$\vec{w} \cdot \vec{x} > c$$

for some threshold constant c, where

$$\vec{w}\ \alpha\ \Sigma^{-1}(\vec{\mu}_1 - \vec{\mu}_0)$$

This means that the criterion of an input $\vec{x}$ being in a class y is purely a function of this linear combination of the known observations.

It is often useful to see this conclusion in geometrical terms : the criterion of an input $\vec{x}$ being in a class y is purely a function of projection of multi-dimensional-space point $\vec{x}$ onto vector $\vec{w}$ (thus, we only consider its direction). In other words, the observation belongs to y if corresponding $\vec{x}$ is located on a certain side of a hyperplane perpendicular to $\vec{w}$. The location of the plane is defined by the threshold c.

Canonical Discriminant Analysis for k Classes

Canonical discriminant analysis (CDA) finds axes (k – 1 canonical co-ordinates, k being the number of classes) that best separate the categories. These linear functions are uncorrelated and define, in effect, an optimal $k - 1$ space through the n-dimensional cloud of data that best separates (the projections in that space of) the k groups.

Fisher's Linear Discriminant

The terms *Fisher's linear discriminant* and *LDA* are often used interchangeably, although Fisher's original article actually describes a slightly different discriminant, which does not make some of the assumptions of LDA such as normally distributed classes or equal class covariances.

Suppose two classes of observations have means $\vec{\mu}_{y=0}, \vec{\mu}_{y=1}$ and covariances $\Sigma_{y=0}, \Sigma_{y=1}$. Then the linear combination of features $\vec{w} \cdot \vec{x}$ will have means $\vec{w} \cdot \vec{\mu}_{y=i}$ and variances $\vec{w}^T \Sigma_{y=i} \vec{w}$ for $i = 0, 1$. Fisher defined the separation between these two distributions to be the ratio of the variance between the classes to the variance within the classes :

$$S = \frac{\sigma^2_{\text{between}}}{\sigma^2_{\text{within}}} = \frac{(\vec{w}\cdot\vec{\mu}_{y=1} - \vec{w}\cdot\vec{\mu}_{y=0})^2}{\vec{w}^T\Sigma_{y=1}\vec{w} + \vec{w}^T\Sigma_{y=0}\vec{w}} = \frac{(\vec{w}\cdot(\vec{\mu}_{y=1} - \vec{\mu}_{y=0}))^2}{\vec{w}^T(\Sigma_{y=0} + \Sigma_{y=1})\vec{w}}$$

This measure is, in some sense, a measure of the signal-to-noise ratio for the class labelling. It can be shown that the maximum separation occurs when

$$\vec{w} \propto (\Sigma_{y=0} + \Sigma_{y=1})^{-1}(\vec{\mu}_{y=1} - \vec{\mu}_{y=0})$$

When the assumptions of LDA are satisfied, the above equation is equivalent to LDA.

Be sure to note that the vector $\vec{w}$ is the normal to the discriminant hyperplane. As an example, in a two dimensional problem, the line that best divides the two groups is perpendicular to $\vec{w}$.

Generally, the data points to be discriminated are projected onto $\vec{w}$; then the threshold that best separates the data is chosen from analysis of the one-dimensional distribution. There is no general rule for the threshold. However, if projections of points from both classes exhibit approximately the same distributions, a good choice would be the hyperplane between projections of the two means, $\vec{w}\cdot\vec{\mu}_{y=0}$ and $\vec{w}\cdot\vec{\mu}_{y=1}$. In this case the parameter c in threshold condition $\vec{w}\cdot\vec{x} > c$ can be found explicitly :

$$c = \vec{w}\cdot\frac{1}{2}(\vec{\mu}_{y=0} + \vec{\mu}_{y=1}) = \frac{1}{2}\vec{\mu}^t_{y=1}\Sigma^{-1}\vec{\mu}_{y=1} - \frac{1}{2}\vec{\mu}^t_{y=0}\Sigma^{-1}\vec{\mu}_{y=0}.$$

Multiclass LDA

In the case where there are more than two classes, the analysis used in the derivation of the Fisher discriminant can be extended to find a sub-space which appears to contain all of the class variability. This generalization is due to C.R. Rao. Suppose that each of C classes has a mean μ_i and the same covariance Σ. Then the between class variability may be defined by the sample covariance of the class means

$$\Sigma_b = \frac{1}{C}\sum_{i=1}^{C}(\mu_i - \mu)(\mu_i - \mu)^T$$

where μ is the mean of the class means. The class separation in a direction $\vec{w}$ in this case will be given by

$$S = \frac{\vec{w}^T\Sigma_b\vec{w}}{\vec{w}^T\Sigma\vec{w}}$$

This means that when $\vec{w}$ is an eigenvector of $\Sigma^{-1}\Sigma_b$ the separation will be equal to the corresponding eigenvalue.

If $\Sigma^{-1}\Sigma_b$ is diagonalizable, the variability between features will be contained in the sub-space spanned by the eigenvectors corresponding to the $C-1$ largest eigenvalues (since Σ_b is of rank $C-1$ at most). These eigenvectors are primarily used in feature reduction, as in PCA. The eigenvectors corresponding to the smaller eigenvalues will tend to be very sensitive to the exact choice of training data, and it is often necessary to use regularisation as described in the next section.

If classification is required, instead of dimension reduction, there are a number of alternative techniques available. For instance, the classes may be partitioned, and a standard Fisher discriminant or LDA used to classify each partition. A common example of this is "one against the rest" where the points from one class are put in one group, and everything else in the other, and then LDA applied. This will result in C classifiers, whose results are combined. Another common method is pairwise classification, where a new classifier is created for each pair of classes (giving $C(C-1)/2$ classifiers in total), with the individual classifiers combined to produce a final classification.

Practical Use

In practice, the class means and covariances are not known. They can, however, be estimated from the training set. Either the maximum likelihood estimate or the maximum a posteriori estimate may be used in place of the exact value in the above equations. Although the estimates of the covariance may be considered optimal in some sense, this does not mean that the resulting discriminant obtained by substituting these values is optimal in any sense, even if the assumption of normally distributed classes is correct.

Another complication in applying LDA and Fisher's discriminant to real data occurs when the number of observations of each sample does not exceed the number of samples. In this case, the covariance estimates do not have full rank, and so cannot be inverted. There are a number of ways to deal with this. One is to use a pseudo inverse instead of the usual matrix inverse in the above formulae. However, better numeric stability may be achieved by first projecting the problem onto the sub-space spanned by Σ_b. Another strategy to deal with small sample size is to use a shrinkage estimator of the covariance matrix, which can be expressed mathematically as

$$\Sigma = (1 - \lambda)\Sigma + \lambda I$$

where I is the identity matrix, and λ is the *shrinkage intensity* or *regularisation parameter*. This leads to the framework of regularized discriminant analysis or shrinkage discriminant analysis.

Also, in many practical cases linear discriminants are not suitable. LDA and Fisher's discriminant can be extended for use in non-linear classification via the kernel trick. Here, the original observations are effectively mapped into a higher dimensional non-linear space. Linear classification in this non-linear space is then equivalent to non-linear classification in the original space. The most commonly used example of this is the kernel Fisher discriminant.

LDA can be generalized to multiple discriminant analysis, where c becomes a categorical variable with N possible states, instead of only two. Analogously, if the class-conditional densities $p(\vec{x} \mid c = i)$ are normal with shared covariances, the sufficient statistic for $P(c \mid \vec{x})$ are the values of N projections, which are the sub-space spanned by the N means, affine projected by the inverse covariance matrix. These projections can be found by solving a generalized eigenvalue problem, where the numerator is the covariance matrix formed by treating the means as the samples, and the denominator is the shared covariance matrix.

Applications

In addition to the examples given below, LDA is applied in positioning and product management.

Bankruptcy Prediction

In bankruptcy prediction based on accounting ratios and other financial variables, linear discriminant analysis was the first statistical method applied to systematically explain which firms entered bankruptcy vs. survived. Despite limitations including known non-conformance of accounting ratios to the normal distribution assumptions of LDA, Edward Altman's 1968 model is still a leading model in practical applications.

Face Recognition

In computerised face recognition, each face is represented by a large number of pixel values. Linear discriminant analysis is primarily used here to reduce the number of features to a more manageable number before classification. Each of the new dimensions is a linear combination of pixel values, which form a template. The linear combinations obtained using Fisher's linear discriminant are called *Fisher faces*, while those obtained using the related principal component analysis are called *eigenfaces*.

Marketing

In marketing, discriminant analysis was once often used to determine the factors which distinguish different types of customers and/or products on the basis of surveys or other forms of collected data. Logistic regression or other methods are now more commonly used. The use of discriminant analysis in marketing can be described by the following steps :

1. Formulate the problem and gather data—Identify the salient attributes consumers use to evaluate products in this category—Use quantitative marketing research techniques (such as surveys) to collect data from a sample of potential customers concerning their ratings of all the product attributes. The data collection stage is usually done by marketing research professionals. Survey questions ask the respondent to rate a product from one to five (or 1 to 7, or 1 to 10) on a range of attributes chosen by the researcher. Anywhere from five to twenty attributes are chosen. They could include things like : ease of use, weight, accuracy, durability, colourfulness, price, or size. The attributes chosen will vary depending on the product being studied. The same question

is asked about all the products in the study. The data for multiple products is codified and input into a statistical program such as R, SPSS or SAS. (This step is the same as in Factor analysis).

2. Estimate the Discriminant Function Coefficients and determine the statistical significance and validity—Choose the appropriate discriminant analysis method. The direct method involves estimating the discriminant function so that all the predictors are assessed simultaneously. The stepwise method enters the predictors sequentially. The two-group method should be used when the dependent variable has two categories or states. The multiple discriminant method is used when the dependent variable has three or more categorical states. Use Wilks's Lambda to test for significance in SPSS or F stat in SAS. The most common method used to test validity is to split the sample into an estimation or analysis sample, and a validation or holdout sample. The estimation sample is used in constructing the discriminant function. The validation sample is used to construct a classification matrix which contains the number of correctly classified and incorrectly classified cases. The percentage of correctly classified cases is called the hit ratio.
3. Plot the results on a two dimensional map, define the dimensions, and interpret the results. The statistical program (or a related module) will map the results. The map will plot each product (usually in two-dimensional space). The distance of products to each other indicate either how different they are. The dimensions must be labelled by the researcher. This requires subjective judgement and is often very challenging.

NON-NEGATIVE MATRIX FACTORIZATION

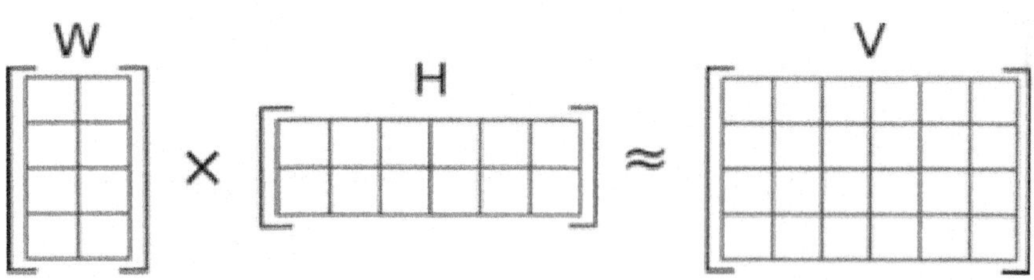

Fig. : Illustration of approximate non-negative matrix factorization : the matrix *V* is represented by the two smaller matrices *W* and *H*, which, when multiplied, approximately reconstruct *V*.

Non-negative matrix factorization (NMF), also **non-negative matrix approximation** is a group of algorithms in multi-variate analysis and linear algebra where a matrixV is factorized into (usually) two matrices **W** and **H**, with the property that all three matrices have no negative elements. This non-negativity makes the resulting matrices easier to inspect. Since the problem is not exactly solvable in general, it is commonly approximated numerically.

NMF finds applications in such fields as computer vision, document clustering, chemometrics and recommender systems.

History

In chemometrics non-negative matrix factorization has a long history under the name "self-modelling curve resolution". In this framework the vectors in the right matrix are continuous curves rather than discrete vectors. Also early work on non-negative matrix factorizations was performed by a Finnish group of researchers in the middle of the 1990s under the name *positive matrix factorization*. It became more widely known as *non-negative matrix factorization* after Lee and Seung investigated the properties of the algorithm and published some simple and useful algorithms for two types of factorizations.

Background

Let matrix V be the product of the matrices W and H such that :

$$WH = V$$

Matrix multiplication can be implemented as linear combinations of column vectors in W with coefficients supplied by cell values in H. Each column in V can be computed as follows :

$$v_i = \sum_{j=1}^{N} H_{ji} w_j$$

where : N is the number of columns in W

v_i is the ith column vector of the product matrix V

H_{ji} is the cell value in the jth row and i^{th} column of the matrix H

w_j is the jth column of the matrix W

When multiplying matrices, the dimensions of the factor matrices may be significantly lower than those of the product matrix and it's this property that forms the basis of NMF. NMF generates factors with significantly reduced dimensions compared to the original matrix. For example, if V is an $m \times n$ matrix, then W is an $m \times p$ matrix, and H is a $p \times n$ matrix. Here p can be significantly less than both m and n.

Here's an example based on a text-mining application :

- Let the input matrix (the matrix to be factored) be V with 10000 rows and 500 columns where words are in rows and documents are in columns. In other words, we have 500 documents indexed by 10000 words. It follows that a column vector v in V represents a document.
- Assume we ask the algorithm to find 10 features in order to generate a features matrix W with 10000 rows and 10 columns and a coefficients matrix H with 10 rows and 500 columns.
- The product of W and H is a matrix with 10000 rows and 500 columns, the same shape as the input matrix V and, if the factorization worked, also a reasonable approximation to the input matrix V.

- From the treatment of matrix multiplication above it follows that each column in the product matrix *WH* is a linear combination of the 10 column vectors in the features matrix *W* with coefficients supplied by the coefficients matrix *H*.

This last point is the basis of NMF because we can consider each original document in our example as being built from a small set of hidden features. NMF generates these features.

It's useful to think of each feature (column vector) in the features matrix *W* as a document archetype comprising a set of words where each word's cell value defines the word's rank in the feature : The higher a word's cell value the higher the word's rank in the feature. A column in the coefficients matrix *H* represents an original document with a cell value defining the document's rank for a feature. This follows because each row in *H* represents a feature. We can now reconstruct a document (column vector) from our input matrix by a linear combination of our features (column vectors in *W*) where each feature is weighted by the feature's cell value from the document's column in *H*.

Types

Approximate Non-negative Matrix Factorization

Usually the number of columns of *W* and the number of rows of *H* in NMF are selected so the product *WH* will become an approximation to *V*. The full decomposition of *V* then amounts to the two non-negative matrices *W* and *H* as well as a residual *U*, such that : $V = WH + U$. The elements of the residual matrix can either be negative or positive.

When *W* and *H* are smaller than *V* they become easier to store and manipulate. Another reason for factorizing *V* into smaller matrices *W* and *H*, is that if one is able to approximately represent the elements of *V* by significantly less data, then one has to infer some latent structure in the data.

Non-negative Rank Factorization

In case the non-negative rank of *V* is equal to its actual rank, $V = WH$ is called a non-negative rank factorization. The problem of finding the NRF of *V*, if it exists, is known to be NP-hard.

Different Cost Functions and Regularizations

There are different types of non-negative matrix factorizations. The different types arise from using different cost functions for measuring the divergence between *V* and *WH* and possibly by regularization of the *W* and/or *H* matrices.

Two simple divergence functions studied by Lee and Seung are the squared error (or Frobenius norm) and an extension of the Kullback-Leibler divergence to positive matrices (the original Kullback-Leibler divergence is defined on probability distributions). Each divergence leads to a different NMF algorithm, usually minimizing the divergence using iterative update rules.

The factorization problem in the squared error version of NMF may be stated as : Given a matrix V find non-negative matrices W and H that minimize the function

$$F(W, H) = ||V - WH||_F^2$$

Another type of NMF for images is based on the total variation norm.

When L1 regularization (akin to Lasso) is added to NMF with the mean squared error cost function, the resulting problem may be called **non-negative sparse coding** due to the similarity to the sparse coding problem, although it may also still be referred to as NMF.

Algorithms

There are several ways in which the W and H may be found : Lee and Seung's multiplicative update rule has been a popular method due to the simplicity of implementation. Since then, a few other algorithmic approaches have been developed.

Some successful algorithms are based on alternating non-negative least squares : in each step of such an algorithm, first H is fixed and W found by a non-negative least squares solver, then W is fixes and H is found analogously. The procedures used to solve for W and H may be the same or different, as some NMF variants regularize one of W and H. Specific approaches include the projected gradient descent methods, the active set method, and the block principal pivoting method among several others.

The currently available algorithms are sub-optimal as they can only guarantee finding a local minimum, rather than a global minimum of the cost function. A provably optimal algorithm is unlikely in the near future as the problem has been shown to generalize the k-means clustering problem which is known to be NP-complete. However, as in many other data-mining applications, a local minimum may still prove to be useful.

Exact NMF

Exact solutions for the variants of NMF can be expected (in polynomial time) when additional constraints hold for matrix V. A polynomial time algorithm for solving non-negative rank factorization if V contains a monomial sub-matrix of rank equal to its rank was given by Campbell and Poole in 1981. Kalofolias and Gallopoulos solved the symmetric counterpart of this problem, where V is symmetric and contains a diagonal principal sub-matrix of rank r. Their algorithm runs in $O(rm^{\wedge 2})$ time in the dense case. Arora, Ge, Kannan, and Moitra give a polynomial time algorithm for exact NMF that works for the case where one of the factors W satisfies the separability condition.

Relation to Other Techniques

In *Learning the parts of objects by non-negative matrix factorization* Lee and Seung proposed NMF mainly for parts-based decomposition of images. It compares

NMF to vector quantization and principal component analysis, and shows that although the three techniques may be written as factorizations, they implement different constraints and therefore produce different results.

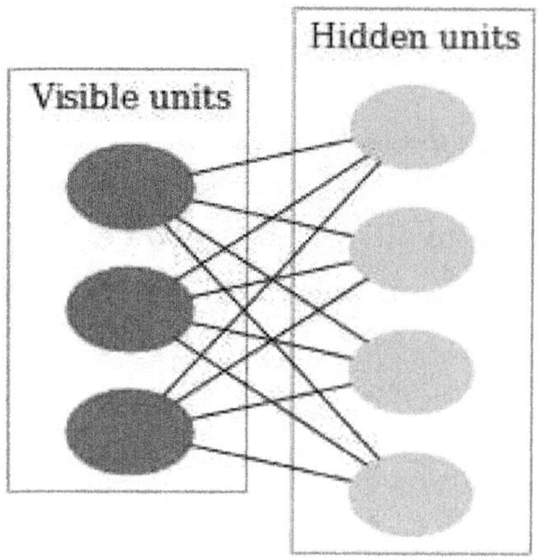

Fig. : NMF as a probabilistic graphical model : visible units (V) are connected to hidden units (H) through weights W, so that V is generated from a probability distribution with mean $\sum_a W_{ia} h_a$.

It was later shown that some types of NMF are an instance of a more general probabilistic model called "multinomial PCA". When NMF is obtained by minimizing the Kullback-Leibler divergence, it is in fact equivalent to another instance of multinomial PCA, probabilistic latent semantic analysis, trained by maximum likelihood estimation. That method is commonly used for analyzing and clustering textual data and is also related to the latent class model.

It has been shown NMF is equivalent to a relaxed form of K-means clustering: matrix factor *W* contains cluster centroids and *H* contains cluster membership indicators, when using the least square as NMF objective. This provides theoretical foundation for using NMF for data clustering.

NMF extends beyond matrices to tensors of arbitrary order. This extension may be viewed as a non-negative version of, *e.g.*, the PARAFAC model.

Other extensions of NMF include joint factorisation of several data matrices and tensors where some factors are shared. Such models are useful for sensor fusion and relational learning.

NMF is an instance of the non-negative quadratic programming (NQP) as well as many other important problems including the support vector machine (SVM). However, SVM and NMF are related at a more intimate level than that of NQP, which allows direct application of the solution algorithms developed for either of the two methods to problems in both domains.

Uniqueness

The factorization is not unique : A matrix and its inverse can be used to transform the two factorization matrices by, *e.g.,*

$$WH = WBB^{-1}H$$

If the two new matrices $\tilde{W} = WB$ and $\tilde{H} = B^{-1}H$ are non-negative they form another parametrization of the factorization.

The non-negativity of $\tilde{W}$ and $\tilde{H}$ applies at least if B is a non-negative monomial matrix. In this simple case it will just correspond to a scaling and a permutation.

More control over the non-uniqueness of NMF is obtained with sparsity constraints.

Applications

Text Mining

NMF can be used for text mining applications. In this process, a *document-term* matrix is constructed with the weights of various terms (typically weighted word frequency information) from a set of documents. This matrix is factored into a *term-feature* and a *feature-document* matrix. The features are derived from the contents of the documents, and the feature-document matrix describes data clusters of related documents.

One specific application used hierarchical NMF on a small subset of scientific abstracts from PubMed. Another research group clustered parts of the Enron email data-set with 65,033 messages and 91,133 terms into 50 clusters. NMF has also been applied to citations data, with one example clustering Wikipedia articles and scientific journals based on the outbound scientific citations in Wikipedia.

Arora, Ge and Moitra have given polynomial-time algorithms to learn topic models using NMF. The algorithm assumes that the topic matrix satisfies a separability condition that is often found to hold in these settings.

Spectral Data Analysis

NMF is also used to analyze spectral data; one such use is in the classification of space objects and debris.

Scalable Internet Distance Prediction

NMF is applied in scalable Internet distance (round-trip time) prediction. For a network with N hosts, with the help of NMF, the distances of all the N^2 end-to-end links can be predicted after conducting only $O(N)$ measurements. This kind of method was firstly introduced in Internet Distance Estimation Service (IDES). Afterwards, as a fully decentralized approach, Phoenix network co-ordinate system is proposed. It achieves better overall prediction accuracy by introducing the concept of weight.

Non-stationary Speech Denoising

Speech denoising has been a long lasting problem in audio signal processing. There are lots of algorithms for denoising if the noise is stationary. For example, the Wiener filter is suitable for additive Gaussian noise. However, if the noise is non-stationary, the classical denoising algorithms usually have poor performance because the statistical information of the non-stationary noise is difficult to estimate. Schmidt *et. al.* use NMF to do speech denoising under non-stationary noise, which is completely different from classical statistical approaches. The key idea is that clean speech signal can be sparsely represented by a speech dictionary, but non-stationary noise cannot. Similarly, non-stationary noise can also be sparsely represented by a noise dictionary, but speech cannot.

The algorithm for NMF denoising goes as follows. Two dictionaries, one for speech and one for noise, need to be trained offline. Once a noisy speech is given, we first calculate the magnitude of the Short-Time-Fourier-Transform. Second, separate it into two parts via NMF, one can be sparsely represented by the speech dictionary, and the other part can be sparsely represented by the noise dictionary. Third, the part that is represented by the speech dictionary will be the estimated clean speech.

Bioinformatics

NMF has been successfully applied to bioinformatics.

Current Research

Current research in non-negative matrix factorization includes, but not limited to,

1. *Algorithmic* : searching for global minima of the factors and factor initialization.
2. *Scalability* : how to factorize million-by-billion matrices, which are commonplace in Web-scale data-mining.
3. *Online* : how to update the factorization when new data comes in without recomputing from scratch.

PRINCIPAL COMPONENT ANALYSIS

Principal component analysis (PCA) is a statistical procedure that uses orthogonal transformation to convert a set of observations of possibly correlated variables into a set of values of linearly uncorrelated variables called **principal components**. The number of principal components is less than or equal to the number of original variables. This transformation is defined in such a way that the first principal component has the largest possible variance (that is, accounts for as much of the variability in the data as possible), and each succeeding component in turn has the highest variance possible under the constraint that it is orthogonal to (*i.e.*, uncorrelated with) the preceding components. Principal components are guaranteed to be independent if the data-set is jointly normally distributed. PCA is sensitive to the relative scaling of the original variables.

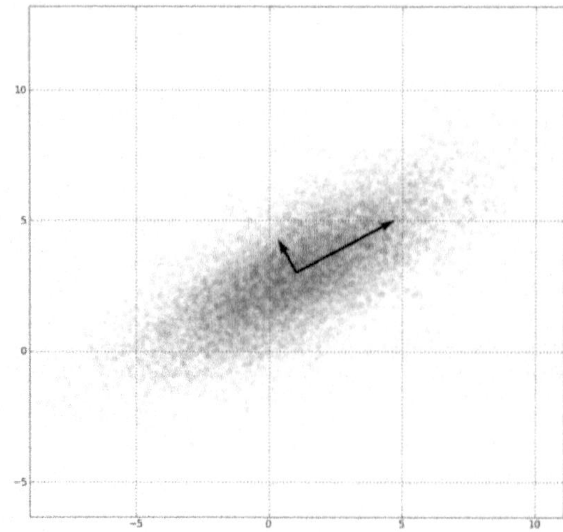

Fig.: PCA of a multi-variate Gaussian distribution centered at (1,3) with a standard deviation of 3 in roughly the (0.878, 0.478) direction and of 1 in the orthogonal direction. The vectors shown are the eigenvectors of the covariance matrix scaled by the square root of the corresponding eigenvalue, and shifted so their tails are at the mean.

Depending on the field of application, it is also named the discrete Karhunen–Loève transform (KLT) in signal processing, the Hotelling transform in multivariate quality control, proper orthogonal decomposition (POD) in mechanical engineering, singular value decomposition (SVD) of **X**, eigenvalue decomposition (EVD) of $\mathbf{X}^T\mathbf{X}$ in linear algebra, factor analysis, Eckart–Young theorem, or Schmidt–Mirsky theorem in psychometrics, empirical orthogonal functions (EOF) in meteorological science, empirical eigenfunction decomposition, empirical component analysis, quasiharmonic modes, spectral decomposition in noise and vibration, and empirical modal analysis in structural dynamics.

PCA was invented in 1901 by Karl Pearson, as an analogue of the principal axes theorem in mechanics; it was later independently developed (and named) by Harold Hotelling in the 1930s. The method is mostly used as a tool in exploratory data analysis and for making predictive models. PCA can be done by eigenvalue decomposition of a data covariance (or correlation) matrix or singular value decomposition of a data matrix, usually after mean centering (and normalizing or using Z-scores) the data matrix for each attribute. The results of a PCA are usually discussed in terms of component scores, sometimes called factor scores (the transformed variable values corresponding to a particular data point), and loadings (the weight by which each standardized original variable should be multiplied to get the component score).

PCA is the simplest of the true eigenvector-based multi-variate analyses. Often, its operation can be thought of as revealing the internal structure of the data in a way that best explains the variance in the data. If a multi-variate data-set is visualised as a set of co-ordinates in a high-dimensional data space (1 axis per variable), PCA can supply the user with a lower-dimensional picture, a projection or "shadow" of this object when viewed from its most informative viewpoint. This is done by using only the first few principal components so that the dimensionality of the transformed data is reduced.

PCA is closely related to factor analysis. Factor analysis typically incorporates more domain specific assumptions about the underlying structure and solves eigenvectors of a slightly different matrix.

PCA is also related to canonical correlation analysis (CCA). CCA defines co-ordinate systems that optimally describe the cross-covariance between two data-sets while PCA defines a new orthogonal co-ordinate system that optimally describes variance in a single data-set.

Details

PCA is mathematically defined as an orthogonallinear transformation that transforms the data to a new co-ordinate system such that the greatest variance by some projection of the data comes to lie on the first co-ordinate (called the first principal component), the second greatest variance on the second co-ordinate, and so on.

Consider a data matrix, **X**, with zero empirical mean (the empirical (sample) mean of the distribution has been subtracted from the data-set), where each of the n rows represents a different repetition of the experiment, and each of the p columns gives a particular kind of datum (say, the results from a particular sensor).

Mathematically, the transformation is defined by a set of p-dimensional vectors of weights or *loadings* $w_{(k)} = (w_1, \ldots, w_p)_{(k)}$ that map each row vector $x(i)$ of **X** to a new vector of principal component *scores* $t_{(i)} = (t_1, \ldots, t_p)_{(i)}$, given by

$$t_{k(i)} = x_{(i)} \cdot w_{(k)}$$

in such a way that the individual variables of **t** considered over the data-set successively inherit the maximum possible variance from **x**, with each loading vector **w** constrained to be a unit vector.

First Component

The first loading vector $\mathbf{w}_{(1)}$ thus has to satisfy

$$w_{(1)} = \arg\max_{\|w\|=1} \left\{ \sum_i (t_1)^2_{(i)} \right\} = \arg\max_{\|w\|=1} \sum_i (x_{(i)} \cdot w)^2$$

Equivalently, writing this in matrix form gives

$$w_{(1)} = \arg\max_{\|w\|=1} \left\{ \|Xw\|^2 \right\} = \arg\max_{\|w\|=1} \left\{ w^T X^T X w \right\}$$

Since $w_{(1)}$ has been defined to be a unit vector, it equivalently also satisfies

$$w_{(1)} = \arg\max \left\{ \frac{w^T X^T X w}{w^T w} \right\}$$

The quantity to be maximised can be recognised as a Rayleigh quotient. A standard result for a symmetric matrix such as $X^T X$ is that the quotient's maximum possible value is the largest eigenvalue of the matrix, which occurs when w is the corresponding eigenvector.

With $w_{(1)}$ found, the first component of a data vector $x_{(i)}$ can then be given as a score $t_{1(i)} = x_{(i)} \cdot w_{(1)}$ in the transformed co-ordinates, or as the corresponding vector in the original variables, $\{x_{(i)} \cdot w_{(1)}\}\, w_{(1)}$.

Further Components

The kth component can be found by subtracting the first $k - 1$ principal components from X :

$$\hat{X}_{k-1} = X - \sum_{s=1}^{k-1} X w_{(s)} w_{(s)}^T$$

and then finding the loading vector which extracts the maximum variance from this new data matrix

$$w_{(k)} = \underset{\|w\|=1}{\arg\max} \left\{ \left\| \hat{X}_{k-1} w \right\|^2 \right\} = \arg\max \left\{ \frac{w^T \hat{X}_{k-1}^T \hat{X}_{k-1} w}{w^T w} \right\}$$

It turns out that this gives the remaining eigenvectors of $X^T X$, with the maximum values for the quantity in brackets given by their corresponding eigenvalues.

The kth principal component of a data vector $x_{(i)}$ can therefore be given as a score $t_{k(i)} = x_{(i)} \cdot w_{(k)}$ in the transformed co-ordinates, or as the corresponding vector in the space of the original variables, $\{x_{(i)} \cdot w_{(k)}\}\, w_{(k)}$, where $w_{(k)}$ is the kth eigenvector of $X^T X$.

The full principal components decomposition of **X** can therefore be given as

$$T = XW$$

where W is a p-by-p matrix whose columns are the eigenvectors of $X^T X$

Covariances

$X^T X$ itself can be recognised as proportional to the empirical sample covariance matrix of the data-set X.

The sample covariance Q between two of the different principal components over the data-set is given by :

$$\begin{aligned} Q(PC_{(j)}), PC_{(k)}) &\propto (Xw_{(j)}) \cdot (Xw_{(k)}) \\ &= w_{(j)}^T X^T X w_{(k)} \end{aligned}$$

$$= w_{(j)}^T \lambda_{(k)} w_{(k)}$$

$$= \lambda_{(k)} w_{(j)}^T w_{(k)}$$

where the eigenvalue property of $w_{(k)}$ has been used to move from line 2 to line 3. However eigenvectors $w_{(j)}$ and $w_{(k)}$ corresponding to eigenvalues of a symmetric matrix are orthogonal (if the eigenvalues are different), or can be orthogonalised (if the vectors happen to share an equal repeated value). The product in the final line is therefore zero; there is no sample covariance between different principal components over the data-set.

Another way to characterise the principal components transformation is therefore as the transformation to co-ordinates which diagonalise the empirical sample covariance matrix.

In matrix form, the empirical covariance matrix for the original variables can be written

$$Q \propto X^T X = W \Lambda W^T$$

The empirical covariance matrix between the principal components becomes

$$W^T Q W \alpha W^T W \Lambda W^T W = \Lambda$$

where $\mathbf{\Lambda}$ is the diagonal matrix of eigenvalues $\lambda_{(k)}$ of $X^T X$

($\lambda_{(k)}$ being equal to the sum of the squares over the data-set associated with each component $k : \lambda_{(k)} = \Sigma_i t_{k\ (i)}^2 = \Sigma_i (\mathbf{x}_{(i)} \cdot w_{(k)})^2$)

Dimensionality reduction

The faithful transformation $T = XW$ maps a data vector $\mathbf{x}_{(i)}$ from an original space of p variables to a new space of p variables which are uncorrelated over the data-set. However, not all the principal components need to be kept. Keeping only the first L principal components, produced by using only the first L loading vectors, gives the truncated transformation

$$T_L = XW_L$$

where the matrix T_L now has n rows but only L columns. By construction, of all the transformed data matrices with only L columns, this score matrix maximises the variance in the original data that has been preserved, while minimising the total squared reconstruction error $\| TW^T - T_L W_L^T \|_2^2$ or $\| X - X_L \|_2^2$.

PCA has successfully found linear combinations of the different markers, that separate out different clusters corresponding to different lines of individuals' Y-chromosomal genetic descent.

Such dimensionality reduction can be a very useful step for visualising and processing high-dimensional data-sets, while still retaining as much of the variance in the data-set as possible. For example, selecting $L = 2$ and keeping only the first two principal components finds the two-dimensional plane through the high-dimensional data-set in which the data is most spread out, so if the data contains clusters these too may be most spread out, and therefore most visible to be plotted out in a two-dimensional diagram; whereas if two directions through the data (or two of the original variables) are chosen at random, the clusters may be much less spread apart from each other, and may in fact be much more likely to substantially overlay each other, making them indistinguishable.

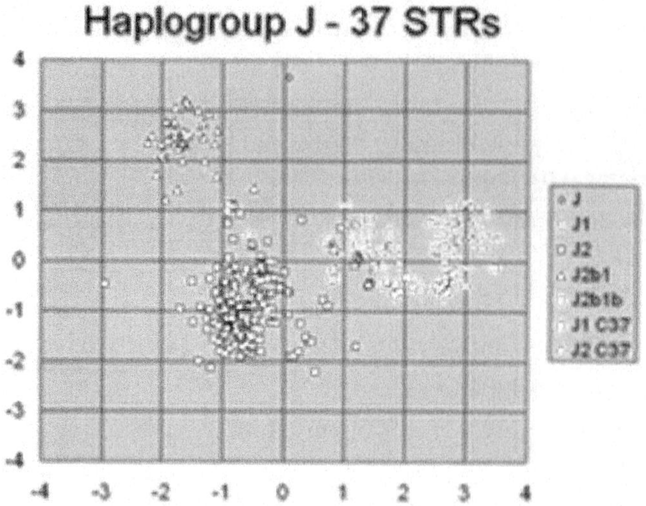

Fig. : A principal components analysis scatterplot of Y-STRhaplotypes calculated from repeat-count values for 37 Y-chromosomal STR markers from 354 individuals.

Similarly, in regression analysis, the larger the number of explanatory variables allowed, the greater is the chance of overfitting the model, producing conclusions that fail to generalise to other data-sets. One approach, especially when there are strong correlations between different possible explanatory variables, is to reduce them to a few principal components and then run the regression against them, a method called principal component regression.

Dimensionality reduction may also be appropriate when the variables in a data-set are noisy. If each column of the data-set contains independent identically distributed Gaussian noise, then the columns of T will also contain similarly identically distributed Gaussian noise (such a distribution is invariant under the effects of the matrix W, which can be thought of as a high-dimensional rotation of the co-ordinate axes). However, with more of the total variance concentrated in the first few principal components compared to the same noise variance, the proportionate effect of the noise is less—the first few components achieve a higher signal-to-noise ratio. PCA thus can have the effect of concentrating much of the signal into the first few principal components, which can usefully be captured by dimensionality reduction; while the later principal components may be dominated by noise, and so disposed of without great loss.

Singular Value Decomposition

The principal components transformation can also be associated with another matrix factorisation, the singular value decomposition (SVD) of X,

$$X = U\Sigma W^T$$

Here $\mathbf{\Sigma}$ is a n-by-p rectangular diagonal matrix of positive numbers $\sigma_{(k)}$, called the singular values of X; U is an n-by-n matrix, the columns of which are orthogonal unit vectors of length n called the left singular vectors of X; and W is a p-by-p whose columns are orthogonal unit vectors of length p and called the right singular vectors of X.

In terms of this factorisation, the matrix X^TX can be written

$$X^TX = W\Sigma U^TU\Sigma W^T$$
$$= W\Sigma^2W^T$$

Comparison with the eigenvector factorisation of X^TX establishes that the right singular vectors W of X are equivalent to the eigenvectors of X^TX, while the singular values $\sigma_{(k)}$ of X are equal to the square roots of the eigenvalues $\lambda_{(k)}$ of X^TX.

Using the singular value decomposition the score matrix **T** can be written

$$T = XW$$
$$= U\Sigma W^TW$$
$$= U\Sigma$$

so each column of T is given by one of the left singular vectors of X multiplied by the corresponding singular value.

Efficient algorithms exist to calculate the SVD of X without having to form the matrix X^TX, so computing the SVD is now the standard way to calculate a principal components analysis from a data matrix, unless only a handful of components are required.

As with the eigen-decomposition, a truncated n-by-L score matrix T_L can be obtained by considering only the first L largest singular values and their singular vectors :

$$T_L = U_L\Sigma_L = XW_L$$

The truncation of a matrix M or T using a truncated singular value decomposition in this way produces a truncated matrix that is the nearest possible matrix of rank L to the original matrix, in the sense of the difference between the two having the smallest possible Frobenius norm, a result known as the Eckart–Young theorem [1936].

Further Considerations

Given a set of points in Euclidean space, the first principal component corresponds to a line that passes through the multi-dimensional mean and minimizes the sum of squares of the distances of the points from the line. The second principal component corresponds to the same concept after all correlation with the first principal component has been subtracted from the points. The singular values (in Σ) are the square roots of the eigenvalues of the matrix X^TX. Each eigenvalue is proportional to the portion of the "variance" (more correctly of the sum of the squared distances of the points from their multi-dimensional mean) that is correlated with each eigenvector. The sum of all the eigenvalues is equal to the sum of the squared distances of the points from their multi-dimensional mean. PCA essentially rotates the set of points around their mean in order to align with the principal components. This moves as much of the variance as possible (using an orthogonal transformation) into the first few dimensions. The values in the remaining dimensions, therefore, tend to be small and may be dropped with minimal loss of information. PCA is often used in this manner for dimensionality reduction. PCA has the distinction of being the optimal orthogonal transformation for keeping the sub-space that has largest "variance". This advantage, however, comes at the price of greater computational requirements if compared, for example and when applicable, to the discrete cosine transform, and in particular to the DCT-II which is simply known as the "DCT". Non-linear dimensionality reduction techniques tend to be more computationally demanding than PCA.

PCA is sensitive to the scaling of the variables. If we have just two variables and they have the same sample variance and are positively correlated, then the PCA will entail a rotation by 45° and the "loadings" for the two variables with respect to the principal component will be equal. But if we multiply all values of the first variable by 100, then the first principal component will be almost the same as that variable, with a small contribution from the other variable, whereas the second component will be almost aligned with the second original variable. This means that whenever the different variables have different units (like temperature and mass), PCA is a somewhat arbitrary method of analysis. (Different results would be obtained if one used Fahrenheit rather than Celsius for example.) Note that Pearson's original paper was entitled "On Lines and Planes of Closest Fit to Systems of Points in Space" — "in space" implies physical Euclidean space where such concerns do not arise. One way of making the PCA less arbitrary is to use variables scaled so as to have unit variance, by standardizing the data and hence use the autocorrelation matrix instead of the autocovariance matrix as a basis for PCA. However, this compresses the fluctuations in all dimensions of the signal space to unit variance.

Mean subtraction (a.k.a. "mean centering") is necessary for performing PCA to ensure that the first principal component describes the direction of maximum variance. If mean subtraction is not performed, the first principal component might instead correspond more or less to the mean of the data. A mean of zero is needed for finding a basis that minimizes the mean square error of the approximation of the data.

PCA is equivalent to empirical orthogonal functions (EOF), a name which is used in meteorology.

An autoencoderneural network with a linear hidden layer is similar to PCA. Upon convergence, the weight vectors of the K neurons in the hidden layer will form a basis for the space spanned by the first K principal components. Unlike PCA, this technique will not necessarily produce orthogonal vectors.

PCA is a popular primary technique in pattern recognition. It is not, however, optimized for class separability. An alternative is the linear discriminant analysis, which does take this into account.

Another application of PCA is reducing the number of parameters in the process of generating computational models of oil reservoirs.

Table of Symbols and Abbreviations

Symbol	*Meaning*	*Dimensions*	*Indices*
$X = \{X[i, j]\}$	data matrix, consisting of the set of all data vectors, one vector per row	$n \times p$	$i = 1 \ldots n$ $j = 1 \ldots p$
n	the number of row vectors in the data-set	1×1	*scalar*
p	the number of elements in each row vector (dimension)	1×1	*scalar*
L	the number of dimensions in the dimensionally reduced sub-space, $1 \le L \le p$	1×1	*scalar*
$u = \{u[j]\}$	vector of empirical means, one mean for each column j of the data matrix	$p \times 1$	$j = 1 \ldots p$
$s = \{s[j]\}$	vector of empirical standard deviations, one standard deviation for each column j of the data matrix	$p \times 1$	$j = 1 \ldots p$
$h = \{h[i]\}$	vector of all 1's	$1 \times n$	$i = 1 \ldots n$
$B = \{B[i, j]\}$	deviations from the mean of each column j of the data matrix	$n \times p$	$i = 1 \ldots n$ $j = 1 \ldots p$
$Z = \{Z[m, n]\}$	z-scores, computed using the mean and standard deviation for each row m of the data matrix	$n \times p$	$i = 1 \ldots n$ $j = 1 \ldots p$
$C = \{C[k, l]\}$	covariance matrix	$p \times p$	$k = 1 \ldots p$ $l = 1 \ldots p$
$R = \{R[k, l]\}$	correlation matrix	$p \times p$	$k = 1 \ldots p$ $l = 1 \ldots p$
$V = \{V[j, k]\}$	matrix consisting of the set of all eigenvectors of C, one eigenvector per column	$p \times p$	$j = 1 \ldots p$ $k = 1 \ldots p$
$D = \{D[k, l]\}$	diagonal matrix consisting of the set of all eigenvalues of C along its principal diagonal, and 0 for all other elements	$p \times p$	$k = 1 \ldots p$ $l = 1 \ldots p$
$W = \{W[j, k]\}$	matrix of basis vectors, one vector per column, where each basis vector is one of the eigenvectors of C, and where the vectors in W are a sub-set of those in V	$p \times L$	$j = 1 \ldots p$ $k = 1 \ldots L$
$T = \{T[i, k]\}$	matrix consisting of n row vectors, where each vector is the projection of the corresponding data vector from matrix X onto the basis vectors contained in the columns of matrix W.	$n \times L$	$i = 1 \ldots n$ $k = 1 \ldots L$

Properties and Limitations of PCA

Properties

Property 1 : For any integer q, $1 \leq q \leq p$, consider the orthogonal linear transformation

$$y = B'x$$

where y is a *q-element* vector and B' is a $(q \times p)$ matrix, and let $\Sigma_y = B'\Sigma B$ be the variance-covariance matrix for y. Then the trace of Σ_y, denoted $tr(\Sigma_y)$, is maximized by taking $B = A_q$, where A_q consists of the first q columns of A(B' is the transposition of B).

Property 2 : Consider again the orthonormal transformation

$$y = B'x$$

with x, B, A and Σ_y defined as before. Then $tr(\Sigma_y)$ is minimized by taking $B = A^*_q$, where A^*_q consists of the last q columns of A.

The statistical implication of this property is that the last few PCs are not simply unstructured left-overs after removing the important PCs. Because these last PCs have variances as small as possible they are useful in their own right. They can help to detect unsuspected near-constant linear relationships between the elements of x, and they may also be useful in regression, in selecting a sub-set of variables from x, and in outlier detection.

Property 3 : (*the Spectral Decomposition of* Σ)

$$\Sigma = \lambda_1\alpha_1\alpha'_1 + \lambda_2\alpha_2\alpha'_2 + \cdots + \lambda_p\alpha_p\alpha'_p$$

Before we look at its usage, we first look at diagonal elements,

$$\text{var}(x_j) = \sum_{k=1}^{P} \lambda_k \alpha_{kj}^2$$

Then, perhaps the main statistical implication of the result is that not only can we decompose the combined variances of all the elements of x into decreasing contributions due to each PC, but we can also decompose the whole covariance matrix into contributions $\lambda_k\alpha_k\alpha'_k$ from each PC. Although not strictly decreasing, the elements of $\lambda_k\alpha_k\alpha'_k$ will tend to become smaller as k increases, as $\lambda_k\alpha_k\alpha'_k$ decreases for increasing k, whereas the elements of α_k tend to stay 'about the same size'because of the normalization constraints : $\alpha'_k\alpha_k = 1, k = 1, 2,..., p$

Limitations

As noted above, the results of PCA depend on the scaling of the variables. A scale-invariant form of PCA has been developed.

The applicability of PCA is limited by certain assumptions made in its derivation.

PCA and Information Theory

The claim that the PCA used for dimensionality reduction preserves most of the information of the data is misleading. Indeed, without any assumption on the signal model, PCA cannot help to reduce the amount of information lost during dimensionality reduction, where information was measured using Shannon entropy.

Under the assumption that

$$x = s + n$$

i.e., that the data vector x is the sum of the desired information-bearing signal s and a noise signal n one can show that PCA can be optimal for dimensionality reduction also from an information-theoretic point-of-view.

In particular, Linsker showed that if s is Gaussian and n is Gaussian noise with a covariance matrix proportional to the identity matrix, the PCA maximizes the mutual information $I(y; s)$ between the desired information s and the dimensionality-reduced output $y = W_L^T x$.

If the noise is still Gaussian and has a covariance matrix proportional to the identity matrix (*i.e.*, the components of the vector n are iid), but the information-bearing signal s is non-Gaussian (which is a common scenario), PCA at least minimizes an upper bound on the *information loss*, which is defined as

$$I(x; s) - I(y; s).$$

The optimality of PCA is also preserved if the noise n is iid and at least more Gaussian (in terms of the Kullback–Leibler divergence) than the information-bearing signal s. In general, even if the above signal model holds, PCA loses its information-theoretic optimality as soon as the noise n becomes dependent.

Computing PCA Using the Covariance Method

The following is a detailed description of PCA using the covariance method. But note that it is better to use the singular value decomposition (using standard software).

The goal is to transform a given data-set X of dimension p to an alternative data-set Y of smaller dimension L. Equivalently, we are seeking to find the matrix Y, where Y is the Karhunen–Loève transform (KLT) of matrix X:

$$Y = KLT\{X\}$$

Organize the Data-set

Suppose you have data comprising a set of observations of p variables, and you want to reduce the data so that each observation can be described with only L variables, $L<p$. Suppose further, that the data are arranged as a set of n data vectors $x_1 \ldots x_n$ with each x_i representing a single grouped observation of the p variables.

- Write $x_1 \ldots x_n$ as row vectors, each of which has p columns.
- Place the row vectors into a single matrix X of dimensions $n \times p$.

Calculate the Empirical Mean

- Find the empirical mean along each dimension $j = 1, ..., p$.
- Place the calculated mean values into an empirical mean vector u of dimensions $p \times 1$.

$$u[j] = \frac{1}{n}\sum_{i=1}^{n} X[i,j]$$

Calculate the Deviations from the Mean

Mean subtraction is an integral part of the solution towards finding a principal component basis that minimizes the mean square error of approximating the data. Hence we proceed by centering the data as follows :

- Subtract the empirical mean vector u from each row of the data matrix X.
- Store mean-subtracted data in the $n \times p$ matrix B.

$$B = X - hu^T$$

where h is an $n \times 1$ column vector of all 1s :

$$h[i] = 1 \qquad \text{for } i = 1,..., n$$

Find the Covariance Matrix

- Find the $p \times p$ empirical covariance matrix C from the outer product of matrix B with itself :

$$C = \frac{1}{n-1} B^* \cdot B$$

where

*is the conjugate transpose operator. Note that if B consists entirely of real numbers, which is the case in many applications, the "conjugate transpose" is the same as the regular transpose.

- Please note that outer products apply to vectors. For tensor cases we should apply tensor products, but the covariance matrix in PCA is a sum of outer products between its sample vectors; indeed, it could be represented as B*.
- The reasoning behind using N-1 instead of N to calculate the covariance is Bessel's correction.

Find the Eigenvectors and Eigenvalues of the Covariance Matrix

- Compute the matrix V of eigenvectors which diagonalizes the covariance matrix **C** :

$$V^{-1}CV = D$$

where D is the diagonal matrix of eigenvalues of C. This step will typically involve the use of a computer-based algorithm for computing eigenvectors and eigenvalues. These algorithms are readily available as sub-components of most matrix algebra systems, such as R, MATLAB, Mathematica, SciPy, IDL (Interactive Data Language), or GNU Octave as well as OpenCV.

- Matrix **D** will take the form of an $M \times M$ diagonal matrix, where

  $D[k, l] = \lambda_k \qquad \text{for } k = l$

  is the jth eigenvalue of the covariance matrix C, and

  $D[k, l] = 0 \qquad \text{for } k \neq l.$
- Matrix V, also of dimension $p \times p$, contains p column vectors, each of length p, which represent the p eigenvectors of the covariance matrix C.
- The eigenvalues and eigenvectors are ordered and paired. The jth eigenvalue corresponds to the jth eigenvector.

Rearrange the Eigenvectors and Eigenvalues

- Sort the columns of the eigenvector matrix V and eigenvalue matrix D in order of *decreasing* eigenvalue.
- Make sure to maintain the correct pairings between the columns in each matrix.

Compute the Cumulative Energy Content for Each Eigenvector

- The eigenvalues represent the distribution of the source data's energy among each of the eigenvectors, where the eigenvectors form a basis for the data. The cumulative energy content g for the jth eigenvector is the sum of the energy content across all of the eigenvalues from 1 through j:

$$g[j] = \sum_{k=1}^{j} D[k,k] \qquad \text{for} \qquad j = 1, \ldots, p$$

Select a Sub-set of the Eigenvectors as Basis Vectors

- Save the first L columns of V as the $p \times L$ matrix W:

  $W[k, l] = V[k, l] \qquad \text{for} \qquad k = 1, \ldots, p \qquad l = 1, \ldots, L$

  where

  $1 \leq L \leq p.$
- Use the vector g as a guide in choosing an appropriate value for L. The goal is to choose a value of L as small as possible while achieving a reasonably high value of g on a percentage basis. For example, you may want to choose L so that the cumulative energy g is above a certain threshold, like 90 per cent. In this case, choose the smallest value of L such that

$$\frac{g[L]}{g[p]} \geq 0.9$$

Convert the Source Data to z-scores (Optional)

- Create an $p \times 1$ empirical standard deviation vector **s** from the square root of each element along the main diagonal of the diagonalized covariance matrix

C. (Note, that scaling operations do not commute with the KLT thus we must scale by the variances of the already-decorrelated vector, which is the diagonal of C) :

$$s = \{s[j]\} = \left\{\sqrt{C[j,j]}\right\} \quad \text{for } j = 1,\ldots,p$$

- Calculate the $n \times p$z-score matrix :

$$Z = \frac{B}{h \cdot s^T} \text{ (divide element-by-element)}$$

- *Note :* While this step is useful for various applications as it normalizes the data-set with respect to its variance, it is not integral part of PCA/KLT.

Project the z-scores of the Data onto the New Basis

- The projected vectors are the columns of the matrix $T = Z \cdot W = KLT\{X\}$.
- The rows of matrix T represent the Karhunen–Loeve transforms (KLT) of the data vectors in the rows of matrix X.

Derivation of PCA Using the Covariance Method

Let X be a d-dimensional random vector expressed as column vector. Without loss of generality, assume X has zero mean.

We want to find (*) a$d \times d$ orthonormal transformation matrix P so that PX has a diagonal covariant matrix (*i.e.* PX is a random vector with all its distinct components pairwise uncorrelated).

A quick computation assuming P were unitary yields :

$$\begin{aligned} \text{var}(PX) &= E[PX(PX)^\dagger] \\ &= E[PX\,X^\dagger P^\dagger] \\ &= P\,E[X\,X^\dagger]P^\dagger \\ &= P\,\text{var}(X)P^{-1} \end{aligned}$$

Hence (*) holds if and only if var(X) were diagonalisable by P.

This is very constructive, as var(X) is guaranteed to be a non-negative definite matrix and thus is guaranteed to be diagonalisable by some unitary matrix.

Iterative Computation

In practical implementations especially with high dimensional data (large p), the covariance method is rarely used because it is not efficient. One way to compute the first principal component efficiently is shown in the following pseudo-code, for a data matrix X with zero mean, without ever computing its covariance matrix

r = a random vector of length p

doc times :

$s = 0$ (a vector of length p)

for each row $x \in X$

$s = s + (x \cdot r)x$

$$r = \frac{s}{|s|}$$

return r

This algorithm is simply an efficient way of calculating $X^T X\, r$, normalizing, and placing the result back in r (power iteration). It avoids the np^2 operations of calculating the covariance matrix. r will typically get close to the first principal component of X within a small number of iterations, c. (The magnitude of s will be larger after each iteration. Convergence can be detected when it increases by an amount too small for the precision of the machine.)

Subsequent principal components can be computed by subtracting component r from X and then repeating this algorithm to find the next principal component. However this simple approach is not numerically stable if more than a small number of principal components are required, because imprecisions in the calculations will additively affect the estimates of subsequent principal components. More advanced methods build on this basic idea, as with the closely related Lanczos algorithm.

One way to compute the eigenvalue that corresponds with each principal component is to measure the difference in mean-squared-distance between the rows and the centroid, before and after subtracting out the principal component. The eigenvalue that corresponds with the component that was removed is equal to this difference.

The NIPALS Method

For very-high-dimensional data-sets, such as those generated in the *omics sciences (*e.g.*, genomics, metabolomics) it is usually only necessary to compute the first few PCs. The non-linear iterative partial least squares (NIPALS) algorithm calculates t_1 and w_1^T from X. The outer product, $t_1 w_1^T$ can then be subtracted from X leaving the residual matrix $\mathbf{E}_1$. This can be then used to calculate subsequent PCs. This results in a dramatic reduction in computational time since calculation of the covariance matrix is avoided.

However, for large data matrices, or matrices that have a high degree of column collinearity, NIPALS suffers from loss of orthogonality due to machine precision limitations accumulated in each iteration step. A Gram–Schmidt (GS) re-orthogonalization algorithm is applied to both the scores and the loadings at each iteration step to eliminate this loss of orthogonality.

Online/Sequential Estimation

In an "online" or "streaming" situation with data arriving piece by piece rather than being stored in a single batch, it is useful to make an estimate of the PCA projection that can be updated sequentially. This can be done efficiently, but requires different algorithms.

Applications

Neuroscience

A variant of principal components analysis is used in neuroscience to identify the specific properties of a stimulus that increase a neuron's probability of generating an action potential. This technique is known as spike-triggered covariance analysis. In a typical application an experimenter presents a white noise process as a stimulus (usually either as a sensory input to a test subject, or as a current injected directly into the neuron) and records a train of action potentials, or spikes, produced by the neuron as a result. Presumably, certain features of the stimulus make the neuron more likely to spike. In order to extract these features, the experimenter calculates the covariance matrix of the *spike-triggered ensemble*, the set of all stimuli (defined and discretized over a finite time window, typically on the order of 100 ms) that immediately preceded a spike. The eigenvectors of the difference between the spike-triggered covariance matrix and the covariance matrix of the *prior stimulus ensemble* (the set of all stimuli, defined over the same length time window) then indicate the directions in the space of stimuli along which the variance of the spike-triggered ensemble differed the most from that of the prior stimulus ensemble. Specifically, the eigenvectors with the largest positive eigenvalues correspond to the directions along which the variance of the spike-triggered ensemble showed the largest positive change compared to the variance of the prior. Since these were the directions in which varying the stimulus led to a spike, they are often good approximations of the sought after relevant stimulus features.

In neuroscience, PCA is also used to discern the identity of a neuron from the shape of its action potential. Spike sorting is an important procedure because extracellular recording techniques often pick up signals from more than one neuron. In spike sorting, one first uses PCA to reduce the dimensionality of the space of action potential waveforms, and then performs clustering analysis to associate specific action potentials with individual neurons.

Relation Between PCA and K-means Clustering

It has been shown recently (2001,2004) that the relaxed solution of K-means clustering, specified by the cluster indicators, is given by the PCA principal components, and the PCA sub-space spanned by the principal directions is identical to the cluster centroid sub-space specified by the between-class scatter matrix. Thus PCA automatically projects to the sub-space where the global solution of K-means clustering lies, and thus facilitates K-means clustering to find near-optimal solutions.

Relation Between PCA and Factor Analysis

Principal components creates variables that are linear combinations of the original variables. The new variables have the property that the variables are all orthogonal. The principal components can be used to find clusters in a set of data. PCA is a variance-focused approach seeking to reproduce the total variable variance, in which components reflect both common and unique variance of the variable. PCA is generally preferred for purposes of data reduction (*i.e.*, translating variable space into optimal factor space) but not when detect the latent construct or factors.

Factor analysis is similar to principal component analysis, in that factor analysis also involves linear combinations of variables. Different from PCA, factor analysis is a correlation-focused approach seeking to reproduce the inter-correlations among variables, in which the factors "represent the common variance of variables, excluding unique variance". Factor analysis is generally used when the research purpose is detecting data structure (*i.e.*, latent constructs or factors) or causal modelling.

Correspondence Analysis

Correspondence analysis (CA) was developed by Jean-Paul Benzécri and is conceptually similar to PCA, but scales the data (which should be non-negative) so that rows and columns are treated equivalently. It is traditionally applied to contingency tables. CA decomposes the chi-squared statistic associated to this table into orthogonal factors. Because CA is a descriptive technique, it can be applied to tables for which the chi-squared statistic is appropriate or not. Several variants of CA are available including detrended correspondence analysis and canonical correspondence analysis. One special extension is multiple correspondence analysis, which may be seen as the counterpart of principal component analysis for categorical data.

Generalizations

Non-linear Generalizations

Linear PCA versus non-linear Principal Manifolds for visualization of breast cancermicroarray data : (a) Configuration of nodes and 2D Principal Surface in the 3D PCA linear manifold. The data-set is curved and cannot be mapped adequately on a 2D principal plane; (b) The distribution in the internal 2D non-linear principal surface co-ordinates (ELMap2D) together with an estimation of the density of points; (c) The same as b), but for the linear 2D PCA manifold (PCA2D). The "basal" breast cancer sub-type is visualized more adequately with ELMap2D and some features of the distribution become better resolved in comparison to PCA2D. Principal manifolds are produced by the elastic maps algorithm. Data are available for public competition. Software is available for free non-commercial use.

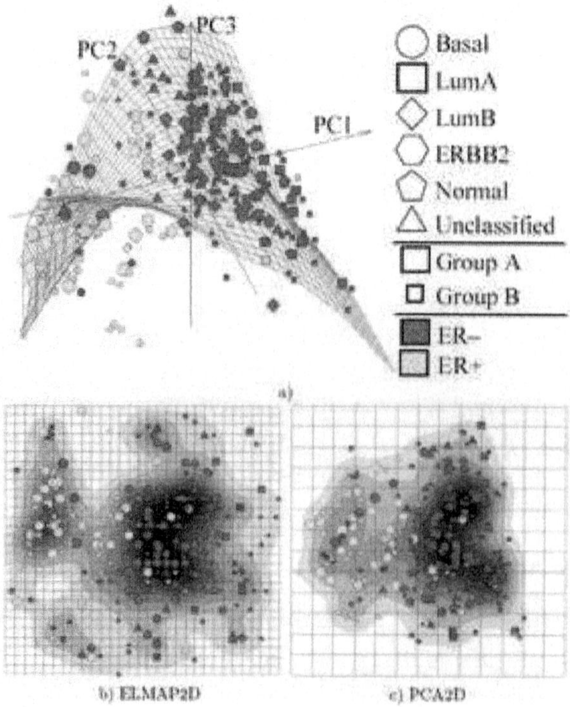

Most of the modern methods for non-linear dimensionality reduction find their theoretical and algorithmic roots in PCA or K-means. Pearson's original idea was to take a straight line (or plane) which will be "the best fit" to a set of data points. **Principal curves and manifolds** give the natural geometric framework for PCA generalization and extend the geometric interpretation of PCA by explicitly constructing an embedded manifold for data approximation, and by encoding using standard geometric projection onto the manifold. Another popular generalization is kernel PCA, which corresponds to PCA performed in a reproducing kernel Hilbert space associated with a positive definite kernel.

Multi-linear Generalizations

In multi-linear sub-space learning, PCA is generalized to multi-linear PCA (MPCA) that extracts features directly from tensor representations. MPCA is solved by performing PCA in each mode of the tensor iteratively. MPCA has been applied to face recognition, gait recognition, etc. MPCA is further extended to uncorrelated MPCA, non-negative MPCA and robust MPCA.

Higher Order

N-way principal component analysis may be performed with models such as Tucker decomposition, PARAFAC, multiple factor analysis, co-inertia analysis, STATIS, and DISTATIS.

Robustness-weighted PCA

While PCA finds the mathematically optimal method (as in minimizing the squared error), it is sensitive to outliers in the data that produce large errors PCA tries to avoid. It therefore is common practice to remove outliers before computing PCA. However, in some contexts, outliers can be difficult to identify. For example, in data-mining algorithms like correlation clustering, the assignment of points to clusters and outliers is not known beforehand. A recently proposed generalization of PCA based on a **weighted PCA** increases robustness by assigning different weights to data objects based on their estimated relevancy.

Software/Source Code

- Mathematica—Implements principal component analysis with the PrincipalComponents command using both covariance and correlation methods.
- NAG Library—Principal components analysis is implemented via the g03aa routine (available in both the Fortran and the C versions of the Library).
- SIMCA—Commercial software package available to perform PCA analysis.
- MATLAB Statistics Toolbox—The functions princomp and pca (R2012b) give the principal components, while the function pcares gives the residuals and reconstructed matrix for a low—rank PCA approximation. An example MATLAB implementation of PCA is available.
- GNU Octave—Free software computational environment mostly compatible with MATLAB, the function princomp gives the principal component.
- R—Free statistical package, the functions princomp and prcomp can be used for principal component analysis; prcomp uses singular value decomposition which generally gives better numerical accuracy. Some packages that implement PCA in R, include, but are not limited to : ade4, vegan, ExPosition, and FactoMineR
- SAS, PROC FACTOR—Offers principal components analysis.
- MLPACK—Provides an implementation of principal component analysis in C++.
- XLMiner—The principal components tab can be used for principal component analysis.
- Stata—The pca command provides principal components analysis.
- Cornell Spectrum Imager—Open—source toolset built on ImageJ, enables PCA analysis for 3D datacubes.
- ImDEV—Free Excel addon to calculate principal components using R package.
- *ViSta* : The Visual Statistics System—Free software that provides principal components analysis, simple and multiple correspondence analysis.

- Spectramap—Software to create a biplot using principal components analysis, correspondence analysis or spectral map analysis.
- FinMath—.NET numerical library containing an implementation of PCA.
- The Unscrambler—Multi-variate analysis software enabling Principal Component Analysis (PCA) with PCA Projection.
- OpenCV
- NMath—Proprietary numerical library containing PCA for the NET Framework.
- IDL—The principal components can be calculated using the function pcomp.
- Weka—Computes principal components.
- Qlucore—Commercial software for analyzing multi-variate data with instant response using PCA.
- Orange (software)—Supports PCA through its Linear Projection widget.
- EIGENSOFT—Provides a version of PCA adapted for population genetics analysis.
- Partek Genomics Suite—Statistical software able to perform PCA.
- libpca C++ library—Offers PCA and corresponding transformations.
- Origin—Contains PCA in its Pro version.
- Scikit-learn—Python library for machine learning which contains PCA, Probabilistic PCA, Kernel PCA, Sparse PCA and other techniques in the decomposition module.

T-DISTRIBUTED STOCHASTIC NEIGHBOUR EMBEDDING

t-Distributed Stochastic Neighbour Embedding (t-SNE) is a machine learning algorithm for dimensionality reduction developed by Laurens van der Maaten and Geoffrey Hinton. It is a non-linear dimensionality reduction technique that is particularly well suited for embedding high-dimensional data into a space of two or three dimensions, which can then be visualized in a scatter plot. Specifically, it models each high-dimensional object by a two-or three-dimensional point in such a way that similar objects are modelled by nearby points and dissimilar objects are modelled by distant points.

The t-SNE algorithms comprises two main stages. First, t-SNE constructs a probability distribution over pairs of high-dimensional objects in such a way that similar objects have a high probability of being picked, whilst dissimilar points have an infinitesimal probability of being picked. Second, t-SNE defines a similar probability distribution over the points in the low-dimensional map, and it minimizes the Kullback–Leibler divergence between the two distributions with respect to the locations of the points in the map.

t-SNE has been used in a wide range of applications, including computer security research, music analysis, cancer research and bio-informatics.

Details

Given a set of N high-dimensional objects $x_1, \ldots, x_N$, t-SNE first computes probabilities p_{ij} that are proportional to the similarity of objects x_i and x_j, as follows :

$$p_{j|i} = \frac{\exp\left(-\left\|x_i - x_j\right\|^2 / 2\sigma_i^2\right)}{\sum_{k \neq i} \exp\left(-\left\|x_i - x_k\right\|^2 / 2\sigma_i^2\right)},$$

$$p_{ij} = \frac{p_{j|i} + p_{i|j}}{2N}$$

The bandwidth of the Gaussian kernels σ_i, is set in such a way that the perplexity of the conditional distribution equals a predefined perplexity using a binary search. As a result, the bandwidth is adapted to the density of the data : smaller values of σ_i are used in denser parts of the data space.

t-SNE aims to learn a d-dimensional map $y_1, \ldots, y_N$ (with $y_i \in R^d$) that reflects the similarities p_{ij} as well as possible. To this end, it measures similarities q_{ij} between two points in the map y_i and y_j, using a very similar approach. Specifically, q_{ij} is defined as :

$$q_{ij} = \frac{\left(1 + \left\|y_i - y_j\right\|^2\right)^{-1}}{\sum_{k \neq l}\left(1 + \left\|y_k - y_l\right\|^2\right)^{-1}}$$

Herein a heavy-tailed Student-t distribution is used to measure similarities between low-dimensional points in order to allow dissimilar objects to be modelled far apart in the map CITATION.

The locations of the points y_i in the map are determined by minimizing the Kullback–Leibler divergence between the two distributions P and Q :

$$KL(P\,||\,Q) = \sum_{i \neq j} p_{ij} \log \frac{p_{ij}}{q_{ij}}$$

The minimization of the Kullback–Leibler divergence with respect to the points y_i is performed using gradient descent. The result of this optimization is a map that reflects the similarities between the high-dimensional inputs well.

Data-mining Methods Make Sense out of Millions of Seemingly Random Data Points

Data-mining methods have many origins, including drawing on insights into learning as it naturally occurs in humans (cognitive science), and advances in computer science and algorithm design on how to best detect patterns in unstructured data. Although traditional statistical methods for analyzing data, based on statistical theories and models, are now widely accepted throughout various industries, data-mining methods have only been widely embraced in business for a decade or two. However, their effectiveness for root cause analysis, and for modelling, optimizing and improving complex processes, are making data-mining increasingly popular—and even necessary—in many real-world discrete manufacturing, batch manufacturing, and continuous-process applications.

There is no single, generally agreed-upon definition of data-mining. As a practical matter, whenever data describing a process are available, in manufacturing for example, then any systematic review of those data to identify useful patterns, correlations, trends, and so forth, could be called "data-mining." Put simply, data-mining uncovers nuggets of information from a sometimes vast repository of data describing the process of interest.

Hypothesis Testing vs. Pattern Recognition

Typical statistical methods used in Six Sigma projects often involve analysis of variance (ANOVA), multiple regression, quality control charting, process capability analysis, and so forth. All of these methods are derived from statistical theory and involve hypothesis testing. For example, in designed experiments (DOE), factors are systematically varied to test the hypothesis that a specific factor or combination of factors has a statistically significant effect on the outcome variable or process of interest.

To summarize, traditional statistical data analysis methods involve hypothesis testing to identify statistically significant factors, points, deviations etc., or the testing of expectations about the process under consideration.

Data-mining methods fundamentally do not rely on or test any *a priori* expectations. Instead, the data-mining process can best be described as pattern recognition : A typically large amount of data are available to describe a process or specific problem, and the goal is to detect in the data the relevant patterns that allow us to improve the process. For example, instead of considering and varying systematically a fixed number of factors in a manufacturing process as you would in a designed experiment, one can extract all available historical data describing the process to identify reliable patterns and factor combinations associated with a problem or optimal process performance.

Following are some examples from different industries and domains where the data-mining approach and methods have become increasingly popular and successful.

Semiconductor Industry

The manufacture of semiconductors involves hundreds of process steps through various semiconductor processing tools. Any problems in a single tool or combination of tools can lead to serious and costly yield problems.

A typical data-set describing the successive processing steps applied to wafer lots will record the specific machine or tool that was used, the time when it was used, and the specific processing that was applied. Each lot goes through hundreds of steps, so when a yield problem occurs, there are almost an infinite number of potential root causes, *i.e.*, individual tools or interactions between specific tools that could be responsible for the decrease in yield.

A first step is to "sift" through all tools to identify those that appear to discriminate between low-yield and high-yield lots. Put another way, the question is which specific tools ("features") or processing steps are common to low-yield vs. high-yield lots.

The graph seen in figure, below, summarizes such a feature-selection graph. The longer the bar for a particular tool shown in this plot, the greater the likelihood that the respective tool is related to the yield problem. That is, the tools associated with the longest bars are the most likely root causes.

Figure 1: Feature-Selection Bar Graph

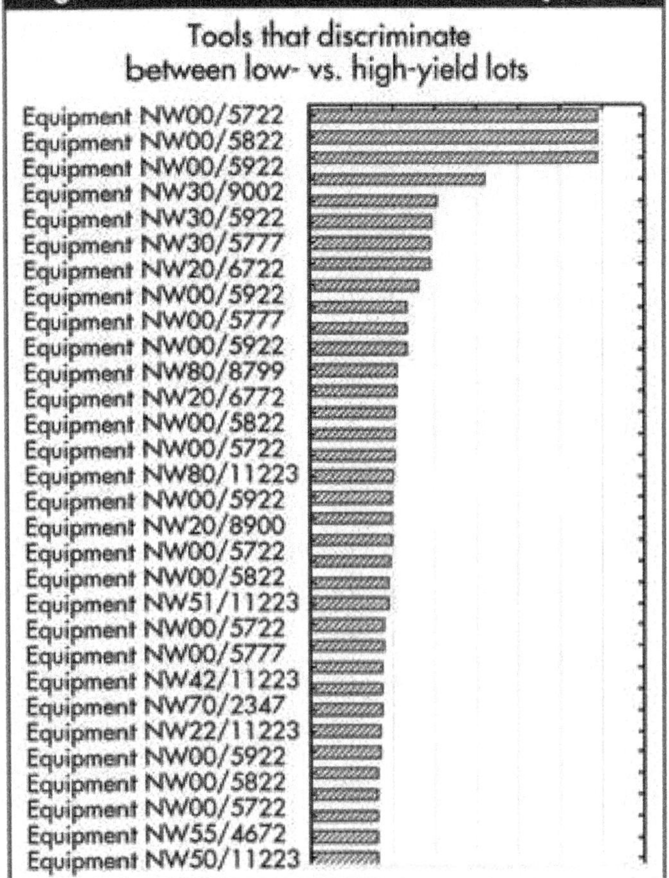

This graph is very similar to the common Pareto chart, except that instead of identifying the frequency of known failures attributable to a particular tool, part, or processing step, the data-mining algorithms are used to identify the specific tools that appear to provide the best discrimination between high-and low-yield lots.

Of course, problems often cannot be traced to a single tool but are rather caused by the combination or interaction between different factors or tools. Now the possible list of combinations of tools to consider becomes extremely large and impossible to review without the help of data-mining algorithms.

Recursive Partitioning

One class of data-mining algorithms that is particularly well-suited to identify interactions between inputs (tools, predictors, etc.) to a process that cause undesirable outputs are the so-called recursive-partitioning or tree algorithms. There are a number of such algorithms commonly implemented in software for data-mining. All algorithms will partition the data, based on the available inputs, into sub-groups of observations that are increasingly homogeneous with respect to the outcome, *e.g.*, into sub-groups that are all of high yield or low yield. In effect, these algorithms will create decision trees that will identify and represent interactions between inputs in the data.

The illustration in figure below shows a simple tree where each split produces two child nodes. The bars inside each node (box) show the proportions of lots that are classified as high yield vs. low yield; at the root node there is approximately an equal number of each, representing the entire data-set for this study. As you traverse down the tree, you can see that a relatively greater number of low-yield cases are partitioned on the left side of the tree (the initial split).

Figure 2: Tree Graph

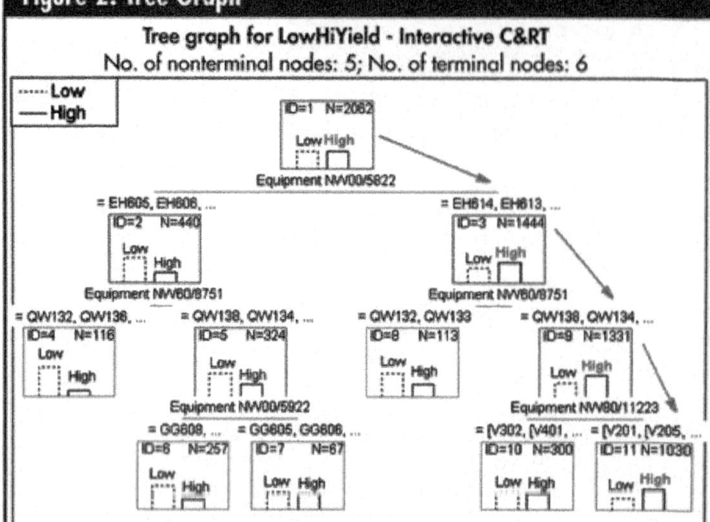

For example, if Equipment NW00/5822 applied the process EH605 or EH606 and Equipment NW60/8751 applied process QW132 or QW136, then many more lots were in the low-yield category than the high-yield category. On the other hand, if you follow the right-hand branch in the illustration down to the final node ID=11, then the respective split ("decision") criteria associated with each split identify the combinations of tool steps ("commonalities") associated with a greater chance of high yield. In node partition ID=11 there are clearly more high-yield lots than low-yield lots. Hence that branch of the tree identifies an interaction between these three pieces of equipment.

These data-mining algorithms are, however, extremely useful for identifying and exploring complex factor interactions in complex high-dimensional data. They are therefore very useful for root cause analysis when all simpler methods have been exhausted.

General Applicability

Most automated manufacturing involves the application of a large number of processing steps to move from raw materials to final product. Because data collection and storage has become very inexpensive during the last few years, comprehensive historical data typically exists describing exactly the steps that were applied to each batch, part, lot, or machine. Data-mining methods for feature selection, as described above, and recursive partitioning are generally very useful for finding the needle in the haystack, that is, identifying the specific tool or processing step, or interaction between two or more steps or tools, responsible for quality problems.

Optimize Product Quality

In short, to control vibration problems during final testing, the manufacture of gas turbines requires that very tight tolerances be carefully observed. In this case, the goal was not only to identify specific root causes and problems during the complex manufacturing process, which could lead to problems in final product testing, but also to build comprehensive data-mining prediction models that would provide a clear line-of-site from the individual manufacturing steps to final product quality.

Caterpillar engineers used data-mining modelling algorithms to build accurate predictive models of key performance and quality parameters, based on manufacturing data. Such prediction models—accurately describing the relationship between the inputs to the manufacturing process to the quality outcomes of the process—can then be used in computer simulation and optimization to identify specific opportunities for improving final product quality. For example, Caterpillar was able to reduce the occurrence of trim balance problems by nearly 45 per cent using these data-mining methods.

Chemical and Pharmaceutical Manufacturing

The manufacture of chemicals or pharmaceuticals often involves monitoring continuous processes of raw materials batches through a lengthy and complex manufacturing process. Multiple process parameters are repeatedly measured over time as the raw materials slowly mature, combine, react, ferment, etc. into the final product.

It is obviously desirable to detect any problems in the maturation process well before final product testing. At that point, repairs, in the discrete manufacturing sense, are usually not possible, and an entire batch of vaccine must be discarded at great cost. The problem, however, is that the maturation process is reflected in the process data as gradual and non-linear changes over time, often simultaneously in different parameters. Therefore, standard control-charting methods are not suitable for quality control because parameters are expected to be correlated and drift over time, so no fixed control or process capability limits can be set.

Tracking "Distance from Model"

An increasingly popular and successful method of implementing effective process monitoring methods for continuous and batch processes is based on a fundamentally simple approach : First, using successful product runs or batches and the data collected during their manufacture, build accurate data-mining prediction models that summarize the relationship between the process parameters and maturation (*e.g.*, time). Then use this prediction model of time (or a proxy-measure of time) for process monitoring by charting deviations or distances of the predictions from the actual observed maturation.

Figure 3: Trajectory of Parameter 5 and 6

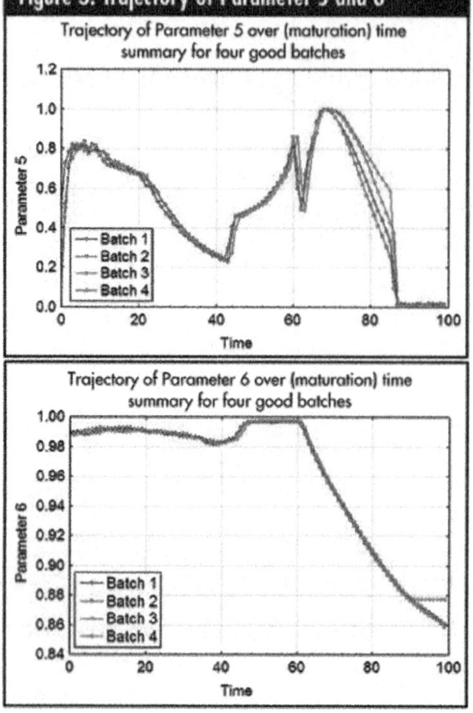

The following example is based on a data-set describing an industrial batch polymerization reactor. The batch duration is approximately two hours, divided into 100 equally spaced time intervals each. At each time interval, 10 process variables were measured.

For example, the graphs in figure show the trajectories over time for Parameter 5 and Parameter 6, for four good batches.

Obviously, standard control charting is not applicable here. In addition, the parameter trajectories are correlated, *i.e.*, the maturation process happens simultaneously for all parameters, each moving according to its own trajectory.

Using time as a proxy for batch maturity over time, we can use data-mining methods to build a prediction model of maturity (time), that is, a model that, given the values of all the parameters, predicts at which stage of maturation the product should be. If that prediction is "bad," *i.e.*, if the actual time at which the measurements were taken is entirely different than that predicted according to the data-mining model, then probably something is wrong with the batch, and its maturation is not compatible with that observed in the successful batches from which the model was built.

A simple control chart to implement this type of process-monitoring method would be to chart the deviations of the predictions from the observed maturation (*i.e.*, time at which the measurements were taken). If a batch of product is compatible with the maturation process of the successful batches from which the models were built, then the prediction residuals (predicted *vs.* residual time) will vary around a mean of zero, and standard X-and-MR charts or cumulative sum (CUSUM) charts can be used to detect when the process is shifting or drifting.

Continuous Processes : Power Generation

The methods described here are applicable to practically all continuous-or batch-process manufacturing, when detailed data are collected describing the process. Another good example where data-mining methods have been shown to be very successful is in the area of furnace and power plant optimization. Power generation from coal-fired furnaces requires the co-ordination of numerous complex subsystems, from coal preparation, through combustion, to the complex treatment of exhaust gases and waste materials to minimize or eliminate harmful emissions. It is not uncommon for a single unit in a power generating station to collect several thousand parameters every minute into a detailed data history, and often years of historical data are available describing all operations.

A study provides a good illustration of how data-mining models can be used to optimize simultaneously the combustion processes and emission control systems for significantly improved overall process performance.

Figure 4: NOX and Flame Temperature

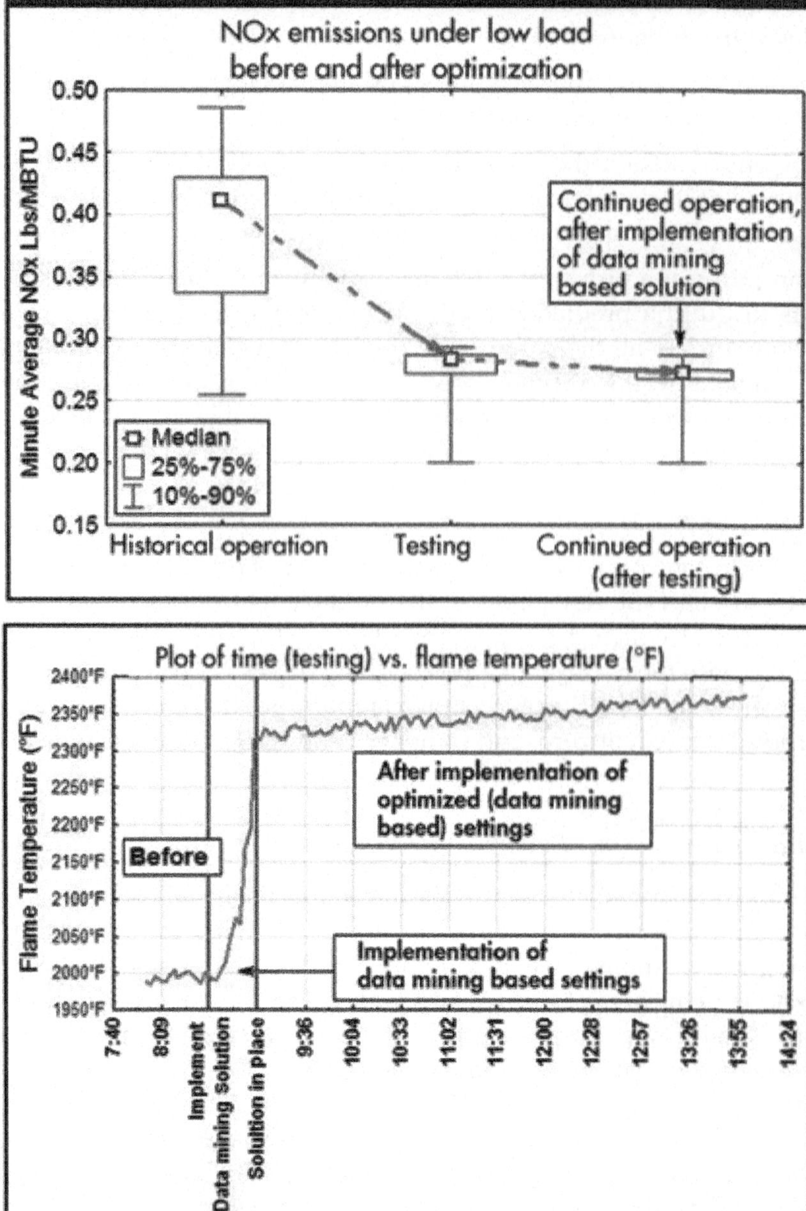

Whenever a complex continuous process is well instrumented, providing detailed data of all parameters controlling the process, data-mining methods can provide effective tools for optimization and control.

Chapter 10

Business Intelligence

Business intelligence (BI) is a set of theories, methodologies, architectures, and technologies that transform raw data into meaningful and useful information for business purposes. BI can handle enormous amounts of unstructured data to help identify, develop and otherwise create new opportunities. BI, in simple words, makes interpreting voluminous data friendly. Making use of new opportunities and implementing an effective strategy can provide a competitive market advantage and long-term stability.

Generally, Business Intelligence is made up of an increasing number of components, these are :

- Multi-dimensional aggregation and allocation
- *Denormalization*, tagging and standardization
- Real-time reporting with analytical alert
- Interface with *unstructured data* source
- Group consolidation, budgeting and rolling forecast
- *Statistical inference* and probabilistic simulation
- Key performance indicators optimization
- Version control and process management
- Open item management.

BI technologies provide historical, current and predictive views of business operations. Common functions of business intelligence technologies are *reporting, online analytical processing, analytics, data-mining, process mining, complex event processing, business performance management, benchmarking, text mining, predictive analytics* and *prescriptive analytics*.

Though the term business intelligence is sometimes a synonym for *competitive intelligence* (because they both support *decision-making*), BI uses technologies, processes, and applications to analyze mostly internal, structured data and business processes while competitive intelligence gathers, analyzes and disseminates information with a topical focus on company competitors. If understood broadly, business intelligence can include the sub-set of competitive intelligence.

HISTORY

The term Business Intelligence was originally first phrased by Richard Millar Devens' in the 'Cyclopædia of Commercial and Business Anecdotes' from 1865. Devens used the term to describe how the banker Sir Henry Furnese, gained profit by receiving and acting upon information about his environment, prior to his competitors. "*Throughout Holland, Flanders, France, and Germany, he maintained a complete and perfect train of business intelligence. The news of the many battles fought was thus received first by him, and the fall of Namur added to his profits, owing to his early receipt of the news*". The ability to collect and react accordingly based on the information retrieved, an ability that Furnese excelled in, is today still at the very heart of BI.

In a 1958 article, IBM researcher Hans Peter Luhn used the term business intelligence. He employed the Webster's dictionary definition of intelligence : "the ability to apprehend the interrelationships of presented facts in such a way as to guide action towards a desired goal."

Business intelligence as it is understood today is said to have evolved from the decision support systems (DSS) that began in the 1960s and developed throughout the mid-1980s. DSS originated in the computer-aided models created to assist with decision-making and planning. From DSS, data warehouses, Executive Information Systems, OLAP and business intelligence came into focus beginning in the late 80s.

In 1988, an Italian-Dutch-French-English consortium organized an international meeting on the Multiway Data Analysis in Rome. The ultimate goal is to reduce the multiple dimensions down to one or two (by detecting the patterns within the data) that can then be presented to human decision-makers.

In 1989, Howard Dresner (later a Gartner Group analyst) proposed "business intelligence" as an umbrella term to describe "concepts and methods to improve business decision-making by using fact-based support systems." It was not until the late 1990s that this usage was widespread.

BUSINESS INTELLIGENCE AND DATA WAREHOUSING

Often BI applications use data gathered from a data warehouse or a data mart. A data warehouse is a copy of analytical data that facilitates decision support. However, not all data warehouses are used for business intelligence, nor do all business intelligence applications require a data warehouse.

To distinguish between the concepts of business intelligence and data warehouses, Forrester Research often defines business intelligence in one of two ways :

Using a broad definition : "Business Intelligence is a set of methodologies, processes, architectures, and technologies that transform raw data into meaningful and useful information used to enable more effective strategic, tactical, and operational insights and decision-making." When using this definition, business intelligence also includes technologies such as data integration, data quality, data warehousing, master data management, text and content analytics, and many others that the market sometimes lumps into the Information Management segment. Therefore, Forrester refers to *data preparation* and *data usage* as two separate, but closely linked segments of the business intelligence architectural stack.

Forrester defines the latter, narrower business intelligence market as, "...referring to just the top layers of the BI architectural stack such as reporting, analytics and dashboards."

BUSINESS INTELLIGENCE AND BUSINESS ANALYTICS

Thomas Davenport argues that business intelligence should be divided into querying, reporting, OLAP, an "alerts" tool, and business analytics. In this definition, business analytics is the sub-set of BI based on statistics, prediction, and optimization.

APPLICATIONS IN AN ENTERPRISE

Business intelligence can be applied to the following business purposes, in order to drive business value.

1. Measurement-program that creates a hierarchy of performance metrics and benchmarking that informs business leaders about progress towards business goals (business process management).
2. Analytics-program that builds quantitative processes for a business to arrive at optimal decisions and to perform business knowledge discovery. Frequently involves : data-mining, process mining, statistical analysis, predictive analytics, predictive modelling, business process modelling, complex event processing and prescriptive analytics.
3. Reporting/enterprise reporting-program that builds infrastructure for strategic reporting to serve the strategic management of a business, not operational reporting. Frequently involves data visualization, executive information system and OLAP.
4. Collaboration/collaboration platform-program that gets different areas (both inside and outside the business) to work together through data sharing and electronic data interchange.
5. Knowledge management-program to make the company data driven through strategies and practices to identify, create, represent, distribute, and enable adoption of insights and experiences that are true business knowledge. Knowledge management leads to learning management and regulatory compliance.

In addition to the above, business intelligence can provide a pro-active approach, such as alert functionality that immediately notifies the end-user if certain conditions are met. For example, if some business metric exceeds a pre-defined threshold, the metric will be highlighted in standard reports, and the business analyst may be alerted via email or another monitoring service. This end-to-end process requires data governance, which should be handled by the expert.

PRIORITIZATION OF BUSINESS INTELLIGENCE PROJECTS

It is often difficult to provide a positive business case for business intelligence initiatives and often the projects must be prioritized through strategic initiatives. Here are some hints and advantages to increase the benefits for a BI project.

- As described by Kimball you must determine the tangible benefits such as eliminated cost of producing legacy reports.
- Enforce access to data for the entire organization. In this way even a small benefit, such as a few minutes saved, makes a difference when multiplied by the number of employees in the entire organization.
- As described by Ross, Weil & Roberson for Enterprise Architecture, consider letting the BI project be driven by other business initiatives with excellent business cases. To support this approach, the organization must have enterprise architects who can identify suitable business projects.
- Use a structured and quantitative methodology to create defensible prioritization in line with the actual needs of the organization, such as a weighted decision matrix.

SUCCESS FACTORS OF IMPLEMENTATION

Before implementing a BI solution, it is worth taking different factors into consideration before proceeding. According to Kimball *et. al.*, these are the three critical areas that you need to assess within your organization before getting ready to do a BI project :

1. The level of commitment and sponsorship of the project from senior management
2. The level of business need for creating a BI implementation
3. The amount and quality of business data available.

Business Sponsorship

The commitment and sponsorship of senior management is according to Kimball *et. al.*, the most important criteria for assessment. This is because having strong management backing helps overcome shortcomings elsewhere in the project. However, as Kimball *et. al.* state : "even the most elegantly designed DW/BI system cannot overcome a lack of business [management] sponsorship".

It is important that personnel who participate in the project have a vision and an idea of the benefits and drawbacks of implementing a BI system. The best business sponsor should have organizational clout and should be well connected within the organization. It is ideal that the business sponsor is demanding but also able to be realistic and supportive if the implementation runs into delays or drawbacks. The management sponsor also needs to be able to assume accountability and to take responsibility for failures and setbacks on the project. Support from multiple members of the management ensures the project does not fail if one person leaves the steering group. However, having many managers work together on the project can also mean that there are several different interests that attempt to pull the project in different directions, such as if different departments want to put more emphasis on their usage. This issue can be countered by an early and specific analysis of the business areas that benefit the most from the implementation. All stakeholders in project should participate in this analysis in order for them to feel ownership of the project and to find common ground.

Another management problem that should be encountered before start of implementation is if the business sponsor is overly aggressive. If the management individual gets carried away by the possibilities of using BI and starts wanting the DW or BI implementation to include several different sets of data that were not included in the original planning phase. However, since extra implementations of extra data may add many months to the original plan, it's wise to make sure the person from management is aware of his actions.

Business Needs

Because of the close relationship with senior management, another critical thing that must be assessed before the project begins is whether or not there is a business need and whether there is a clear business benefit by doing the implementation. The needs and benefits of the implementation are sometimes driven by competition and the need to gain an advantage in the market. Another reason for a business-driven approach to implementation of BI is the acquisition of other organizations that enlarge the original organization it can sometimes be beneficial to implement DW or BI in order to create more oversight.

Companies that implement BI are often large, multi-national organizations with diverse subsidiaries. A well-designed BI solution provides a consolidated view of key business data not available anywhere else in the organization, giving management visibility and control over measures that otherwise would not exist.

Amount and Quality of Available Data

Without good data, it does not matter how good the management sponsorship or business-driven motivation is. Without proper data, or with too little quality data, any BI implementation fails. Before implementation it is a good idea to do data profiling. This analysis identifies the "content, consistency and structure [..]" of the data. This should be done as early as possible in the process and if the analysis shows that data is lacking, put the project on the shelf temporarily while the IT department figures out how to properly collect data.

When planning for business data and business intelligence requirements, it is always advisable to consider specific scenarios that apply to a particular organization, and then select the business intelligence features best suited for the scenario.

Often, scenarios revolve around distinct business processes, each built on one or more data sources. These sources are used by features that present that data as information to knowledge workers, who subsequently act on that information. The business needs of the organization for each business process adopted correspond to the essential steps of business intelligence. These essential steps of business intelligence include but are not limited to :

1. Go through business data sources in order to collect needed data
2. Convert business data to information and present appropriately
3. Query and analyze data
4. Act on those data collected.

The **quality aspect** in business intelligence should cover all the process from the source data to the final reporting. At each step, the **quality gates** are different :

1. *Source Data :*
   - *Data Standardization :* Make data comparable (same unit, same pattern..)
   - *Master Data Management :* Unique referential
2. *Operational Data Store (ODS) :*
   - *Data Cleansing :* Detect & correct inaccurate data
   - *Data Profiling :* Check inappropriate value, null/empty
3. *Datawarehouse :*
   - *Completeness :* Check that all expected data are loaded
   - *Referential Integrity :* Unique and existing referential over all sources
   - *Consistency Between Sources :* Check consolidated data vs sources
4. *Reporting :*
   - *Uniqueness of Indicators :* Only one share dictionary of indicators
   - *Formula Accuracy :* Local reporting formula should be avoided or checked.

User Aspect

Some considerations must be made in order to successfully integrate the usage of business intelligence systems in a company. Ultimately the BI system must be accepted and utilized by the users in order for it to add value to the organization. If the usability of the system is poor, the users may become frustrated and spend a considerable amount of time figuring out how to use the system or may not be able to really use the system. If the system does not add value to the users´ mission, they simply don't use it.

To increase user acceptance of a BI system, it can be advisable to consult business users at an early stage of the DW/BI lifecycle, for example at the requirements gathering phase. This can provide an insight into the business process and what the users need from the BI system. There are several methods for gathering this information, such as questionnaires and interview sessions.

When gathering the requirements from the business users, the local IT department should also be consulted in order to determine to which degree it is possible to fulfill the business's needs based on the available data.

Taking on a user-centered approach throughout the design and development stage may further increase the chance of rapid user adoption of the BI system.

Besides focusing on the user experience offered by the BI applications, it may also possibly motivate the users to utilize the system by adding an element of competition. Kimball suggests implementing a function on the Business Intelligence portal website where reports on system usage can be found. By doing so, managers can see how well their departments are doing and compare themselves to others and this may spur them to encourage their staff to utilize the BI system even more.

In a 2007 article, H. J. Watson gives an example of how the competitive element can act as an incentive. Watson describes how a large call centre implemented performance dashboards for all call agents, with monthly incentive bonuses tied to performance metrics. Also, agents could compare their performance to other team members. The implementation of this type of performance measurement and competition significantly improved agent performance.

BI chances of success can be improved by involving senior management to help make BI a part of the organizational culture, and by providing the users with necessary tools, training, and support. Training encourages more people to use the BI application.

Providing user support is necessary to maintain the BI system and resolve user problems. User support can be incorporated in many ways, for example by creating a website. The website should contain great content and tools for finding the necessary information. Furthermore, helpdesk support can be used. The help desk can be manned by power users or the DW/BI project team.

BI Portals

A **Business Intelligence portal** (BI portal) is the primary access interface for Data Warehouse (DW) and Business Intelligence (BI) applications. The BI portal is the users first impression of the DW/BI system. It is typically a browser application, from which the user has access to all the individual services of the DW/BI system, reports and other analytical functionality. The BI portal must be implemented in such a way that it is easy for the users of the DW/BI application to call on the functionality of the application.

The BI portal's main functionality is to provide a navigation system of the DW/BI application. This means that the portal has to be implemented in a way that the user has access to all the functions of the DW/BI application.

The most common way to design the portal is to custom fit it to the business processes of the organization for which the DW/BI application is designed, in that way the portal can best fit the needs and requirements of its users.

The BI portal needs to be easy to use and understand, and if possible have a look and feel similar to other applications or web content of the organization the DW/BI application is designed for (consistency).

The following is a list of desirable features for web portals in general and BI portals in particular :

Usable

User should easily find what they need in the BI tool.

Content Rich

The portal is not just a report printing tool, it should contain more functionality such as advice, help, support information and documentation.

Clean

The portal should be designed so it is easily understandable and not over complex as to confuse the users.

Current

The portal should be updated regularly.

Interactive

The portal should be implemented in a way that makes it easy for the user to use its functionality and encourage them to use the portal. Scalability and customization give the user the means to fit the portal to each user.

Value Oriented

It is important that the user has the feeling that the DW/BI application is a valuable resource that is worth working on.

Marketplace

There are a number of business intelligence vendors, often categorized into the remaining independent "pure-play" vendors and consolidated "megavendors" that have entered the market through a recent trend of acquisitions in the BI industry.

Some companies adopting BI software decide to pick and choose from different product offerings (best-of-breed) rather than purchase one comprehensive integrated solution.

Industry-specific

Specific considerations for business intelligence systems have to be taken in some sectors such as governmental banking regulations. The information collected by banking institutions and analyzed with BI software must be protected from some groups or individuals, while being fully available to other groups or individuals. Therefore BI solutions must be sensitive to those needs and be flexible enough to adapt to new regulations and changes to existing law.

Semi-structured or Unstructured Data

Businesses create a huge amount of valuable information in the form of e-mails, memos, notes from call-centers, news, user groups, chats, reports, web-pages, presentations, image-files, video-files, and marketing material and news. According to Merrill Lynch, more than 85% of all business information exists in these forms. These information types are called either *semi-structured* or *unstructured* data. However, organizations often only use these documents once.

The management of semi-structured data is recognized as a major unsolved problem in the information technology industry. According to projections from Gartner, white collar workers spend anywhere from 30 to 40 per cent of their time searching, finding and assessing unstructured data. BI uses both structured and unstructured data, but the former is easy to search, and the latter contains a large quantity of the information needed for analysis and decision-making. Because of the difficulty of properly searching, finding and assessing unstructured or semi-structured data, organizations may not draw upon these vast reservoirs of information, which could influence a particular decision, task or project. This can ultimately lead to poorly informed decision-making.

Therefore, when designing a business intelligence/DW-solution, the specific problems associated with semi-structured and unstructured data must be accommodated for as well as those for the structured data.

Unstructured Data *vs.* Semi-structured Data

Unstructured and semi-structured data have different meanings depending on their context. In the context of relational database systems, unstructured data cannot be stored in predictably ordered columns and rows. One type of unstructured data is typically stored in a BLOB (binary large object), a catch-all data type available in most relational database management systems. Unstructured data may also refer to irregularly or randomly repeated column patterns that vary from row to row within each file or document.

Many of these data types, however, like e-mails, word processing text files, PPTs, image-files, and video-files conform to a standard that offers the possibility of metadata. Metadata can include information such as author and time of creation, and this can be stored in a relational database. Therefore it may be more accurate to talk about this as semi-structured documents or data, but no specific consensus seems to have been reached.

Unstructured data can also simply be the knowledge that business users have about future business trends. Business forecasting naturally aligns with the BI system because business users think of their business in aggregate terms. Capturing the business knowledge that may only exist in the minds of business users provides some of the most important data points for a complete BI solution.

Problems with Semi-structured or Unstructured Data

There are several challenges to developing BI with semi-structured data. According to Inmon & Nesavich, some of those are :

1. Physically accessing unstructured textual data-unstructured data is stored in a huge variety of formats.
2. Terminology—Among researchers and analysts, there is a need to develop a standardized terminology.
3. Volume of data—As stated earlier, up to 85% of all data exists as semi-structured data. Couple that with the need for word-to-word and semantic analysis.
4. Searchability of unstructured textual data—A simple search on some data, *e.g.* apple, results in links where there is a reference to that precise search term. gives an example : "a search is made on the term felony. In a simple search, the term felony is used, and everywhere there is a reference to felony, a hit to an unstructured document is made. But a simple search is crude. It does not find references to crime, arson, murder, embezzlement, vehicular homicide, and such, even though these crimes are types of felonies."

The use of Metadata

To solve problems with searchability and assessment of data, it is necessary to know something about the content. This can be done by adding context through the use of metadata. Many systems already capture some metadata (*e.g.* file name, author, size, etc.), but more useful would be metadata about the actual content *e.g.* summaries, topics, people or companies mentioned. Two technologies designed for generating metadata about content are automatic categorization and information extraction.

FUTURE

A 2009 Gartner paper predicted these developments in the business intelligence market :

- Because of lack of information, processes, and tools, through 2012, more than 35 per cent of the top 5,000 global companies regularly fail to make insightful decisions about significant changes in their business and markets.
- By 2012, business units will control at least 40 per cent of the total budget for business intelligence.
- By 2012, one-third of analytic applications applied to business processes will be delivered through coarse-grained application mashups.

A 2009 *Information Management* special report predicted the top BI trends : "green computing, social networking services, data visualization, mobile BI, predictive analytics, composite applications, cloud computing and multitouch."

Other business intelligence trends include the following :

- Third party SOA-BI products increasingly address ETL issues of volume and throughput.

- Companies embrace in-memory processing, 64-bit processing, and pre-packaged analytic BI applications.
- Operational applications have callable BI components, with improvements in response time, scaling, and concurrency.
- Near or real time BI analytics is a baseline expectation.
- Open source BI software replaces vendor offerings.

Other lines of research include the combined study of business intelligence and uncertain data. In this context, the data used is not assumed to be precise, accurate and complete. Instead, data is considered uncertain and therefore this uncertainty is propagated to the results produced by BI.

According to a study by the Aberdeen Group, there has been increasing interest in Software-as-a-Service (SaaS) business intelligence over the past years, with twice as many organizations using this deployment approach as one year ago-15% in 2009 compared to 7% in 2008.

An article by InfoWorld's Chris Kanaracus points out similar growth data from research firm IDC, which predicts the SaaS BI market will grow 22 per cent each year through 2013 thanks to increased product sophistication, strained IT budgets, and other factors.

DATA-MINING ANALYTICS FOR BUSINESS INTELLIGENCE AND DECISION SUPPORT

Knowledge discovery and data-mining (KDD) techniques are used for analyzing and discovering actionable insights from data. Today, data-mining is being increasingly used and embedded in vertical solutions for business intelligence and decision support, such as platforms for data management, e-commerce and critical large-scale solutions (*e.g.* CRM).

Background

The use of database technology to manage large volumes of business data is pervasive and growing at an explosive rate. The traditional use of database systems—to support applications that employ query-based report generation—continues to be the primary target of the technology. However, the size and volume of data being managed raises new and interesting issues : Can we utilize methods wherein the data can help businesses achieve competitive advantage? Can the data be used to model underlying business processes? Can we gain insights from the data to help improve business processes? These are the goals of business intelligence (BI) systems, and data-mining is the set of embeddable (in BI systems) analytic methods that provide the capabilities to explore, summarize and model the data.

Before applying these methods to data, the data typically has to be organized into history repositories, known as data warehouses. Data warehousing may require integration of multiple sources of data, which may involve dealing with multiple formats, multiple database systems, distributed databases, cleaning the data, and creating unified logical view of the underlying non-homogeneous data.

Online analytics processing (OLAP) is an extension of the structured query language (SQL) framework to accommodate queries that would otherwise have been computationally impossible on a relational database management system. This is achieved by utilizing and storing pre-computed aggregates (*e.g.* credit-card sales in a certain geographic region over a certain time period) that are automatically updated as the underlying data changes. Deciding upon which aggregates to pre-compute is determined by the business end-user. Providing technical capabilities for automatic computation and updating of aggregates is the strength of OLAP analytics.

Data-mining analytics try to go beyond OLAP by providing abilities for discovering insights that are computer driven and not end-user driven. Data size is increasing at a rate far exceeding any rates that end-users can cope with. Providing solutions when end-users cannot reasonably supply all possible aggregates to pre-compute, or when it is not possible to express an insight as a pre-computed aggregate, is the goal of data-mining analytics.

A collection of early foundational papers on data-mining appear in "Advances in Knowledge Discovery and Data-mining", and a recent special issue of the Communications of the ACM contains a collection of papers that discuss current accomplishments and trends.

Data-Mining

Data-mining may be viewed as automated search procedures for discovering credible and actionable insights from large volumes of high-dimensional data. Often, the process emphases symbolic learning and modelling methods (*i.e.*, techniques that produce interpretable results) and data management methods (for providing scalable techniques for large data volumes). Data-mining employs techniques from statistics, pattern recognition and machine learning. Many of these methods are also frequently used in vision, speech recognition, image processing, handwriting recognition and natural language understanding. However, the issues of scalability and automated business intelligence solutions differentiate data-mining from the other applications of machine learning and statistical modelling.

Typical business intelligence applications of data-mining include :

- ***Risk analysis*** (given a set of current customers and their finance/insurance history data, build a predictive model that can be used to classify a new customer into a risk category);
- ***Targeted marketing*** (given a set of current customers and history on their purchases and their responses to promotions, target new promotions to those most likely to respond);
- ***Customer retention*** (given a set of past customers and their behaviour prior to leaving, predict who is most likely to leave and take proactive action); and

- *Fraud detection* (detect fraudulent activities either proactively or on-line real-time).

Many other new application domains are surfacing as we continue to explore and expand upon new data-mining opportunities.

There is more to building a successful business intelligence solution than just data-mining. The process of identifying valid, novel, potentially useful and understandable patterns in data may require selecting or sampling data from a data warehouse, cleaning or pre-processing it, transforming or reducing it, applying a data-mining component to extract models or patterns, and evaluating the derived structure. Data-mining is a key component in this methodology that is concerned with the algorithmic means by which structures are extracted from data while meeting computational efficiency constraints.

Key algorithm families used in data-mining include :

- *Predictive modelling* (predict a specific attribute [database field] based upon the other attributes [fields] in the data);
- *Clustering* (also known as segmentation, which groups data records into sub-sets where items in sub-sets are more similar to each other than to items in other sub-sets);
- *Frequent patterns* (find interesting similarities between a few attributes in sub-sets of the data);
- *Change and deviation* (detect and account for interesting sequence of information in data records); and
- *Dependencies* (generate the joint probability density function that might have generated the data).

Predictive modelling may be defined as the estimation of a function f that maps points from an input space X to an output space Y given only a finite sampling of the mapping. Typically this translates into predicting the value of a field (Y) in a database based on the other fields (X) in the database. Predictive modelling algorithms are designed to accurately construct an estimator f′ of f from a typically finite sample of the data known as the training set. Training data is assumed to be potentially corrupted (*i.e.*, noisy), and appropriate handling schemes need to be built into the modelling algorithm. If the predicted quantity is numeric, then the prediction problem is that of regression modelling. If the predicted quantity is discrete, then the prediction problem is that of classification modelling. In business intelligence applications, the more generalized form of predictive modelling—probabilistic modelling—is also frequently used since many decision-support systems can work better with ranked (by probability) predictions, since they are more easily amenable to optimization and constraint satisfaction analytics.

A number of interesting technical issues are addressed in the design of predictive modelling algorithms. Transformations on the input space X to improve estimation capability are performed by feature extraction, construction and selection methods. Evaluating the estimate *f′* in terms of how well it performs on data not present in the training set permits the maximization of prediction accuracy by avoiding under-fitting or over-fitting. Trading off model complexity versus model accuracy is addressed by methods such as bias-variance tradeoff, penalized likelihood, minimum message length (MML) and minimum description length (MDL) encoding.

Classification modelling enables the prediction of the most likely state of a categorical variable (the class) given the values of other variables. This can be viewed as a density estimation problem; *i.e.*, deriving the value of Y given X from the joint density on Y and X. There are several approaches to building classification modelling algorithms, including : kernel density estimators, metric space-based methods (*e.g.* k-nearest neighbour) and projection into decision regions (*i.e.*, dividing attribute space into decision regions and associating a prediction with each region). Projection methods are by far the most practical for data-mining, and these include linear classifiers, neural networks, decision trees and disjunctive normal form (DNF) rule-based classifiers.

Regression modelling enables the prediction of the most likely value of a numerical variable (the target column) given the values of other variables. This can be viewed as a numerical function approximation problem; *i.e.*, deriving the value of Y given X from the joint probability distribution on Y and X. There are several approaches to regression modelling algorithms, including : statistical probability models (*e.g.* linear regression), projection into decision regions (dividing attribute space into decision regions and associating a constant value with each region) and hybrid methods (coupling projection methods with statistical methods). Projection and hybrid methods are by far the most practical data-mining methods, and these include neural networks, decision trees and disjunctive normal form (DNF) rule-based classifiers.

Predictive modelling is the most frequently used data-mining technique for building business intelligence and decision-support solutions. Essentially, it provides a robust and automated mechanism for building forecasting systems in data rich environments. There are three major steps in this data-mining process. Historical data is first mined to train patterns/models for predicting future behaviour. These behaviors can include typical business goals such as predicting response to direct mail, defects in manufactured parts, declining activity, credit risk, delinquency, likelihood to buy specific products, profitability, etc. These patterns/models are then used to score new transactions to determine their likelihood to exhibit the modeled behaviour. These scores are then used to act upon for optimizing a business objective.

Clustering is a data-mining technique (also known as segmentation) in which a finite sampling of points are grouped into sets of similar points. Points with common characteristics are essentially "clustered." While predictive modelling requires that the target class (or value) membership be known in the training data, in clustering this knowledge is not known *a priori*, and is potentially being discovered by the clustering or segmentation process.

Typical clustering techniques involve a two-stage approach : an outer loop to determine the optimal cluster number k, and an inner loop to fit data points to clusters. Several different algorithmic techniques are available for fitting data points to clusters. Metric distance-based methods find best k-way partitions so that points in a partition are closer to each other than to points in other partitions. Model-based methods hypothesize a best-fit (very typically probabilistic) model for each cluster.

Frequent patterns data-mining extracts compact patterns that describe subsets of data. These patterns could be either row-wise or column-wise. A widely popular column-wise pattern detection technique is the "association" rules data-mining technique. Associations mining detects combinations of attribute values that occur with a minimum level of frequency (support) and certainty (confidence). Data-mining specific associations algorithms provide scalable capability to find all such patterns in linear time under certain conditions of data sparseness. It should be pointed out that these patterns are not strictly statements about causal effects amongst attributes, but can still provide useful insights in existing large volumes of high dimensional data.

Change and deviation techniques can be used in detecting sequence information in data, where the sequential ordering could be either temporal in nature or some other ordering. The ordering information in the transactions is utilized for computing, under certain conditions of data sparseness, sequences with desired levels of frequency and certainty.

Dependency modelling techniques can be used for detecting causal structure within data. These causal models can be either in the form of probabilistic distributions governing the data, or functional dependencies between attributes in the data. Techniques for discovering causal structures include density estimation methods (expectation maximization) and explicit causal modelling methods.

Data-mining techniques can be potentially applied to such key business areas as profitability, customer relationships and business process efficiency. Identifying the most likely potential customers for a particular product, discovering the most effective market segments for a business, determining how to increase market share of products, reducing costs without impacting production, and optimizing inventories are all typical instances of recent applications of data-mining.

Predictive modelling can be used in assigning risk levels to new insurance and financial contracts. Clustering/segmentation can be used in identifying distinct market groups in customer populations. Frequent patterns can be used for market basket analysis (what gets shopped together in a supermarket). Change and deviation methods can be used for fraud discovery in health claim data and discovering shopping patterns over time. There appear to be numerous business application opportunities in diverse domains, including retail/distribution, health care, manufacturing, utilities, telecommunications, transportation, government, cross industry and financial services. Comprehensive textbooks on data-mining algorithms and analytics include "Data-mining : Concepts and Techniques" and "Principles of Data-mining".

Current Status and Future Directions

Data-mining is currently considered an enabler for business intelligence systems. Data-mining algorithm suites are available as software packages, some loosely coupled with database technology. Building a data-mining application usually begins with a heavy emphasis on data warehousing followed by exploratory data-mining. The analysis and application building is typically conducted by consultants or in-house analytic teams. The key challenges to the successful completion of a data-mining project are the data warehousing requirements, and the sophisticated analytics requirements.

To address these challenges, key research trends in data-mining include systems research, for enabling transparent and pervasive usage; algorithms research, for providing scalable, optimized, and robust mining; and solutions research, for embedding data-mining into vertically integrated applications.

The key goal in systems research is to enable transparent usage of data-mining in environments where data typically resides and is being managed. This involves :

- ***Building database extenders*** (*e.g.* user-defined functions for model training/ scoring and sufficient statistics like histograms, counts, samples, etc.);
- ***Parallel and distributed data-mining*** (for supporting scalability via parallelization and inbuilt sampling);
- ***XML based APIs for database coupling and application embedding*** (to enable interoperability and training/scoring in different environments); and
- ***Intelligent or semi-automated data warehousing for mining*** (by providing industry specific templates and meta-data-mining).

The key goal in algorithms research is to enable robust and automated data-mining, thereby making it easier for non-experts to conduct and run data-mining applications. This involves :

- Building better techniques for automated evaluation metrics;
- Automated feature extraction/transformation/selection
- Discovering relational and hierarchical structures amongst attributes;
- Incorporating prior knowledge to account for costs/benefits/uncertainty/ missing values;
- Incremental and online mining;
- Privacy-preserving data-mining; and
- Heterogeneous data-mining.

The key goal in solutions research is to develop solution specific data-mining components that are optimized to the application at hand and can be embedded into a vertically integrated application. Some key application areas that are driving this research include business processes such as risk management, targeted marketing and portfolio management; systems processes such as computer and network performance management; and Internet processes such as site profiling and performance tuning, and user personalization.

We see two areas in which data-mining and operations research (and optimization techniques) will begin to intersect and interact more frequently as the data-mining technology matures. While data-mining can assist in the automated discovery of actionable insights from data, the efficient execution of the actions can only be effected by coupling the output of data-mining with optimization methods. Very often the actionable insights need to be acted upon, taking into account business constraints such as budgets and schedules. This can be effectively done only by applying optimization techniques to the outputs of data-mining.

A classic example is the mail stream optimization (or horizontal marketing) solution. A more recent example of a solution that integrates predictive modelling and optimization techniques appears in the *International Journal of Production Reasearch*. As the size of available data becomes large, we expect traditional statistical modelling techniques to be replaced by data-mining techniques for building more effective solutions.

A second area where the two disciplines will start interacting more frequently and productively is in the actual design of the data-mining algorithms. Many data-mining algorithms rely upon heuristic search techniques that are trying to optimize some objective function (*e.g.* minimizing predictive error on hold-out data). It is natural that optimization methods can contribute here to designing more effective data-mining algorithms.

In conclusion, given the current research and solution development directions, we see two key trends emerging. First, data-mining will join mathematical programming and optimization as a key scientific technology for building decision support systems. Second, using data-mining should eventually become as easy and pervasive as working with databases and spreadsheets is today.

PERFORMANCE METRIC

A **performance metric** is that which determines an organization's behaviour and performance. Performance metrics measure of an organization's activities and performance. It should support a range of *stakeholder* needs from *customers, shareholders* to *employees*. While traditionally many metrics are finance based, inwardly focusing on the performance of the organization, metrics may also focus on the performance against customer requirements and *value*. In *project management*, performance metrics are used to assess the health of the project and consist of the measuring of seven criteria : safety, time, cost, resources, scope, quality, and actions. In *call centres*, performance metrics help capture internal performance and can include productivity measurements and the quality of service provided by the *customer service advisor*. These metrics can include : Calls Answered, Calls Abandoned, Average Handle Time and Average Wait Time.

Developing performance metrics usually follows a process of :

1. Establishing critical processes/customer requirements
2. Identifying specific, quantifiable outputs of work
3. Establishing *targets* against which results can be scored.

A criticism of performance metrics is that when the value of information is computed using mathematical methods, it shows that even performance metrics professionals choose measures that have little value. This is referred to as the "measurement inversion". For example, metrics seem to emphasize what organizations find immediately measurable—even if those are low value—and tend to ignore high value measurements simply because they seem harder to measure (whether they are or not).

To correct for the measurement inversion other methods, like *applied information economics*, introduce the "value of information analysis" step in the process so that metrics focus on high-value measures. Organizations where this has been applied find that they define completely different metrics than they otherwise would have and, often, fewer metrics. For projects, the effort to collect a metric has to be weighed against its value as projects are temporary endeavours performed with finite resources.

There are a variety of ways in which organizations may react to results. This may be to trigger specific activity relating to *performance* (*i.e.*, an improvement plan) or to use the data merely for statistical information. Often closely tied in with outputs, performance metrics should usually encourage improvement, effectiveness and appropriate levels of control.

Performance metrics are often linked in with corporate *strategy* and are often derived in order to measure performance against a *critical success factor*.

METRICS REFERENCE MODEL

The **Metrics Reference Model (MRM)** is the reference model created by the Consortium for Advanced Management-International (CAM-I) to be a single reference library of performance metrics. This library is useful for accelerating to development of and improving the content of any organization's business intelligence solution.

Background

The MRM was created by the Business Intelligence Working Group of CAM-I. Substantial secondary research performed by the group revealed that there was no single, central location for performance metrics information. All published material focused on a single dimension of performance metrics (*e.g.*, financial), or focused on how to use such metrics to affect the performance of an organization. Exhaustive lists of performance metrics that were generic to nearly any organization across all aspects of the business (*e.g.*, customer, product/service, employee) were not found. Furthermore, without a single reference point, no research was located where the relationships between these performance metrics had been documented and described.

Overview

MRM is a technology agnostic reference model based on the basic elements of a generic organization for which cost, process, and performance improvements may be desired.

In keeping with CAM-I's generic model of an organization, these elements include :

- Customers (or business sustaining forces)
- Products and services
- Processes and activities
- Employees
- Finance
- Research and development
- Infrastructure
- Resources
- Suppliers.

The figure below illustrates the flow between these components in a typical organization :

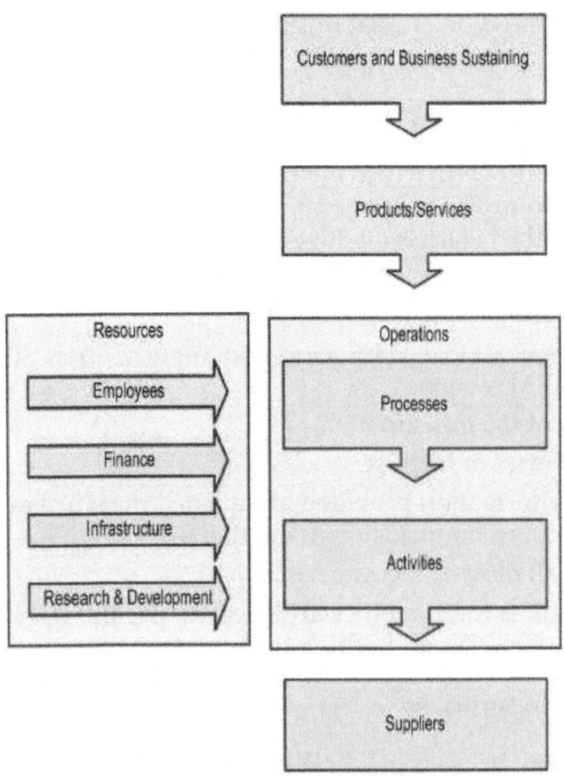

Fig. : CAM-I's Generic Model of an Organization.

For each of the organizational elements, the MRM provides a comprehensive list of cost, process, and performance metrics of possible interest to many organizations. For example, some of the performance metrics provided for the Employee organizational element include Average Customer Review, Customer Retention Rate, Sales per Customer/Spending Rate, Average Quantity per Customer Order, Order Frequency, Average Number of Orders per Customer, and Average Customer Lifetime Value.

For each performance metric, this information is also provided :

- *Type* : Provides further levels of suborganization within the organizational element where sensible (*e.g.*, Financial).
- *Definition/Calculation* : A generally accepted means of calculating the metric from measures is provided.
- *Strengths/Opportunities* : Known advantages for using the metric.
- *Weaknesses/Problems/Risks* : Known disadvantages for using the metric.
- *Target Setting* : A discussion of the desired outcome.
- *Possible Correlations/Market Basket of Metrics* : Consideration of other metrics and correlations of environmental factors are provided. This area explores the relationship of the measure against every other metric in the MRM, an indication of the strength of the relationship (*e.g.*, weak, moderate, strong), and an indication to the type of relationship (*e.g.*, negatively correlated, positively correlated).

Value and Use of the MRM

The benefit of the MRM is that each metric has already been identified, defined, and considered by a group of experienced business intelligence professionals. The idea of the MRM is for the business intelligence (BI) practitioner to follow these steps :

- Assess the requirements of the BI solution and determine which organization components apply.
- For each organizational component that applies, open this document to the associated MRM section.
- Review each of the measures.
- Identify measures of interest.
- Review the information provided about each measure and either accept or reject the measure for inclusion in the BI tool.
- Accumulate all desired measures in a list.
- Use the measures (or, if metrics are selected, use the measures in the Definition/Calculation column) to drive the BI data model.

Using the MRM in Industry

How the MRM may be used practically in the industry depends on whether an individual or an organization is considered.

Individual interest in the MRM :

- *Program analysts :* Use the MRM to tailor to a specific program/organization.
- *Managers :* Use the MRM model to validate that the metrics selected represent the needs of their business.
- *Strategic planners :* Use the MRM to determine metric relationships and correlations in order to determine strategic decisions.
- *Operational planners :* Use the MRM to understand operational data points that you have control over to affect the metrics and understand the components of the metrics.
- *BI practitioners :* Use the MRM to define the BI requirements and calculations, as well as the interpretation of the data. Help the business owner determine the appropriate display of metrics.
- *Vendors :* Use the MRM to provide relevant products/services that answer clients' needs out-of-the-box.
- *Educators :* Use the MRM to discuss the benefits and drawbacks of the various measures.

Organizational interest in the MRM :

- New organizations/agencies, mergers & acquisitions
- Existing organizations that are starting performance measurement efforts
- Consulting organizations for use in client situations
- Software vendors
- Monitoring organizations
- Those with performance challenges (may have metrics, but not enough to interpret that metric)
- Those who do not wish to go through a large strategic planning exercise but want to start using best practice metrics for common functions
- Small budgets or time constraints or those that do not have the expertise to create metrics

Use Cases

The MRM to be used in a number of different ways, including :

- *Use as part of planning process :* Rather than starting from a blank slate, go through each and add those that are specific to their organization.* Use to generate ideas. Customize existing, add specific to your organization.
- *Use as a cookbook :* Select ingredients (metrics) of interest from each area when there are known objectives but want to save time.
- Use to refine existing program (rare), compare metrics vs. existing (completeness review).
- *BI practitioner :* Use for developing the calculations, finding the data needed to run the calculations, and building appropriate data structures.

- For consulting organizations to provide more thorough results quickly (competitive advantage).
- For software vendors to provide specific applications.
- Education tool (understanding strengths/weaknesses of metrics, etc.) for staff at all levels to understand what's important.

PROCESS MINING

Process mining is a process management technique that allows for the analysis of business processes based on event logs. The basic idea is to extract knowledge from event logs recorded by an information system. Process mining aims at improving this by providing techniques and tools for discovering process, control, data, organizational, and social structures from event logs.

Overview

Process mining techniques are often used when no formal description of the process can be obtained by other approaches, or when the quality of an existing documentation is questionable. For example, the audit trails of a workflow management system, the transaction logs of an enterprise resource planning system, and the electronic patient records in a hospital can be used to discover models describing processes, organizations, and products. Moreover, such event logs can also be used to compare event logs with some *prior* model to see whether the observed reality conforms to some prescriptive or descriptive model.

Contemporary management trends such as BAM (Business Activity Monitoring), BOM (Business Operations Management), BPI (business process intelligence) illustrate the interest in supporting the diagnosis functionality in the context of Business Process Management technology (*e.g.*, Workflow Management Systems but also other process-aware information systems).

Application

Process mining follows the options known from business process engineering and goes beyond with feedback to business process modelling :

- **Process analysis** filters, orders and compresses logfiles for further insight into the connex of process operations.
- **Process design** may be supported by feedback from process **monitoring**, which means basically action or event logging
- **Process enactment** uses results from process mining based on logging for triggering further process operation

Classification

There are three classes of process mining techniques. This classification is based on whether there is a prior model and, if so, how it is used.

- *Discovery* : There is no *a priori* model, *i.e.*, based on an event log some model is constructed a process model can be discovered based on low-level events. For example, using the alpha algorithm, which is a didactically driven approach, where the authors state the lack of analytic capability for large event data volumes with such simple method. There exist many techniques to automatically construct process models (*e.g.*, in terms of a Petri net) based some event log. Recently, process mining research also started to target the other perspectives (*e.g.*, data, resources, time, etc.). For example, the technique described in (Aalst, Reijers, & Song, 2005) can be used to construct a social network.
- *Conformance analysis* : There is an *a priori* model. This model is compared with the event log and discrepancies between the log and the model are analyzed. For example, there may be a process model indicating that purchase orders of more than 1 million euro require two checks. Another example is the checking of the so-called "four-eyes" principle. Conformance checking may be used to detect deviations to enrich the model. An example is the extension of a process model with performance data, *i.e.*, some *a priori* process model is used to project the bottlenecks on. Another example is the decision miner described in which takes an *a priori* process model and analyzes every choice in the process model. For each choice the event log is consulted to see which information is typically available the moment the choice is made. Then classical data-mining techniques are used to see which data elements influence the choice. As a result, a decision tree is generated for each choice in the process.
- *Extension* : There is a prior model also. This model is extended with a new aspect or perspective, *i.e.*, the goal is not to check conformance but to enrich the model. An example is the extension of a process model with performance data, *i.e.*, some prior process model dynamically annotated with performance data (*e.g.*, bottlenecks are shown by colouring parts of the process model).

Software for Process Mining

A software framework for the evaluation of process mining algorithms has been developed at the Eindhoven University of Technology by Wil van der Aalst and others, and is available as an open source toolkit.

- Process Mining
- ProM Framework
- ProM Import Framework.

Process Mining functionality is also offered by the following commercial vendors :

- Futura Reflect, a Process Mining and Process Intelligence suite developed by Futura Technology. Acquired by Perceptive Software.
- Interstage Automated Process Discovery, a Process Mining service offered by Fujitsu, Ltd. as part of the Interstage Integration Middleware Suite.
- Celonis Process Mining, the Process Mining solution offered by Celonis.
- BPMone, offering both basic process mining functionality as well as a more comprehensive process mining module as part of the Pallas Athena BPMone

software suite. Acquired by Perceptive Software and rebranded Perceptive Reflect.

- Nitro is a tool by Fluxicon for easily converting CSV and XLS event logs for ProM.
- ARIS Process Performance Manager, a Process Mining and Process Intelligence Tool offered by Software AG as part of the Process Intelligence Solution.
- QPR Process Analyzer, a tool for Automated Business Process Discovery, and QPR Process Analysis, a service based on the aforementioned, offered by QPR Software Plc.
- Disco is a complete process mining software by Fluxicon.

BUSINESS PROCESS MODELLING

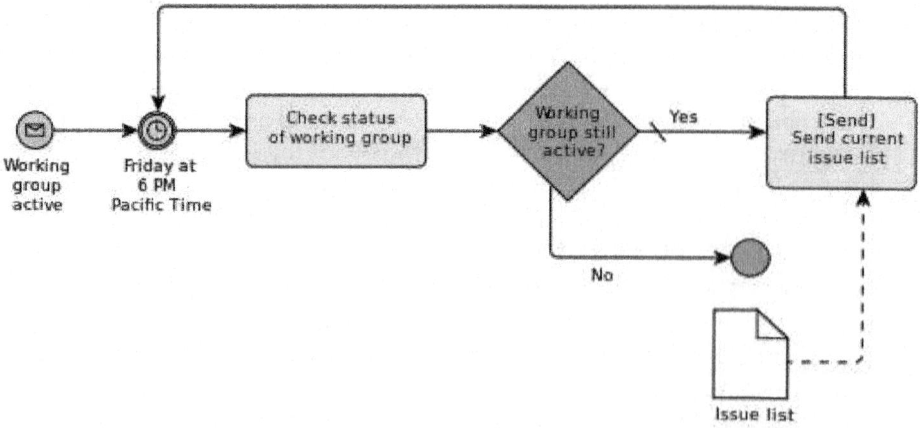

Fig. : Example of business process modelling of a process with a normal flow with the *Business Process Modelling Notation.*

Business process modelling (BPM) in *systems engineering* is the activity of *representing processes* of an enterprise, so that the current process may be analyzed and improved. BPM is typically performed by business analysts and managers who are seeking to improve process efficiency and quality. The process improvements identified by BPM may or may not require *information technology* involvement, although that is a common driver for the need to model a business process, by creating a process master. Business process modelling results in the improvement of the way tasks performed by the business. They can pick up errors or cons about the way processes are currently being performed and model an improved way of carrying out these processes.

Change management programs are typically involved to put the improved business processes into practice. With advances in technology from large platform vendors, the vision of BPM models becoming fully executable (and capable of simulations and round-trip engineering) is coming closer to reality.

History

Techniques to model business process such as the flow chart, functional flow block diagram, control flow diagram, Gantt chart, PERT diagram, and IDEF have emerged since the beginning of the 20th century. The Gantt charts were among the first to arrive around 1899, the flow charts in the 1920s, Functional Flow Block Diagram and PERT in the 1950s, Data Flow Diagrams and IDEF in the 1970s. Among the modern methods are Unified Modelling Language and Business Process Model and Notation. Still, these represent just a fraction of the methodologies used over the years to document business processes. The term "business process modelling" itself was coined in the 1960s in the field of systems engineering by S. Williams in his 1967 article "Business Process Modelling Improves Administrative Control". His idea was that techniques for obtaining a better understanding of physical control systems could be used in a similar way for business processes. It was not until the 1990s that the term became popular.

In the 1990s the term "process" became a new productivity paradigm. Companies were encouraged to think in *processes* instead of *functions* and *procedures.* Process thinking looks at the chain of events in the company from purchase to supply, from order retrieval to sales, etc. The traditional modelling tools were developed to illustrate time and cost, while modern tools focus on cross-functional activities. These cross-functional activities have increased significantly in number and importance, due to the growth of complexity and dependence. New methodologies include business process redesign, business process innovation, business process management, integrated business planning, among others, all "aiming at improving processes across the traditional functions that comprise a company".

In the field of software engineering, the term "business process modelling" opposed the common software process modelling, aiming to focus more on the state of the practice during software development. In that time all existing and new modelling techniques to illustrate business processes were consolidated as "business process modelling languages." In the Object Oriented approach, it was considered to be an essential step in the specification of business application systems. Business process modelling became the base of new methodologies, for instance those that supported data collection, data flow analysis, process flow diagrams and reporting facilities. Around 1995, the first visually oriented tools for business process modelling and implementation were being presented.

BPM Topics

Business Model

A business model is a framework for creating economic, social, and/or other forms of value. The term 'business model' is thus used for a broad range of informal and formal descriptions to represent core aspects of a business, including purpose, offerings, strategies, infrastructure, organizational structures, trading practices, and operational processes and policies.

In the most basic sense, a business model is the method of doing business by which a company can sustain itself. That is, generate revenue. The business model spells-out how a company makes money by specifying where it is positioned in the value chain.

Business Process

A business process is a collection of related, structured activities or tasks that produce a specific service or product (serve a particular goal) for a particular customer or customers. There are three main types of business processes :

1. Management processes, that govern the operation of a system. Typical management processes include corporate governance and strategic management.
2. Operational processes, that constitute the core business and create the primary value stream. Typical operational processes are purchasing, manufacturing, marketing, and sales.
3. Supporting processes, that support the core processes. Examples include accounting, recruitment, and technical support.

A business process can be decomposed into several sub-processes, which have their own attributes, but also contribute to achieving the goal of the super-process. The analysis of business processes typically includes the mapping of processes and sub-processes down to activity level. A business process model is a model of one or more business processes, and defines the ways in which operations are carried out to accomplish the intended objectives of an organization. Such a model remains an abstraction and depends on the intended use of the model. It can describe the workflow or the integration between business processes. It can be constructed in multiple levels.

A workflow is a depiction of a sequence of operations, declared as work of a person, of a simple or complex mechanism, of a group of persons, of an organization of staff, or of machines. Workflow may be seen as any abstraction of real work, segregated into workshare, work split or other types of ordering. For control purposes, workflow may be a view of real work under a chosen aspect.

Artifact-centric Business Process

The artifact-centric business process model has emerged as a new approach for modelling business processes, as it provides a highly flexible solution to capture operational specifications of business processes. It particularly focuses on describing the data of business processes, known as "artifacts", by characterizing business-relevant data objects, their lifecycles, and related services. The artifact-centric process modelling approach fosters the automation of the business operations and supports the flexibility of the workflow enactment and evolution.

Business Process Modelling Tools

Business process modelling tools provide business users with the ability to model their business processes, implement and execute those models, and refine the models based on as-executed data. As a result, business process modelling tools can provide transparency into business processes, as well as the centralization of corporate business process models and execution metrics.

Modelling and Simulation

Modelling and simulation functionality allows for pre-execution "what-if" modelling and simulation. Post-execution optimization is available based on the analysis of actual as-performed metrics.

- Use case diagrams created by Ivar Jacobson, 1992. Currently integrated in UML.
- Activity diagrams, also currently adopted by UML.

Some business process modelling techniques are :

- Business Process Model and Notation (BPMN).
- Cognition enhanced Natural language Information Analysis Method (CogNIAM).
- Extended Business Modelling Language (xBML).
- Event-driven process chain (EPC).
- ICAM Definition (IDEF0).
- Unified Modelling Language (UML), extensions for business process such as Eriksson-Penker's.

Programming Language Tools for BPM

BPM suite software provides programming interfaces (web services, application program interfaces (APIs)) which allow enterprise applications to be built to leverage the BPM engine. This component is often referenced as the *engine* of the BPM suite.

Programming languages that are being introduced for BPM include :

- BPMN.
- Business Process Execution Language (BPEL).
- Web Services Choreography Description Language (WS-CDL).
- XML Process Definition Language (XPDL).

Some vendor-specific languages :

- Architecture of Integrated Information Systems (ARIS) supports EPC,
- Java Process Definition Language (JBPM).

Other technologies related to business process modelling include model-driven architecture and service-oriented architecture.

Related Topics

Business Reference Model

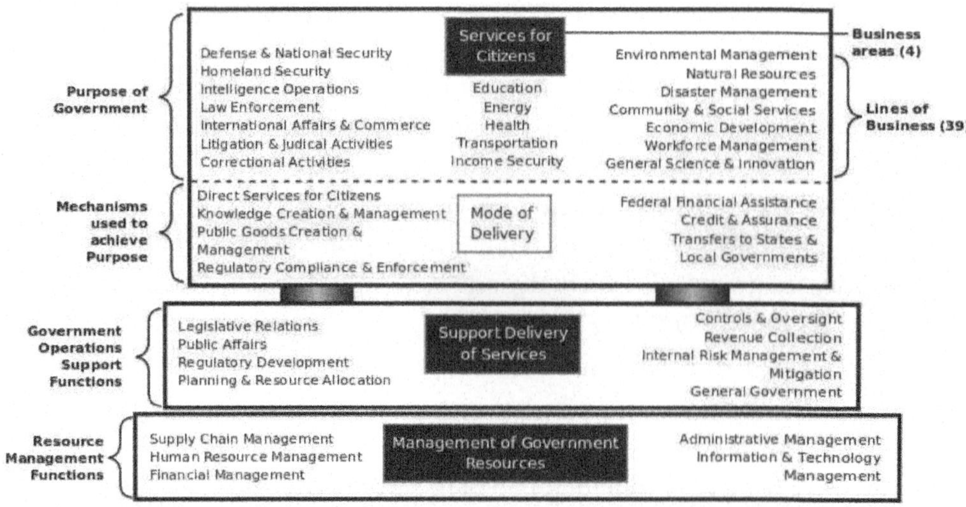

Fig. : Example of the US Federal Government Business Reference Model.

A business reference model is a reference model, concentrating on the functional and organizational aspects of an enterprise, service organization or government agency. In general a reference model is a model of something that embodies the basic goal or idea of something and can then be looked at as a reference for various purposes. A business reference model is a means to describe the business operations of an organization, independent of the organizational structure that perform them. Other types of business reference model can also depict the relationship between the business processes, business functions, and the business area's business reference model. These reference models can be constructed in layers, and offer a foundation for the analysis of service components, technology, data, and performance.

The most familiar business reference model is the Business Reference Model of the US Federal Government. That model is a function-driven framework for describing the business operations of the Federal Government independent of the agencies that perform them. The Business Reference Model provides an organized, hierarchical construct for describing the day-to-day business operations of the Federal government. While many models exist for describing organizations-organizational charts, location maps, etc. – this model presents the business using a functionally driven approach.

Business Process Integration

A business model, which may be considered an elaboration of a business process model, typically shows business data and business organizations as well as business processes. By showing business processes and their information flows a business model allows business stakeholders to define, understand, and validate their business enterprise. The data model part of the business model shows how business information is stored, which is useful for developing software code.

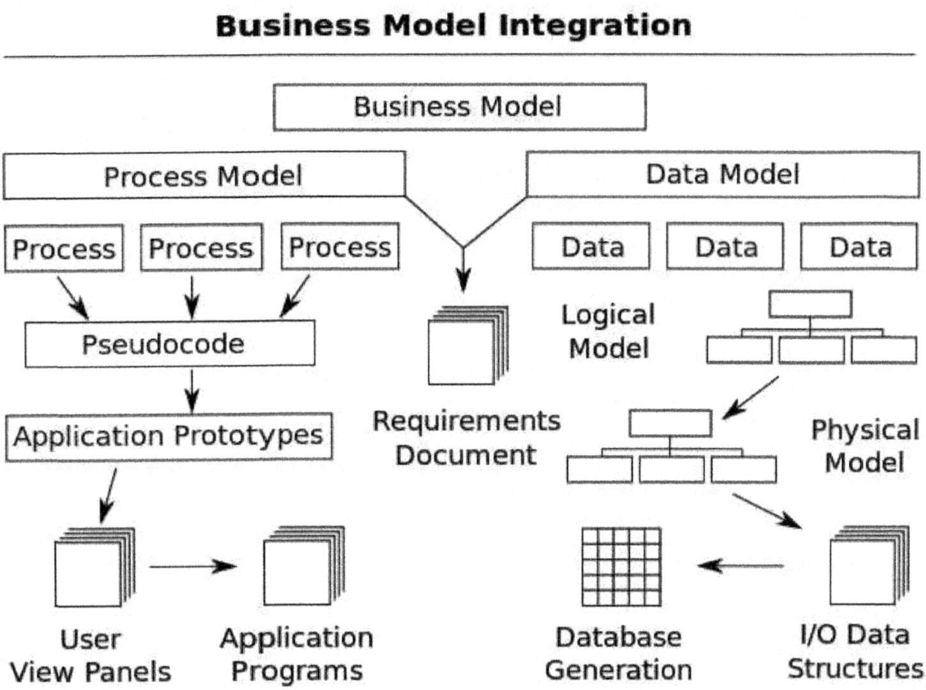

Fig. : Example of the interaction between business process and data models.

Usually a business model is created after conducting an interview, which is part of the business analysis process. The interview consists of a facilitator asking a series of questions to extract information about the subject business process. The interviewer is referred to as a facilitator to emphasize that it is the participants, not the facilitator, who provide the business process information. Although the facilitator should have some knowledge of the subject business process, but this is not as important as the mastery of a pragmatic and rigorous method interviewing business experts. The method is important because for most enterprises a team of facilitators is needed to collect information across the enterprise, and the findings of all the interviewers must be compiled and integrated once completed.

Business models are developed as defining either the current state of the process, in which case the final product is called the "as is" snapshot model, or a concept of what the process should become, resulting in a "to be" model. By comparing and contrasting "as is" and "to be" models the business analysts can determine if the existing business processes and information systems are sound and only need minor modifications, or if re-engineering is required to correct problems or improve efficiency. Consequently, business process modelling and subsequent analysis can be used to fundamentally reshape the way an enterprise conducts its operations.

Business Process Reengineering

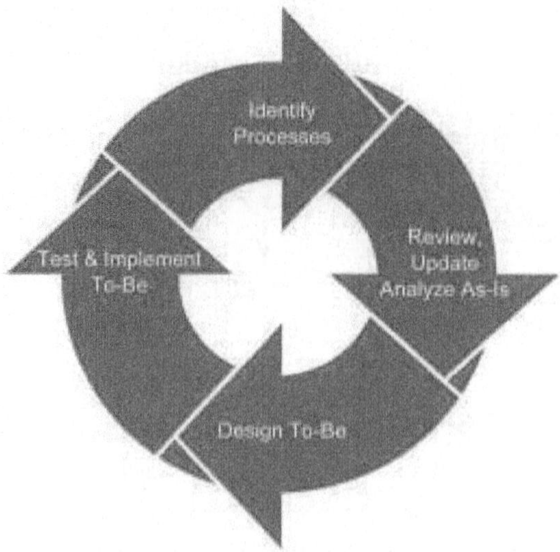

Fig. : Business Process Reengineering Cycle.

Business process re-engineering (BPR) aims to improve the efficiency and effectiveness of the processes that exist within and across organizations. It examines business processes from a "clean slate" perspective to determine how best to construct them.

Business process re-engineering (BPR) began as a private sector technique to help organizations fundamentally rethink how they do their work. A key stimulus for re-engineering has been the development and deployment of sophisticated information systems and networks. Leading organizations use this technology to support innovative business processes, rather than refining current ways of doing work.

Business Process Management

Business process management is a field of management focused on aligning organizations with the wants and needs of clients. It is a holistic management approach that promotes business effectiveness and efficiency while striving for innovation, flexibility and integration with technology. As organizations strive for attainment of their objectives, business process management attempts to continuously improve processes—the process to define, measure and improve your processes—a "process optimization" process.

COMPLEX EVENT PROCESSING

Event processing is a method of tracking and analyzing (processing) streams of information (data) about things that happen (events), and deriving a conclusion from them. **Complex event processing**, or CEP, is event processing that combines data from multiple sources to infer events or patterns that suggest more complicated circumstances. The goal of complex event processing is to identify meaningful events (such as opportunities or threats) and respond to them as quickly as possible.

These events may be happening across the various layers of an organization as sales leads, orders or customer service calls. Or, they may be news items, text messages, social media posts, stock market feeds, traffic reports, weather reports, or other kinds of data. An event may also be defined as a "change of state," when a measurement exceeds a predefined threshold of time, temperature, or other value. Analysts suggest that CEP will give organizations a new way to analyze patterns in real-time, and help the business side communicate better with IT and service departments.

The vast amount of information available about events is sometimes referred to as the event cloud.

Conceptual Description

Among thousands of incoming events, a monitoring system may for instance receive the following three from the same source :

1. Church bells ringing.
2. The appearance of a man in a tuxedo with a woman in a flowing white gown.
3. Rice flying through the air.

From these events the monitoring system may infer a *complex event* : a wedding. CEP as a technique helps discover complex events by analyzing and correlating other events : the bells, the man and woman in wedding attire and the rice flying through the air.

CEP relies on a number of techniques, including :

- Event-pattern detection
- Event abstraction
- Event filtering
- Event aggregation and transformation
- Modelling event hierarchies
- Detecting relationships (such as causality, membership or timing) between events
- Abstracting event-driven processes.

Commercial applications of CEP exist in variety of industries and include algorithmic stock-trading, the detection of credit-card fraud, business activity monitoring, and security monitoring.

History

The CEP area has roots in Discrete event simulation, active database area and some programming languages. The activity in the industry was preceded by a wave of research projects in the 1990s. According to the first project that paved the way to a generic CEP language and execution model was the Rapide project in Stanford University, directed by David Luckham, in parallel there have been three other research projects : Infospheres in California Institute of Technology, directed by K. Mani Chandy, Apama in University of Cambridge directed by John Bates, and Amit in IBM Haifa Research Laboratory, directed by Opher Etzion. The commercial products were dependents of the concepts developed in these and some later research projects. Community efforts started in a series of event processing symposiums organized by the Event Processing Technical Society, and later by the ACM DEBS conference series. One of the community effort was in producing the event processing manifesto

Related Concepts

CEP is used in Operational Intelligence (OI) solutions to provide insight into business operations by running query analysis against live feeds and event data. OI solutions use real-time data to collect and correlate against historical data to provide insight into and analysis of the current situation. Multiple sources of data can be combined from different organizational silos to provide a common operating picture that uses current information. Wherever real-time insight has the greatest value, OI solutions can be applied to deliver the information and need.

In network management, systems management, application management and service management, people usually refer instead to event correlation. As CEP engines, event correlation engines (*event correlators*) analyze a mass of events, pinpoint the most significant ones, and trigger actions. However, most of them do not produce new inferred events. Instead, they relate high-level events with low-level events.

Inference engines, *e.g.* rule-based reasoning engines typically produce inferred information in artificial intelligence. However, they do not usually produce new information in the form of complex (*i.e.,* inferred) events.

Example

A more systemic example of CEP involves a car, some sensors and various events and reactions. Imagine that a car has several sensors—one that measures tire pressure, one that measures speed, and one that detects if someone sits on a seat or leaves a seat.

In the first situation, the car is moving and the pressure of one of the tires moves from 45 psi (pound per square inch) to 41 psi over 15 minutes. As the pressure in the tire is decreasing, a series of events containing the tire pressure is generated. In addition, a series of events containing the speed of the car is generated. The car's Event Processor may detect a situation whereby a loss of tire pressure over a relatively long period of time results in the creation of the "loss of Tire Pressure" event. This new event may trigger a reaction process to note the pressure loss into the car's maintenance log, and alert the driver via the car's portal that the tire pressure has reduced.

In the second situation, the car is moving and the pressure of one of the tires drops from 45 psi to 20 psi in 5 seconds. A different situation is detected—perhaps because the loss of pressure occurred over a shorter period of time, or perhaps because the difference in values between each event were larger than a predefined limit. The different situation results in a new event "blow Out Tire" being generated. This new event triggers a different reaction process to immediately alert the driver and to initiate on board computer routines to assist the driver in bringing the car to a stop without losing control through skidding.

In addition, events that represent detected situations can also be combined with other events in order to detect more complex situations. For example, in the final situation the car was moving normally but suffers a blown tire which results in the car leaving the road and striking a tree and the driver is thrown from the car. A series of different situations are rapidly detected. The combination of "blow Out Tire", "zero Speed" and "driver Left Seat" within a very short space of time results in a new situation being detected : "occupant Thrown Accident". Even though there is no direct measurement that can determine conclusively that the driver was thrown, or that there was an accident, the combination of events allows the situation to be detected and a new event to be created to signify the detected situation. This is the essence of a complex (or composite) event. It is complex because one cannot directly detect the situation; one has to infer or deduce that the situation has occurred from a combination of other events.

Types

Most CEP solutions and concepts can be classified into two main categories :

1. Aggregation-oriented CEP
2. Detection-oriented CEP.

An aggregation-oriented CEP solution is focused on executing on-line algorithms as a response to event data entering the system. A simple example is to continuously calculate an average based in data on the inbound events.

Detection-oriented CEP is focused on detecting combinations of events called events patterns or situations. A simple example of detecting a situation is to look for a specific sequence of events.

Currently many application use hybrid of the two approaches.

Integration with Business Process Management

Of course, rarely does the application of a new technology exist in isolation. A natural fit for CEP has been with business process management, or BPM. BPM very much focuses on end-to-end business processes, in order to continuously optimize and align for its operational environment.

However, the optimization of a business does not rely solely upon its individual, end-to-end processes. Seemingly disparate processes can affect each other significantly. Consider this scenario : In the aerospace industry, it is good practice to monitor breakdowns of vehicles to look for trends (determine potential weaknesses in manufacturing processes, material, etc.). Another separate process monitors current operational vehicles' life cycles and decommissions them when appropriate. Now one use for CEP is to link these separate processes, so that in the case of when the initial process (breakdown monitoring) discovers a malfunction based on metal fatigue (a significant event) an action can be created to exploit the second process (life cycle) to issue a recall on vehicles using the same batch of metal discovered as faulty in the initial process.

The integration of CEP and BPM must exist at two levels, both at the business awareness level (users must understand the potential holistic benefits of their individual processes) and also at the technological level (there needs to be a method by which CEP can interact with BPM implementation).

Computation-oriented CEP's role can arguably be seen to overlap with Business Rule technology.

For example, customer service centers are using CEP for click-stream analysis and customer experience management. CEP software can factor real-time information about millions of events (clicks or other interactions) per second into business intelligence and other decision-support applications. These "recommendation applications" help agents provide personalized service based on each customer's experience. The CEP application may collect data about what customers on the phone are currently doing, or how they have recently interacted with the company in other various channels, including in-branch, or on the Web via self-service features, instant messaging and email. The application then analyzes the total customer experience and recommends scripts or next steps that guide the agent on the phone, and hopefully keep the customer happy.

Another example of CEP in practice is in the healthcare industry. The HyReminder system, developed by the Worcester Polytechnic Institute and UMass Medical School, continually tracks healthcare workers for hygiene compliance (*e.g.* sanitizing hands and wearing masks), reminding them to perform hygiene when appropriate to prevent the spread of infectious disease. Each worker wears an RFID badge that displays a green (safe), yellow (warning) or red (violation) light, depending on what behaviour the RFID chip has observed.

In Financial Services

The financial services industry was an early adopter of CEP technology, using complex event processing to structure and contextualize available data so that it could inform trading behaviour, specifically algorithmic trading, by identifying opportunities or threats that indicate traders (or automatic trading systems) should buy or sell. Algorithmic trading is already a practice in stock trading. It is estimated that around 60% of Equity trading in the United States is by way of algo trades. CEP is expected to continue to help financial institutions improve their algorithms and be more efficient.

Recent improvements in CEP technologies have made it more affordable, helping smaller firms to create trading algorithms of their own and compete with larger firms. CEP has evolved from an emerging technology to an essential platform of many capital markets. The technology's most consistent growth has been in banking, serving fraud detection, online banking, and multi-channel marketing initiatives.

Today, a wide variety of financial applications use CEP, including profit, loss, and risk management systems, order and liquidity analysis, quantitative trading and signal generation systems, and others.

Integration with Time Series Databases

A time series database is a software system that is optimized for the handling of data organized by time. Time series are finite or infinite sequences of data items, where each item has an associated timestamp and the sequence of timestamps is non-decreasing. Elements of a time series are often called ticks. The timestamps are not required to be ascending (merely non-decreasing) because in practice the time resolution of some systems such as financial data sources can be quite low (milliseconds, micro-seconds or even nano-seconds), so consecutive events may carry equal timestamps.

Time series data provides a historical context to the analysis typically associated with complex event processing. This can apply to any vertical industry such as finance and co-operatively with other technologies such as BPM as described elsewhere in this document.

Consider the scenario in finance where there is a need to understand historic price volatility to determine statistical thresholds of future price movements. This is helpful for both trade models and transaction cost analysis.

The ideal case for CEP analysis is to view historical time series and real-time streaming data as a single time continuum. What happened yesterday, last week or last month is simply an extension of what is occurring today and what may occur in the future. An example may involve comparing current market volumes to historic volumes, prices and volatility for trade execution logic. Or the need to act upon live market prices may involve comparisons to benchmarks that include sector and index movements, whose intra-day and historic trends gauge volatility and smooth outliers.

PRESCRIPTIVE ANALYTICS

Prescriptive analytics is the third and final phase of *business analytics (BA)* which includes descriptive, *predictive* and prescriptive analytics.

Prescriptive analytics automatically synthesizes *big data*, multiple disciplines of *mathematical sciences* and *computational sciences*, and *business rules*, to make *predictions* and then suggests decision options to take advantage of the predictions. The first stage of business analytics is descriptive analytics, which still accounts for the majority of all business analytics today. Descriptive analytics answers the questions what happened and why did it happen. Descriptive analytics looks at past performance and understands that performance by mining historical data to look for the reasons behind past success or failure. Most management reporting—such as *sales*, *marketing*, *operations*, and *finance*—uses this type of post-mortem analysis.

The next phase is *predictive analytics*. Predictive analytics answers the question what will happen. This is when historical performance data is combined with rules, *algorithms*, and occasionally external data to determine the probable future outcome of an event or the likelihood of a situation occurring. The final phase is prescriptive analytics, which goes beyond predicting future outcomes by also suggesting actions to benefit from the predictions and showing the implications of each decision option.

Prescriptive analytics not only anticipates what will happen and when it will happen, but also why it will happen. Further, prescriptive analytics suggests decision options on how to take advantage of a future opportunity or mitigate a future risk and shows the implication of each decision option. Prescriptive analytics can continually take in new data to re-predict and re-prescribe, thus automatically improving prediction accuracy and prescribing better decision options. Prescriptive analytics ingests hybrid data, a combination of structured (numbers, categories) and unstructured data (videos, images, sounds, texts), and business rules to predict what lies ahead and to prescribe how to take advantage of this predicted future without compromising other priorities.

All three phases of analytics can be performed through professional services or technology or a combination. In order to scale, prescriptive analytics technologies need to be adaptive to take into account the growing volume, velocity, and variety of data that most mission critical processes and their environments may produce.

History

Prescriptive analytics has been around since about 2003. The technology behind prescriptive analytics synergistically combines hybrid data, business rules with mathematical models and computational models. The data inputs to prescriptive analytics may come from multiple sources : internal, such as inside a corporation; and external, also known as environmental data. The data may be structured, which includes numbers and categories, as well as unstructured data, such as texts, images, sounds, and videos. Unstructured data differs from structured data in that its format varies widely and cannot be stored in traditional relational databases without significant effort at data transformation. More than 80% of the world's data today is unstructured, according to IBM. In addition to this variety of data types and growing data volume, incoming data can also evolve with respect to velocity, that is, more data being generated at a faster or a variable pace. Business rules define the business process and include objectives constraints, preferences, policies, best practices, and boundaries. Mathematical models and computational models are techniques derived from mathematical sciences, computer science and related disciplines such as applied statistics, machine learning, operations research, natural language processing, computer vision, pattern recognition, image processing, speech recognition, and signal processing.

Applications in Healthcare

Multiple factors are driving healthcare providers to dramatically improve business processes and operations as the United States healthcare industry embarks on the necessary migration from a largely fee-for service, volume-based system to a fee-for-performance, value-based system. Prescriptive analytics is playing a key role to help improve the performance in a number of areas involving various stakeholders : payers, providers and pharmaceutical companies.

Prescriptive analytics can help providers improve effectiveness of their clinical care delivery to the population they manage and in the process achieve better patient satisfaction and retention. Providers can do better population health management by identifying appropriate intervention models for risk stratified population combining data from the in-facility care episodes and home based telehealth.

Prescriptive analytics can also benefit healthcare providers in their capacity planning by using analytics to leverage operational and usage data combined with data of external factors such as economic data, population demographic trends and population health trends, to more accurately plan for future capital investments such as new facilities and equipment utilization as well as understand the trade-offs between adding additional beds and expanding an existing facility *versus* building a new one.

Prescriptive analytics can help pharmaceutical companies to expedite their drug development by identifying patient cohorts that are most suitable for the clinical trials worldwide-patients who are expected to be compliant and will not drop out of the trial due to complications. Analytics can tell companies how much time and money they can save if they choose one patient cohort in a specific country *vs.* another.

In provider-payer negotiations, providers can improve their negotiating position with health insurers by developing a robust understanding of future service utilization. By accurately predicting utilization, providers can also better allocate personnel.

Applications in Oil and Gas

Energy is the largest industry in the world ($6 trillion in size). The processes and decisions related to oil and natural gas exploration, development and production generate large amounts of data. Many types of captured data are used to create models and images of the Earth's structure and layers 5,000-35,000 feet below the surface and to describe activities around the wells themselves, such as depositional characteristics, machinery performance, oil flow rates, reservoir temperatures and pressures. Prescriptive analytics software can help with both finding and producing oil and gas. It can take in seismic data, well log data, production data, and other related data sets to prescribe where to drill, how to drill (to maximize production, and minimize cost and environmental impact).

Prescriptive analytics software can accurately predict production issues by modelling numerous internal and external variables simultaneously. Prescriptive analytics software can also provide decision options and show the impact of each decision option so the operations managers can proactively take appropriate actions, on time, to guarantee future exploration and production performance. In the area of Health, Safety, and Environment, prescriptive analytics can predict and preempt incidents that can lead to reputational and financial loss for oil and gas companies.

Pricing is another area of focus. Natural gas prices fluctuate dramatically depending upon supply, demand, econometrics, geopolitics, and weather conditions. Gas producers, pipeline transmission companies and utility firms have a keen interest in more accurately predicting gas prices so that they can lock in favourable terms while hedging downside risk. Prescriptive analytics software can accurately predict prices by modelling internal and external variables simultaneously and also provide decision options and show the impact of each decision option.

EXECUTIVE INFORMATION SYSTEM

An **executive information system (EIS)** is a type of *management information system* that facilitates and supports senior executive information and *decision-making* needs. It provides easy access to internal and external information relevant to *organizational* goals. It is commonly considered a specialized form of *decision support system* (DSS).

EIS emphasizes graphical displays and easy-to-use *user interfaces*. They offer strong reporting and *drill-down* capabilities. In general, EIS are enterprise-wide DSS that help top-level executives analyze, compare, and highlight trends in important *variables* so that they can monitor performance and identify opportunities and problems. EIS and *data warehousing* technologies are converging in the marketplace.

In recent years, the term EIS has lost popularity in favour of *business intelligence* (with the sub-areas of reporting, *analytics*, and *digital dashboards*).

History

Traditionally, executive information systems were mainframe computer-based programs. The purpose was to package a company's data and to provide sales performance or market research statistics for decision makers, such as financial officers, marketing directors, and chief executive officers, who were not necessarily well acquainted with computers. The objective was to develop computer applications that highlighted information to satisfy senior executives' needs. Typically, an EIS provides only data that supported executive level decisions, not all company data.

Today, the application of EIS is not only in typical corporate hierarchies, but also at personal computers on a local area network. EIS now cross computer hardware platforms and integrate information stored on mainframes, personal computer systems, and mini-computers. As some client service companies adopt the latest enterprise information systems, employees can use their personal computers to get access to the company's data and identify information relevant to their decision-making. This arrangement lets all users customize their access to company data, and provides relevant information to upper and lower corporate levels.

Components

EIS components can typically be classified as :

- Hardware
- Software
- User interface
- Telecommunications.

Hardware

When talking about computer hardware for an EIS environment, we should focus on the hardware that meet the executive's needs. The executive must be put first and the executive's needs must be defined before the hardware can be selected. The basic hardware needed for a typical EIS includes four components :

1. *Input data-entry devices :* These devices allow the executive to enter, verify, and update data immediately.
2. The central processing unit (CPU), which is the kernel because it controls the other computer system components.
3. *Data storage files :* The executive can use this part to save useful business information, and this part also help the executive to search historical business information easily.
4. Output devices, which provide a visual or permanent record for the executive to save or read. This device refers to the visual output device such as monitor or printer.

In addition, with the advent of local area networks (LAN), several EIS products for networked workstations became available. These systems require less support and less expensive computer hardware. They also increase EIS information access to more company users.

Software

Choosing the appropriate software is vital to an effective EIS. Therefore, the software components and how they integrate the data into one system are important. A typical EIS includes four software components :

1. *Text-handling software* : Documents are typically text-based.
2. *Database* : Heterogeneous databases on a range of vendor-specific and open computer platforms help executives access both internal and external data.
3. *Graphic base* : Graphics can turn volumes of text and statistics into visual information for executives. Typical graphic types are : time series charts, scatter diagrams, maps, motion graphics, sequence charts, and comparison-oriented graphs (*i.e.*, bar charts).
4. *Model base* : EIS models contain routine and special statistical, financial, and other quantitative analysis.

User Interface

An EIS must be efficient to retrieve relevant data for decision makers, so the user interface is very important. Several types of interfaces can be available to the EIS structure, such as scheduled reports, questions/answers, menu driven, command language, natural language, and input/output.

Telecommunication

As decentralizing is becoming the current trend in companies, telecommunications will play a pivotal role in networked information systems. Transmitting data from one place to another has become crucial for establishing a reliable network. In addition, telecommunications within an EIS can accelerate the need for access to distributed data.

Applications

EIS helps executives find data according to user-defined criteria and promote information-based insight and understanding. Unlike a traditional management information system presentation, EIS can distinguish between vital and seldom-used data, and track different key critical activities for executives, both which are helpful in evaluating if the company is meeting its corporate objectives. After realizing its advantages, people have applied EIS in many areas, especially, in manufacturing, marketing, and finance areas.

Manufacturing

Manufacturing is the transformation of raw materials into finished goods for sale, or intermediate processes involving the production or finishing of semi-manufactures. It is a large branch of industry and of secondary production. Manufacturing operational control focuses on day-to-day operations, and the central idea of this process is effectiveness and efficiency.

Marketing

In an organization, marketing executives' duty is managing available marketing resources to create a more effective future. For this, they need make judgments about risk and uncertainty of a project and its impact on the company in short-term and long-term. To assist marketing executives in making effective marketing decisions, an EIS can be applied. EIS provides sales forecasting, which can allow the market executive to compare sales forecast with past sales. EIS also offers an approach to product price, which is found in venture analysis. The market executive can evaluate pricing as related to competition along with the relationship of product quality with price charged. In summary, EIS software package enables marketing executives to manipulate the data by looking for trends, performing audits of the sales data, and calculating totals, averages, changes, variances, or ratios.

Financial

Financial analysis is one of the most important steps to companies today. Executives needs to use financial ratios and cash flow analysis to estimate the trends and make capital investment decisions. An EIS integrates planning or budgeting with control of performance reporting, and it can be extremely helpful to finance executives. EIS focuses on financial performance accountability, and recognizes the importance of cost standards and flexible budgeting in developing the quality of information provided for all executive levels.

Advantages and Disadvantages

Advantages of EIS

- Easy for upper-level executives to use, extensive computer experience is not required in operations.
- Provides timely delivery of company summary information.
- Information that is provided is better understood.
- EIS provides timely delivery of information. Management can make decisions promptly.
- Improves tracking information.
- Offers efficiency to decision makers.

Disadvantages of EIS

- System dependent.
- Limited functionality, by design.
- Information overload for some managers.
- Benefits hard to quantify.
- High implementation costs.
- System may become slow, large, and hard to manage.

- Need good internal processes for data management.
- May lead to less reliable and less secure data.

Future Trends

The future of executive info systems is not bound by mainframe computer systems. This trend free executives from learning different computer operating systems, and substantially decreases implementation costs. Because this trend includes using existing software applications, executives don't need to learn a new or special language for the EIS package.

ONLINE ANALYTICAL PROCESSING

In computing, **online analytical processing**, or OLAP/'oʊlæp/, an approach to answering multi-dimensional analytical (MDA) queries swiftly. OLAP is part of the broader category of business intelligence, which also encompasses relational database, report writing and data-mining. Typical applications of OLAP include business reporting for sales, marketing, management reporting, business process management (BPM), budgeting and forecasting, financial reporting and similar areas, with new applications coming up, such as agriculture. The term *OLAP* was created as a slight modification of the traditional database term OLTP (Online Transaction Processing).

OLAP tools enable users to analyze multi-dimensional data interactively from multiple perspectives. OLAP consists of three basic analytical operations : consolidation (roll-up), drill-down, and slicing and dicing. Consolidation involves the aggregation of data that can be accumulated and computed in one or more dimensions. For example, all sales offices are rolled up to the sales department or sales division to anticipate sales trends. By contrast, the drill-down is a technique that allows users to navigate through the details. For instance, users can view the sales by individual products that make up a region's sales. Slicing and dicing is a feature whereby users can take out (slicing) a specific set of data of the OLAP cube and view (dicing) the slices from different viewpoints.

Databases configured for OLAP use a multi-dimensional data model, allowing for complex analytical and *ad hoc* queries with a rapid execution time. They borrow aspects of navigational databases, hierarchical databases and relational databases.

Overview of OLAP Systems

The core of any OLAP system is an OLAP cube (also called a 'multi-dimensional cube' or a hypercube). It consists of numeric facts called *measures* which are categorized by *dimensions*. The measures are placed at the intersections of the hypercube, which is spanned by the dimensions as a Vector space. The usual interface to manipulate an OLAP cube is a matrix interface like Pivot tables in a spreadsheet program, which performs projection operations along the dimensions, such as aggregation or averaging.

The cube metadata is typically created from a star schema or snowflake schema or fact constellation of tables in a relational database.

Each *measure* can be thought of as having a set of *labels*, or meta-data associated with it. A *dimension* is what describes these *labels*; it provides information about the *measure*.

A simple example would be a cube that contains a store's sales as a *measure*, and Date/Time as a *dimension*. Each Sale has a Date/Time *label* that describes more about that sale.

Any number of *dimensions* can be added to the structure such as Store, Cashier, or Customer by adding a foreign key column to the fact table. This allows an analyst to view the *measures* along any combination of the *dimensions*.

For example :

```
  Sales Fact Table
+-------------+----------+
| sale_amount | time_id  |
+-------------+----------+          Time Dimension
|      2008.10|     1234 |---+     +---------+-------------------+
+-------------+----------+   |     | time_id | timestamp         |
                             |     +---------+-------------------+
                             +---->|   1234  | 20080902 12:35:43 |
                                   +---------+-------------------+
```

Multi-dimensional Databases

Multi-dimensional structure is defined as "a variation of the relational model that uses multi-dimensional structures to organize data and express the relationships between data". The structure is broken into cubes and the cubes are able to store and access data within the confines of each cube. "Each cell within a multi-dimensional structure contains aggregated data related to elements along each of its dimensions". Even when data is manipulated it remains easy to access and continues to constitute a compact database format. The data still remains interrelated. Multi-dimensional structure is quite popular for analytical databases that use online analytical processing (OLAP) applications. Analytical databases use these databases because of their ability to deliver answers to complex business queries swiftly. Data can be viewed from different angles, which gives a broader perspective of a problem unlike other models.

Aggregations

It has been claimed that for complex queries OLAP cubes can produce an answer in around 0.1% of the time required for the same query on OLTP relational data. The most important mechanism in OLAP which allows it to achieve such performance is the use of *aggregations*. Aggregations are built from the fact table by changing the granularity on specific dimensions and aggregating up data along these dimensions. The number of possible aggregations is determined by every possible combination of dimension granularities.

The combination of all possible aggregations and the base data contains the answers to every query which can be answered from the data.

Because usually there are many aggregations that can be calculated, often only a predetermined number are fully calculated; the remainder are solved on demand. The problem of deciding which aggregations (views) to calculate is known as the view selection problem. View selection can be constrained by the total size of the selected set of aggregations, the time to update them from changes in the base data, or both. The objective of view selection is typically to minimize the average time to answer OLAP queries, although some studies also minimize the update time. View selection is NP-Complete. Many approaches to the problem have been explored, including greedy algorithms, randomized search, genetic algorithms and A* search algorithm.

Types

OLAP systems have been traditionally categorized using the following taxonomy.

Multi-dimensional

MOLAP is a "multi-dimensional online analytical processing". 'MOLAP' is the 'classic' form of OLAP and is sometimes referred to as just OLAP. MOLAP stores this data in an optimized multi-dimensional array storage, rather than in a relational database. Therefore it requires the pre-computation and storage of information in the cube-the operation known as processing. MOLAP tools generally utilize a pre-calculated data set referred to as a data cube. The data cube contains all the possible answers to a given range of questions. MOLAP tools have a very fast response time and the ability to quickly write back data into the data set.

Advantages of MOLAP

- Fast query performance due to optimized storage, multi-dimensional indexing and caching.
- Smaller on-disk size of data compared to data stored in relational database due to compression techniques.
- Automated computation of higher level aggregates of the data.
- It is very compact for low dimension data sets.
- Array models provide natural indexing.
- Effective data extraction achieved through the pre-structuring of aggregated data.

Disadvantages of MOLAP

- Within some MOLAP Solutions the processing step (data load) can be quite lengthy, especially on large data volumes. This is usually remedied by doing only incremental processing, *i.e.*, processing only the data which have changed (usually new data) instead of reprocessing the entire data set.

- MOLAP tools traditionally have difficulty querying models with dimensions with very high cardinality (*i.e.*, millions of members).
- Some MOLAP products have difficulty updating and querying models with more than ten dimensions. This limit differs depending on the complexity and cardinality of the dimensions in question. It also depends on the number of facts or measures stored. Other MOLAP products can handle hundreds of dimensions.
- Some MOLAP methodologies introduce data redundancy.

Relational

ROLAP works directly with relational databases. The base data and the dimension tables are stored as relational tables and new tables are created to hold the aggregated information. Depends on a specialized schema design. This methodology relies on manipulating the data stored in the relational database to give the appearance of traditional OLAP's slicing and dicing functionality. In essence, each action of slicing and dicing is equivalent to adding a "WHERE" clause in the SQL statement. ROLAP tools do not use pre-calculated data cubes but instead pose the query to the standard relational database and its tables in order to bring back the data required to answer the question. ROLAP tools feature the ability to ask any question because the methodology does not limit to the contents of a cube. ROLAP also has the ability to drill down to the lowest level of detail in the database.

Hybrid

There is no clear agreement across the industry as to what constitutes "Hybrid OLAP", except that a database will divide data between relational and specialized storage. For example, for some vendors, a HOLAP database will use relational tables to hold the larger quantities of detailed data, and use specialized storage for at least some aspects of the smaller quantities of more-aggregate or less-detailed data. HOLAP addresses the shortcomings of MOLAP and ROLAP by combining the capabilities of both approaches. HOLAP tools can utilize both pre-calculated cubes and relational data sources.

Comparison

Each type has certain benefits, although there is disagreement about the specifics of the benefits between providers.

- Some MOLAP implementations are prone to database explosion, a phenomenon causing vast amounts of storage space to be used by MOLAP databases when certain common conditions are met : high number of dimensions, pre-calculated results and sparse multi-dimensional data.
- MOLAP generally delivers better performance due to specialized indexing and storage optimizations. MOLAP also needs less storage space compared to ROLAP because the specialized storage typically includes compression techniques.

- ROLAP is generally more scalable. However, large volume pre-processing is difficult to implement efficiently so it is frequently skipped. ROLAP query performance can therefore suffer tremendously.
- Since ROLAP relies more on the database to perform calculations, it has more limitations in the specialized functions it can use.
- HOLAP encompasses a range of solutions that attempt to mix the best of ROLAP and MOLAP. It can generally pre-process swiftly, scale well, and offer good function support.

Other Types

The following acronyms are also sometimes used, although they are not as widespread as the ones above :

- **WOLAP**-Web-based OLAP.
- **DOLAP-Desktop** OLAP.
- **RTOLAP**-Real-Time OLAP.

APIs and Query Languages

Unlike relational databases, which had SQL as the standard query language, and widespread APIs such as ODBC, JDBC and OLEDB, there was no such unification in the OLAP world for a long time. The first real standard API was OLE DB for OLAP specification from Microsoft which appeared in 1997 and introduced the MDX query language. Several OLAP vendors-both server and client-adopted it. In 2001 Microsoft and Hyperion announced the XML for Analysis specification, which was endorsed by most of the OLAP vendors. Since this also used MDX as a query language, MDX became the *de facto* standard. Since September-2011 LINQ can be used to query SSAS OLAP cubes from Microsoft.NET.

Products

History

The first product that performed OLAP queries was *Express,* which was released in 1970 (and acquired by Oracle in 1995 from Information Resources). However, the term did not appear until 1993 when it was coined by Edgar F. Codd, who has been described as "the father of the relational database". Codd's paper resulted from a short consulting assignment which Codd undertook for former Arbor Software (later Hyperion Solutions, and in 2007 acquired by Oracle), as a sort of marketing coup. The company had released its own OLAP product, *Essbase,* a year earlier. As a result Codd's "twelve laws of online analytical processing" were explicit in their reference to Essbase. There was some ensuing controversy and when Computer world learned that Codd was paid by Arbor, it retracted the article. OLAP market experienced strong growth in late 90s with dozens of commercial products going into market. In 1998, Microsoft released its first OLAP

Server-Microsoft Analysis Services, which drove wide adoption of OLAP technology and moved it into mainstream.

Product Comparison

Market Structure

Below is a list of top OLAP vendors in 2006, with figures in millions of US Dollars.

Vendor	*Global Revenue*	*Consolidated company*
Microsoft Corporation	1,806	Microsoft
Hyperion Solutions Corporation	1,077	Oracle
Cognos	735	IBM
Business Objects	416	SAP
MicroStrategy	416	MicroStrategy
SAP AG	330	SAP
Cartesis (SAP)	210	SAP
Applix	205	IBM
Infor	199	Infor
Oracle Corporation	159	Oracle
Others	152	Others
Total	**5,700**	

MASTER DATA MANAGEMENT

In business, **master data management (MDM)** comprises the processes, governance, policies, standards and tools that consistently define and manage the critical data of an *organization* to provide a single point of reference.

The data that is mastered may include :

- *Master data*—the business objects for transactions, and the dimensions for analysis
- Transactional data—supports applications
- Analytical data—supports decision-making.

In *computing*, an MDM tool can be used to support master data management by removing duplicates, standardizing data (mass maintaining), and incorporating rules to eliminate incorrect data from entering the system in order to create an authoritative source of master data. Master data are the products, accounts and parties for which the business transactions are completed. The root cause problem stems from business unit and product line segmentation, in which the same customer will be serviced by different product lines, with redundant data being entered about the customer (aka party in the role of customer) and account in order to process the transaction. The redundancy of party and account data is compounded in the front to back office life cycle, where the authoritative single source for the party, account and product data is needed but is often once again redundantly entered or augmented.

MDM has the objective of providing processes for collecting, aggregating, matching, consolidating, quality-assuring, persisting and distributing such data throughout an organization to ensure consistency and control in the ongoing maintenance and application use of this information.

The term recalls the concept of a *master file* from an earlier computing era.

Issues

At a basic level, MDM seeks to ensure that an organization does not use multiple (potentially inconsistent) versions of the same master data in different parts of its operations, which can occur in large organizations. A common example of poor MDM is the scenario of a bank at which a customer has taken out a mortgage and the bank begins to send mortgage solicitations to that customer, ignoring the fact that the person already has a mortgage account relationship with the bank. This happens because the customer information used by the marketing section within the bank lacks integration with the customer information used by the customer services section of the bank. Thus the two groups remain unaware that an existing customer is also considered a sales lead. The process of record linkage is used to associate different records that correspond to the same entity, in this case the same person.

Other problems include (for example) issues with the quality of data, consistent classification and identification of data, and data-reconciliation issues. Master data management of disparate data systems requires data transformations as the data extracted from the disparate source data system is transformed and loaded into the master data management hub. To synchronize the disparate source master data, the managed master data extracted from the master data management hub is again transformed and loaded into the disparate source data system as the master data is updated. As with other Extract, Transform, Load-based data movement, these processes are expensive and inefficient to develop and to maintain which greatly reduces the return on investment for the master data management product.

One of the most common reasons some large corporations experience massive issues with MDM is growth through mergers or acquisitions. Two organizations which merge will typically create an entity with duplicate master data (since each likely had at least one master database of its own prior to the merger). Ideally, database administrators resolve this problem through deduplication of the master data as part of the merger. In practice, however, reconciling several master data systems can present difficulties because of the dependencies that existing applications have on the master databases. As a result, more often than not the two systems do not fully merge, but remain separate, with a special reconciliation process defined that ensures consistency between the data stored in the two systems. Over time, however, as further mergers and acquisitions occur, the problem multiplies, more and more master databases appear, and data-reconciliation processes become extremely complex, and consequently unmanageable and unreliable. Because of this trend, one can find organizations with 10, 15, or even as many as 100 separate, poorly integrated master databases, which can cause serious operational problems in the areas of customer satisfaction, operational efficiency, decision-support, and regulatory compliance.

Solutions

Processes commonly seen in MDM include source identification, data collection, data transformation, normalization, rule administration, error detection and correction, data consolidation, data storage, data distribution, data classification, taxonomy services, item master creation, schema mapping, product codification, data enrichment and data governance.

The tools include data networks, file systems, a data warehouse, data marts, an operational data store, data-mining, data analysis, data virtualization, data federation and data visualization. One of the newest tools, virtual master data management utilizes data virtualization and a persistent metadata server to implement a multi-level automated MDM hierarchy.

The selection of entities considered for MDM depends somewhat on the nature of an organization. In the common case of commercial enterprises, MDM may apply to such entities as customer (customer data integration), product (product information management), employee, and vendor. MDM processes identify the sources from which to collect descriptions of these entities. In the course of transformation and normalization, administrators adapt descriptions to conform to standard formats and data domains, making it possible to remove duplicate instances of any entity. Such processes generally result in an organizational MDM repository, from which all requests for a certain entity instance produce the same description, irrespective of the originating sources and the requesting destination.

OPERATIONAL DATA STORE

An **operational data store** (or "**ODS**") is a *database* designed to *integrate data* from multiple sources for additional operations on the data. The data is then passed back to *operational systems* for further operations and to the *data warehouse* for reporting.

Because the *data* originates from multiple sources, the integration often involves *cleaning*, resolving redundancy and checking against *business rules* for *integrity*. An ODS is usually designed to contain low-level or atomic (indivisible) data (such as transactions and prices) with limited history that is captured "real time" or "near real time" as opposed to the much greater volumes of data stored in the data warehouse generally on a less-frequent basis.

General Use

The general purpose of an ODS is to integrate data from disparate source systems in a single structure, using data integration technologies like data virtualization, data federation, or extract, transform, and load. This will allow operational access to the data for operational reporting, master data or reference data management.

An ODS is not a replacement or substitute for a data warehouse but in turn could become a source.

REFERENTIAL INTEGRITY

artist_id	artist_name
1	Bono
2	Cher
3	Nuno Bettencourt

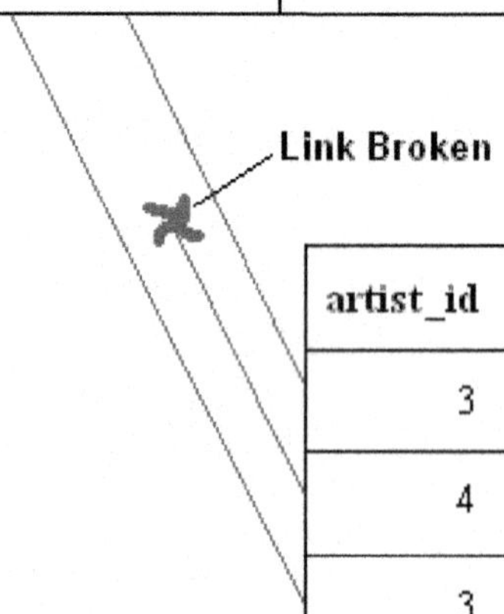

artist_id	album_id	album_name
3	1	Schizophonic
4	2	Eat the rich
3	3	Crave (single)

An example of a database that has not enforced **referential integrity**. In this example, there is a foreign key (artist_id) value in the album table that references a non-existent artist — in other words there is a foreign key value with no corresponding primary key value in the referenced table. What happened here was that there was an artist called "Aerosmith", with an artist_id of 4, which was deleted from the artist table. However, the album "Eat the Rich" referred to this artist. With referential integrity enforced, this would not have been possible.

Referential integrity is a property of data which, when satisfied, requires every value of one attribute (column) of a relation (table) to exist as a value of another attribute in a different (or the same) relation (table).

For referential integrity to hold in a relational database, any field in a table that is declared a foreign key can contain either a null value, or only values from a parent table's primary key or a candidate key. In other words, when a foreign key value is used it must reference a valid, existing primary key in the parent table. For instance, deleting a record that contains a value referred to by a foreign key in another table would break referential integrity. Some relational database management systems (RDBMS) can enforce referential integrity, normally either by deleting the foreign key rows as well to maintain integrity, or by returning an error and not performing the delete. Which method is used may be determined by a referential integrity constraint defined in a data dictionary.

"Referential" the adjective describes the action that a foreign key performs, 'referring' to a link field in another table. In simple terms, 'referential integrity' is a guarantee that the target it 'refers' to will be found. A lack of referential integrity in a database can lead relational databases to return incomplete data, usually with no indication of an error. A common problem occurs with relational database tables linked with an 'inner join' which requires non-NULL values in both tables, a requirement that can only be met through careful design and referential integrity.

Formalization

An **inclusion dependency** over two (possibly identical) predicates R and S from a schema is written $R[A_1,..., A_n] \subseteq S [B_1,..., B_n]$, where the A_i, B_i are distinct attributes (column names) of R and S. It implies that the tuples of values appearing in columns $A_1,..., A_n$ for facts of R must also appear as a tuple of values in columns $B_1,..., B_n$ for some fact of S.

Logical implication between inclusion dependencies can be axiomatized by inference rules and can be decided by a PSPACE algorithm. The problem can be shown to be PSPACE-complete by reduction from the acceptance problem for a linear bounded automaton. However, logical implication between dependencies that can be inclusion dependencies or functional dependencies is undecidable by reduction from the word problem for monoids.

BUSINESS PERFORMANCE MANAGEMENT

Business performance management is a set of *management* and analytic processes that enables the management of an *organization*'s performance to achieve one or more pre-selected goals. Synonyms for "business performance management" include **"corporate performance management (CPM)"** and **"enterprise performance management"**.

Business performance management is contained within approaches to *business process management*.

Business performance management has three main activities :

1. Selection of goals,
2. *Consolidation* of measurement information relevant to an organization's progress against these goals, and
3. Interventions made by managers in light of this information with a view to improving future performance against these goals.

Although presented here sequentially, typically all three activities will run concurrently, with interventions by managers affecting the choice of goals, the measurement information monitored, and the activities being undertaken by the organization.

Because business performance management activities in large organizations often involve the collation and reporting of large volumes of data, many software vendors, particularly those offering *business intelligence* tools, market products intended to assist in this process. As a result of this marketing effort, business performance management is often incorrectly understood as an activity that necessarily relies on software systems to work, and many definitions of business performance management explicitly suggest software as being a definitive component of the approach.

This interest in business performance management from the software community is sales-driven— "The biggest growth area in operational BI analysis is in the area of business performance management."

Since 1992, business performance management has been strongly influenced by the rise of the *balanced scorecard* framework. It is common for managers to use the balanced scorecard framework to clarify the goals of an organization, to identify how to track them, and to structure the mechanisms by which interventions will be triggered. These steps are the same as those that are found in BPM, and as a result balanced scorecard is often used as the basis for business performance management activity with organizations.

In the past, owners have sought to drive strategy down and across their organizations, transform these strategies into actionable metrics and use analytics to expose the cause-and-effect relationships that, if understood, could give insight into decision-making.

History

Reference to non-business performance management occurs in Sun Tzu's *The Art of War*. Sun Tzu claims that to succeed in war, one should have full knowledge of one's own strengths and weaknesses as well as those of one's enemies. Lack of either set of knowledge might result in defeat. Parallels between the challenges in business and those of war include :

- Collecting data-both internal and external
- Discerning patterns and meaning in the data (analyzing)
- Responding to the resultant information.

Prior to the start of the Information Age in the late 20th century, businesses sometimes took the trouble to laboriously collect data from non-automated sources. As they lacked computing resources to properly analyze the data, they often made commercial decisions primarily on the basis of intuition.

As businesses started automating more and more systems, more and more data became available. However, collection often remained a challenge due to a lack of infrastructure for data exchange or due to incompatibilities between systems. Reports on the data gathered sometimes took months to generate. Such reports allowed informed long-term strategic decision-making. However, short-term tactical decision-making often continued to rely on intuition.

In 1989 Howard Dresner, a research analyst at Gartner, popularized "business intelligence" (BI) as an umbrella term to describe a set of concepts and methods to improve business decision-making by using fact-based support systems. Performance management builds on a foundation of BI, but marries it to the planning-and-control cycle of the enterprise—with enterprise planning, consolidation and modelling capabilities.

Increasing standards, automation, and technologies have led to vast amounts of data becoming available. Data warehouse technologies have allowed the building of repositories to store this data. Improved ETL and enterprise application integration tools have increased the timely collecting of data. OLAP reporting technologies have allowed faster generation of new reports which analyze the data. As of 2010, business intelligence has become the art of sieving through large amounts of data, extracting useful information and turning that information into actionable knowledge.

Definition and Scope

Business performance management consists of a set of management and analytic processes, supported by technology, that enable businesses to define strategic goals and then measure and manage performance against those goals. Core business performance management processes include financial planning, operational planning, business modelling, consolidation and reporting, analysis, and monitoring of key performance indicators linked to strategy.

Business performance management involves consolidation of data from various sources, querying, and analysis of the data, and putting the results into practice.

Methodologies

Various methodologies for implementing business performance management exist. The discipline gives companies a top-down framework by which to align planning and execution, strategy and tactics, and business-unit and enterprise objectives. Reactions may include the Six Sigma strategy, balanced scorecard, activity-based costing (ABC), Total Quality Management, economic value-add, integrated strategic measurement and Theory of Constraints.

The balanced scorecard is the most widely adopted performance management methodology.

Methodologies on their own cannot deliver a full solution to an enterprise's CPM needs. Many pure-methodology implementations fail to deliver the anticipated benefits due to lack of integration with fundamental CPM processes.

Metrics and Key Performance Indicators

Some of the areas from which bank management may gain knowledge by using business performance management include :

- *Customer-related numbers :*
    - New customers acquired
    - Status of existing customers
    - Attrition of customers (including breakup by reason for attrition).
- Turnover generated by segments of the customers-possibly using demographic filters
- Outstanding balances held by segments of customers and terms of payment-possibly using demographic filters
- Collection of bad debts within customer relationships
- Demographic analysis of individuals (potential customers) applying to become customers, and the levels of approval, rejections and pending numbers
- Delinquency analysis of customers behind on payments
- Profitability of customers by demographic segments and segmentation of customers by profitability
- Campaign management
- Real-time dashboard on key operational metrics
    - overall equipment effectiveness,
- Clickstream analysis on a website
- Key product portfolio trackers
- Marketing-channel analysis
- Sales-data analysis by product segments
- Call center metrics.

Though the above list describes what a bank might monitor, it could refer to a telephone company or to a similar service-sector company.

Items of generic importance include :

1. Consistent and correct KPI-related data providing insights into operational aspects of a company
2. Timely availability of KPI-related data
3. KPIs designed to directly reflect the efficiency and effectiveness of a business
4. Information presented in a format which aids decision-making for management and decision-makers
5. Ability to discern patterns or trends from organized information.

Business performance management integrates the company's processes with CRM or ERP. Companies should become better able to gauge customer satisfaction, control customer trends and influence shareholder value.

Application Software Types

People working in business intelligence have developed tools that ease the work of business performance management, especially when the business-intelligence task involves gathering and analyzing large amounts of unstructured data.

Tool categories commonly used for business performance management include :

- MOLAP—Multi-dimensional online analytical processing, sometimes simply called "analytics" (based on dimensional analysis and the so-called "hyper-cube" or "cube")
- Scorecarding, dashboarding and data visualization
- Data warehouses
- Document warehouses
- Text mining
- DM—data-mining
- BPO—business performance optimization
- EPM—enterprise performance management
- EIS—executive information systems
- DSS—decision support systems
- MIS—management information systems
- SEMS—strategic enterprise management software.

Design and Implementation

Questions asked when implementing a business performance management program include :

- *Goal-alignment queries :* Determine the short-and medium-term purpose of the program. What strategic goal(s) of the organization will the program address? What organizational mission/vision does it relate to? A hypothesis needs to be crafted that details how this initiative will eventually improve results/ performance (*i.e.* a strategy map).
- *Baseline queries :* Assess current information-gathering competency. Does the organization have the capability to monitor important sources of information? What data is being collected and how is it being stored? What are the statistical parameters of this data, *e.g.*, how much random variation does it contain? Is this being measured?
- *Cost and risk queries :* Estimate the financial consequences of a new BI initiative. Assess the cost of the present operations and the increase in costs associated with the BPM initiative. What is the risk that the initiative will fail? This risk assessment should be converted into a financial metric and included in the planning.

- *Customer and stakeholder queries* : Determine who will benefit from the initiative and who will pay. Who has a stake in the current procedure? What kinds of customers/stakeholders will benefit directly from this initiative? Who will benefit indirectly? What quantitative/qualitative benefits follow? Is the specified initiative the best or only way to increase satisfaction for all kinds of customers? How will customer benefits be monitored? What about employees, shareholders, and distribution channel members?
- *Metrics-related queries* : Information requirements need operationalization into clearly defined metrics. Decide which metrics to use for each piece of information being gathered. Are these the best metrics and why? How many metrics need to be tracked? If this is a large number (it usually is), what kind of system can track them? Are the metrics standardized, so they can be benchmarked against performance in other organizations? What are the industry standard metrics available?
- *Measurement methodology-related queries* : Establish a methodology or a procedure to determine the best (or acceptable) way of measuring the required metrics. How frequently will data be collected? Are there any industry standards for this? Is this the best way to do the measurements? How do we know that?
- *Results-related queries* : Monitor the BPM program to ensure that it meets objectives. The program itself may require adjusting. The program should be tested for accuracy, reliability, and validity. How can it be demonstrated that the BI initiative, and not something else, contributed to a change in results? How much of the change was probably random?

BUSINESS ANALYTICS

Business analytics (BA) refers to the skills, technologies, applications and practices for continuous iterative exploration and investigation of past business performance to gain insight and drive business planning. Business analytics focuses on developing new insights and understanding of business performance based on data and statistical methods. In contrast, business intelligence traditionally focuses on using a consistent set of metrics to both measure past performance and guide business planning, which is also based on data and statistical methods.

Business analytics makes extensive use of data, statistical and quantitative analysis, explanatory and predictive modelling, and fact-based management to drive decision-making. It is therefore closely related to management science. Analytics may be used as input for human decisions or may drive fully automated decisions. Business intelligence is querying, reporting, OLAP, and "alerts."

In other words, querying, reporting, OLAP, and alert tools can answer questions such as what happened, how many, how often, where the problem is, and what actions are needed. Business analytics can answer questions like why is this happening, what if these trends continue, what will happen next (that is, predict), what is the best that can happen (that is, optimize).

Examples of Application

Banks, such as Capital One, use data analysis (or *analytics*, as it is also called in the business setting), to differentiate among customers based on credit risk, usage and other characteristics and then to match customer characteristics with appropriate product offerings. Harrah's, the gaming firm, uses analytics in its customer loyalty programs. E & J Gallo Winery quantitatively analyzes and predicts the appeal of its wines. Between 2002 and 2005, Deere & Company saved more than $1 billion by employing a new analytical tool to better optimize inventory.

Types of Analytics

- *Descriptive Analytics :* Gain insight from historical data with reporting, score-cards, clustering etc.
- Predictive analytics (predictive modelling using statistical and machine learning techniques)
- Prescriptive analytics recommend decisions using optimization, simulation etc.
- *Decisive analytics :* Supports human decisions with visual analytics the user models to reflect reasoning.

Basic Domains Within Analytics

- Retail sales analytics
- Financial services analytics
- Risk & Credit analytics
- Talent analytics
- Marketing analytics
- Behavioural analytics
- Cohort Analysis
- Collections analytics
- Fraud analytics
- Pricing analytics
- Telecommunications
- Supply Chain analytics
- Transportation analytics
- Contextual data modelling-supports the human reasoning that occurs after viewing "executive dashboards" or any other visual analytics

History

Analytics have been used in business since the time management exercises that were initiated by Frederick Winslow Taylor in the late 19th century. Henry Ford measured pacing of assembly line. But analytics began to command more attention in the late 1960s when computers were used in decision support systems. Since then, analytics have evolved with the development of enterprise resource planning (ERP) systems, data warehouses, and a wide variety of other hardware and software tools and applications.

With the recent explosion of big data and intuitive BI tools, data is more accessible to business professionals and managers than ever before. Thus there is a big opportunity to make better decisions using that data to drive incremental revenue, decrease cost and loss by building better products, improving customer experience, catching fraud before it happens, improving customer engagement through targeting and customization—all with the power of data. More and more companies are now equipping their employees with the know—how of Business Analytics to drive efficiency in day-to-day decision-making.

Challenges

Business analytics depends on sufficient volumes of high quality data. The difficulty in ensuring data quality is integrating and reconciling data across different systems, and then deciding what sub-sets of data to make available.

Previously, analytics was considered a type of after-the-fact method of forecasting consumer behaviour by examining the number of units sold in the last quarter or the last year. This type of data warehousing required a lot more storage space than it did speed. Now business analytics is becoming a tool that can influence the outcome of customer interactions. When a specific customer type is considering a purchase, an analytics-enabled enterprise can modify the sales pitch to appeal to that consumer. This means the storage space for all that data must react extremely fast to provide the necessary data in real-time.

Competing on Analytics

Davenport argues that businesses can optimize a distinct business capability via analytics and thus better compete. He identifies these characteristics of an organization that are apt to compete on analytics :

- One or more senior executives who strongly advocate fact-based decision-making and, specifically, analytics.
- Widespread use of not only descriptive statistics, but also predictive modelling and complex optimization techniques.
- Substantial use of analytics across multiple business functions or processes.
- Movement toward an enterprise level approach to managing analytical tools, data, and organizational skills and capabilities.

COMPETITIVE INTELLIGENCE

Competitive intelligence is the action of defining, gathering, analyzing, and distributing *intelligence* about products, customers, competitors, and any aspect of the environment needed to support executives and managers making strategic decisions for an organization.

Competitive intelligence essentially means understanding and learning what's happening in the world outside your business so you can be as competitive as possible. It means learning as much as possible—as soon as possible—about your industry in general, your competitors, or even your county's particular zoning rules. In short, it empowers you to anticipate and face challenges head on.

Key points of this definition :

1. Competitive intelligence is an ethical and legal business practice, as opposed to *industrial espionage*, which is illegal.
2. The focus is on the external business environment.
3. There is a process involved in gathering information, converting it into intelligence and then utilizing this in business decision-making. Some CI professionals erroneously emphasize that if the intelligence gathered is not usable, or actionable, then it is not intelligence.

A more focused definition of CI regards it as the organizational function responsible for the early identification of risks and opportunities in the market before they become *obvious*. Experts also call this process the early signal analysis. This definition focuses attention on the difference between dissemination of widely available factual information (such as market statistics, financial reports, newspaper clippings) performed by functions such as libraries and information centers, and competitive intelligence which is a *perspective* on developments and events aimed at yielding a competitive edge.

The term CI is often viewed as synonymous with *competitor analysis*, but competitive intelligence is more than analyzing competitors—it is about making the organization more competitive relative to its entire environment and stakeholders: customers, competitors, distributors, technologies, and macroeconomic data.

Historic Development

The literature associated with the field of competitive intelligence is best exemplified by the detailed bibliographies that were published in the Society of Competitive Intelligence Professionals' refereed academic journal called *The Journal of Competitive Intelligence and Management*. Although elements of organizational intelligence collection have been a part of business for many years, the history of competitive intelligence arguably began in the U.S. in the 1970s, although the literature on the field pre-dates this time by at least several decades. In 1980, Michael Porter published the study *Competitive-Strategy : Techniques for Analyzing Industries and Competitors* which is widely viewed as the foundation of modern competitive intelligence. This has since been extended most notably by the pair of Craig Fleisher and Babette Bensoussan, who through several popular books on competitive analysis have added 48 commonly applied competitive intelligence analysis techniques to the practitioner's tool box. In 1985, Leonard Fuld published his best selling book dedicated to competitor intelligence. However, the institutionalization of CI as a formal activity among American corporations can be traced to 1988, when Ben and Tamar Gilad published the first organizational model of a formal corporate CI function, which was then adopted widely by US companies. The first professional certification program (CIP) was created in 1996 with the establishment of The Fuld-Gilad-Herring Academy of Competitive Intelligence in Cambridge, Massachusetts.

In 1972 the *Society of Competitive Intelligence Professionals* (SCIP) was founded in the United States and grew in the late 1990s to around 6,000 members worldwide, mainly in the United States and Canada, but with large numbers especially in the UK and Australia. Due to financial difficulties in 2009, the organization merged with Frost & Sullivan under the Frost & Sullivan Institute. SCIP has since been renamed "Strategic & Competitive Intelligence Professionals" to emphasise the strategic nature of the subject, and also to refocus the organisation's general approach, while keeping the existing SCIP brandname and logo. A number of efforts have been made to discuss the field's advances in post-secondary (university) education, covered by several authors including Blenkhorn & Fleisher, Fleisher, Fuld, Prescott, and McGonagle. Although the general view would be that competitive intelligence concepts can be readily found and taught in many business schools around the globe, there are still relatively few dedicated academic programs, majors, or degrees in the field, a concern to academics in the field who would like to see it further researched. These issues were widely discussed by over a dozen knowledgeable individuals in a special edition of the Competitive Intelligence Magazine that was dedicated to this topic. In France, a Specialized Master in Economic Intelligence and Knowledge Management was created in 1995 within the CERAM Business School, now SKEMA Business School, in Paris, with the objective of delivering a full and professional training in Economic Intelligence. A Centre for Global Intelligence and Influence was created in September 2011 in the same School.

On the other hand, practitioners and companies regard professional accreditation as especially important in this field. In 2011, SCIP recognized the Fuld-Gilad-Herring Academy of Competitive Intelligence's CIP certification process as its global, dual-level (CIP-I and CIP-II) certification program.

Global developments have also been uneven in competitive intelligence. Several academic journals, particularly the *Journal of Competitive Intelligence and Management* in its third volume, provided coverage of the field's global development. For example, in 1997 the *Ecole de Guerre Economique* (*School of economic warfare*) was founded in Paris, France. It is the first European institution which teaches the tactics of economic warfare within a globalizing world. In Germany, competitive intelligence was unattended until the early 1990s. The term "competitive intelligence" first appeared in German literature in 1997. In 1995 a German SCIP chapter was founded, which is now second in terms of membership in Europe. In summer 2004 the Institute for Competitive Intelligence was founded, which provides a postgraduate certification program for Competitive Intelligence Professionals. Japan is currently the only country that officially maintains an economic intelligence agency (JETRO). It was founded by the Ministry of International Trade and Industry (MITI) in 1958.

Accepting the importance of competitive intelligence, major multi-national corporations, such as ExxonMobil, Procter & Gamble, and Johnson and Johnson, have created formal CI units. Importantly, organizations execute competitive intelligence activities not only as a safeguard to protect against market threats and changes, but also as a method for finding new opportunities and trends.

Organizations use competitive intelligence to compare themselves to other organizations ("competitive benchmarking"), to identify risks and opportunities in their markets, and to pressure-test their plans against market response (war gaming), which enable them to make informed decisions. Most firms today realize the importance of knowing what their competitors are doing and how the industry is changing, and the information gathered allows organizations to understand their strengths and weaknesses.

One of the major activities involved in corporate competitive intelligence is use of ratio analysis, using key performance indicators (KPI). Organizations compare annual reports of their competitors on certain KPI and ratios, which are intrinsic to their industry. This helps them track their performance, *vis-a-vis* their competitors.

The actual importance of these categories of information to an organization depends on the contestability of its markets, the organizational culture, the personality and biases of its top decision makers, and the reporting structure of competitive intelligence within the company.

Strategic Intelligence (SI) focuses on the longer term, looking at issues affecting a company's competitiveness over the course of a couple of years. The actual time horizon for SI ultimately depends on the industry and how quickly it's changing. The general questions that SI answers are, 'Where should we as a company be in X years?' and 'What are the strategic risks and opportunities facing us?' This type of intelligence work involves among others the identification of weak signals and application of methodology and process called Strategic Early Warning (SEW), first introduced by Gilad, followed by Steven Shaker and Victor Richardson, Alessandro Comai and Joaquin Tena, and others. According to Gilad, 20% of the work of competitive intelligence practitioners should be dedicated to strategic early identification of weak signals within a SEW framework.

Tactical Intelligence : The focus is on providing information designed to improve shorter-term decisions, most often related with the intent of growing market share or revenues. Generally, it is the type of information that you would need to support the sales process in an organization. It investigates various aspects of a product/product line marketing :

- Product—what are people selling?
- Price—what price are they charging?
- Promotion—what activities are they conducting for promoting this product?

- Place—where are they selling this product?
- Other—sales force structure, clinical trial design, technical issues, etc.

With the right amount of information, organizations can avoid unpleasant surprises by anticipating competitors' moves and decreasing response time. Examples of competitive intelligence research is evident in daily newspapers, such as the *Wall Street Journal*, *Business Week*, and *Fortune*. Major airlines change hundreds of fares daily in response to competitors' tactics. They use information to plan their own marketing, pricing, and production strategies.

Resources, such as the Internet, have made gathering information on competitors easy. With a click of a button, analysts can discover future trends and market requirements. However competitive intelligence is much more than this, as the ultimate aim is to lead to competitive advantage. As the Internet is mostly public domain material, information gathered is less likely to result in insights that will be unique to the company. In fact there is a risk that information gathered from the Internet will be misinformation and mislead users, so competitive intelligence researchers are often wary of using such information.

As a result, although the Internet is viewed as a key source, most CI professionals should spend their time and budget gathering intelligence using primary research—networking with industry experts, from trade shows and conferences, from their own customers and suppliers, and so on. Where the Internet is used, it is to gather sources for primary research as well as information on what the company says about itself and its online presence (in the form of links to other companies, its strategy regarding search engines and online advertising, mentions in discussion forums and on blogs, etc.). Also, important are online subscription databases and news aggregation sources which have simplified the secondary source collection process. Social media sources are also becoming important—providing potential interviewee names, as well as opinions and attitudes, and sometimes breaking news (*e.g.*, via Twitter).

Organizations must be careful not to spend too much time and effort on old competitors without realizing the existence of any new competitors. Knowing more about your competitors will allow your business to grow and succeed. The practice of competitive intelligence is growing every year, and most companies and business students now realize the importance of knowing their competitors.

According to Arjan Singh and Andrew Beurschgens in their 2006 article in the *Competitive Intelligence Review*, there are four stages of development of a competitive intelligence capability with a firm. It starts with "stick fetching", where a CI department is very reactive, up to "world class", where it is completely integrated in the decision-making process.

Recent Trends

The technical advances in massive parallel processing offered by the Hadoop "big data" architecture has allowed the creation of multiple platforms for named-entity recognition such as the Apache Projects Open NLP and Apache Stanbol. The former includes pre-trained statistical parsers that can discern elements key to establishing trends and evaluating competitive position and responding appropriately. Public information mining from *SEC.gov*, Federal Contract Awards, social media (Twitter, Reddit, Facebook, and others), vendors, and competitor websites now permit real-time counterintelligence as a strategy for horizontal and vertical market expansion and product positioning. This occurs in an automated fashion on massive marketplaces such as Amazon.com and their classification and prediction of product associations and purchase probability.

Similar Fields

Competitive intelligence has been influenced by national strategic intelligence. Although national intelligence was researched 50 years ago, competitive intelligence was introduced during the 1990s. Competitive-intelligence professionals can learn from national-intelligence experts, especially in the analysis of complex situations. Competitive intelligence may be confused with (or seen to overlap) environmental scanning, business intelligence and market research. Craig Fleisher questions the appropriateness of the term, comparing it to business intelligence, competitor intelligence, knowledge management, market intelligence, marketing research and strategic intelligence.

Fleisher suggests that business intelligence has two forms. Its narrow (contemporary) form is more focused on information technology and internal focus than CI, while its broader (historical) definition is more inclusive than CI. Knowledge management (KM), when improperly achieved, is seen as an information-technology driven organizational practice relying on data-mining, corporate intranets and mapping organizational assets to make it accessible to organization members for decision-making. CI shares some aspects of KM; they are human-intelligence-and experience-based for a more-sophisticated qualitative analysis. KM is essential for effective change. A key effective factor is a powerful, dedicated IT system executing the full intelligence cycle.

Market intelligence (MI) is industry-targeted intelligence developed in real-time aspects of competitive events taking place among the four Ps of the marketing mix (pricing, place, promotion and product) in the product (or service) marketplace to better understand the market's attractiveness. A time-based competitive tactic, MI is used by marketing and sales managers to respond to consumers more quickly in the marketplace. Fleisher suggests it is not distributed as widely as some forms of CI, which are also distributed to non-marketing decision-makers. Market intelligence has a shorter time horizon than other intelligence areas, and is measured in days, weeks, or (in slower-moving industries) months.

Market research is a tactical, method-driven field consisting of neutral, primary research of customer data (beliefs and perceptions) gathered in surveys or focus groups, and is analyzed with statistical-research techniques. CI draws on a wider variety (primary and secondary) of sources from a wider range of stakeholders (suppliers, competitors, distributors, substitutes and media) to answer existing questions, raise new ones and guide action.

Ben Gilad and Jan Herring lay down a set of prerequisites defining CI, distinguishing it from other information-rich disciplines such as market research or business development. They show that a common body of knowledge and a unique set of tools (key intelligence topics, business war games and blindspots analysis) distinguish CI; while other sensory activities in a commercial firm focus on *one* segment of the market (customers, suppliers or acquisition targets), CI is synthesizes data from all high-impact players (HIP).

Gilad later focused his delineation of CI on the difference between information and intelligence. According to him, the common denominator among organizational sensory functions (whether they are called market research, business intelligence or market intelligence) is that they deliver information rather than intelligence. Intelligence, says Gilad, is a perspective on facts rather than the facts themselves. Unique among corporate functions, competitive intelligence has a perspective of risks and opportunities for a firm's performance; as such, it (not information activities) is part of an organization's risk-management activity.

Ethics

Ethics has been a long-held issue of discussion among CI practitioners. Essentially, the questions revolve around what is and is not allowable in terms of CI practitioners' activity. A number of very excellent scholarly treatments have been generated on this topic, most prominently addressed through Society of Competitive Intelligence Professionals publications. The book *Competitive Intelligence Ethics : Navigating the Gray Zone* provides nearly twenty separate views about ethics in CI, as well as another 10 codes used by various individuals or organizations. Combining that with the over two dozen scholarly articles or studies found within the various CI bibliographic entries, it is clear that no shortage of study has gone into better classifying, understanding and addressing CI ethics.

Competitive information may be obtained from public or subscription sources, from networking with competitor staff or customers, disassembly of competitor products or from field research interviews. Competitive intelligence research is distinguishable from industrial espionage, as CI practitioners generally abide by local legal guidelines and ethical business norms.

COMPETITOR ANALYSIS

Competitor analysis in *marketing* and strategic *management* is an assessment of the strengths and weaknesses of current and potential *competitors*. This analysis provides both an offensive and defensive strategic context to identify opportunities and threats. Profiling coalesces all of the relevant sources of competitor analysis into one framework in the support of efficient and effective strategy formulation, implementation, monitoring and adjustment.

Competitor analysis is an essential component of corporate strategy. It is argued that most firms do not conduct this type of analysis systematically enough. Instead, many enterprises operate on what is called "informal impressions, conjectures, and intuition gained through the tidbits of information about competitors every manager continually receives." As a result, traditional environmental scanning places many firms at risk of dangerous competitive blindspots due to a lack of robust competitor analysis.

Competitor Array

One common and useful technique is constructing a *competitor array*. The steps include :

- Define your industry—scope and nature of the industry
- Determine who your competitors are
- Determine who your customers are and what benefits they expect
- Determine what the key success factors are in your industry
- Rank the key success factors by giving each one a weighting—The sum of all the weightings must add up to one.
- Rate each competitor on each of the key success factors
- Multiply each cell in the matrix by the factor weighting.

This can best be displayed on a two dimensional matrix-competitors along the top and key success factors down the side. An example of a competitor array follows :

Key Industry Success Factors	*Weighting*	*Competitor #1 rating*	*Competitor #1 weighted*	*Competitor #2 rating*	*Competitor #2 weighted*
1. Extensive distribution	.4	6	2.4	3	1.2
2. Customer focus	.3	4	1.2	5	1.5
3. Economies of scale	.2	3	.6	3	.6
4. Product innovation	.1	7	.7	4	.4
Totals	**1.0**	**20**	**4.9**	**15**	**3.7**

In this example competitor #1 is rated higher than competitor #2 on product innovation ability (7 out of 10, compared to 4 out of 10) and distribution networks (6 out of 10), but competitor #2 is rated higher on customer focus (5 out of 10). Overall, competitor #1 is rated slightly higher than competitor #2 (20 out of 40 compared to 15 out of 40). When the success factors are weighted according to their importance, competitor #1 gets a far better rating (4.9 compared to 3.7).

Two additional columns can be added. In one column you can rate your own company on each of the key success factors (try to be objective and honest). In another column you can list benchmarks. They are the ideal standards of comparisons on each of the factors. They reflect the workings of a company using all the industry's best practices.

Competitor Profiling

The strategic rationale of competitor profiling is powerfully simple. Superior knowledge of rivals offers a legitimate source of competitive advantage. The raw material of competitive advantage consists of offering superior customer value in the firm's chosen market. The definitive characteristic of customer value is the adjective, superior. Customer value is defined relative to rival offerings making competitor knowledge an intrinsic component of corporate strategy. Profiling facilitates this strategic objective in three important ways. First, profiling can reveal strategic weaknesses in rivals that the firm may exploit. Second, the proactive stance of competitor profiling will allow the firm to anticipate the strategic response of their rivals to the firm's planned strategies, the strategies of other competing firms, and changes in the environment. Third, this proactive knowledge will give the firms strategic agility. Offensive strategy can be implemented more quickly in order to exploit opportunities and capitalize on strengths. Similarly, defensive strategy can be employed more deftly in order to counter the threat of rival firms from exploiting the firm's own weaknesses.

Clearly, those firms practicing systematic and advanced competitor profiling have a significant advantage. As such, a comprehensive profiling capability is rapidly becoming a core competence required for successful competition. An appropriate analogy is to consider this advantage as akin to having a good idea of the next move that your opponent in a chess match will make. By staying one move ahead, checkmate is one step closer. Indeed, as in chess, a good offense is the best defense in the game of business as well.

A common technique is to create detailed profiles on each of your major competitors. These profiles give an in-depth description of the competitor's background, finances, products, markets, facilities, personnel, and strategies. This involves :

- Background
    - Location of offices, plants, and online presences
    - o History-key personalities, dates, events, and trends
    - o Ownership, corporate governance, and organizational structure.
- Financials
    - P-E ratios, dividend policy, and profitability
    - Various financial ratios, liquidity, and cash flow
    - Profit growth profile; method of growth (organic or acquisitive).
- Products
    - Products offered, depth and breadth of product line, and product portfolio balance
    - New products developed, new product success rate, and R&D strengths
    - Brands, strength of brand portfolio, brand loyalty and brand awareness
    - Patents and licenses

- Quality control conformance
- Reverse engineering or deformulation.

- Marketing
    - Segments served, market shares, customer base, growth rate, and customer loyalty
    - Promotional mix, promotional budgets, advertising themes, ad agency used, sales force success rate, online promotional strategy
    - Distribution channels used (direct & indirect), exclusivity agreements, alliances, and geographical coverage
    - Pricing, discounts, and allowances.
- Facilities
    - Plant capacity, capacity utilization rate, age of plant, plant efficiency, capital investment
    - Location, shipping logistics, and product mix by plant.
- Personnel
    - Number of employees, key employees, and skill sets
    - Strength of management, and management style
    - Compensation, benefits, and employee morale & retention rates.
- Corporate and marketing strategies
    - Objectives, mission statement, growth plans, acquisitions, and divestitures
    - Marketing strategies.

Media Scanning

Scanning competitor's ads can reveal much about what that competitor believes about marketing and their target market. Changes in a competitor's advertising message can reveal new product offerings, new production processes, a new branding strategy, a new positioning strategy, a new segmentation strategy, line extensions and contractions, problems with previous positions, insights from recent marketing or product research, a new strategic direction, a new source of sustainable competitive advantage, or value migrations within the industry. It might also indicate a new pricing strategy such as penetration, price discrimination, price skimming, product bundling, joint product pricing, discounts, or loss leaders. It may also indicate a new promotion strategy such as push, pull, balanced, short-term sales generation, long-term image creation, informational, comparative, affective, reminder, new creative objectives, new unique selling proposition, new creative concepts, appeals, tone, and themes, or a new advertising agency. It might also indicate a new distribution strategy, new distribution partners, more extensive distribution, more intensive distribution, a change in geographical focus, or exclusive distribution. Similar techniques can be used by observing a competitor's search engine optimization targets and practices. For example, by conducting keyword research, one may be able to determine a competitor's target market, keywords, or products. Other metrics allow for detection of a competitor's success. Little of this intelligence is definitive : additional information is needed before conclusions should be drawn.

A competitor's media strategy reveals budget allocation, segmentation and targeting strategy, and selectivity and focus. From a tactical perspective, it can also be used to help a manager implement his own media plan. By knowing the competitor's media buy, media selection, frequency, reach, continuity, schedules, and flights, the manager can arrange his own media plan so that they do not coincide.

Other sources of corporate intelligence include trade shows, patent filings, mutual customers, annual reports, and trade associations.

Some firms hire competitor intelligence professionals to obtain this information. The Society of Competitive Intelligence Professionals maintains a listing of individuals who provide these services.

New Competitors

In addition to analysing current competitors, it is necessary to estimate future competitive threats. The most common sources of new competitors are :

- Companies competing in a related product/market
- Companies using related technologies
- Companies already targeting your prime market segment but with unrelated products
- Companies from other geographical areas and with similar products
- New start-up companies organized by former employees and/or managers of existing companies.

The entrance of new competitors is likely when :

- There are high profit margins in the industry
- There is unmet demand (insufficient supply) in the industry
- There are no major barriers to entry
- There is future growth potential
- Competitive rivalry is not intense
- Gaining a competitive advantage over existing firms is feasible.

DASHBOARD

Dashboards often provide at-a-glance views of KPIs (*key performance indicators*) relevant to a particular objective or business process (*e.g. sales, marketing, human resources*, or *production*). The term dashboard originates from the *automobile dashboard* where drivers monitor the major functions at a glance via the instrument cluster. Dashboards give signs about a business letting the user know something is wrong or something is right. The corporate world has tried for years to come up with a solution that would tell them if their business needed maintenance or if the temperature of their business was running above normal. Dashboards typically are limited to show summaries, key *trends*, comparisons, and exceptions. There are four Key elements to a good dashboard.

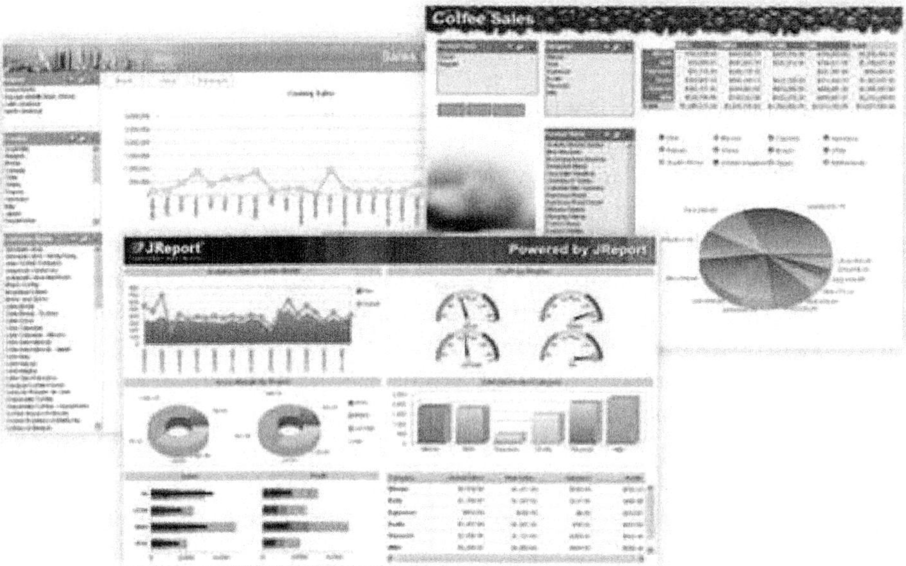

Fig. : Business Dashboards.

1. Simple, communicates easily
2. Minimum distractions...it could cause confusion
3. Supports organized business with meaning and useful data
4. Applies human visual perception to visual presentation of information.

In *management information systems*, a **dashboard** is

"An easy to read, often single page, real-time user interface, showing a graphical presentation of the current status (snapshot) and historical trends of an organization's** key performance indicators **(KPIs) to enable instantaneous and informed decisions to be made at a glance."

History

Early predecessors of the modern business dashboard were first developed in the 1980s in the form of Executive Information Systems (EISs). Due to problems primarily with data refreshing and handling, it was soon realized that the approach wasn't practical as information was often incomplete, unreliable, and spread across too many disparate sources. Thus, EISs hibernated until the 1990s when the information age quickened pace and data warehousing, and online analytical processing (OLAP) allowed dashboards to function adequately. Despite the availability of enabling technologies, the dashboard use didn't become popular until later in that decade, with the rise of key performance indicators (KPIs), and the introduction of Robert S. Kaplan and David P. Norton's Balanced Scorecard. Today, the use of dashboards forms an important part of [Business Performance Management] (BPM).

Classification

Dashboards can be broken down according to role and are either strategic, analytical, operational, or informational. Strategic dashboards support managers at any level in an organization, and provide the quick overview that decision makers need to monitor the health and opportunities of the business. Dashboards of this type focus on high level measures of performance, and forecasts. Strategic dashboards benefit from static snapshots of data (daily, weekly, monthly, and quarterly) that are not constantly changing from one moment to the next. Dashboards for analytical purposes often include more context, comparisons, and history, along with subtler performance evaluators. Analytical dashboards typically support interactions with the data, such as drilling down into the underlying details. Dashboards for monitoring operations are often designed differently from those that support strategic decision-making or data analysis and often require monitoring of activities and events that are constantly changing and might require attention and response at a moment's notice.

Dashboards and Scoreboards

Balanced Scoreboards and Dashboards have been linked together as if they were interchangeable. However, although both visually display critical information, the difference is in the format : Scoreboards can open the quality of an operation while dashboards provide calculated direction. A balanced scoreboard has what they called a "prescriptive" format. It should always contain these components (Active Strategy)...

- Perspectives-groupings of high level strategic areas
- Objectives-verb-noun phrases pulled from a strategy plan
- Measures-also called Metric or Key Performance Indicators (KPIs)
- Spotlight Indicators-red, yellow, or green symbols that provide an at-a-glance view of a measure's performance.

Each of these sections ensures that a Balanced Scorecard is essentially connected to the businesses critical strategic needs.

The design of a dashboard is more loosely defined. Dashboards are usually a series of graphics, charts, gauges and other visual indicators that can be monitored and interpreted. Even when there is a strategic link, on a dashboard, it may not be noticed as such since objectives are not normally present on dashboards. However, dashboards can be customized to link their graphs and charts to strategic objectives.

Design

Dashboards are unique : The design of each dashboard is driven by the business and their needs and culture. What may work for one business may not work for another. There are general guidelines that are available when initially developing the tool or when looking to improve current performance metrics.

Planning and researching a good design is crucial for dashboards : A good information design will clearly communicate key information to users and makes supporting information easily accessible. Setting up a business dashboard can also be fun and worthwhile if done correctly.

LOCATION INTELLIGENCE

Location Intelligence is the capacity to organize and understand complex phenomena through the use of geographical relationships inherent in all information.

The optimal location for any business is one that supports workplace success and addresses a variety of customer and financial objectives. Location Intelligence streamlines business processes and identifies unique geographies for winning market share in the B2C (retail and site selection analytics) and B2B (corporate real estate location advisory) business models. The goal of location intelligence is to select the best "geographies of opportunity" by providing a fit-to-purpose spatial analytic solution.

Deploying location intelligence by analyzing data using a geographical information system (GIS) within business is becoming a critical core strategy for success in an increasingly competitive global economy. Location or GIS tools enable spatial experts to collect, store, analyze and visualize data. Location intelligence experts are defined by their advanced education in spatial technology and applied use of spatial methodologies.

Location intelligence experts can use a variety of spatial and business analytical tools to measure optimal locations for operating a business or providing a service. Location intelligence experts begin with defining the business ecosystem which has many interconnected economic influences. Such economic influences include but are not limited to culture, lifestyle, labour, healthcare, cost of living, crime, economic climate and education.

The term Location Intelligence is often used to describe the people, data and technology employed to geographically "map" information. These mapping applications can transform large amounts of data into colour-coded visual representations that make it easy to see trends and generate meaningful intelligence. The creation of location intelligence is directed by domain knowledge, formal frameworks, and a focus on decision support.

Location Intelligence is also used to describe the integration of a geographical component into Business Intelligence processes and tools, often incorporating Spatial Database and spatial OLAP tools.

In 2012, Dr. Wayne Gearey from the commercial real estate industry was selected to offer the first applied course on Location Intelligence at the University of Texas at Dallas. In this course, Dr. Gearey defines location intelligence as the process for selecting the optimal location that will support workplace success and address a variety of business and financial objectives.

Commercial Applications

Today, Location Intelligence is used by a broad range of industries to improve overall business results. Applications include :

- *Communications & Telecommunications :* Network planning and design, boundary identification, identifying new customer markets.
- *Financial Services :* Optimize branch locations, market analysis, share of wallet and cross-sell activities, mergers & acquisitions, industry sector analysis, risk management.
- *Government :* Census updates, law enforcement crime analysis, emergency response, environmental and land management, electoral redistricting, tax jurisdiction assignment, urban planning.
- *Healthcare :* Site selection, market segmentation, network analysis, growth assessments.
- *Higher Education :* Student Recruitment, Alumni & Donor Tracking, Campus Mapping.
- *Hotels & Restaurants :* Customer profile analysis, site selection, target marketing, expansion planning.
- *Insurance :* Address validation, underwriting and risk management, claims management, marketing and sales analysis, market penetration studies.
- *Media :* Target market identification, subscriber demographics, media planning.
- *Real Estate :* Site reports, comprehensive site analysis, retail modelling, presentation quality maps.
- *Retail :* Site selection, maximize per-store sales, identify under-performing stores, market analysis.
- *Transportation :* Transport planning, route monitoring.
- *K-12 :* School Site Selection, enrollment planning, school attendance area modification (boundary change), school consolidation, district consolidation, student achievement plotting.

MOBILE BUSINESS INTELLIGENCE

Mobile Business Intelligence (Mobile BI or Mobile Intelligence) is defined as "The capability that enables the mobile workforce to gain business insights through information analysis using applications optimized for mobile devices" (Verkooij 2012). *Business intelligence* (BI) refers to computer-based techniques used in spotting, digging-out, and analyzing business data, such as sales revenue by products and/or departments or associated costs and incomes.

Although the concept of mobile computing has been prevalent for over a decade, Mobile BI has shown a momentum/growth only very recently. This change has been partly encouraged by a change from the 'wired world' to a wireless world with the advantage of *smartphones* which has led to a new era of mobile computing, especially in the field of BI.

According to the Aberdeen Group, a large number of companies are rapidly undertaking mobile BI owing to a large number of market pressures such as the need for higher efficiency in business processes, improvement in employee productivity (*e.g.*, time spent looking for information), better and faster decision-making, better customer service, and delivery of real-time bi-directional data access to make decisions anytime and anywhere. But despite the apparent advantages of mobile information delivery, mobile BI is still in the 'early adopter' phase. Some CFOs remain sceptical of the business benefits and with the perceived lack of specific business use cases and tangible ROI, mobile BI adoption is still behind the curve compared with other enterprise mobile applications.

History

Information Delivery to Mobile Devices

The predominant method for accessing BI information is using proprietary software or a Web browser on a personal computer to connect to BI applications. These BI applications request data from databases. Starting in the late 1990s, BI systems offered alternatives for receiving data, including email and mobile devices.

Static Data Push

Initially, mobile devices such as pagers and mobile phones received pushed data using a short message service (SMS) or text messages. These applications were designed for specific mobile devices, contained minimal amounts of information, and provided no data interactivity. As a result, the early mobile BI applications were expensive to design and maintain while providing limited informational value, and garnered little interest.

Data Access via a Mobile Browser

The mobile browser on a smartphone, a handheld computer integrated with a mobile phone, provided a means to read simple tables of data. The small screen space, immature mobile browsers, and slow data transmission could not provide a satisfactory BI experience. Accessibility and bandwidth may be perceived as issues when it comes to mobile technology, but BI solutions provide advanced functionality to predict and outperform such potential challenges. While Web-based mobile BI solutions provide little to no control over the processing of data in a network, managed BI solutions for mobile devices only utilize the server for specific operations. In addition, local reports are compressed both during transmission and on the device, permitting greater flexibility for storage and receipt of these reports. Within a mobile environment, users capitalize on easy access to information because the mobile application operates within a single authoring environment that permits access to all BI content (respecting existing security) regardless of language or locale. Furthermore, the user will not need to build and maintain a separate mobile BI deployment. In addition, mobile BI requires much less bandwidth for functionality. Mobile BI promises a small report footprint on memory, encryption during transmission as well as on the device, and compressed data storage for offline viewing and use.

Mobile Client Application

In 2002, Research in Motion released the first BlackBerry smartphone optimized for wireless e-mail use. Wireless e-mail proved to be the "killer app" that accelerated the popularity of the smartphone market. By the mid-2000s, Research in Motion's BlackBerry had solidified its hold on the smartphone market with both corporate and governmental organizations. The BlackBerry smartphones eliminated the obstacles to mobile business intelligence. The BlackBerry offered a consistent treatment of data across its many models, provided a much larger screen for viewing data, and allowed user interactivity via the thumbwheel and keyboard. BI vendors re-entered the market with offerings spanning different mobile operating systems (BlackBerry, Windows, Symbian) and data access methods. The two most popular data access options were :

- To use the mobile browser to access data, similar to desktop computer, and
- To create a native application designed specifically for the mobile device.

Research in Motion is continuing to lose market share to Apple and Android smartphones. In the first three months of 2011 Google's Android OS gained 7 points of market share. During the same time period RIM's market share collapsed and dropped almost 5 points.

Purpose-built Mobile BI Apps

Apple quickly set the standard for mobile devices with the introduction of the iPhone. In the first three years, Apple sold over 33.75 million units. Similarly, in 2010, Apple sold over 1 million iPads in just under three months. Both devices feature an interactive touchscreen display that is the *de facto* standard on many mobile phones and tablet computers.

In 2008, Apple published the SDK for which developers can build applications that run natively on the iPhone and iPad instead of Safari-based applications. These native applications can give the user a robust, easier-to-read and easier-to-navigate experience. The Apple App Store now has over 750,000 apps, giving the Apple mobile devices a decided advantage.

Others were quick to replicate Apple's success. The Android Market now has over 700,000 apps available for the mobile devices running the Android operating system.

More importantly, the advent of the Apple iPhone has radically changed the way people use data on their mobile devices. This includes mobile BI. Business intelligence applications can be used to transform reports and data into mobile dashboards, and have them instantly delivered to any iPhone or iPad.

Google Inc.'s Android has overtaken Apple Inc.'s iOS in the wildly growing arena of app downloads. In the second quarter of 2011, 44% of all apps downloaded from app marketplaces across the web were for Android devices while 31% were for Apple devices, according to new data from ABI Research. The remaining apps were for various other mobile operating systems, including BlackBerry and Windows Phone 7.

Mobile BI applications have evolved from being a client application for viewing data to a purpose-built application designed to provide information and workflows necessary to quickly make business decisions and take action.

Web Applications vs. Device-specific Applications for Mobile BI

In early 2011, as the mobile BI software market started to mature and adoption started to grow at a significant pace in both small and large enterprises, most vendors adopted either a purpose-built, device-specific application strategy (*e.g.* iPhone or Android apps, downloaded from iTunes or the Android Market) or a web application strategy (browser-based, works on most devices without an application being installed on the device). This debate continues and there are benefits and drawbacks to both methods. One potential solution will be the wider adoption of HTML5 on mobile devices which will give web applications many of the characteristics of dedicated applications while still allowing them to work on many devices without an installed application.

Microsoft has announced their mobile BI strategy. Microsoft plans to support browser-based applications such as Reporting Services and Performance Point on iOS in the first half of 2012 and touch-based applications on iOS and Android by the second half of 2012. Despite popular perception that Microsoft only acknowledges its own existence, recent moves suggest the company is aware that it is not the only player in the technology ecosystem. Instead of attempting to squelch competition or suggesting new technology developments were ridiculous, the company has instead decided to make its technology accessible to a wider audience.

There are many mobile devices and platforms available today. The list is constantly growing and so is the platform support. There are hundreds of models available today, with multiple hardware and software combinations. The enterprise must select a device very carefully. The target devices will impact the mobile BI design itself because the design for a smartphone will be different than for a tablet. The screen size, processor, memory, etc. all vary. The mobile BI program must account for lack of device standardization from the providers by constantly testing devices for the mobile BI apps. Some best practices can always be followed. For example, a smartphone is a good candidate for operational mobile BI. However, for analytics and what-if analysis, tablets are the best option. Hence, the selection or availability of the device plays a big role in the implementation.

Demand

Gartner analyst Ted Friedman believes that mobile delivery of BI is all about practical, tactical information needed to make immediate decisions—"The biggest value is in operational BI—information in the context of applications—not in pushing lots of data to somebody's phone."

Accessing the Internet through a mobile device such as a smartphone is also known as the mobile Internet or mobile Web. IDC expects the US mobile workforce to increase by 73% in 2011. Morgan Stanley reports the mobile Internet is ramping up faster than its predecessor, the desktop Internet, enabling companies to deliver knowledge to their mobile workforce to help them make more profitable decisions.

Michael Cooney from Gartner has identified bring-your-own-technology at work as becoming the norm, not the exception. By 2015 media tablet shipments will reach around 50% of laptop shipments and Windows 8 will likely be in third place behind Android and Apple. The net result is that Microsoft's share of the client platform, be it PC, tablet or smartphone, will likely be reduced to 60% and it could fall below 50%.

Business Benefits

In its latest Magic Quadrant for Business Intelligence Platforms, Gartner examines whether the platform enables users to "fully interact with BI content delivered to mobile devices." The phrase "fully interact" is the key. The ability to send alerts embedded in email or text messages, or links to static content in email messages hardly represents sophistication in mobile analytics. For users to benefit from mobile BI, they must be able to navigate dashboard and guided analytics comfortably—or as comfortably as the mobile device will allow, which is where devices with high-resolution screens and touch interfaces (like the iPhone and Android-based phones) have a clear edge over, say, earlier editions of BlackBerry. It is equally important to take a step back to define your purpose and adoption patterns. Which business users stand to benefit the most from mobile analytics—and what, exactly, is their requirement? You don't need mobile analytics to send a few alerts or summary reports to their handhelds—without interactivity, mobile BI is indistinguishable from merely informative email or text messages.

Applications

Similar to consumer applications, which have shown an ever increasing growth over the past few years, a constant demand for anytime, anywhere access to BI is leading to a number of custom mobile application development. Businesses have also started adopting mobile solutions for their workforce and are soon becoming key components of core business processes. In an Aberdeen survey conducted in May 2010, 23% of companies participating indicated that they now have a mobile BI app or dashboard in place, while another 31% indicated that they plan to implement some form of mobile BI in the next year.

Definitions

Mobile BI applications can be defined/segregated as follows :

- *Mobile Browser Rendered App :* Almost any mobile device enables Web-based, thin client, HTML-only BI applications. However, these apps are static and provide little data interactivity. Data is viewed just as it would be over a browser from a personal computer. Little additional effort is required to display data but mobile browsers can typically only support a small sub-set of the interactivity of a web browser.
- *Customized App :* A step up from this approach is to render each (or all) reports and dashboards in device-specific format. In other words provide information specific to the screen size, optimize usage of screen real estate, and enable device-specific navigation controls. Examples of these include thumb wheel or thumb button for BlackBerry, up/down/left/right arrows for Palm, gestural manipulation for iPhone. This approach requires more effort than the previous but no additional software.
- *Mobile Client App :* The most advanced, the client app provides full interactivity with the BI content viewed on the device. In addition, this approach provides periodic caching of data which can be viewed and analyzed even offline.

Companies across all verticals, from retail to even non-profit organizations are realizing the value of purpose-specific mobile applications suited for their mobile workforce.

Development

Developing a native mobile BI app poses challenges, especially concerning data display rendering and user interactivity. Mobile BI App development has traditionally been a time-consuming and expensive effort requiring businesses to justify the investment for the mobile workforce. They do not only require texting and alerts, they need information customized for their line of work which they can interact with and analyze to gain deeper information.

Custom-coded Mobile BI Apps

Mobile BI applications are often custom-coded apps specific to the underlying mobile operating system. For example, the iPhone apps require coding in Objective-C while Android apps require coding in Java. In addition to the user functionality of the app, the app must be coded to work with the supporting server infrastructure required to serve data to the mobile BI app. While custom-coded apps offer near limitless options, the specialized software coding expertise and infrastructure can be expensive to develop, modify, and maintain.

Fixed-form Mobile BI Applications

Business data can be displayed in a mobile BI client (or web browser) that serves as a user interface to existing BI platforms or other data sources, eliminating the need for new master sources of data and specialized server infrastructure. This option offers fixed and configurable data visualizations such as charts, tables, trends, KPIs, and links, and can usually be deployed quickly using existing data sources. However, the data visualizations are not limitless and cannot always be extended to beyond what is available from the vendor.

Graphical Tool-developed Mobile BI Apps

Mobile BI apps can also be developed using the graphical, drag-and-drop development environments of BI platforms. The advantages including the following :

1. Apps can be developed without coding,
2. Apps can be easily modified and maintained using the BI platform change management tools,
3. Apps can use any range of data visualizations and not be limited to just a few,
4. Apps can incorporate specific business workflows, and
5. The BI platform provides the server infrastructure.

Using graphical BI development tools can allow faster mobile BI app development when a custom application is required.

Security Considerations for Mobile BI Apps

High adoption rates and reliance on mobile devices makes safe mobile computing a critical concern. The Mobile Business Intelligence Market Study discovered that security is the number one issue (63%) for organizations.

A comprehensive mobile security solution must provide security at these levels :

- Device
- Transmission
- Authorization, Authentication, and Network Security.

Device Security

A senior analyst at the Burton Group research firm recommends that the best way to ensure data will not be tampered with is to not store it on the client device (mobile device). As such, there is no local copy to lose if the mobile device is stolen and the data can reside on servers within the data center with access permitted only over the network. Most smartphone manufacturers provide a complete set of security features including full-disk encryption, email encryption, as well as remote management which includes the ability to wipe contents if device is lost or stolen. Also, some devices have embedded third-party antivirus and firewall software such as RIM's BlackBerry.

Transmission Security

Transmission security refers to measures that are designed to protect data from unauthorized interception, traffic analysis, and imitative deception. These measures include Secure Sockets Layer (SSL), iSeries Access for Windows, and virtual private network (VPN) connections. A secure data transmission should enable the identity of the sender and receiver to be verified by using a cryptographic shared key system as well as protect the data to be modified by a third party when it crosses the network. This can be done using AES or Triple DES with an encrypted SSL tunnel.

Authorization, Authentication, and Network Security

Authorization refers to the act of specifying access rights to control access of information to users. Authentication refers to the act of establishing or confirming the user as true or authentic. Network security refers to all the provisions and policies adopted by the network administrator to prevent and monitor unauthorized access, misuse, modification, or denial of the computer network and network-accessible resources. The mobility adds to unique security challenges. As data is trafficked beyond the enterprise firewall towards unknown territories, ensuring that it is handled safely is of paramount importance. Towards this, proper authentication of user connections, centralized access control (like LDAP Directory), encrypted data transfer mechanisms can be implemented.

Role of BI for Securing Mobile Apps

To ensure high security standards, BI software platforms must extend the authentication options and policy controls to the mobile platform. Business intelligence software platforms need to ensure a secure encrypted key-chain for storage of credentials. Administrative control of password policies should allow creation of security profiles for each user and seamless integration with centralized security directories to reduce administration and maintenance of users.

PERFORMANCE INDICATOR

A **performance indicator** or **key performance indicator** (**KPI**) is a type of performance measurement. An organization may use KPIs to evaluate its success, or to evaluate the success of a particular activity in which it is engaged. Sometimes success is defined in terms of making progress toward strategic goals, but often success is simply the repeated, periodic achievement of some level of operational goal (*e.g.* zero defects, 10/10 customer satisfaction, etc.). Accordingly, choosing the right KPIs relies upon a good understanding of what is important to the organization. 'What is important' often depends on the department measuring the performance—*e.g.* the KPIs useful to finance will be quite different from the KPIs assigned to sales. Since there is a need to understand well what is important (to an organization), various techniques to assess the present state of the business, and its key activities, are associated with the selection of performance indicators.

These assessments often lead to the identification of potential improvements, so performance indicators are routinely associated with 'performance improvement' initiatives. A very common way to choose KPIs is to apply a management framework such as the balanced scorecard.

Categorization of Indicators

Key performance indicators define a set of values used to measure against. These raw sets of values, which are fed to systems in charge of summarizing the information, are called **indicators**. Indicators identifiable and marked as possible candidates for KPIs can be summarized into the following sub-categories :

- **Quantitative indicators** that can be presented with a number.
- **Qualitative indicators** that can't be presented as a number.
- **Leading indicators** that can predict the outcome of a process.
- **Lagging indicators** that present the success or failure *post hoc*.
- **Input indicators** that measure the amount of resources consumed during the generation of the outcome.
- **Process indicators** that represent the efficiency or the productivity of the process.
- **Output indicators** that reflect the outcome or results of the process activities.
- **Practical indicators** that interface with existing company processes.
- **Directional indicators** specifying whether or not an organization is getting better.
- **Actionable indicators** are sufficiently in an organization's control to effect change.
- **Financial indicators** used in performance measurement and when looking at an operating index.

Key performance indicators, in practical terms and for strategic development, are *objectives* to be targeted that will add the most *value* to the business. These are also referred to as 'key success indicators'.

Identifying Indicators of Organization

Performance indicators differ from business drivers and aims (or goals). A school might consider the failure rate of its students as a key performance indicator which might help the school understand its position in the educational community, whereas a business might consider the percentage of income from returning customers as a potential KPI.

The key stages in identifying KPIs are :

- Having a pre-defined business process (BP).
- Having requirements for the BPs.
- Having a quantitative/qualitative measurement of the results and comparison with set goals.

- Investigating variances and tweaking processes or resources to achieve short-term goals.

Key performance indicators (KPIs) are ways to periodically assess the performances of organizations, business units, and their division, departments and employees. Accordingly, KPIs are most commonly defined in a way that is understandable, meaningful, and measurable. They are rarely defined in such a way such that their fulfillment would be hampered by factors seen as non-controllable by the organizations or individuals responsible. Such KPIs are usually ignored by organizations.

A KPI can follow the SMART criteria. This means the measure has a Specific purpose for the business, it is Measurable to really get a value of the KPI, the defined norms have to be Achievable, the improvement of a KPI has to be Relevant to the success of the organization, and finally it must be Time phased, which means the value or outcomes are shown for a predefined and relevant period.

In order to be evaluated, KPIs are linked to target values, so that the value of the measure can be assessed as meeting expectations or not.

KPI Examples

Marketing

Some examples are :

1. New customers acquisition.
2. Demographic analysis of individuals (potential customers) applying to become customers, and the levels of approval, rejections, and pending numbers.
3. Status of existing customers.
4. Customer attrition.
5. Turnover (*i.e.*, revenue) generated by segments of the customer population.
6. Outstanding balances held by segments of customers and terms of payment
7. Collection of bad debts within customer relationships.
8. Profitability of customers by demographic segments and segmentation of customers by profitability.

// Many of these customer KPIs are developed and managed with customer relationship management software.

Faster availability of data is a competitive issue for most organizations. For example, businesses which have higher operational/credit risk (involving for example credit cards or wealth management) may want weekly or even daily availability of KPI analysis, facilitated by appropriate IT systems and tools.

Manufacturing

Overall equipment effectiveness, is a set of broadly accepted non-financial metrics which reflect manufacturing success.

- *Cycle Time :* Cycle time is the total time from the beginning to the end of your process, as defined by you and your customer. Cycle time includes process time, during which a unit is acted upon to bring it closer to an output, and delay time, during which a unit of work is spent waiting to take the next action.
- **Cycle Time Ratio** (CTR)—CTR = Standard Cycle Time/Real Cycle Time
- Utilization
- Rejection rate.

IT Operations

- Availability/Uptime
- Mean time between failure
- Mean time to repair
- Unplanned availability.

IT Project Execution

- Earned Value
- Estimate to Complete
- Labour Spent/Month
- Dollars Spent/Month
- Planned Dollars/Month
- Planned Labour/Month.

Supply Chain Management

Businesses can utilize KPIs to establish and monitor progress toward a variety of goals, including lean manufacturing objectives, minority business enterprise and diversity spending, environmental "green" initiatives, cost avoidance programs and low-cost country sourcing targets.

Any business, regardless of size, can better manage supplier performance with the help of KPIs robust capabilities, which include :

- Automated entry and approval functions
- On-demand, real-time scorecard measures
- Rework on procured inventory
- Single data repository to eliminate inefficiencies and maintain consistency
- Advanced workflow approval process to ensure consistent procedures
- Flexible data-input modes and real-time graphical performance displays
- Customized cost savings documentation
- Simplified setup procedures to eliminate dependence upon IT resources.

Main SCM KPIs will detail the following processes :

- Sales forecasts
- Inventory
- Procurement and suppliers
- Warehousing
- Transportation
- Reverse logistics.

Suppliers can implement KPIs to gain an advantage over the competition. Suppliers have instant access to a user-friendly portal for submitting standardized cost savings templates. Suppliers and their customers exchange vital supply chain performance data while gaining visibility to the exact status of cost improvement projects and cost savings documentation.

Government

The provincial government of Ontario, Canada has been using KPIs since 1998 to measure the performance of higher education institutions in the province. All post secondary schools collect and report performance data in five areas-graduate satisfaction, student satisfaction, employer satisfaction, employment rate, and graduation rate.

Further Performance Indicators

- Duration of a stockout situation
- Customer order waiting time.

Problems

In practice, overseeing key performance indicators can prove expensive or difficult for organizations. Some indicators such as staff morale may be impossible to quantify. As such dubious KPIs can be adopted that can be used as a rough guide rather than a precise benchmark.

Key performance indicators can also lead to perverse incentives and unintended consequences as a result of employees working to the specific measurements at the expense of the actual quality or value of their work. For example, measuring the productivity of a software development team in terms of source lines of code encourages copy and paste code and over-engineered design, leading to bloated code bases that are particularly difficult to maintain, understand and modify.

PERFORMANCE INTELLIGENCE

Performance Intelligence refers to technologies that utilize standard *Business Intelligence* techniques on performance-related IT metrics. Traditionally, the performance metrics are loaded into a set of tables that resemble the table structure of a data warehouse. Analysis is performed with standard *Business Intelligence* techniques such as slicing, dicing, drilling down and the trend analysis.

This type of analysis is useful, because it allows the user to quickly look at the data from several perspectives to see trends or anomalies that they might miss while looking at data in a more granular form.

PERVASIVE BUSINESS INTELLIGENCE

Pervasive Business intelligence is an architecture or framework for working with Business Intelligence. It combines the practices of Enterprise Architecture with Service Oriented Architecture and Active Data Warehousing to provide enterprises with a more holistic approach to Business Intelligence. The aim is to facilitate tactical decisions by establishing a data warehouse which stores both historical data and real time data that is available instantly and accessible 24/7/365. It could be valuable to share access to the data warehouse with various stakeholders, clients and suppliers as they might gain and provide insight.

It is however important to note that the goal with Pervasive Business Intelligence is not only to provide near realtime data. The goal is to reduce lag or latency between events and the subsequent action.

Active Data Warehousing

An Active Data Warehouse is an important part of pervasive BI as it provides historical data infused with real time data on the fly either pushed or pulled to the user. The difference between a traditional data warehouse and one that is active is in the way it handles events. An active data warehouse will have pre-defined rules, schedules or event triggers which cause a certain automatic action such as notifying the user, giving a discount to a certain customer, etc.

The definitions of the ADW active elements are quoted from Teradata : Active Data Warehousing :

- *Active Access :* Front-line users access the data warehouse for operational decision-making with a service level agreement of five seconds or less (also known as "web speed").
- *Active Load :* Near-real-time data enters the data warehouse via mini-batch loading, replication services or continuous streams of data from message queuing systems.
- *Active Events :* Event-driven architectures and business activity monitoring detect significant business events and issue alerts for timely, informed decisions.
- *Active Workload Management :* Dynamic priority management inside the data warehouse ensures service levels are achieved across multiple user communities and workload types.

- *Active Enterprise Integration* Integration tools and designs connect the data warehouse to web sites, portals, SOA, web services, enterprise service busses, workflow and batch systems.
- *Active Availability* : Policies, procedures and redundant hardware ensure the entire information supply chain is protected when subsystem failure occurs.

Importance of Service Level Agreements

In order to take full advantage of Pervasive BI, Teradata suggests that the data and architecture is held up against a Service Level Agreements (SLA) which guarantees that the system lives up to certain minimal expectations. The categories for the Data SLA are;

- Data Freshness
- Data Completeness
- Data Cleansing and Accuracy

The categories for the BI SLA are;

- Scalability
- Performance
- Reliability.

Usage

Pervasive Business Intelligence has been used in the following tasks :

- Inbound and outbound cross-selling by Customer Service.

Representatives

- Point-of-sale fraud detection
- Automated insurance claims triage and fraud detection
- Personalized next best offer on web sites, ATMs, and POS.

Devices

- Supply chain business activity monitoring
- Retail out-of-stock monitoring on promotional items
- Labour and crew scheduling
- Optimizing trailer and container loading and routing
- Anti-money laundering
- Dynamic pricing and yield management
- Real-time manufacturing line quality alerts.

QUANTITATIVE MARKETING RESEARCH

Quantitative marketing research is the application of quantitative research techniques to the field of marketing. It has roots in both the *positivist* view of the world, and the modern marketing viewpoint that marketing is an interactive process in which both the buyer and seller reach a satisfying agreement on the "*four Ps*" of marketing : Product, Price, Place (location) and Promotion.

As a *social research* method, it typically involves the construction of *questionnaires* and *scales*. People who respond (respondents) are asked to complete the *survey*. *Marketers* use the information so obtained to understand the needs of individuals in the marketplace, and to create *strategies* and *marketing plans*.

Typical General Procedure

Simply put, there are five major and important steps involved in the research process :

1. Defining the problem.
2. Research design.
3. Data collection.
4. Data analysis.
5. Report Writing & presentation.

A brief discussion on these steps is :

1. *Problem audit and problem definition :* What is the problem? What are the various aspects of the problem? What information is needed?
2. *Conceptualization and operationalization :* How exactly do we define the concepts involved? How do we translate these concepts into observable and measurable behaviours?
3. *Hypothesis specification :* What claim(s) do we want to test?
4. *Research design specification :* What type of methodology to use?—examples : questionnaire, survey.
5. *Question specification :* What questions to ask? In what order?
6. *Scale specification :* How will preferences be rated?
7. *Sampling design specification :* What is the total population? What sample size is necessary for this population? What sampling method to use?—examples : *Probability Sampling :* (cluster sampling, stratified sampling, simple random sampling, multi-stage sampling, systematic sampling) & *Non-probability sampling :* (Convenience Sampling, Judgement Sampling, Purposive Sampling, Quota Sampling, Snowball Sampling, etc.)
8. *Data collection :* Use mail, telephone, internet, mall intercepts.
9. *Codification and re-specification :* Make adjustments to the raw data so it is compatible with statistical techniques and with the objectives of the research-examples : assigning numbers, consistency checks, substitutions, deletions, weighting, dummy variables, scale transformations, scale standardization.
10. *Statistical analysis :* Perform various descriptive and inferential techniques on the raw data. Make inferences from the sample to the whole population. Test the results for statistical significance.

11. *Interpret and integrate findings* : What do the results mean? What conclusions can be drawn? How do these findings relate to similar research?
12. *Write the research report* : Report usually has headings such as : (1) executive summary; (2) objectives; (3) methodology; (4) main findings; (5) detailed charts and diagrams. Present the report to the client in a 10 minute presentation. Be prepared for questions.

The design step may involve a pilot study in order to discover any hidden issues. The codification and analysis steps are typically performed by computer, using statistical software. The data collection steps, can in some instances be automated, but often require significant manpower to undertake. Interpretation is a skill mastered only by experience.

Statistical Analysis

The data acquired for quantitative marketing research can be analysed by almost any of the range of techniques of statistical analysis, which can be broadly divided into descriptive statistics and statistical inference. An important set of techniques is that related to statistical surveys. In any instance, an appropriate type of statistical analysis should take account of the various types of error that may arise, as outlined below.

Reliability and Validity

Research should be tested for reliability, generalizability, and validity.

Generalizability is the ability to make inferences from a sample to the population.

Reliability is the extent to which a measure will produce consistent results.

- *Test-retest reliability* checks how similar the results are if the research is repeated under similar circumstances. Stability over repeated measures is assessed with the Pearson coefficient.
- *Alternative forms reliability* checks how similar the results are if the research is repeated using different forms.
- *Internal consistency reliability* checks how well the individual measures included in the research are converted into a composite measure. Internal consistency may be assessed by correlating performance on two halves of a test (split-half reliability). The value of the Pearson product-moment correlation coefficient is adjusted with the Spearman–Brown prediction formula to correspond to the correlation between two full-length tests. A commonly used measure is Cronbach's α, which is equivalent to the mean of all possible split-half coefficients. Reliability may be improved by increasing the sample size.

Validity asks whether the research measured what it intended to.

- *Content validation* (also called face validity) checks how well the content of the research are related to the variables to be studied; it seeks to answer whether the research questions are representative of the variables being researched.

It is a demonstration that the items of a test are drawn from the domain being measured.

- *Criterion validation* checks how meaningful the research criteria are relative to other possible criteria. When the criterion is collected later the goal is to establish predictive validity.
- *Construct validation* checks what underlying construct is being measured. There are three variants of construct validity : *convergent validity* (how well the research relates to other measures of the same construct), *discriminant validity* (how poorly the research relates to measures of opposing constructs), and *nomological validity* (how well the research relates to other variables as required by theory).
- *Internal validation*, used primarily in experimental research designs, checks the relation between the dependent and independent variables (*i.e.* Did the experimental manipulation of the independent variable actually cause the observed results?)
- *External validation* checks whether the experimental results can be generalized.

Validity implies reliability : A valid measure must be reliable. Reliability does not necessarily imply validity, however : A reliable measure does not imply that it is valid.

Types of Errors

Random sampling errors :

- Sample too small
- Sample not representative
- Inappropriate sampling method used
- Random errors.

Research design errors :

- Bias introduced
- Measurement error
- Data analysis error
- Sampling frame error
- Population definition error
- Scaling error
- Question construction error.

Interviewer errors :

- Recording errors
- Cheating errors
- Questioning errors
- Respondent selection error.

Respondent errors :

- Non-response error
- Inability error
- Falsification error.

Hypothesis errors :

- Type I error (also called alpha error)
  - The study results lead to the rejection of the null hypothesis even though it is actually true.
- Type II error (also called beta error)
  - The study results lead to the acceptance (non-rejection) of the null hypothesis even though it is actually false.

REACTIVE BUSINESS INTELLIGENCE

Reactive business intelligence (RBI) advocates an holistic approach that integrates data-mining, modelling and interactive visualization, into an end-to-end discovery and continuous innovation process powered by human and automated learning.

In the area of decision-making this approach has been used to adapt the decision method to the knowledge which is progressively acquired from the decision-maker.

Relationships with reactive search optimization (RSO)

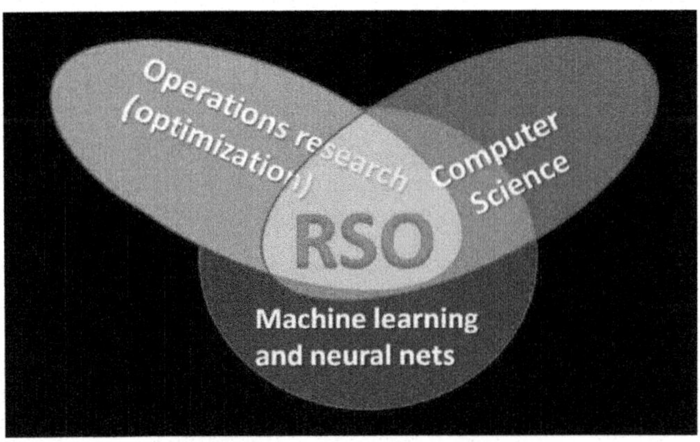

Fig. : RSO is multi-disciplinary.

RSO is a multi-disciplinary research area between operations research (optimization), computer science, machine learning and neural networks that studies online learning schemes applied to problem-solving and optimization, according to a *learning while optimizing* principle. The word **reactive** hints at a ready response to events during the search through an internal feedback loop for online self-tuning and dynamic adaptation. Online adaptation and model revision is used in **reactive business intelligence** to adapt the data-mining and interactive visualization techniques to the knowledge derived from a user in real time.

REAL-TIME BUSINESS INTELLIGENCE

Real-time business intelligence (RTBI) is the process of delivering information about business operations as they occur. Real time means near to zero latency and access to information whenever it is required.

The speed of today's processing systems has moved classical data warehousing into the realm of real-time. The result is real-time business intelligence. Business transactions as they occur are fed to a real-time business intelligence system that maintains the current state of the enterprise. The RTBI system not only supports the classic strategic functions of data warehousing for deriving information and knowledge from past enterprise activity, but it also provides real-time tactical support to drive enterprise actions that react immediately to events as they occur. As such, it replaces both the classic data warehouse and the enterprise application integration (EAI) functions. Such event-driven processing is a basic tenet of real-time business intelligence.

In this context, *real-time* means a range from milliseconds to a few seconds after the business event has occurred. While traditional business intelligence presents historical data for manual analysis, *real-time business intelligence* compares current business events with historical patterns to detect problems or opportunities automatically. This automated analysis capability enables corrective actions to be initiated and/or business rules to be adjusted to optimize business processes.

RTBI is an approach in which up-to-a-minute data is analyzed, either directly from Operational sources or feeding business transactions into a real time data warehouse and Business Intelligence system. RTBI analyzes real time data.

Real-time business intelligence makes sense for some applications but not for others—a fact that organizations need to take into account as they consider investments in real-time BI tools. Key to deciding whether a real-time BI strategy would pay dividends is understanding the needs of the business and determining whether end users require immediate access to data for analytical purposes, or if something less than real time is fast enough.

Evolution of RTBI

In today's competitive environment with high consumer expectation, decisions that are based on the most current data available to improve customer relationships, increase revenue, maximize operational efficiencies, and yes—even save lives. This technology is real-time business intelligence. Real-time business intelligence systems provide the information necessary to strategically improve an enterprise's processes as well as to take tactical advantage of events as they occur.

Latency

All real-time business intelligence systems have some latency, but the goal is to minimize the time from the business event happening to a corrective action or notification being initiated. Analyst Richard Hackathorn describes three types of latency :

- Data latency; the time taken to collect and store the data
- Analysis latency; the time taken to analyze the data and turn it into actionable information
- Action latency; the time taken to react to the information and take action.

Real-time business intelligence technologies are designed to reduce all three latencies to as close to zero as possible, whereas traditional business intelligence only seeks to reduce data latency and does not address analysis latency or action latency since both are governed by manual processes.

Some commentators have introduced the concept of *right time business intelligence* which proposes that information should be delivered just before it is required, and not necessarily in real-time.

Architectures

Event-based

Real-time Business Intelligence systems are event driven, and may use Event Stream Processing and Mashup (web application hybrid) techniques to enable events to be analysed without being first transformed and stored in a database. These in-memory techniques have the advantage that high rates of events can be monitored, and since data does not have to be written into databases data latency can be reduced to milli-seconds.

Data Warehouse

An alternative approach to event driven architectures is to increase the refresh cycle of an existing data warehouse to update the data more frequently. These real-time data warehouse systems can achieve near real-time update of data, where the data latency typically is in the range from minutes to hours. The analysis of the data is still usually manual, so the total latency is significantly different from event driven architectural approaches.

Server-less technology

The latest alternative innovation to "real-time" event driven and/or "real-time" data warehouse architectures is MSSO Technology (Multiple Source Simple Output) which removes the need for the data warehouse and intermediary servers altogether since it is able to access live data directly from the source (even from multiple, disparate sources). Because live data is accessed directly by server-less means, it provides the potential for zero-latency, real-time data in the truest sense.

Process-aware

This is sometimes considered a sub-set of Operational intelligence and is also identified with Business Activity Monitoring. It allows entire processes (transactions, steps) to be monitored, metrics (latency, completion/failed ratios, etc.) to be viewed, compared with warehoused historic data, and trended in real-time. Advanced implementations allow threshold detection, alerting and providing feedback to the process execution systems themselves, thereby 'closing the loop'.

Technologies That Support Real-time Analytics

Technologies that can be supported to enable real-time business intelligence are data visualization, data federation, enterprise information integration, enterprise application integration and service oriented architecture. Complex event processing tools can be used to analyze data streams in real time and either trigger automated actions or alert workers to patterns and trends.

Data warehouse appliance : Data warehouse appliance is a combination of hardware and software product which was designed exclusively for analytical processing. In data warehouse implementation, tasks that involve tuning, adding or editing structure around the data, data migration from other databases, reconciliation of data are done by DBA. Another task for DBA was to make the database to perform well for large sets of users. Whereas with data warehouse appliances, it is the vendor responsibility of the physical design and tuning the software as per hardware requirements. Data warehouse appliance package comes with its own operating system, storage, DBMS, software, and required hardware. If required data warehouse appliances can be easily integrated with other tools.

Mobile technology : There are very limited vendors for providing Mobile business intelligence; MBI is integrated with existing BI architecture. MBI is a package that uses existing BI applications so people can use on their mobile phone and make informed decision in real time.

Application Areas

- Algorithmic trading
- Fraud detection
- Systems monitoring
- Application performance monitoring
- Customer Relationship Management
- Demand sensing
- Dynamic pricing and yield management
- Data validation
- Operational intelligence and risk management
- Payments & cash monitoring
- Data security monitoring
- Supply chain optimization
- RFID/sensor network data analysis
- Workstreaming

- Call center optimization
- Enterprise Mashups and Mashup Dashboards
- Transportation industry.

Transportation industry can be benefited by using real-time analytics. For an example railroad network. Depending on the results provided by the real-time analytics, dispatcher can make a decision on what kind of train he can dispatch on the track depending on the train traffic and commodities shipped.

SALES INTELLIGENCE

Sales intelligence (SI) refers to technologies, applications and practices for the collection, integration, analysis, and presentation of information to help salespeople keep up to date with clients, prospect data and drive business. In addition to providing metrics for win-loss and sales confidence, SI can present contextually relevant customer and product information.

The 2008 survey of 300 companies by the Aberdeen Group show that the recent economic downturn has lengthened traditional sales cycles. As businesses have been forced to reduce spending, sales representatives have been challenged to meet quotas. Top performing companies have implemented sales intelligence programs to improve the quality and quantity of sales leads. SI contextualizes opportunities by providing relevant industry, corporate and personal information. Frequently SI's fact-based information is integrated or includes Customer Relationship Management (CRM).

Although some aspects of sales intelligence overlaps business intelligence (BI), SI is specifically designed for the use of salespeople and sales managers. Unlike *customer relationship management* (CRM) and traditional *business intelligence* (BI) applications, SI provides real-time analysis of current sales data and assists with suggesting and delivering actionable, relevant information.

Sales intelligence solutions are predominantly designed for companies in the manufacturing, distribution and wholesale sectors. These are highly competitive markets, where volumes are high, margins are low. (SI) solutions provide unique insight into customer buying patterns. By automatically analysing and evaluating these patterns, Sales Intelligence pro-actively identifies and delivers up-sell, cross-sell and switch-sell opportunities.

UTILITY-BUSINESS INTELLIGENCE

Utility-Business Intelligence (UBI) is an extension of the term *Business Intelligence*. The *public utility* industry in the United States is governed by specific sets of regulations which leads to a drastically enhanced set of reporting and analysis requirements. The additional requirements far exceed those addressed by traditional BI technologies. These needs include an additional ledger of *accounts*, known as *FERC* Accounts, special treatment of *inventory*, the tying of regulated and *GAAP* accounting, the rapid development of specific reports for *rate cases*, and other regulatory characteristics that make utility industry analysis unique.

The utility industry's additional reporting and accounting requirements make traditional BI technologies insufficient to meet the needs of regulated utilities. Utility-Business Intelligence represents an enhanced set of analysis and reporting technologies which meet these specific needs.

Handbook of
Data Mining for Design and Manufacturing

Handbook of Data Mining for Design and Manufacturing

Contributors :
Fernando Sánchez Lasheras,
Paulino José García Nieto, *et al.*

Volume 2

AURIS Reference
AURIS REFERENCE LTD.
London, UK

Handbook of Data Mining for Design and Manufacturing
Contributors : Fernando Sánchez Lasheras *and* Paulino José García Nieto, *et al.*

Auris Reference Ltd., UK

www.aurisreference.com

United Kingdom

Printed in 2017 for Sale in the Indian Subcontinent

Handbook of Data Mining for Design and Manufacturing

ISBN: 978-1-78154-484-6

British Library Cataloguing in Publication Data
A CIP record for this book is available from the British Library

Exclusively distributed by CBS Publishers & Distributors Pvt. Ltd.

Sales & Distribution Rights only for India, Pakistan, Bangladesh, Sri Lanka, Nepal and Bhutan.This book is not to be sold outside these territories.

Preface

Data mining uses a relatively large amount of computing power operating on a large set of data to determine regularities and connections between data points. Algorithms that employ techniques from statistics, machine learning and pattern recognition are used to search large databases automatically. Data mining is also known as Knowledge-Discovery in Databases (KDD).

Like the term artificial intelligence, data mining is an umbrella term that can be applied to a number of varying activities. In the corporate world, data mining is used most frequently to determine the direction of trends and predict the future. It is employed to build models and decision support systems that give people information they can use. Data mining takes a front-line role in the battle against terrorism. It was supposedly used to determine the leader of the 9/11 attacks.

Data miners are statisticians who use techniques with names like near-neighbor models, k-means clustering, holdout method, k-fold cross validation, the leave-one-out method, and so on. Regression techniques are used to subtract irrelevant patterns, leaving only useful information. The term Bayesian is seen frequently in the field, referring to a class of inference techniques that predict the likelihood of future events by combining prior probabilities and probabilities based on conditional events. Spam filtering is arguably a form of data mining, which automatically brings relevant messages to the surface from a chaotic sea of phishing attempts and Viagra pitches.

Decision trees are used to filter mountains of data. In a decision tree, all data passes through an entrance node, where it faces a filter that separates the data into streams depending on its characteristics. For example, data about consumer behavior is likely to be filtered based on demographic factors. Data mining is not primarily about fancy graphs and visualization techniques, but it does employ them to show what it has found. It is known that we can absorb more statistical information visually than verbally and this format for presentation can be very persuasive and powerful if used in the right context.

As our civilization becomes increasingly data-saturated and sensors are distributed en masse into our local environments, we will inadvertently discover things that might be missed on the first pass over. Data mining will let us correct these mistakes and discover new insights based on past data, giving us more bang for our data storage buck.

CONTENTS

This page left intentionally blank.

List of Contributors

Fernando Sánchez Lasheras

Department of Construction and Manufacturing Engineering, University of Oviedo, Gijón 33204, Spain

Paulino José García Nieto

Department of Mathematics, University of Oviedo, Oviedo 33007, Spain; E-Mail: lato@orion.ciencias.uniovi.es

Francisco Javier de Cos Juez

Department of Mining Engineering and Exploitation, University of Oviedo, Oviedo 33004, Spain; E-Mail: fjcos@uniovi.es

Ricardo Mayo Bayón

Department of Electrical Engineering, University of Oviedo, Gijón 33204, Spain; E-Mails: rmayo@uniovi.es (R.M.B.); vmsuarez@uniovi.es (V.M.G.S.)

Victor Manuel González Suárez

Department of Electrical Engineering, University of Oviedo, Gijón 33204, Spain; E-Mails: rmayo@uniovi.es (R.M.B.); vmsuarez@uniovi.es (V.M.G.S.)

This page left intentionally blank.

Chapter 11

PREDICTIVE ANALYTICS

Predictive analytics encompasses a variety of statistical techniques from modelling, machine learning, and data-mining that analyze current and historical facts to make predictions about future, or otherwise unknown, events.

In business, predictive models exploit patterns found in historical and transactional data to identify risks and opportunities. Models capture relationships among many factors to allow assessment of risk or potential associated with a particular set of conditions, guiding decision-making for candidate transactions.

Predictive analytics is used in actuarial science, marketing, financial services, insurance, telecommunications, retail, travel, healthcare, pharmaceuticals and other fields.

One of the most well known applications is credit scoring, which is used throughout financial services. Scoring models process a customer's credit history, loan application, customer data, etc., in order to rank-order individuals by their likelihood of making future credit payments on time. A well-known example is the FICO score.

DEFINITION

Predictive analytics is an area of data-mining that deals with extracting information from data and using it to predict trends and behaviour patterns. Often the unknown event of interest is in the future, but predictive analytics can be applied to any type of unknown whether it be in the past, present or future. For example, identifying suspects after a crime has been committed, or credit card fraud as it occurs. The core of predictive analytics relies on capturing relationships between explanatory variables and the predicted variables from past occurrences, and exploiting them to predict the unknown outcome. It is important to note, however, that the accuracy and usability of results will depend greatly on the level of data analysis and the quality of assumptions.

TYPES

Generally, the term predictive analytics is used to mean predictive modelling, "scoring" data with predictive models, and forecasting. However, people are increasingly using the term to refer to related analytical disciplines, such as descriptive modelling and decision modelling or optimization. These disciplines also involve rigorous data analysis, and are widely used in business for segmentation and decision-making, but have different purposes and the statistical techniques underlying them vary.

PREDICTIVE MODELS

Predictive models are models of the relation between the specific performance of a unit in a sample and one or more known attributes or features of the unit. The objective of the model is to assess the likelihood that a similar unit in a different sample will exhibit the specific performance. This category encompasses models that are in many areas, such as marketing, where they seek out subtle data patterns to answer questions about customer performance, such as fraud detection models. Predictive models often perform calculations during live transactions, for example, to evaluate the risk or opportunity of a given customer or transaction, in order to guide a decision. With advancements in computing speed, individual agent modelling systems have become capable of simulating human behaviour or reactions to given stimuli or scenarios.

The available sample units with known attributes and known performances is referred to as the "training sample." The units in other sample, with known attributes but un-known performances, are referred to as "out of [training] sample" units. The out of sample bare no chronological relation to the training sample units. For example, the training sample may consists of literary attributes of writings by Victorian authors, with known attribution, and the out-of sample unit may be newly found writing with unknown authorship; a predictive model may aid the attribution of the unknown author. Another example is given by analysis of blood splatter in simulated crime scenes in which the out-of sample unit is the actual blood splatter pattern from a crime scene. The out of sample unit may be from the same time as the training units, from a previous time, or from a future time.

DESCRIPTIVE MODELS

Descriptive models quantify relationships in data in a way that is often used to classify customers or prospects into groups. Unlike predictive models that focus on predicting a single customer behaviour (such as credit risk), descriptive models identify many different relationships between customers or products. Descriptive models do not rank-order customers by their likelihood of taking a particular action the way predictive models do. Instead, descriptive models can be used, for example, to categorize customers by their product preferences and life stage. Descriptive modelling tools can be utilized to develop further models that can simulate large number of individualized agents and make predictions.

DECISION MODELS

Decision models describe the relationship between all the elements of a decision—the known data (including results of predictive models), the decision, and the forecast results of the decision—in order to predict the results of decisions involving many variables. These models can be used in optimization, maximizing certain outcomes while minimizing others. Decision models are generally used to develop decision logic or a set of business rules that will produce the desired action for every customer or circumstance.

APPLICATIONS

Although predictive analytics can be put to use in many applications, we outline a few examples where predictive analytics has shown positive impact in recent years.

Analytical Customer Relationship Management (CRM)

Analytical Customer Relationship Management is a frequent commercial application of Predictive Analysis. Methods of predictive analysis are applied to customer data to pursue CRM objectives, which involve constructing a holistic view of the customer no matter where their information resides in the company or the department involved. CRM uses predictive analysis in applications for marketing campaigns, sales, and customer services to name a few. These tools are required in order for a company to posture and focus their efforts effectively across the breadth of their customer base. They must analyze and understand the products in demand or have the potential for high demand, predict customers' buying habits in order to promote relevant products at multiple touch points, and proactively identify and mitigate issues that have the potential to lose customers or reduce their ability to gain new ones. Analytical Customer Relationship Management can be applied throughout the customers lifecycle (acquisition, relationship growth, retention, and win-back). Several of the application areas described below (direct marketing, cross-sell, customer retention) are part of Customer Relationship Managements.

Clinical Decision Support Systems

Experts use predictive analysis in health care primarily to determine which patients are at risk of developing certain conditions, like diabetes, asthma, heart disease, and other lifetime illnesses. Additionally, sophisticated clinical decision support systems incorporate predictive analytics to support medical decision-making at the point of care. A working definition has been proposed by Robert Hayward of the Centre for Health Evidence : "Clinical Decision Support Systems link health observations with health knowledge to influence health choices by clinicians for improved health care."

Collection Analytics

Every portfolio has a set of delinquent customers who do not make their payments on time. The financial institution has to undertake collection activities on these customers to recover the amounts due. A lot of collection resources are wasted on customers who are difficult or impossible to recover. Predictive analytics can help optimize the allocation of collection resources by identifying the most effective collection agencies, contact strategies, legal actions and other strategies to each customer, thus significantly increasing recovery at the same time reducing collection costs.

Cross-sell

Often corporate organizations collect and maintain abundant data (*e.g.* customer records, sale transactions) as exploiting hidden relationships in the data can provide a competitive advantage. For an organization that offers multiple products, predictive analytics can help analyze customers' spending, usage and other behaviour, leading to efficient cross sales, or selling additional products to current customers. This directly leads to higher profitability per customer and stronger customer relationships.

Customer Retention

With the number of competing services available, businesses need to focus efforts on maintaining continuous consumer satisfaction, rewarding consumer loyalty and minimizing customer attrition. Businesses tend to respond to customer attrition on a reactive basis, acting only after the customer has initiated the process to terminate service. At this stage, the chance of changing the customer's decision is almost impossible. Proper application of predictive analytics can lead to a more proactive retention strategy. By a frequent examination of a customer's past service usage, service performance, spending and other behaviour patterns, predictive models can determine the likelihood of a customer terminating service sometime soon. An intervention with lucrative offers can increase the chance of retaining the customer. Silent attrition, the behaviour of a customer to slowly but steadily reduce usage, is another problem that many companies face. Predictive analytics can also predict this behaviour, so that the company can take proper actions to increase customer activity.

Direct Marketing

When marketing consumer products and services, there is the challenge of keeping up with competing products and consumer behaviour. Apart from identifying prospects, predictive analytics can also help to identify the most effective combination of product versions, marketing material, communication channels and timing that should be used to target a given consumer. The goal of predictive analytics is typically to lower the cost per order or cost per action.

Fraud Detection

Fraud is a big problem for many businesses and can be of various types : inaccurate credit applications, fraudulent transactions (both offline and online), identity thefts and false insurance claims. These problems plague firms of all sizes in many industries. Some examples of likely victims are credit card issuers, insurance companies, retail merchants, manufacturers, business-to-business suppliers and even services providers. A predictive model can help weed out the "bads" and reduce a business's exposure to fraud.

Predictive modelling can also be used to identify high-risk fraud candidates in business or the public sector. Mark Nigrini developed a risk-scoring method to identify audit targets. He describes the use of this approach to detect fraud in the franchisee sales reports of an international fast-food chain. Each location is scored using 10 predictors. The 10 scores are then weighted to give one final overall risk score for each location. The same scoring approach was also used to identify high-risk check kiting accounts, potentially fraudulent travel agents, and questionable vendors. A reasonably complex model was used to identify fraudulent monthly reports submitted by divisional controllers.

The Internal Revenue Service (IRS) of the United States also uses predictive analytics to mine tax returns and identify tax fraud.

Recent advancements in technology have also introduced predictive behaviour analysis for web fraud detection. This type of solution utilizes heuristics in order to study normal web user behaviour and detect anomalies indicating fraud attempts.

Portfolio, Product or Economy-level Prediction

Often the focus of analysis is not the consumer but the product, portfolio, firm, industry or even the economy. For example, a retailer might be interested in predicting store-level demand for inventory management purposes. Or the Federal Reserve Board might be interested in predicting the unemployment rate for the next year. These types of problems can be addressed by predictive analytics using time series techniques. They can also be addressed via machine learning approaches which transform the original time series into a feature vector space, where the learning algorithm finds patterns that have predictive power.

Risk Management

When employing risk management techniques, the results are always to predict and benefit from a future scenario. The Capital asset pricing model (CAP-M) "predicts" the best portfolio to maximize return, Probabilistic Risk Assessment (PRA)—when combined with mini-Delphi Techniques and statistical approaches yields accurate forecasts and RiskAoA is a stand-alone predictive tool. These are three examples of approaches that can extend from project to market, and from near to long-term. Underwriting and other business approaches identify risk management as a predictive method.

Underwriting

Many businesses have to account for risk exposure due to their different services and determine the cost needed to cover the risk. For example, auto insurance providers need to accurately determine the amount of premium to charge to cover each automobile and driver. A financial company needs to assess a borrower's potential and ability to pay before granting a loan. For a health insurance provider, predictive analytics can analyze a few years of past medical claims data, as well as lab, pharmacy and other records where available, to predict how expensive an enrollee is likely to be in the future. Predictive analytics can help underwrite these quantities by predicting the chances of illness, default, bankruptcy, etc. Predictive analytics can streamline the process of customer acquisition by predicting the future risk behaviour of a customer using application level data. Predictive analytics in the form of credit scores have reduced the amount of time it takes for loan approvals, especially in the mortgage market where lending decisions are now made in a matter of hours rather than days or even weeks. Proper predictive analytics can lead to proper pricing decisions, which can help mitigate future risk of default.

TECHNOLOGY AND BIG DATA INFLUENCES

Big data is a collection of data sets that are so large and complex that they become awkward to work with using traditional database management tools. The volume, variety and velocity of big data have introduced challenges across the board for capture, storage, search, sharing, analysis, and visualization. Examples of big data sources include web logs, RFID and sensor data, social networks, Internet search indexing, call detail records, military surveillance, and complex data in astronomic, biogeochemical, genomics, and atmospheric sciences. Thanks to technological advances in computer hardware—faster CPUs, cheaper memory, and MPP architectures—and new technologies such as Hadoop, Map Reduce, and in-database and text analytics for processing big data, it is now feasible to collect, analyze, and mine massive amounts of structured and unstructured data for new insights. Today, exploring big data and using predictive analytics is within reach of more organizations than ever before and new methods that are capable for handling such data-sets are proposed.

ANALYTICAL TECHNIQUES

The approaches and techniques used to conduct predictive analytics can broadly be grouped into regression techniques and machine learning techniques.

Regression Techniques

Regression models are the mainstay of predictive analytics. The focus lies on establishing a mathematical equation as a model to represent the interactions between the different variables in consideration. Depending on the situation, there is a wide variety of models that can be applied while performing predictive analytics. Some of them are briefly discussed below.

Linear Regression Model

The linear regression model analyzes the relationship between the response or dependent variable and a set of independent or predictor variables. This relationship is expressed as an equation that predicts the response variable as a linear function of the parameters. These parameters are adjusted so that a measure of fit is optimized. Much of the effort in model fitting is focused on minimizing the size of the residual, as well as ensuring that it is randomly distributed with respect to the model predictions.

The goal of regression is to select the parameters of the model so as to minimize the sum of the squared residuals. This is referred to as **ordinary least squares** (OLS) estimation and results in best linear unbiased estimates (BLUE) of the parameters if and only if the Gauss-Markov assumptions are satisfied.

Once the model has been estimated we would be interested to know if the predictor variables belong in the model—*i.e.* is the estimate of each variable's contribution reliable? To do this we can check the statistical significance of the model's coefficients which can be measured using the t-statistic. This amounts to testing whether the coefficient is significantly different from zero. How well the model predicts the dependent variable based on the value of the independent variables can be assessed by using the R^2 statistic. It measures predictive power of the model *i.e.* the proportion of the total variation in the dependent variable that is "explained" (accounted for) by variation in the independent variables.

Discrete Choice Models

Multivariate regression (above) is generally used when the response variable is continuous and has an unbounded range. Often the response variable may not be continuous but rather discrete. While mathematically it is feasible to apply multivariate regression to discrete ordered dependent variables, some of the assumptions behind the theory of multivariate linear regression no longer hold, and there are other techniques such as discrete choice models which are better suited for this type of analysis. If the dependent variable is discrete, some of those superior methods are logistic regression, multi-nomial logit and probit models. Logistic regression and probit models are used when the dependent variable is binary.

Logistic Regression

In a classification setting, assigning outcome probabilities to observations can be achieved through the use of a logistic model, which is basically a method which transforms information about the binary dependent variable into an unbounded continuous variable and estimates a regular multivariate model.

The Wald and likelihood-ratio test are used to test the statistical significance of each coefficient *b* in the model. A test assessing the goodness-of-fit of a classification model is the "percentage correctly predicted".

Multi-nomial Logistic Regression

An extension of the binary logit model to cases where the dependent variable has more than 2 categories is the multi-nomial logit model. In such cases collapsing the data into two categories might not make good sense or may lead to loss in the richness of the data. The multi-nomial logit model is the appropriate technique in these cases, especially when the dependent variable categories are not ordered (for examples colours like red, blue, green). Some authors have extended multi-nomial regression to include feature selection/importance methods such as Random multi-nomial logit.

Probit Regression

Probit models offer an alternative to logistic regression for modelling categorical dependent variables. Even though the outcomes tend to be similar, the underlying distributions are different. Probit models are popular in social sciences like economics.

A good way to understand the key difference between probit and logit models is to assume that there is a latent variable z.

We do not observe z but instead observe y which takes the value 0 or 1. In the logit model we assume that y follows a logistic distribution. In the probit model we assume that y follows a standard normal distribution. Note that in social sciences (*e.g.* economics), probit is often used to model situations where the observed variable y is continuous but takes values between 0 and 1.

Logit *versus* Probit

The Probit model has been around longer than the logit model. They behave similarly, except that the logistic distribution tends to be slightly flatter tailed. One of the reasons the logit model was formulated was that the probit model was computationally difficult due to the requirement of numerically calculating integrals. Modern computing however has made this computation fairly simple. The coefficients obtained from the logit and probit model are fairly close. However, the odds ratio is easier to interpret in the logit model.

Practical reasons for choosing the probit model over the logistic model would be :

- There is a strong belief that the underlying distribution is normal
- The actual event is not a binary outcome (*e.g.*, bankruptcy status) but a proportion (*e.g.*, proportion of population at different debt levels).

Time Series Models

Time series models are used for predicting or forecasting the future behaviour of variables. These models account for the fact that data points taken over time may have an internal structure (such as autocorrelation, trend or seasonal variation) that should be accounted for. As a result standard regression techniques cannot be applied to time series data and methodology has been developed to decompose the trend, seasonal and cyclical component of the series. Modelling the dynamic path of a variable can improve forecasts since the predictable component of the series can be projected into the future.

Time series models estimate difference equations containing stochastic components. Two commonly used forms of these models are autoregressive models (AR) and moving average (MA) models. The Box-Jenkins methodology developed by George Box and G.M. Jenkins combines the AR and MA models to produce the ARMA (autoregressive moving average) model which is the cornerstone of stationary time series analysis. ARIMA(autoregressive integrated moving average models) on the other hand are used to describe non-stationary time series. Box and Jenkins suggest differencing a non stationary time series to obtain a stationary series to which an ARMA model can be applied. Non-stationary time series have a pronounced trend and do not have a constant long-run mean or variance.

Box and Jenkins proposed a three stage methodology which includes : model identification, estimation and validation. The identification stage involves identifying if the series is stationary or not and the presence of seasonality by examining plots of the series, autocorrelation and partial autocorrelation functions. In the estimation stage, models are estimated using non-linear time series or maximum likelihood estimation procedures. Finally the validation stage involves diagnostic checking such as plotting the residuals to detect outliers and evidence of model fit.

In recent years time series models have become more sophisticated and attempt to model conditional heteroskedasticity with models such as ARCH (autoregressive conditional heteroskedasticity) and GARCH (generalized autoregressive conditional heteroskedasticity) models frequently used for financial time series. In addition time series models are also used to understand inter-relationships among economic variables represented by systems of equations using VAR (vector autoregression) and structural VAR models.

Survival or Duration Analysis

Survival analysis is another name for time to event analysis. These techniques were primarily developed in the medical and biological sciences, but they are also widely used in the social sciences like economics, as well as in engineering (reliability and failure time analysis).

Censoring and non-normality, which are characteristic of survival data, generate difficulty when trying to analyze the data using conventional statistical models such as multiple linear regression. The normal distribution, being a symmetric distribution, takes positive as well as negative values, but duration by its very nature cannot be negative and therefore normality cannot be assumed when dealing with duration/survival data. Hence the normality assumption of regression models is violated.

The assumption is that if the data were not censored it would be representative of the population of interest. In survival analysis, censored observations arise whenever the dependent variable of interest represents the time to a terminal event, and the duration of the study is limited in time.

An important concept in survival analysis is the hazard rate, defined as the probability that the event will occur at time t conditional on surviving until time t. Another concept related to the hazard rate is the survival function which can be defined as the probability of surviving to time t.

Most models try to model the hazard rate by choosing the underlying distribution depending on the shape of the hazard function. A distribution whose hazard function slopes upward is said to have positive duration dependence, a decreasing hazard shows negative duration dependence whereas constant hazard is a process with no memory usually characterized by the exponential distribution. Some of the distributional choices in survival models are : F, gamma, Weibull, log normal, inverse normal, exponential etc. All these distributions are for a non-negative random variable.

Duration models can be parametric, non-parametric or semi-parametric. Some of the models commonly used are Kaplan-Meier and Cox proportional hazard model (non-parametric).

Hierarchical Optimal Discriminant Analysis (HODA), (also called classification tree analysis) is a generalization of Optimal discriminant analysis that may be used to identify the statistical model that has maximum accuracy for predicting the value of a categorical dependent variable for a data-set consisting of categorical and continuous variables. The output of HODA is a non-orthogonal tree that combines categorical variables and cut points for continuous variables that yields maximum predictive accuracy, an assessment of the exact Type I error rate, and an evaluation of potential cross-generalizability of the statistical model. Hierarchical Optimal Discriminant analysis may be thought of as a generalization of Fisher's linear discriminant analysis. Optimal discriminant analysis is an alternative to ANOVA (analysis of variance) and regression analysis, which attempt to express one dependent variable as a linear combination of other features or measurements. However, ANOVA and regression analysis give a dependent variable that is a numerical variable, while hierarchical optimal discriminant analysis gives a dependent variable that is a class variable.

Classification and regression trees (CART) is a non-parametric decision tree learning technique that produces either classification or regression trees, depending on whether the dependent variable is categorical or numeric, respectively.

Decision trees are formed by a collection of rules based on variables in the modelling data-set :

- Rules based on variables' values are selected to get the best split to differentiate observations based on the dependent variable
- Once a rule is selected and splits a node into two, the same process is applied to each "child" node (*i.e.* it is a recursive procedure)

- Splitting stops when CART detects no further gain can be made, or some pre-set stopping rules are met. (Alternatively, the data are split as much as possible and then the tree is later pruned.)

Each branch of the tree ends in a terminal node. Each observation falls into one and exactly one terminal node, and each terminal node is uniquely defined by a set of rules.

A very popular method for predictive analytics is Leo Breiman's Random forests or derived versions of this technique like Random multi-nomial logit.

Multivariate Adaptive Regression Splines

Multivariate adaptive regression splines (MARS) is a non-parametric technique that builds flexible models by fitting piece-wise linear regressions.

An important concept associated with regression splines is that of a knot. Knot is where one local regression model gives way to another and thus is the point of intersection between two splines.

In multivariate and adaptive regression splines, basis functions are the tool used for generalizing the search for knots. Basis functions are a set of functions used to represent the information contained in one or more variables. Multivariate and Adaptive Regression Splines model almost always creates the basis functions in pairs.

Multivariate and adaptive regression spline approach deliberately overfits the model and then prunes to get to the optimal model. The algorithm is computationally very intensive and in practice we are required to specify an upper limit on the number of basis functions.

Machine Learning Techniques

Machine learning, a branch of artificial intelligence, was originally employed to develop techniques to enable computers to learn. Today, since it includes a number of advanced statistical methods for regression and classification, it finds application in a wide variety of fields including medical diagnostics, credit card fraud detection, face and speech recognition and analysis of the stock market. In certain applications it is sufficient to directly predict the dependent variable without focusing on the underlying relationships between variables. In other cases, the underlying relationships can be very complex and the mathematical form of the dependencies unknown. For such cases, machine learning techniques emulate human cognition and learn from training examples to predict future events.

A brief discussion of some of these methods used commonly for predictive analytics is provided below. A detailed study of machine learning can be found in Mitchell.

Neural Networks

Neural networks are non-linear sophisticated modelling techniques that are able to model complex functions. They can be applied to problems of prediction, classification or control in a wide spectrum of fields such as finance, cognitive psychology/neuroscience, medicine, engineering, and physics.

Neural networks are used when the exact nature of the relationship between inputs and output is not known. A key feature of neural networks is that they learn the relationship between inputs and output through training. There are three types of training in neural networks used by different networks, supervised and unsupervised training, reinforcement learning, with supervised being the most common one.

Some examples of neural network training techniques are back propagation, quick propagation, conjugate gradient descent, projection operator, Delta-Bar-Delta etc. Some unsupervised network architectures are multi-layer perceptrons, Kohonen networks, Hopfield networks, etc.

Multi-layer Perceptron (MLP)

The Multi-layer Perceptron (MLP) consists of an input and an output layer with one or more hidden layers of non-linearly-activating nodes or sigmoid nodes. This is determined by the weight vector and it is necessary to adjust the weights of the network. The back propogation employs gradient fall to minimize the squared error between the network output values and desired values for those outputs. The weights adjusted by an iterative process of repetitive present of attributes. Small changes in the weight to get the desired values are done by the process called training the net and is done by the training set (learning rule).

Radial Basis Functions

A radial basis function (RBF) is a function which has built into it a distance criterion with respect to a center. Such functions can be used very efficiently for interpolation and for smoothing of data. Radial basis functions have been applied in the area of neural networks where they are used as a replacement for the sigmoidal transfer function. Such networks have 3 layers, the input layer, the hidden layer with the RBF non-linearity and a linear output layer. The most popular choice for the non-linearity is the Gaussian. RBF networks have the advantage of not being locked into local minima as do the feed-forward networks such as the multi-layer perceptron.

Support Vector Machines

Support Vector Machines (SVM) are used to detect and exploit complex patterns in data by clustering, classifying and ranking the data. They are learning machines that are used to perform binary classifications and regression estimations. They commonly use kernel based methods to apply linear classification techniques to non-linear classification problems. There are a number of types of SVM such as linear, polynomial, sigmoid etc.

Naïve Bayes

Naïve Bayes based on Bayes conditional probability rule is used for performing classification tasks. Naïve Bayes assumes the predictors are statistically independent which makes it an effective classification tool that is easy to interpret. It is best employed when faced with the problem of 'curse of dimensionality' *i.e.* when the number of predictors is very high.

k-nearest Neighbours

The nearest neighbour algorithm (KNN) belongs to the class of pattern recognition statistical methods. The method does not impose a priori any assumptions about the distribution from which the modelling sample is drawn. It involves a training set with both positive and negative values. A new sample is classified by calculating the distance to the nearest neighbouring training case. The sign of that point will determine the classification of the sample. In the k-nearest neighbour classifier, the k nearest points are considered and the sign of the majority is used to classify the sample. The performance of the kNN algorithm is influenced by three main factors : (1) the distance measure used to locate the nearest neighbours; (2) the decision rule used to derive a classification from the k-nearest neighbours; and (3) the number of neighbours used to classify the new sample. It can be proved that, unlike other methods, this method is universally asymptotically convergent, *i.e.* : as the size of the training set increases, if the observations are independent and identically distributed (i.i.d.), regardless of the distribution from which the sample is drawn, the predicted class will converge to the class assignment that minimizes misclassification error.

Geospatial Predictive Modelling

Conceptually, geospatial predictive modelling is rooted in the principle that the occurrences of events being modelled are limited in distribution. Occurrences of events are neither uniform nor random in distribution—there are spatial environment factors (infrastructure, socio-cultural, topographic, etc.) that constrain and influence where the locations of events occur. Geospatial predictive modelling attempts to describe those constraints and influences by spatially correlating occurrences of historical geospatial locations with environmental factors that represent those constraints and influences. Geospatial predictive modelling is a process for analyzing events through a geographic filter in order to make statements of likelihood for event occurrence or emergence.

Tools

Historically, using predictive analytics tools—as well as understanding the results they delivered—required advanced skills. However, modern predictive analytics tools are no longer restricted to IT specialists. As more organizations adopt predictive analytics into decision-making processes and integrate it into their operations, they are creating a shift in the market toward business users as the primary consumers of the information. Business users want tools they can use on their own. Vendors are responding by creating new software that removes the mathematical complexity, provides user-friendly graphic interfaces and/or builds in short cuts that can, for example, recognize the kind of data available and suggest an appropriate predictive model. Predictive analytics tools have become sophisticated enough to adequately present and dissect data problems, so that any data-savvy information worker can utilize them to analyze data and retrieve meaningful, useful results. For example, modern tools present findings using simple charts, graphs, and scores that indicate the likelihood of possible outcomes.

PREDICTIVE MODELLING

Predictive modelling is the process by which a model is created or chosen to try to best predict the probability of an outcome. In many cases the model is chosen on the basis of detection theory to try to guess the probability of an outcome given a set amount of input data, for example given an e-mail determining how likely that it is spam.

Models can use one or more classifiers in trying to determine the probability of a set of data belonging to another set, say spam or 'ham'.

Models

Nearly any regression model can be used for prediction purposes. Broadly speaking, there are two classes of predictive models : parametric and non-parametric. A third class, semi-parametric models, includes features of both. Parametric models make "specific assumptions with regard to one or more of the population parameters that characterize the underlying distribution(s)", while non-parametric regressions make fewer assumptions than their parametric counterparts.

Group Method of Data Handling

Ordinary Least Square

Ordinary least squares is a method that minimizes the sum of squared distance between an observed and its predicted value.

Generalized Linear Models (GLM)

Generalized linear model is a flexible family of models that are unified under a single method. Logistic regression is a notable special case of GLM. Other types of GLM include Poisson regression, Gamma regression, Multi-nomial regression.

Logistic Regression

Logistic regression is a technique in which unknown values of a discrete variable are predicted based on known values of one or more continuous and/or discrete variables. Logistic regression differs from ordinary least squares (OLS) regression in that the dependent variable is binary in nature. This procedure has many applications. In bio-statistics, the researcher may be interested in trying to model the probability of a patient being diagnosed with a certain type of cancer based on knowing, say, the incidence of that cancer in his or her family. In business, the marketer may be interested in modelling the probability of an individual purchasing a product based on the price of that product. Both of these are examples of a simple, binary logistic regression model. The model is "simple" in that each has only one independent, or predictor, variable, and it is "binary" in that the dependent variable can take on only one of two values : cancer or no cancer, and purchase or does not purchase.

Generalized Additive Models

Generalized additive model is a smoothing method for multiple predictors that allows for non-parametric predictions.

Robust Regression

Robust regression include a number of modelling approaches to handle high leverage observations or violation of assumptions. Models can be both parametric (*e.g.* regression with Huber, White, Sandwich variance estimators) as well as non-parametric(*e.g.* quantile regression).

Semiparametric Regression

Semiparametric regression include proportional odds model and the Cox proportional hazard model where the response is a rank.

Presenting and Using the Results of a Predictive Model

Predictive models can either be used directly to estimate a response (output) given a defined set of characteristics (input), or indirectly to drive the choice of decision rules.

Depending on the methodology employed for the prediction, it is often possible to derive a formula that may be used in a **spreadsheet software**. This has some advantages for end users or decision-makers, the main one being familiarity with the software itself, hence a lower barrier to adoption.

Nomograms are useful graphical representation of a predictive model. As in spreadsheet software, their use depends on the methodology chosen. The advantage of nomograms is the immediacy of computing predictions without the aid of a computer.

Point estimates tables are one of the simplest form to represent a predictive tool. Here combination of characteristics of interests can either be represented via a table or a graph and the associated prediction read off the y-axis or the table itself.

Tree based methods (*e.g.* CART, survival trees) provide one of the most graphically intuitive ways to present predictions. However, their usage is limited to those methods that use this type of modelling approach which can have several drawbacks. Trees can also be employed to represent decision rules graphically.

Score charts are graphical tabular or graphical tools to represent either predictions or decision rules.

A new class of **modern tools** are represented by web based applications. For example, Shiny is a web based tool developed by Rstudio, an R IDE. With a Shiny app, a modeller has the advantage to represent any which way he or she chooses to represent the predictive model while allowing the user some control. A user can choose a combination of characteristics of interest via sliders or input boxes and results can be generated, from graphs to confidence intervals to tables and various statistics of interests. However, these tools often require a server installation of Rstudio.

Applications

Uplift Modelling

Uplift Modelling is a technique for modelling the *change in probability* caused by an action. Typically this is a marketing action such as an offer to buy a product, to use a product more or to re-sign a contract. For example in a retention campaign you wish to predict the change in probability that a customer will remain a customer if they are contacted. A model of the change in probability allows the retention campaign to be targeted at those customers on whom the change in probability will be beneficial. This allows the retention programme to avoid triggering unnecessary churn or customer attrition without wasting money contacting people who would act anyway.

Archaeology

Predictive modelling in archaeology gets its foundations from Gordon Willey's mid-fifties work in the Virú Valley of Peru. Complete, intensive surveys were performed then covariability between cultural remains and natural features such as slope, and vegetation were determined. Development of quantitative methods and a greater availability of applicable data led to growth of the discipline in the 1960s and by the late 1980s, substantial progress had been made by major land managers worldwide.

Generally, predictive modelling in archaeology is establishing statistically valid causal or covariable relationships between natural proxies such as soil types, elevation, slope, vegetation, proximity to water, geology, geomorphology, etc., and the presence of archaeological features. Through analysis of these quantifiable attributes from land that has undergone archaeological survey, sometimes the "archaeological sensitivity" of unsurveyed areas can be anticipated based on the natural proxies in those areas. Large land managers in the United States, such as the Bureau of Land Management (BLM), the Department of Defense (DOD), and numerous highway and parks agencies, have successfully employed this strategy. By using predictive modelling in their cultural resource management plans, they are capable of making more informed decisions when planning for activities that have the potential to require ground disturbance and subsequently affect archaeological sites.

Customer Relationship Management

Predictive modelling is used extensively in analytical customer relationship management and data-mining to produce customer-level models that describe the likelihood that a customer will take a particular action. The actions are usually sales, marketing and customer retention related.

For example, a large consumer organisation such as a mobile telecommunications operator will have a set of predictive models for product cross-sell, product deep-sell and churn. It is also now more common for such an organisation to have a model of savability using an uplift model. This predicts the likelihood that a customer can be saved at the end of a contract period (the change in churn probability) as opposed to the standard churn prediction model.

Auto Insurance

Predictive Modelling is utilised in vehicle insurance to assign risk of incidents to policy-holders from information obtained from policy-holders. This is extensively employed in usage-based insurance solutions where predictive models utilise telemetry based data to build a model of predictive risk for claim likelihood. Black-box auto insurance predictive models utilise GPS or accelerometer sensor input only. Some models include a wide range of predictive input beyond basic telemetry including advanced driving behaviour, independent crash records, road history, and user profiles to provide improved risk models.

Health Care

In 2009 Parkland Health & Hospital System began analyzing electronic medical records in order to use predictive modelling to help identify patients at high risk of readmission. Initially the hospital focused on patients with congestive heart failure, but the program has expanded to include patients with diabetes, acute myocardial infarction, and pneumonia.

Notable Failures of Predictive Modelling

Although not widely discussed by the mainstream predictive modelling community, predictive modelling is a methodology that has been widely used in the financial industry in the past and some of the spectacular failures have contributed to the financial crisis of 2008. These failures exemplify the danger of relying blindly on models that are essentially backforward looking in nature. The following examples are by no mean a complete list :

1. *Bond rating* : S&P, Moody's and Fitch quantify the probability of default of bonds with discrete variables called rating. The rating can take on discrete values from AAA down to D. The rating is a predictor of the risk of default

based on a variety of variables associated with the borrower and macro-economic data that are drawn from historicals. The rating agencies failed spectacularly with their ratings on the 600 billion USD mortgage backed CDO market. Almost the entire AAA sector (and the super-AAA sector, a new rating the rating agencies provided to represent super safe investment) of the CDO market defaulted or severely downgraded during 2008, many of which obtained their ratings less than just a year ago.

2. *Statistical models that attempt to predict equity market prices based on historical data*: So far, no such model is considered to consistently make correct predictions over the long-term. One particularly memorable failure is that of Long-term Capital Management, a fund that hired highly qualified analysts, including a Nobel Prize winner in economics, to develop a sophisticated statistical model that predicted the price spreads between different securities. The models produced impressive profits until a spectacular debacle that caused the then Federal Reserve chairman Alan Greenspan to step in to broker a rescue plan by the wall street broker dealers in order to prevent a meltdown of the bond market.

Possible Fundamental Limitations of Predictive Model Based on Data Fitting

1. *History cannot always predict future* : Using relations derived from historical data to predict the future implicitly assumes there are certain steady-state conditions or constants in the complex system. This is almost always wrong when the system involves people.
2. *The issue of unknown unknowns* : In all data collection, the collector first defines the set of variables for which data is collected. However, no matter how extensive the collector considers his selection of the variables, there is always the possibility of new variables that have not been considered or even defined, yet critical to the outcome.
3. *Self-defeat of an algorithm* : After an algorithm becomes an accepted standard of measurement, it can be taken advantage of by people who understand the algorithm and have the incentive to fool or manipulate the outcome. This is what happened to the CDO rating. The CDO dealers actively fulfilled the rating agencies input to reach an AAA or super-AAA on the CDO they are issuing by cleverly manipulating variables that were "unknown" to the rating agencies' "sophisticated" models.

ARTIFICIAL NEURAL NETWORK

In computer science and related fields, **artificial neural networks** are computational models inspired by animals' central nervous systems (in particular the brain) that are capable of machine learning and pattern recognition. They are usually presented as systems of interconnected "neurons" that can compute values from inputs by feeding information through the network.

For example, in a neural network for handwriting recognition, a set of input neurons may be activated by the pixels of an input image representing a letter or digit. The activations of these neurons are then passed on, weighted and transformed by some function determined by the network's designer, to other neurons, etc., until finally an output neuron is activated that determines which character was read.

Like other machine learning methods, neural networks have been used to solve a wide variety of tasks that are hard to solve using ordinary rule-based programming, including computer vision and speech recognition.

Background

The inspiration for the neural networks came from examination of central nervous systems. In an artificial neural network, simple artificial nodes, called "neurons", "neurodes", "processing elements" or "units", are connected together to form a network which mimics a biological neural network.

There is no single formal definition of what an artificial neural network is. Commonly, a class of statistical models may be called "neural" if they :

1. Consist of sets of adaptive weights, *i.e.* numerical parameters that are tuned by a learning algorithm, and
2. Are capable of approximating non-linear functions of their inputs.

The adaptive weights are conceptually connection strengths between neurons, which are activated during training and prediction.

Neural networks are also similar to biological neural networks in performing functions collectively and in parallel by the units, rather than there being a clear delineation of sub-tasks to which various units are assigned. The term "neural network" usually refers to models employed in statistics, cognitive psychology and artificial intelligence. Neural network models which emulate the central nervous system are part of theoretical neuroscience and computational neuroscience.

In modern software implementations of artificial neural networks, the approach inspired by biology has been largely abandoned for a more practical approach based on statistics and signal processing. In some of these systems, neural networks or parts of neural networks (like artificial neurons) form components in larger systems that combine both adaptive and non-adaptive elements. While the more general approach of such systems is more suitable for real-world problem solving, it has little to do with the traditional artificial intelligence connectionist models. What they do have in common, however, is the principle of non-linear, distributed, parallel and local processing and adaptation. Historically, the use of neural networks models marked a paradigm shift in the late eighties from high-level (symbolic) artificial intelligence, characterized by expert systems with knowledge embodied in *if-then* rules, to low-level (sub-symbolic) machine learning, characterized by knowledge embodied in the parameters of a dynamical system.

History

Warren McCulloch and Walter Pitts created a computational model for neural networks based on mathematics and algorithms. They called this model threshold logic. The model paved the way for neural network research to split into two distinct approaches. One approach focused on biological processes in the brain and the other focused on the application of neural networks to artificial intelligence.

In the late 1940s psychologist Donald Hebb created a hypothesis of learning based on the mechanism of neural plasticity that is now known as Hebbian learning. Hebbian learning is considered to be a 'typical' unsupervised learning rule and its later variants were early models for long-term potentiation. These ideas started being applied to computational models in 1948 with Turing's B-type machines.

Farley and Wesley A. Clark first used computational machines, then called calculators, to simulate a Hebbian network at MIT. Other neural network computational machines were created by Rochester, Holland, Habit, and Duda.

Frank Rosenblatt created the perceptron, an algorithm for pattern recognition based on a two-layer learning computer network using simple addition and subtraction. With mathematical notation, Rosenblatt also described circuitry not in the basic perceptron, such as the exclusive-or circuit, a circuit whose mathematical computation could not be processed until after the back propagation algorithm was created by Paul Werbos.

Neural network research stagnated after the publication of machine learning research by Marvin Minsky and Seymour Papert. They discovered two key issues with the computational machines that processed neural networks. The first issue was that single-layer neural networks were incapable of processing the exclusive-or circuit. The second significant issue was that computers were not sophisticated enough to effectively handle the long-run time required by large neural networks. Neural network research slowed until computers achieved greater processing power. Also key later advances was the back propagation algorithm which effectively solved the exclusive-or problem.

The parallel distributed processing of the mid-1980s became popular under the name connectionism. The text by David E. Rumelhart and James McClelland provided a full exposition on the use of connectionism in computers to simulate neural processes.

Neural networks, as used in artificial intelligence, have traditionally been viewed as simplified models of neural processing in the brain, even though the relation between this model and brain biological architecture is debated, as it is not clear to what degree artificial neural networks mirror brain function.

In the 1990s, neural networks were overtaken in popularity in machine learning by support vector machines and other, much simpler methods such as linear classifiers. Renewed interest in neural nets was sparked in the 2000s by the advent of deep learning.

Recent Improvements

Biophysical models, such as BCM theory, have been important in understanding mechanisms for synaptic plasticity, and have had applications in both computer science and neuroscience. Research is ongoing in understanding the computational algorithms used in the brain, with some recent biological evidence for radial basis networks and neural back propagation as mechanisms for processing data.

Computational devices have been created in CMOS, for both biophysical simulation and neuromorphic computing. More recent efforts show promise for creating nano-devices for very large scale principal components analyses and convolution. If successful, these efforts could usher in a new era of neural computing that is a step beyond digital computing, because it depends on learning rather than programming and because it is fundamentally analog rather than digital even though the first instantiations may in fact be with CMOS digital devices.

Between 2009 and 2012, the recurrent neural networks and deep feedforward neural networks developed in the research group of Jürgen Schmidhuber at the Swiss AI Lab IDSIA have won eight international competitions in pattern recognition and machine learning. For example, multi-dimensional long-short-term memory (LSTM) won three competitions in connected handwriting recognition at the 2009 International Conference on Document Analysis and Recognition (ICDAR), without any prior knowledge about the three different languages to be learned.

Variants of the back-propagation algorithm as well as unsupervised methods by Geoff Hinton and colleagues at the University of Toronto can be used to train deep, highly non-linear neural architectures similar to the 1980 Neocognitron by Kunihiko Fukushima, and the "standard architecture of vision", inspired by the simple and complex cells identified by David H. Hubel and Torsten Wiesel in the primary visual cortex.

Deep learning feedforward networks, such as convolutional neural networks, alternate convolutional layers and max-pooling layers, topped by several pure classification layers. Fast GPU-based implementations of this approach have won several pattern recognition contests, including the IJCNN 2011 Traffic Sign Recognition Competition and the ISBI 2012 Segmentation of Neuronal Structures in Electron Microscopy Stacks challenge. Such neural networks also were the first artificial pattern recognizers to achieve human-competitive or even superhuman performance on benchmarks such as traffic sign recognition, or the MNIST handwritten digits problem of Yann LeCun and colleagues at NYU.

Successes in Pattern Recognition Contests Since 2009

Between 2009 and 2012, the recurrent neural networks and deep feedforward neural networks developed in the research group of Jürgen Schmidhuber at the Swiss AI Lab IDSIA have won eight international competitions in pattern recognition and machine learning. For example, the bi-directional and multi-dimensional long short term memory (LSTM) of Alex Graves *et. al.* won three competitions in connected handwriting recognition at the 2009 International Conference on Document Analysis and Recognition (ICDAR), without any prior knowledge about the three different languages to be learned. Fast GPU-based implementations of this approach by Dan Ciresan and colleagues at IDSIA have won several pattern recognition contests, including the IJCNN 2011 Traffic Sign Recognition Competition, the ISBI 2012 Segmentation of Neuronal Structures in Electron Microscopy Stacks challenge, and others. Their neural networks also were the first artificial pattern recognizers to achieve human-competitive or even superhuman performance on important benchmarks such as traffic sign recognition, or the MNIST handwritten digits problem of Yann LeCun at NYU. Deep, highly non-linear neural architectures similar to the 1980 neocognitron by Kunihiko Fukushima and the "standard architecture of vision" can also be pre-trained by unsupervised methods of Geoff Hinton's lab at University of Toronto. A team from this lab won a 2012 contest sponsored by Merck to design software to help find molecules that might lead to new drugs.

Models

Neural network models in artificial intelligence are usually referred to as artificial neural networks (ANNs); these are essentially simple mathematical models defining a function $f : X \rightarrow Y$ or a distribution over X or both X and Y, but sometimes models are also intimately associated with a particular learning algorithm or learning rule. A common use of the phrase ANN model really means the definition of a *class* of such functions (where members of the class are obtained by varying parameters, connection weights, or specifics of the architecture such as the number of neurons or their connectivity).

Network Function

The word *network* in the term 'artificial neural network' refers to the inter-connections between the neurons in the different layers of each system. An example system has three layers. The first layer has input neurons which send data via synapses to the second layer of neurons, and then via more synapses to the third layer of output neurons. More complex systems will have more layers of neurons with some having increased layers of input neurons and output neurons. The synapses store parameters called "weights" that manipulate the data in the calculations.

An ANN is typically defined by three types of parameters :

1. The interconnection pattern between the different layers of neurons
2. The learning process for updating the weights of the interconnections
3. The activation function that converts a neuron's weighted input to its output activation.

Mathematically, a neuron's network function $f(x)$ is defined as a composition of other functions $g_i(x)$, which can further be defined as a composition of other functions. This can be conveniently represented as a network structure, with arrows depicting the dependencies between variables. A widely used type of composition is the *non-linear weighted sum*, where $f(x) = K\left(\Sigma_i w_i g_i(x)\right)$, where K (commonly referred to as the activation function) is some pre-defined function, such as the hyperbolic tangent. It will be convenient for the following to refer to a collection of functions g_i as simply a vector $g = (g_1, g_2, \ldots, g_n)$.

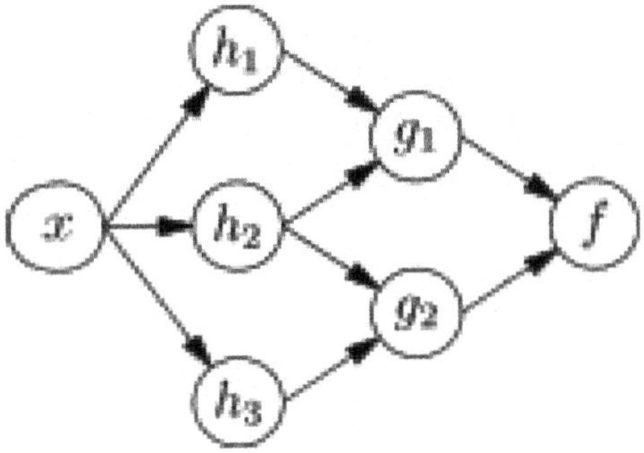

Fig. : ANN dependency graph.

This figure depicts such a decomposition of f, with dependencies between variables indicated by arrows. These can be interpreted in two ways.

The first view is the functional view : the input x is transformed into a 3-dimensional vector h, which is then transformed into a 2-dimensional vector g, which is finally transformed into f. This view is most commonly encountered in the context of optimization.

The second view is the probabilistic view : the random variable $F = f(G)$ depends upon the random variable $G = g(H)$, which depends upon $H = h(X)$, which depends upon the random variable X. This view is most commonly encountered in the context of graphical models.

The two views are largely equivalent. In either case, for this particular network architecture, the components of individual layers are independent of each other (*e.g.*, the components of g are independent of each other given their input h). This naturally enables a degree of parallelism in the implementation.

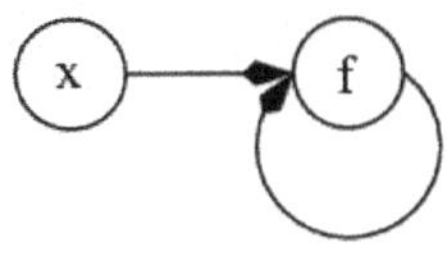

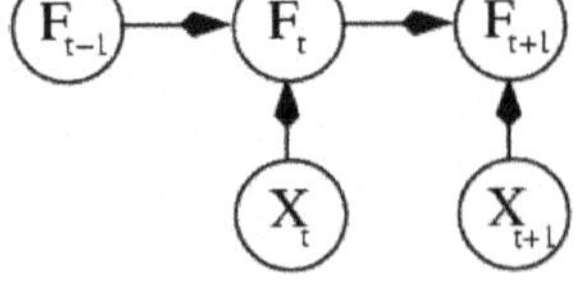

Fig. : Two separate depictions of the recurrent ANN dependency graph.

Networks such as the previous one are commonly called feedforward, because their graph is a directed acyclic graph. Networks with cycles are commonly called recurrent. Such networks are commonly depicted in the manner shown at the top of the figure, where f is shown as being dependent upon itself. However, an implied temporal dependence is not shown.

Learning

What has attracted the most interest in neural networks is the possibility of *learning*. Given a specific *task* to solve, and a *class* of functions F, learning means using a set of *observations* to find $f^* \in F$ which solves the task in some *optimal* sense.

This entails defining a cost function $C : F \rightarrow R$ such that, for the optimal solution f^*, $C(f^*) \leq C(f)\ \forall f \in F$ – *i.e.*, no solution has a cost less than the cost of the optimal solution.

The cost function C is an important concept in learning, as it is a measure of how far away a particular solution is from an optimal solution to the problem to be solved. Learning algorithms search through the solution space to find a function that has the smallest possible cost.

For applications where the solution is dependent on some data, the cost must necessarily be a *function of the observations*, otherwise we would not be modelling anything related to the data. It is frequently defined as a statistic to which only approximations can be made. As a simple example, consider the problem of finding the model f, which minimizes $C = E[f(x) - y)^2]$, for data pairs (x, y) drawn from some distribution D. In practical situations we would only have N samples from D and thus, for the above example, we would only minimize $\hat{C} = \frac{1}{N}\sum_{i=1}^{N}(f(x_i) - y_i)^2$.

Thus, the cost is minimized over a sample of the data rather than the entire dataset.

When $N \rightarrow \infty$ some form of online machine learning must be used, where the cost is partially minimized as each new example is seen. While online machine learning is often used when D is fixed, it is most useful in the case where the distribution changes slowly over time. In neural network methods, some form of online machine learning is frequently used for finite datasets.

Choosing a Cost Function

While it is possible to define some arbitrary ad hoc cost function, frequently a particular cost will be used, either because it has desirable properties (such as convexity) or because it arises naturally from a particular formulation of the problem (*e.g.*, in a probabilistic formulation the posterior probability of the model can be used as an inverse cost). Ultimately, the cost function will depend on the desired task. An overview of the three main categories of learning tasks is provided below :

Learning Paradigms

There are three major learning paradigm, each corresponding to a particular abstract learning task. These are supervised learning, unsupervised learning and reinforcement learning.

Supervised Learning

In supervised learning, we are given a set of example pairs (x, y), $x \in X$, $y \in Y$ and the aim is to find a function $f : X \rightarrow Y$ in the allowed class of functions that matches the examples. In other words, we wish to *infer* the mapping implied by the data; the cost function is related to the mismatch between our mapping and the data and it implicitly contains prior knowledge about the problem domain.

A commonly used cost is the mean-squared error, which tries to minimize the average squared error between the network's output, $f(x)$, and the target value y over all the example pairs. When one tries to minimize this cost using gradient descent for the class of neural networks called multi-layer perceptrons, one obtains the common and well-known back propagation algorithm for training neural networks.

Tasks that fall within the paradigm of supervised learning are pattern recognition (also known as classification) and regression (also known as function approximation). The supervised learning paradigm is also applicable to sequential data (*e.g.*, for speech and gesture recognition). This can be thought of as learning with a "teacher," in the form of a function that provides continuous feedback on the quality of solutions obtained thus far.

Unsupervised Learning

In unsupervised learning, some data x is given and the cost function to be minimized, that can be any function of the data x and the network's output, f.

The cost function is dependent on the task (what we are trying to model) and our *a priori* assumptions (the implicit properties of our model, its parameters and the observed variables).

As a trivial example, consider the model $f(x) = a$ where a is a constant and the cost $C = E[x - f(x))^2]$. Minimizing this cost will give us a value of a that is equal to the mean of the data. The cost function can be much more complicated. Its form depends on the application : for example, in compression it could be related to the mutual information between x and $f(x)$, whereas in statistical modelling, it could be related to the posterior probability of the model given the data. (Note that in both of those examples those quantities would be maximized rather than minimized).

Tasks that fall within the paradigm of unsupervised learning are in general estimation problems; the applications include clustering, the estimation of statistical distributions, compression and filtering.

Reinforcement Learning

In reinforcement learning, data x are usually not given, but generated by an agent's interactions with the environment. At each point in time t, the agent performs an action y_t and the environment generates an observation x_t and an instantaneous cost c_t, according to some (usually unknown) dynamics. The aim is to discover a *policy* for selecting actions that minimizes some measure of a long-term cost; *i.e.*, the expected cumulative cost. The environment's dynamics and the long-term cost for each policy are usually unknown, but can be estimated.

More formally the environment is modelled as a Markov decision process (MDP) with states $s_1,..., s_n \in S$ and actions $a_1,..., a_m \in A$ with the following probability distributions : the instantaneous cost distribution $P(c_t \mid s_t)$, the observation distribution $P(x_t \mid s_t)$ and the transition $P(s_t + 1 \mid s_t, a_t)$, while a policy is defined as conditional distribution over actions given the observations. Taken together, the two then define a Markov chain (MC). The aim is to discover the policy that minimizes the cost; *i.e.*, the MC for which the cost is minimal.

ANNs are frequently used in reinforcement learning as part of the overall algorithm. Dynamic programming has been coupled with ANNs (Neuro dynamic programming) by Bertsekas and Tsitsiklis and applied to multi-dimensional non-linear problems such as those involved in vehicle routing, natural resources management or medicine because of the ability of ANNs to mitigate losses of accuracy even when reducing the discretization grid density for numerically approximating the solution of the original control problems.

Tasks that fall within the paradigm of reinforcement learning are control problems, games and other sequential decision-making tasks.

Learning Algorithms

Training a neural network model essentially means selecting one model from the set of allowed models (or, in a Bayesian framework, determining a distribution over the set of allowed models) that minimizes the cost criterion. There are numerous algorithms available for training neural network models; most of them can be viewed as a straightforward application of optimization theory and statistical estimation.

Most of the algorithms used in training artificial neural networks employ some form of gradient descent. This is done by simply taking the derivative of the cost function with respect to the network parameters and then changing those parameters in a gradient-related direction.

Evolutionary methods, gene expression programming, simulated annealing, expectation-maximization, non-parametric methods and particle swarm optimization are some commonly used methods for training neural networks.

Employing Artificial Neural Networks

Perhaps the greatest advantage of ANNs is their ability to be used as an arbitrary function approximation mechanism that 'learns' from observed data. However, using them is not so straightforward, and a relatively good understanding of the underlying theory is essential.

- *Choice of model :* This will depend on the data representation and the application. Overly complex models tend to lead to problems with learning.
- *Learning algorithm :* There are numerous trade-offs between learning algorithms. Almost any algorithm will work well with the *correct hyperparameters* for training on a particular fixed data-set. However, selecting and tuning an algorithm for training on unseen data requires a significant amount of experimentation.
- *Robustness :* If the model, cost function and learning algorithm are selected appropriately the resulting ANN can be extremely robust.

With the correct implementation, ANNs can be used naturally in online learning and large data-set applications. Their simple implementation and the existence of mostly local dependencies exhibited in the structure allows for fast, parallel implementations in hardware.

Applications

The utility of artificial neural network models lies in the fact that they can be used to infer a function from observations. This is particularly useful in applications where the complexity of the data or task makes the design of such a function by hand impractical.

Real-life Applications

The tasks artificial neural networks are applied to tend to fall within the following broad categories :

- Function approximation, or regression analysis, including time series prediction, fitness approximation and modelling.
- Classification, including pattern and sequence recognition, novelty detection and sequential decision-making.
- Data processing, including filtering, clustering, blind source separation and compression.
- Robotics, including directing manipulators, prosthesis.
- Control, including Computer numerical control.

Application areas include the system identification and control (vehicle control, process control, natural resources management), quantum chemistry, game-playing and decision-making (backgammon, chess, poker), pattern recognition (radar systems, face identification, object recognition and more), sequence recognition (gesture, speech, handwritten text recognition), medical diagnosis, financial applications (*e.g.* automated trading systems), data-mining (or knowledge discovery in databases, "KDD"), visualization and e-mail spam filtering.

Artificial neural networks have also been used to diagnose several cancers. An ANN based hybrid lung cancer detection system named HLND improves the accuracy of diagnosis and the speed of lung cancer radiology. These networks have also been used to diagnose prostate cancer. The diagnoses can be used to make specific models taken from a large group of patients compared to information of one given patient. The models do not depend on assumptions about correlations of different variables. Colourectal cancer has also been predicted using the neural networks. Neural networks could predict the outcome for a patient with colourectal cancer with more accuracy than the current clinical methods. After training, the networks could predict multiple patient outcomes from unrelated institutions.

Neural Networks and Neuroscience

Theoretical and computational neuroscience is the field concerned with the theoretical analysis and the computational modelling of biological neural systems. Since neural systems are intimately related to cognitive processes and behaviour, the field is closely related to cognitive and behavioural modelling.

The aim of the field is to create models of biological neural systems in order to understand how biological systems work. To gain this understanding, neuroscientists strive to make a link between observed biological processes (data), biologically plausible mechanisms for neural processing and learning (biological neural network models) and theory (statistical learning theory and information theory).

Types of Models

Many models are used in the field, defined at different levels of abstraction and modelling different aspects of neural systems. They range from models of the short-term behaviour of individual neurons, models of how the dynamics of neural circuitry arise from interactions between individual neurons and finally to models of how behaviour can arise from abstract neural modules that represent complete sub-systems. These include models of the long-term, and short-term plasticity, of neural systems and their relations to learning and memory from the individual neuron to the system level.

Neural Network Software

Neural network software is used to simulate, research, develop and apply artificial neural networks, biological neural networks and, in some cases, a wider array of adaptive systems.

Types of Artificial Neural Networks

Artificial neural network types vary from those with only one or two layers of single direction logic, to complicated multi-input many directional feedback loops and layers. On the whole, these systems use algorithms in their programming to determine control and organization of their functions. Some may be as simple as a one-neuron layer with an input and an output, and others can mimic complex systems such as dANN, which can mimic chromosomal DNA through sizes at the cellular level, into artificial organisms and simulate reproduction, mutation and population sizes.

Most systems use "weights" to change the parameters of the throughput and the varying connections to the neurons. Artificial neural networks can be autonomous and learn by input from outside "teachers" or even self-teaching from written-in rules.

Theoretical Properties

Computational Power

The multi-layer perceptron (MLP) is a universal function approximator, as proven by the Cybenko theorem. However, the proof is not constructive regarding the number of neurons required or the settings of the weights.

Work by Hava Siegelmann and Eduardo D. Sontag has provided a proof that a specific recurrent architecture with rational valued weights (as opposed to full precision real number-valued weights) has the full power of a Universal Turing Machine using a finite number of neurons and standard linear connections. They have further shown that the use of irrational values for weights results in a machine with super-Turing power.

Capacity

Artificial neural network models have a property called 'capacity', which roughly corresponds to their ability to model any given function. It is related to the amount of information that can be stored in the network and to the notion of complexity.

Convergence

Nothing can be said in general about convergence since it depends on a number of factors. Firstly, there may exist many local minima. This depends on the cost function and the model. Secondly, the optimization method used might not be guaranteed to converge when far away from a local minimum. Thirdly, for a very large amount of data or parameters, some methods become impractical. In general, it has been found that theoretical guarantees regarding convergence are an unreliable guide to practical application.

Generalization and Statistics

In applications where the goal is to create a system that generalizes well in unseen examples, the problem of over-training has emerged. This arises in convoluted or over-specified systems when the capacity of the network significantly exceeds the needed free parameters. There are two schools of thought for avoiding this problem : The first is to use cross-validation and similar techniques to check for the presence of overtraining and optimally select hyperparameters such as to minimize the generalization error. The second is to use some form of *regularization*. This is a concept that emerges naturally in a probabilistic (Bayesian) framework, where the regularization can be performed by selecting a larger prior probability over simpler models; but also in statistical learning theory, where the goal is to minimize over two quantities : the 'empirical risk' and the 'structural risk', which roughly corresponds to the error over the training set and the predicted error in unseen data due to overfitting.

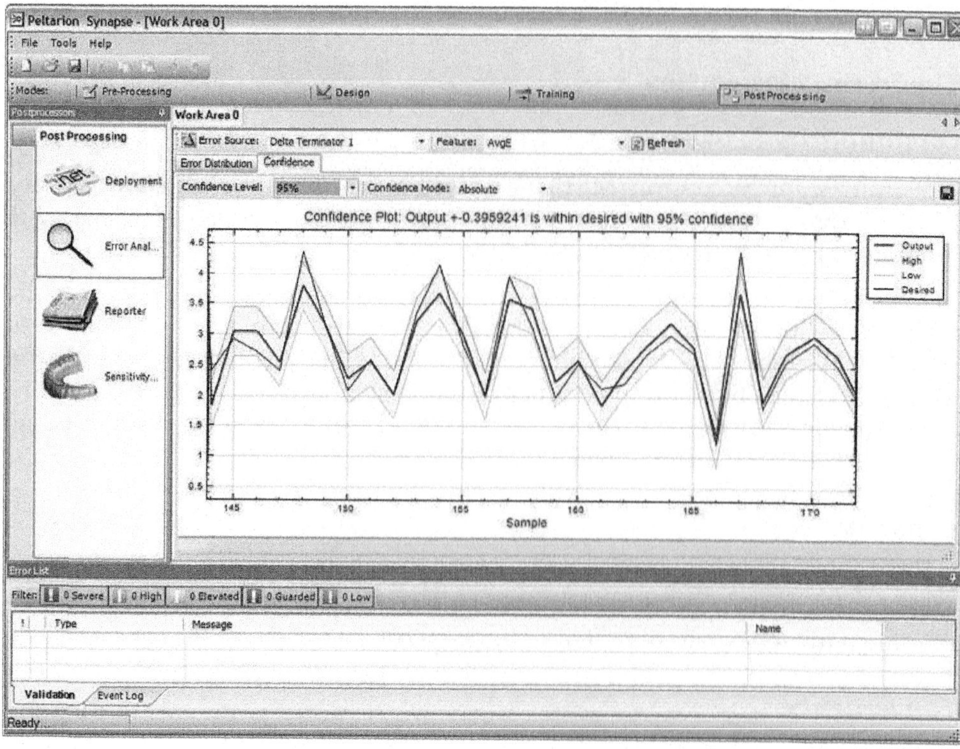

Fig. : Confidence analysis of a neural network.

Supervised neural networks that use an MSE cost function can use formal statistical methods to determine the confidence of the trained model. The MSE on a validation set can be used as an estimate for variance. This value can then be used to calculate the confidence interval of the output of the network, assuming a normal distribution. A confidence analysis made this way is statistically valid as long as the output probability distribution stays the same and the network is not modified.

By assigning a softmax activation function, a generalization of the logistic function, on the output layer of the neural network (or a softmax component in a component-based neural network) for categorical target variables, the outputs can be interpreted as posterior probabilities. This is very useful in classification as it gives a certainty measure on classifications.

The softmax activation function is :

$$y_i = \frac{e^{x_i}}{\sum_{j=1}^{c} e^{x_j}}$$

Dynamic Properties

Various techniques originally developed for studying disordered magnetic systems (*i.e.*, the spin glass) have been successfully applied to simple neural network architectures, such as the Hopfield network. Influential work by E. Gardner and B. Derrida has revealed many interesting properties about perceptrons with real-valued synaptic weights, while later work by W. Krauth and M. Mezard has extended these principles to binary-valued synapses.

Criticism

A common criticism of neural networks, particularly in robotics, is that they require a large diversity of training for real-world operation. This is not surprising, since any learning machine needs sufficient representative examples in order to capture the underlying structure that allows it to generalize to new cases. Dean Pomerleau, in his research presented in the paper "Knowledge-based Training of Artificial Neural Networks for Autonomous Robot Driving," uses a neural network to train a robotic vehicle to drive on multiple types of roads (single lane, multi-lane, dirt, etc.). A large amount of his research is devoted to (1) extrapolating multiple training scenarios from a single training experience, and (2) preserving past training diversity so that the system does not become overtrained (if, for example, it is presented with a series of right turns—it should not learn to always turn right). These issues are common in neural networks that must decide from amongst a wide variety of responses, but can be dealt with in several ways, for example by randomly shuffling the training examples, by using a numerical optimization algorithm that does not take too large steps when changing the network connections following an example, or by grouping examples in so-called mini-batches.

A. K. Dewdney, a former *Scientific American* columnist, wrote in 1997, "Although neural nets do solve a few toy problems, their powers of computation are so limited that I am surprised anyone takes them seriously as a general problem-solving tool."

Arguments for Dewdney's position are that to implement large and effective software neural networks, much processing and storage resources need to be committed. While the brain has hardware tailored to the task of processing signals through a graph of neurons, simulating even a most simplified form on Von Neumann technology may compel a neural network designer to fill many millions of database rows for its connections—which can consume vast amounts of computer memory and hard disk space. Furthermore, the designer of neural network systems will often need to simulate the transmission of signals through many of these connections and their associated neurons—which must often be matched with incredible amounts of CPU processing power and time.

While neural networks often yield *effective* programs, they too often do so at the cost of *efficiency* (they tend to consume considerable amounts of time and money).

Arguments against Dewdney's position are that neural nets have been successfully used to solve many complex and diverse tasks, ranging from autonomously flying aircraft to detecting credit card fraud.

Technology writer Roger Bridgman commented on Dewdney's statements about neural nets :

Neural networks, for instance, are in the dock not only because they have been hyped to high heaven, (what hasn't?) but also because you could create a successful net without understanding how it worked : the bunch of numbers that captures its behaviour would in all probability be "an opaque, unreadable table... valueless as a scientific resource."

In spite of his emphatic declaration that science is not technology, Dewdney seems here to pillory neural nets as bad science when most of those devising them are just trying to be good engineers. An unreadable table that a useful machine could read would still be well worth having.

In response to this kind of criticism, one should note that although it is true that analyzing what has been learned by an artificial neural network is difficult, it is much easier to do so than to analyze what has been learned by a biological neural network. Furthermore, researchers involved in exploring learning algorithms for neural networks are gradually uncovering generic principles which allow a learning machine to be successful. For example, Bengio and LeCun wrote an article regarding local vs non-local learning, as well as shallow vs deep architecture.

Some other criticisms came from believers of hybrid models (combining neural networks and symbolic approaches). They advocate the intermix of these two approaches and believe that hybrid models can better capture the mechanisms of the human mind.

SELF-ORGANIZING MAP

A **self-organizing map** (**SOM**) or **self-organizing feature map** (**SOFM**) is a type of *artificial neural network* (ANN) that is trained using *unsupervised learning* to produce a low-dimensional (typically two-dimensional), discretized representation of the input space of the training samples, called a **map**. Self-organizing maps are different from other artificial neural networks in the sense that they use a neighbourhood function to preserve the *topological* properties of the input space.

This makes SOMs useful for *visualizing* low-dimensional views of high-dimensional data, akin to *multi-dimensional scaling*. The model was first described as an artificial neural network by the *Finnish* professor *Teuvo Kohonen*, and is sometimes called a **Kohonen map** or **network**.

Like most artificial neural networks, SOMs operate in two modes : training and mapping. "Training" builds the map using input examples (a *competitive process*, also called *vector quantization*), while "mapping" automatically classifies a new input vector.

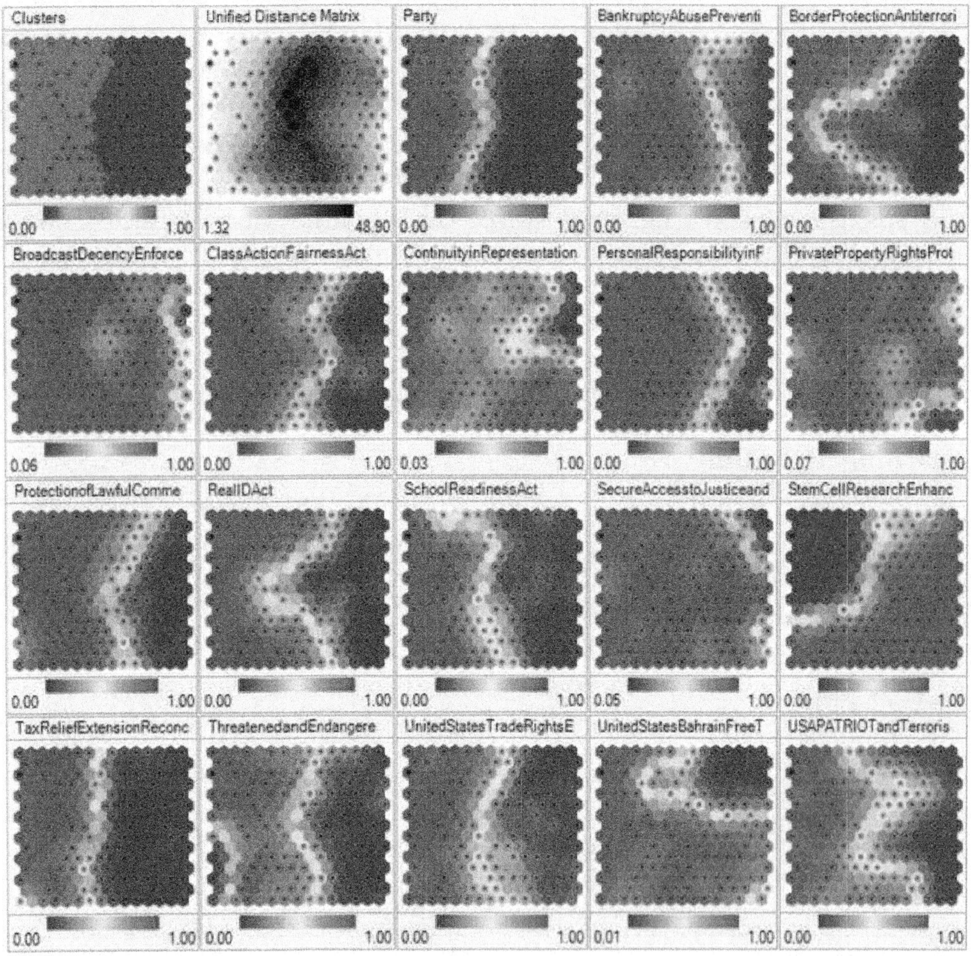

Fig. : A self-organizing map showing *U.S. Congress* voting patterns visualized in *Synapse*. The first two boxes show clustering and distances while the remaining ones show the component planes. Red means a yes vote while blue means a no vote in the component planes (except the party component where red is *Republican* and blue is *Democratic*).

A self-organizing map consists of components called nodes or neurons. Associated with each node is a weight vector of the same dimension as the input data vectors and a position in the map space. The usual arrangement of nodes is a two-dimensional regular spacing in a *hexagonal* or *rectangular* grid. The self-organizing map describes a mapping from a higher-dimensional input space to a lower-dimensional map space. The procedure for placing a vector from data space onto the map is to find the node with the closest (smallest distance metric) weight vector to the data space vector.

While it is typical to consider this type of network structure as related to *feedforward networks* where the nodes are visualized as being attached, this type of architecture is fundamentally different in arrangement and motivation.

Useful extensions include using *toroidal* grids where opposite edges are connected and using large numbers of nodes.

It has been shown that while self-organizing maps with a small number of nodes behave in a way that is similar to *K-means,* larger self-organizing maps rearrange data in a way that is fundamentally topological in character.

It is also common to use the *U-Matrix*. The U-Matrix value of a particular node is the average distance between the node and its closest neighbours. In a square grid, for instance, we might consider the closest 4 or 8 nodes (the *Von Neumann* and *Moore neighbourhoods*, respectively), or six nodes in a hexagonal grid.

Large SOMs display emergent properties. In maps consisting of thousands of nodes, it is possible to perform cluster operations on the map itself.

Learning Algorithm

The goal of learning in the self-organizing map is to cause different parts of the network to respond similarly to certain input patterns. This is partly motivated by how visual, auditory or other sensory information is handled in separate parts of the cerebral cortex in the human brain.

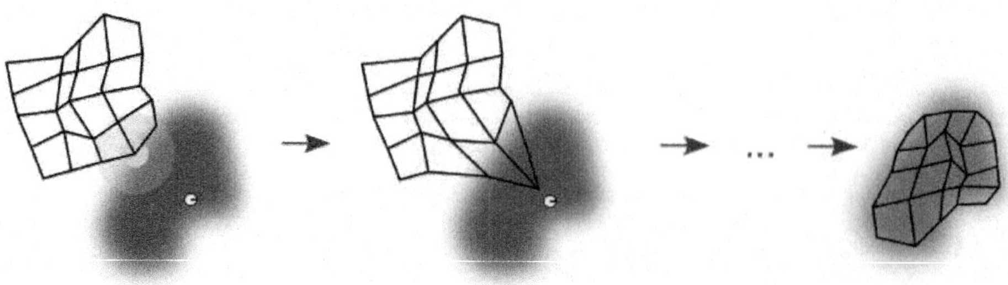

Fig.: An illustration of the training of a self-organizing map. The blue blob is the distribution of the training data, and the small white disc is the current training datum drawn from that distribution. At first (left) the SOM nodes are arbitrarily positioned in the data space. The node (highlighted in yellow) which is nearest to the training datum is selected. It is moved towards the training datum, as (to a lesser extent) are its neighbours on the grid. After many iterations the grid tends to approximate the data distribution (right).

The weights of the neurons are initialized either to small random values or sampled evenly from the sub-space spanned by the two largest principal component eigenvectors. With the latter alternative, learning is much faster because the initial weights already give a good approximation of SOM weights.

The network must be fed a large number of example vectors that represent, as close as possible, the kinds of vectors expected during mapping. The examples are usually administered several times as iterations.

The training utilizes competitive learning. When a training example is fed to the network, its Euclidean distance to all weight vectors is computed. The neuron whose weight vector is most similar to the input is called the best matching unit (BMU). The weights of the BMU and neurons close to it in the SOM lattice are adjusted towards the input vector. The magnitude of the change decreases with time and with distance (within the lattice) from the BMU. The update formula for a neuron with weight vector $Wv(s)$ is

$$Wv(s + 1) = Wv(s) + \Theta(u, v, s)\, \alpha(s)(D(t) - Wv(s)),$$

where s is the step index, t an index into the training sample, u is the index of the BMU for $D(t)$, $\alpha(s)$ is a monotonically decreasing learning coefficient and $D(t)$ is the input vector; v is assumed to visit all neurons for every value of s and t. Depending on the implementations, t can scan the training data-set systematically (t is 0, 1, 2...T-1, then repeat, T being the training sample's size), be randomly drawn from the data-set (bootstrap sampling), or implement some other sampling method (such as jackknifing).

The neighbourhood function $\Theta(u, v, s)$ depends on the lattice distance between the BMU (neuron u) and neuron v. In the simplest form it is 1 for all neurons close enough to BMU and 0 for others, but a Gaussian function is a common choice, too. Regardless of the functional form, the neighbourhood function shrinks with time. At the beginning when the neighbourhood is broad, the self-organizing takes place on the global scale. When the neighbourhood has shrunk to just a couple of neurons, the weights are converging to local estimates. In some implementations the learning coefficient α and the neighbourhood function Θ decrease steadily with increasing s, in others (in particular those where t scans the training data-set) they decrease in step-wise fashion, once every T steps.

This process is repeated for each input vector for a (usually large) number of cycles λ. The network winds up associating output nodes with groups or patterns in the input data-set. If these patterns can be named, the names can be attached to the associated nodes in the trained net.

During mapping, there will be one single *winning* neuron : the neuron whose weight vector lies closest to the input vector. This can be simply determined by calculating the Euclidean distance between input vector and weight vector.

It should be noted that any kind of object which can be represented digitally, which has an appropriate distance measure associated with it, and in which the necessary operations for training are possible can be used to construct a self-organizing map. This includes matrices, continuous functions or even other self-organizing maps.

Preliminary Definitions

Consider an $n \times m$ array of nodes, each of which contains a weight vector and is aware of its location in the array. Each weight vector is of the same dimension as the node's input vector. The weights may initially be set to random values.

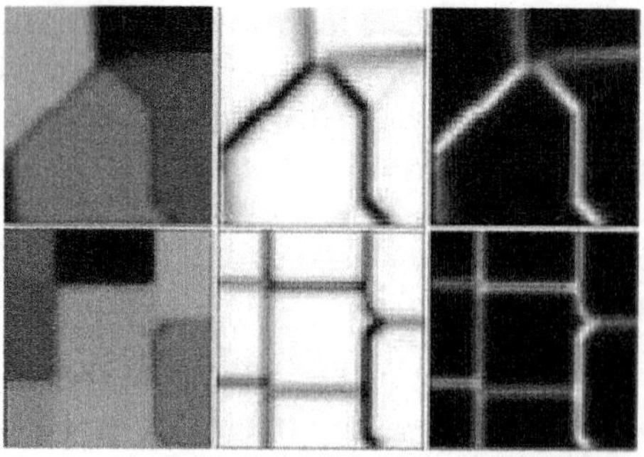

Fig. : Self organizing maps (SOM) of three and eight colours with U-Matrix.

Now we need input to feed the map—The generated map and the given input exist in separate sub-spaces. We will create three vectors to represent colours. Colours can be represented by their red, green, and blue components. Consequently our input vectors will have three components, each corresponding to a colour space. The input vectors will be :

R = <255, 0, 0>

G = <0, 255, 0>

B = <0, 0, 255>

The colour training vector data sets used in SOM :

three Colours = [255, 0, 0], [0, 255, 0], [0, 0, 255]

eight Colours = [0, 0, 0], [255, 0, 0], [0, 255, 0], [0, 0, 255], [255, 255, 0], [0, 255, 255], [255, 0, 255], [255, 255, 255]

The data vectors should preferably be normalized (vector length is equal to one) before training the SOM.

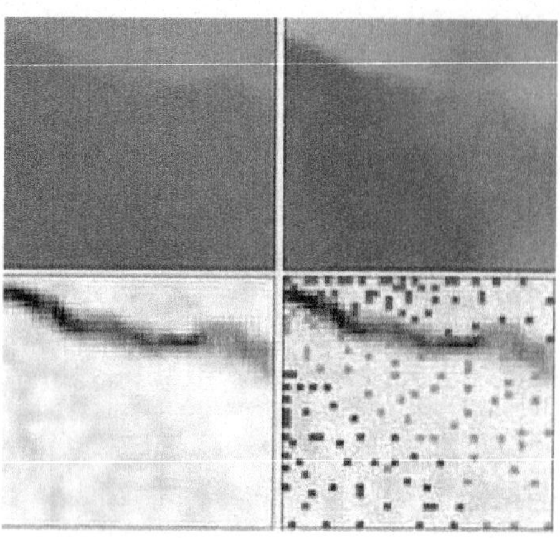

Fig. : Self organizing map of Fisher's Iris flower data.

Neurons (40×40 square grid) are trained for 250 iterations with a learning rate of 0.1 using the normalized Iris flower data-set which has four-dimensional data vectors. Shown are : a colour image formed by the first three dimensions of the four-dimensional SOM weight vectors (top left), a pseudo-colour image of the magnitude of the SOM weight vectors (top right), a U-Matrix (Euclidean distance between weight vectors of neighbouring cells) of the SOM (bottom left), and an overlay of data points (red : *I. setosa*, green : *I. versicolour* and blue : *I. virginica*) on the U-Matrix based on the minimum Euclidean distance between data vectors and SOM weight vectors.

Variables

These are the variables needed, with vectors in bold,

- s is the current iteration
- λ is the iteration limit
- t is the index of the target input data vector in the input data-set D
- $D(t)$ is a target input data vector
- v is the index of the node in the map
- Wv is the current weight vector of node v
- u is the index of the best matching unit (BMU) in the map
- $\Theta(u, v, s)$ is a restraint due to distance from BMU, usually called the neighbourhood function, and
- $\alpha(s)$ is a learning restraint due to iteration progress.

Algorithm

1. Randomize the map's nodes' weight vectors
2. Grab an input vector $D(t)$
3. Traverse each node in the map

1. Use the Euclidean distance formula to find the similarity between the input vector and the map's node's weight vector
2. Track the node that produces the smallest distance (this node is the best matching unit, BMU)

4. Update the nodes in the neighbourhood of the BMU (including the BMU itself) by pulling them closer to the input vector

   1. $Wv(s + 1) = Wv(s) + \Theta(u, v, s)\ \alpha(s)(\mathbf{D}(t)\text{-}Wv(s))$

5. Increase s and repeat from step 2 while $s < \lambda$

A variant algorithm :

1. Randomize the map's nodes' weight vectors
2. Traverse each input vector in the input data-set

1. Traverse each node in the map

1. Use the Euclidean distance formula to find the similarity between the input vector and the map's node's weight vector

2. Track the node that produces the smallest distance (this node is the best matching unit, BMU).

   2. Update the nodes in the neighbourhood of the BMU (including the BMU itself) by pulling them closer to the input vector.

1. $Wv(s + 1) = Wv(s) + \Theta(u, v, s)\, \alpha(s)(D(t)\text{-}Wv(s))$

   3. Increase s and repeat from step 2 while $s < \lambda$

Interpretation

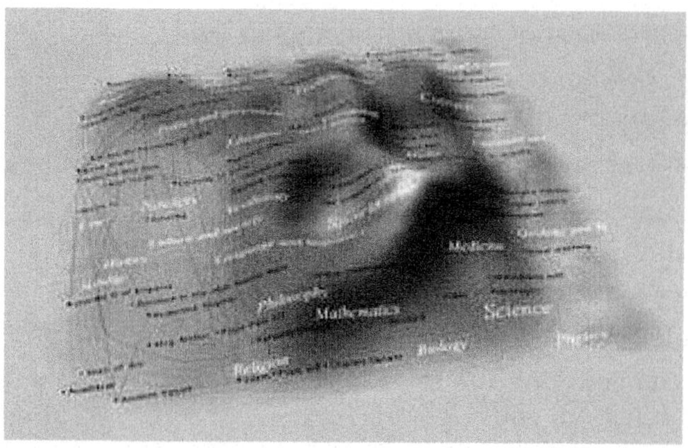

Fig. : Cartographical representation of a self-organizing map (U-Matrix). Distance is inversely proportional to similarity. The "mountains" are edges between clusters. The red lines are links between articles.

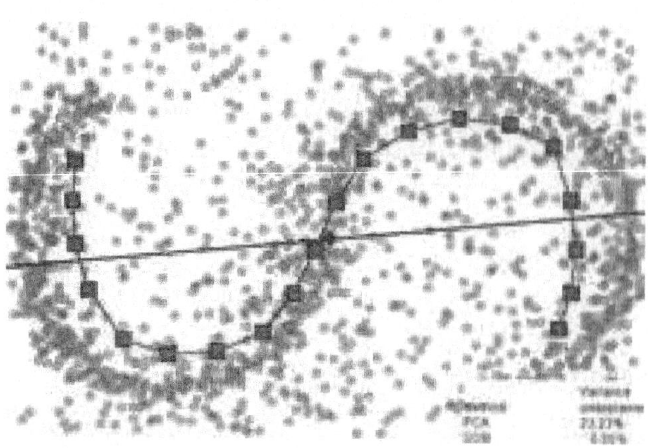

Fig. : One-dimensional SOM *versus* principal component analysis (PCA) for data approximation. SOM is a red broken line with squares, 20 nodes. The first principal component is presented by a blue line. Data points are the small grey circles. For PCA, the fraction of variance unexplained in this example is 23.23%, for SOM it is 6.86%.

There are two ways to interpret a SOM. Because in the training phase weights of the whole neighbourhood are moved in the same direction, similar items tend to excite adjacent neurons. Therefore, SOM forms a semantic map where similar samples are mapped close together and dissimilar ones apart. This may be visualized by a U-Matrix (Euclidean distance between weight vectors of neighbouring cells) of the SOM.

The other way is to think of neuronal weights as pointers to the input space. They form a discrete approximation of the distribution of training samples. More neurons point to regions with high training sample concentration and fewer where the samples are scarce.

SOM may be considered a non-linear generalization of Principal components analysis (PCA). It has been shown, using both artificial and real geophysical data, that SOM has many advantages over the conventional feature extraction methods such as Empirical Orthogonal Functions (EOF) or PCA.

Originally, SOM was not formulated as a solution to an optimisation problem. Nevertheless, there have been several attempts to modify the definition of SOM and to formulate an optimisation problem which gives similar results. For example, Elastic maps use the mechanical metaphor of elasticity to approximate principal manifolds : the analogy is an elastic membrane and plate.

Alternatives

- The **generative topographic map** (GTM) is a potential alternative to SOMs. In the sense that a GTM explicitly requires a smooth and continuous mapping from the input space to the map space, it is topology preserving. However, in a practical sense, this measure of topological preservation is lacking.
- The **time adaptive self-organizing map** (TASOM) network is an extension of the basic SOM. The TASOM employs adaptive learning rates and neighbourhood functions. It also includes a scaling parameter to make the network invariant to scaling, translation and rotation of the input space. The TASOM and its variants have been used in several applications including adaptive clustering, multi-level thresholding, input space approximation, and active contour modelling. Moreover, a Binary Tree TASOM or BTASOM, resembling a binary natural tree having nodes composed of TASOM networks has been proposed where the number of its levels and the number of its nodes are adaptive with its environment.
- The **growing self-organizing map** (GSOM) is a growing variant of the self-organizing map. The GSOM was developed to address the issue of identifying a suitable map size in the SOM. It starts with a minimal number of nodes (usually four) and grows new nodes on the boundary based on a heuristic. By using a value called the *spread factor*, the data analyst has the ability to control the growth of the GSOM.
- The **elastic maps** approach borrows from the spline interpolation the idea of minimization of the elastic energy. In learning, it minimizes the sum of quadratic bending and stretching energy with the least squares approximation error.

- The conformal approach that uses conformal mapping to interpolate each training sample between grid nodes in a continuous surface. An one-to-one smooth mapping is possible in this approach.

SUPPLY CHAIN DESIGN : A NECESSARY CORE COMPETENCY TO BUILD SUSTAINABLE COMPETITIVE ADVANTAGE

Introduction to Supply Chain Management

Definition of Supply Chain Management : A number of definitions have been proposed for supply chain management. Perhaps the best is simply, "an integrative approach to managing supply and distribution networks." The key word is "integrative," making the chain work better and at lower cost than would be possible by managing each segment of the chain independently. Another way of defining supply chain management is to state that it is the management of physical materials, services, information, money, and time across and between organizations in a business relationship in a manner that achieves the objectives of all the organizations at the least total cost.

Objectives of Supply Chain Management : Some common goals of supply chain management include : Reduce waste and nonvalue-added activities (*i.e.* cost reduction) including excess inventory, increase customer service/responsiveness, improve supply chain communication (speed/timeliness, accuracy of information, information sharing), reduce cycle time (*e.g.* new product development, supply leadtime), improve co-ordination of efforts (continuous improvement, understanding of goals).

Results of Applying Supply Chain Management : The most-sought and most common result of applying supply chain management is reduced costs in the supply chain. A cost that exists throughout most supply chains and therefore is often the target of supply chain management is the cost of inventory, managing it and carrying it. Many supply chain management efforts are aimed at reducing inventory costs. Other results of supply chain management include improved quality, more dependable quality, more reliable supplier performance, *e.g.* in order fulfillment and delivery, improved transportation service, reduced packaging costs, elimination or combination of steps in the supply chain, faster cycle times and more satisfied customers. It is by nature an integrative approach, as much entrepreneurial as logistical, and results come from the following :

- Having knowledge of a variety of methods, processes, techniques, and systems that can be used to manage supply chains.
- Studying particular supply chains to identify areas of potential improvement.
- Applying and implementing methods, processes, techniques, and systems as appropriate to improve supply chain performance.
- Evaluating changes, revising as needed, and practicing of continuous improvement through periodic performance reviews and value analysis.

Example of how Supply Chains Rather than Individual Firms Compete : The final product or service offered and delivered by any firm is the sum total of its own efforts and those of all of its supply chains. A recent example is the Ford Explorer-Firestone tire recall situation. The vehicle sold by Ford included the tires provided by a member of its supply chain—Firestone. Firestone's apparent failure to provide an adequate product is affecting the entire supply chain's ability to sell the product. Not only are Ford and Firestone affected, but so are all the other members of Ford's supply chain—the fewer Explorers that are sold, the fewer parts they can sell to Ford. The discussion reported in the press subsequent to the recall of tires indicates that Ford and Firestone were not operating together as ideal supply chain partners. Firestone recommended a higher tire pressure but, to get a softer ride, Ford recommended a lower pressure. Instead of working together—as if they were one supply chain—to solve the problem, they each proceeded separately with a disastrous result.

Development of Competitive Advantage for Supply Chains

The necessity to consider the supply chain as the competitive entity results from the integrative nature of supply chain management that, in turn, produces a high degree of interdependence among the members of a supply chain. The timeliness, quality and cost of each member of the chain combine to produce the final product, it is only partly a result of the actions of the chain member that produces the final product. Thus, a supply chain that produces a product at the level of quality demanded by the consumer, but has high-cost upstream supply chain members or uses high-cost transportation to deliver the product will be at a cost disadvantage to a chain that uses lower-cost alternatives.

Examples of Competitive Advantage for a Supply Chain : Competitive advantage accrues to supply chains in a variety of ways. Some of these include :

- Ability to maximize leverage with other supply chain members
- Ability to identify and manage cost drivers
- Optimization of supply chain capacities/release of unused capacity/no added capital—balanced supply/demand throughout the chain
- Improvement in total supply chain response time
- Leverage technology—adoption of current technology by other supply chain members
- Improved sourcing processes that enable optimal commodity strategies
- Improved cash flow throughout the chain
- Benchmarks and performance measures that represent the integrated operation of chain members rather than individual company results
- Improved competitiveness of the supply base resulting from upgrades necessary to participate in the supply chain
- Ability to tier strategies and implement enhanced business strategies
- Lower total cost of ownership for purchased materials and services

- Improved economic value added (EVA) throughout the chain
- Reduced cycle times across the chain
- Reduced transaction costs for supply requirements
- Reduced supply process costs
- Ability to charge lower prices for products
- Ability to provide higher quality products
- Improved ability to consistently meet customer requirements for quality
- Improved ability to meet customer delivery requirements
- Improved Ability to provide improved customer service before, during, and after the sale transaction
- Ability to more effectively communicate with customers to determine their requirements.

Concept of Core Competency : Supply chains, not surprisingly, often develop competitive advantages around their core competencies. Core competencies can be defined as those activities that a firm does best and most cost-effectively, and which are central or "core" to success in its business. Thus there are some firms that only market, outsourcing all manufacturing, production, and sourcing activities, because their core competency is marketing. The core competency concept must be kept in mind when designing supply chains because, in an ideal supply chain, each member will perform those things that are their core competencies and assign to other chain members those things that are not core competencies.

Supply Chain Design and Implementation

Supply chains should not "just happen," they should be deliberately designed and developed the same as individual supplier members of supply chains are selected and developed.

Supply Chain Design – Importance of Design : Supply chains determine the ability of the firms included in them to compete in the marketplace. How supply chains are designed will affect their ability to compete. A firm that is attempting to compete in a market where low cost determines who gets the business will have difficulty if it includes high cost suppliers in its supply chain. The characteristics of the end-market in which a firm is competing must be considered when designing supply chains.

Supply Chain Design—Design Process

1. *Select a chain* : Criteria for selection may include : What are the largest products or purchased items in terms of sales dollars or spend, products/materials/services that need improvement in cycle or response time, problem products (any type problem that affects customer satisfaction and cost effectiveness), products/items critical to company mission or goals.

2. *Form a design team* : Include all affected parties inside and outside your company, especially other members of the supply chain such as suppliers, customers, third party service providers.
3. *Map the chain as a team* : Get a common understanding of what you are working to improve. The entire chain does not need to be mapped in detail if you are focusing only on particular parts of it but the entire chain should be mapped generally to assure that all members who could be affected by changes are identified.
4. *Determine supply chain performance criteria* : Where or in what activities or results must the supply chain develop and maintain a competitive advantage. At what level must criteria be set, *e.g.* what level of quality or service is necessary to compete in the final consumer market for the product in question. Performance criteria will drive the metrics, or performance measures that will be used to determine if the supply chain and its individual members are performing well enough to compete. Some areas of performance criteria and metrics include :
   - Total cost Quality
   - Cycle time
   - Reliability
   - Specific product performance characteristics.
5. *Analyze each step of the selected supply chain* : Select best way of performing that step. Use benchmarking and best practice studies as references and sources of improved business methods. Identify specific measurement criteria and levels to be used in determining metrics and levels of performance that the supply chain must achieve to compete.
6. Evaluate the impact of changes from existing practice needed to perform each step in the supply chain as a whole. Identify and evaluate tradeoffs of changes at each step to determine the overall best way to perform that step to meet the performance required of the overall chain.
7. Reiterate the preceding process until the combination of practices that best meets the supply chain performance criteria is determined.

Critical Areas of Supply Chain Design

Supplier Selection : Suppliers must be carefully selected based on the particular combination of attributes they must provide to the supply chain. There may be one criterion of primary importance but others must not be overlooked. For example, a low-cost supplier that has low delivery reliability, or poor financial condition may not be a good choice even if low cost is the principle attribute that is needed.

Supplier Development : To get the exact "fit" or combination of attributes desired, it may be necessary to work with suppliers to help them develop exactly the right capabilities to enhance the supply chain.

Value Added : A supply chain should (must) enhance the customers' ability to compete when the customer is not the final consumer. For example, the supply chain may be required to provide services such as quick replenishment or training, or results such as overall cycle time reduction. Each chain member must add some value to the ultimate final product. Each chain member must provide flexibility to innovate and help to bring unique value to the chain.

Supply Chain Tier Integration, Reduction, Expansion, Cooperation, Communication: This is determining how the chain members fit together. Get the best combination of members, functions, and responsibilities to meet customer requirements. Determine in detail with affected chain members how they will work together initially and ongoing. Use written understandings and procedures to minimize misunderstandings and problems.

Change Management : Decide how to implement changes necessary to initially launch the supply chain relationship and how to agree on and implement subsequent changes that may be necessary.

Relationship Between Information and Process or Activity Cycles : Determine how information will trigger processing or activity. Determine how activity will be recorded and reported to affect future processing or activity. Determine how information systems will be adapted to the needs of the supply chain.

Communication Within the Supply Chain : What, when, where, how, who, must be planned in detail.

Goals of Particular Supply Chain : Specifically what must the supply chain achieve? Quality, performance, cost? Other goals?

Goals vs. Vision for Supply Chain : The initial development of a supply chain should come from a vision of what it must accomplish. This vision must be translated into specific goals that can be operationalized.

Match (Congruence) Between Supply Chain Goals and the Goals/Objectives/Mission/Vision of the Members of the Chain : Any significant mismatch will result in an ineffective and (probably) non-competitive supply chain.

Co-ordination of Supply Chain Strategies, Planning, Operations : Use the team approach, involve all affected parties and entities, use written procedures and guides.

Performance Measures (Metrics) : Absolutely vital to know if the supply chain is performing as needed and to determine where improvements or changes may be needed. Must be agreed upon as valid and necessary by all affected supply chain members.

Identify and Use Best Practices to determine supply chain processes and procedures.

Supply Chain Implementation

Leadership to Drive Change : There must be credible leadership for the supply chain implementation to be effective. For any significant implementation effort, corporate management must visibly endorse and support the effort through a mandate or other visible means.

Team Approach : Although leadership is essential, a team approach is another vital element of successful supply chain implementation. Use the design team, extend and change it as necessary, but be certain to include members from all entities in the supply chain that are affected and/or must do something to successfully implement and operate the supply chain.

Consider Ease of Implementation and Likelihood of Success : Pick your battles. Evaluate implementation alternatives and choose those with the highest likelihood of implementation success even if they do not offer the greatest potential improvement. Implementation success with a lesser degree of improvement is still improvement where a failure of a high potential implementation is no improvement at all.

Training : Cannot be overemphasized. Absolutely vital to successful implementation. If a person is to do something different than they have previously done, they must know how to do it.

Sources of Resistance : The organizational behaviour literature identifies eight common sources of resistance to change : (Schermerhorn, Hunt, Osborn, 1997)

- Fear of the unknown
- Disrupted habits
- Loss of confidence Loss of control
- Poor timing
- Work overload
- Loss of face
- Lack of purpose.

The same reference identifies six ways to deal with resistance to change :

- Education and communication
- Participation and involvement
- Facilitation and support
- Facilitation and agreement
- Manipulation and co-optation
- Explicit and implicit coercion.

Obviously the last two ways should be avoided or used only if the others fail. The team approach to supply chain design and implementation coupled with leadership and training utilize the first four methods and will usually result in success.

Evaluation and Review

Performance Measurement Review : Perform periodic measurement and review as determined by the design team. Revise, change, and improve the supply chain design as needed to achieve the desired results or as a result of changes in the goals, objectives, or environment of the supply chain.

Practice Continuous Improvement : Competitive advantage gained from supply chain design must be sustained by a planned continuous improvement program and supported by a vigorous, pro-active change management initiative.

SUPPLY CHAIN NETWORK DESIGN : EMBRACE THE CULTURE OF MODELLING

The cost savings of "re-shoring," or shifting production to North America, was not apparent until a number of complex and interconnected factors were considered, namely increasing ocean freight rates, the shrinking differential of labour and other manufacturing costs, as well as the true cost of in-transit and safety stock inventory.

Today, there are a number of "new normals" that are causing transportation and logistics managers to rethink their supply chain networks. The top five are listed below :

1. *Re-shoring :* Macro-economic shifts like the ones described above coupled with increased pressure to position production and inventory closer to points of market consumption, are making re-shoring and near-shoring a more viable supply chain configuration.
2. *Global transportation infrastructure :* Transportation infrastructure investments like the forthcoming *expansion of the Panama Canal* are changing the cost and speed of global freight movements and introducing new logistics options.
3. *Carrier capacity and freight rate trends :* Transportation regulations aimed at safety, like the new *hours-of-service (HOS) rules* issued by the U.S. Federal Motor Carrier Safety Administration, are affecting carrier operations.

4. Energy : Continued fuel cost fluctuations are affecting modal choice and spurring investment in technology innovations. Alternative fuels like natural gas are becoming more widely available and redefining the operational landscape.

5. *Omni-channel fulfillment and the "Amazon Effect :"* Customer-focused fulfillment initiatives that are designed to make products available to buy or return anywhere in the distribution or retail networks are introducing new drop-ship, direct-ship, replenishment, and *reverse logistics* paradigms.

A FRESH PERSPECTIVE

An attempt to internalize these realities within a specific industry or business environment quickly reveals what we intuitively know : We are not very good at predicting seemingly straightforward changes and their effects on our supply chains.

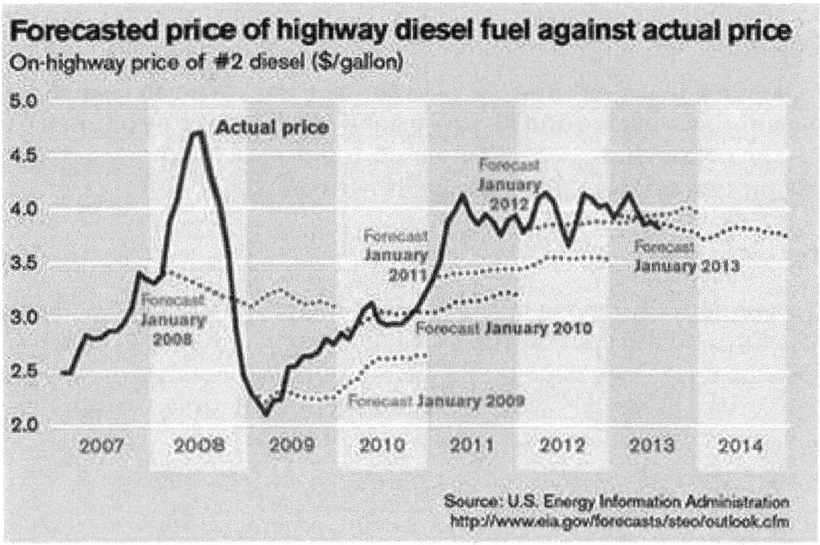

Consider, for instance, the new HOS rules. It was widely touted by the carrier market that reducing the number of on-duty hours would tighten capacity and increase transportation costs by as much as 17 per cent. According to an analysis conducted by Chainalytics' Freight Market Intelligence Consortium (FMIC), rates have not increased much if at all since the rule went into effect earlier this year, and have even decreased in some markets.

The study examined actual freight costs of more than 100 shippers who altogether purchase $18 billion in freight each year. Any incremental change can largely be attributed to inflation and fuel surcharges.

And let's not forget the fool's game that is forecasting fuel prices. The chart below that tracks the forecast for the price of highway diesel fuel against the actual shows the best efforts of the U.S. Department of Energy over the last five years. Simply put, the system is too complex to get it right.

To make matters worse, changes don't happen in isolation—they occur in conjunction with other complex changes. For this reason, supply chain and logistics managers should focus not on predicting the future, but instead preparing their network for a range of possible outcomes.

Think about the forthcoming Panama Canal expansion and its effect on global transportation movements. How much the new Asia to Eastern U.S. port rates and canal transit fees will be—which are still unknown and subject to market fluctuations—are not the only questions worth considering. When will the various port-deepening projects be ready? How will it affect the price of alternative modes or routes? What will the overall economy look like?

The future as it relates to online and omni-channel commerce holds just as many hard to answer questions. Consumers continue to demand customized products and fulfillment services, and retailers are rapidly experimenting with and deploying alternative supply chain networks to meet this new demand.

We assume that these forces will eventually converge on a small number of dominant designs, but we don't know what this new era of e-commerce will look like yet. The bottom line : A company must establish an ongoing, scenario-based business process to effectively manage its network.

How to Get Started

Supply chain and logistics managers should be constantly reviewing their network infrastructure and transportation flows and testing them in a modelling environment. The goal is to develop an agile network that is robust, provides low cost and high service under the widest possible range of business conditions and can be easily transformed once a key tipping point is reached.

We recommend the following steps to get started :

1. *Set yourself up for success :* Consider using an independent partner, one that is deeply familiar with supply chain design, your industry challenges, and the various tools available in today's technology marketplace. Strategic and tactical planning is too important to skimp on.
2. *Build an ongoing, stable competency :* Whether it's in-house or outsourced, the team needs to have enough critical mass to support organizational change management, training, and mentoring. Generally, internal groups are sustainable with five or more people. Groups with fewer than that have a much higher risk of being set back with the promotion or departure of key team members.
3. *Use high quality design data :* This type of data describes new options in the network. It is data that did not exist historically, such as the costs and capacities of new warehouses and the freight costs of new lanes. Despite the fact that freight accounts for nearly 80 per cent of the costs in a supply chain network analysis, many companies still lack the ability to accurately estimate these costs on a large number of lanes. Freight rate benchmarks, like those provided by the FMIC, are increasingly being used as reliable estimates for this critical input.
4. *Generate executive support and buy-in :* Build confidence in the process, data, and models so executives and managers alike will come to trust and rely on the dynamic tool to support strategic and tactical decision-making. The best modelling and analytics money can buy are of little value if the management team lacks confidence in the recommendations.
5. *Keep the process and supporting models current with frequent refreshes :* The rule of inertia that says "objects at rest tend to stay at rest, and those in motion tend to stay in motion" applies to supply chain and transportation network modelling. Also, keep in mind that jump-starting a discarded model and process is much more difficult and costly than maintaining one. An ongoing, refreshed model can be used to answer *ad-hoc* questions as well as conduct periodic reviews

and planned network analyses. It can also provide differentiated support to sales and operations planning (S&OP) and other cross-functional planning teams.

Case Study : **From silos to a culture of modelling** In the absence of a standard, repeatable process for supply chain design, decisions are made in decentralized silos, often with sub-optimal results. One manufacturer of consumer goods realized that it lacked cross-silo decision-making. Its procurement, manufacturing, and distribution decisions were being made independently, each with a different set of constraints and data assumptions.

As an example of this separation, one plant in the company's network had negotiated a tiered pricing arrangement for its inbound resin products. While the top-most tiers provided substantial discounts, the associated quantities were far above the individual plant's historic volumes. We worked with the manufacturer in building a repeatable process and culture of modelling that incorporated all the relevant stakeholders from procurement, manufacturing, and distribution.

The network model demonstrated that it was beneficial to leverage the top-tier of resin pricing even if it meant substantially increasing the plant's production and service territory and scaling back the other plants in the network. This originally seemed counter-intuitive to the planners within each silo, but it made sense when the entire network was considered. The savings from material procurement more than offset the increase in freight costs due to the reconfiguration of the manufacturing network.

Today, the company uses supply chain modelling to frequently analyze network alternatives. It can shift production and distribution on a monthly or quarterly basis as necessary to take advantage of cost and service opportunities. And its in-house modelling team has demonstrated the importance of cross-silo analysis and how a culture of modelling can support and inform strategic and tactical decisions.

Facing the New Realities

The realities of the new global economy are a reminder that we operate in a complex and dynamic environment. The future is unpredictable, and the best way a company can manage uncertainty is to continuously assess its supply chain and transportation networks. By establishing an ongoing business process that evaluates a supply chain's design and its ability to respond to changing market forces, a company can have confidence in its plans, even under uncertain conditions.

NEW PRODUCT DEVELOPMENT

In business and engineering, **new product development** (NPD) is the complete process of bringing a new product to market. A product is a set of benefits offered for exchange and can be tangible (that is, something physical you can touch) or intangible (like a service, experience, or belief). There are two parallel paths involved in the NPD process : one involves the idea generation, product design and detail engineering; the other involves market research and marketing analysis. Companies typically see new product development as the first stage in generating and commercializing new product within the overall strategic process of product life cycle management used to maintain or grow their market share.

Marketing Considerations

There have been a number of approaches proposed for analyzing and responding to the marketing challenges of new product development. Two of these are *the eight stages* process of Koen and a process known as *the fuzzy front end.*

The Eight Stages

1. **Idea Generation** is often called the "NPD" of the NPD process.
   - Ideas for new products can be obtained from basic research using a SWOT analysis (Strengths, Weaknesses, Opportunities & Threats). Market and consumer trends, company's R&D department, competitors, focus groups, employees, salespeople, corporate spies, trade shows, or ethnographic discovery methods (searching for user patterns and habits) may also be used to get an insight into new product lines or product features.
   - Lots of ideas are generated about the new product. Out of these ideas many are implemented. The ideas are generated in many forms. Many reasons are responsible for generation of an idea.
   - Idea Generation or Brainstorming of new product, service, or store concepts-idea generation techniques can begin when you have done your OPPORTUNITY ANALYSIS to support your ideas in the **Idea Screening Phase** (shown in the next development step).
2. *Idea Screening*
   - The object is to eliminate unsound concepts prior to devoting resources to them.
   - The screeners should ask several questions :
   - Will the customer in the target market benefit from the product?
   - What is the size and growth forecasts of the market segment/target market?
   - What is the current or expected competitive pressure for the product idea?
   - What are the industry sales and market trends the product idea is based on?
   - Is it technically feasible to manufacture the product?
   - Will the product be profitable when manufactured and delivered to the customer at the target price?
3. *Concept Development and Testing*
   - Develop the marketing and engineering details
   - Investigate intellectual property issues and search patent databases

- Who is the target market and who is the decision maker in the purchasing process?
- What product features must the product incorporate?
- What benefits will the product provide?
- How will consumers react to the product?
- How will the product be produced most cost effectively?
- Prove feasibility through virtual computer aided rendering and rapid prototyping.
- What will it cost to produce it?
- Testing the Concept may involve asking a number of prospective customers to evaluate the idea.

4. *Business Analysis*
   - Estimate likely selling price based upon competition and customer feedback
   - Estimate sales volume based upon size of market and such tools as the Fourt-Woodlock equation.
   - Estimate profitability and break-even point.
5. *Beta Testing and Market Testing*
   - Produce a physical prototype or mock-up.
   - Test the product (and its packaging) in typical usage situations.
   - Conduct focus group customer interviews or introduce at trade show.
   - Make adjustments where necessary.
   - Produce an initial run of the product and sell it in a test market area to determine customer acceptance.
6. *Technical Implementation*
   - New program initiation
   - Finalize Quality management system
   - Resource estimation
   - Requirement publication
   - Publish technical communications such as data sheets
   - Engineering operations planning
   - Department scheduling
   - Supplier collabouration
   - Logistics plan
   - Resource plan publication
   - Program review and monitoring
   - Contingencies-what-if planning
7. *Commercialization* (often considered post-NPD)
   - Launch the product

- Produce and place advertisements and other promotions
- Fill the distribution pipeline with product
- Critical path analysis is most useful at this stage.

8. *New Product Pricing*
    - Impact of new product on the entire product portfolio
    - Value Analysis (internal & external)
    - Competition and alternative competitive technologies
    - Differing value segments (price, value and need)
    - Product Costs (fixed & variable)
    - Forecast of unit volumes, revenue, and profit.

These steps may be iterated as needed. Some steps may be eliminated. To reduce the time that the NPD process takes, many companies are completing several steps at the same time (referred to as **concurrent engineering** or **time to market**). Most industry leaders see new product development as a *proactive* process where resources are allocated to identify market changes and seize upon new product opportunities before they occur (in contrast to a *reactive strategy* in which nothing is done until problems occur or the competitor introduces an innovation). Many industry leaders see new product development as an ongoing process (referred to as *continuous development*) in which the entire organization is always looking for opportunities.

For the more innovative products indicated on the diagram above, great amounts of uncertainty and change may exist which makes it difficult or impossible to plan the complete project before starting it. In this case, a more flexible approach may be advisable.

Because the NPD process typically requires both engineering and marketing expertise, cross-functional teams are a common way of organizing projects. The team is responsible for all aspects of the project, from initial idea generation to final commercialization, and they usually report to senior management (often to a vice-president or Program Manager). In those industries where products are technically complex, development research is typically expensive and product life cycles are relatively short, strategic alliances among several organizations helps to spread the costs, provide access to a wider skill set and speeds up the overall process.

Because both engineering and marketing expertise are usually critical to the process, choosing an appropriate blend of the two is important.

A new product pricing process is important to reduce risk and increase confidence in the pricing and marketing decisions to be made. Processes have been proposed to break down the complex task of new product pricing into more manageable elements.

The Path to Developing Successful New Products points out three key processes that can play critical role in product development : Talk to the customer; Nurture a project culture; Keep it focused.

Fuzzy Front End

The Fuzzy Front End is the messy "getting started" period of new product engineering development processes. It is in the front end where the organization formulates a concept of the product to be developed and decides whether or not to invest resources in the further development of an idea. It is the phase between first consideration of an opportunity and when it is judged ready to enter the structured development process (Kim and Wilemon, 2007; Koen *et. al.*, 2001). It includes all activities from the search for new opportunities through the formation of a germ of an idea to the development of a precise concept. The Fuzzy Front End ends when an organization approves and begins formal development of the concept.

Although the Fuzzy Front End may not be an expensive part of product development, it can consume 50% of development time, and it is where major commitments are typically made involving time, money, and the product's nature, thus setting the course for the entire project and final end product. Consequently, this phase should be considered as an essential part of development rather than something that happens "before development," and its cycle time should be included in the total development cycle time.

Koen distinguishes five different front-end elements (not necessarily in a particular order) : *et. al.*

1. Opportunity Identification
2. Opportunity Analysis
3. Idea Genesis
4. Idea Selection
5. Concept and Technology Development.
    - The first element is the opportunity identification. In this element, large or incremental business and technological chances are identified in a more or less structured way. Using the guidelines established here, resources will eventually be allocated to new projects... which then lead to a structured NPPD (New Product & Process Development) strategy.
    - The second element is the opportunity analysis. It is done to translate the identified opportunities into implications for the business and technology specific context of the company. Here extensive efforts may be made to align ideas to target customer groups and do market studies and/or technical trials and research.
    - The third element is the idea genesis, which is described as evolutionary and iterative process progressing from birth to maturation of the opportunity into a tangible idea. The process of the idea genesis can be made internally

or come from outside inputs, *e.g.* a supplier offering a new material/technology or from a customer with an unusual request.

- The fourth element is the idea selection. Its purpose is to choose whether to pursue an idea by analyzing its potential business value.
- The fifth element is the concept and technology development. During this part of the front-end, the business case is developed based on estimates of the total available market, customer needs, investment requirements, competition analysis and project uncertainty. Some organizations consider this to be the first stage of the NPPD process (*i.e.*, Stage 0).

The Fuzzy Front End is also described in literature as "Front End of Innovation", "Phase 0", "Stage 0" or "Pre-Project-Activities".

A universally acceptable definition for Fuzzy Front End or a dominant framework has not been developed so far. In a glossary of PDMA, it is mentioned that the Fuzzy Front End generally consists of three tasks : strategic planning, concept generation, and, especially, pre-technical evaluation. These activities are often chaotic, unpredictable, and unstructured. In comparison, the subsequent new product development process is typically structured, predictable, and formal. The term *Fuzzy Front End* was first popularized by Smith and Reinertsen. R.G.Cooper describes the early stages of NPPD as a four step process in which ideas are generated (I), subjected to a preliminary technical and market assessment (II) and merged to coherent product concepts (III) which are finally judged for their fit with existing product strategies and portfolios (IV).

Other Approaches

Others have divided pre-development product development activities differently :

1. Preliminary market assessment
2. Technical assessment
3. Source-of-supply-assessment : suppliers and partners or alliances
4. Market research : market size and segmentation analysis, VoC (voice of the customer) research
5. Product concept testing
6. Value-to-the customer assessment
7. Product definition
8. Business and financial analysis.

These activities yield vital information to make a Go/No-Go to Development decision.

One of the earliest studies using the case study method defined the front-end to include the interrelated activities of :

- Product strategy formulation and communication
- Opportunity identification and assessment
- Idea generation

- Product definition
- Project planning
- Executive reviews.

Economical analysis, benchmarking of competitive products and modelling and prototyping are also important activities during the front-end activities. The outcomes of FFE are the :

- Mission statement
- Customer needs
- Details of the selected concept
- Product definition and specifications
- Economic analysis of the product
- The development schedule
- Project staffing and the budget
- A business plan aligned with corporate strategy.

A conceptual model of Front-End Process was proposed which includes early phases of the innovation process. This model is structured in three phases and three gates :

- *Phase 1 :* Environmental screening or opportunity identification stage in which external changes will be analysed and translated into potential business opportunities.
- *Phase 2 :* Preliminary definition of an idea or concept.
- *Phase 3 :* Detailed product, project or concept definition, and Business planning.

The gates are :

- Opportunity screening
- Idea evaluation
- Go/No-Go for development.

The final gate leads to a dedicated new product development project. Many professionals and academics consider that the general features of Fuzzy Front End (fuzziness, ambiguity, and uncertainty) make it difficult to see the FFE as a structured process, but rather as a set of interdependent activities. However, Husig *et. al.*, 2005 argue that front-end not need to be fuzzy, but can be handled in a structured manner. In fact Carbone showed that when using the front end success factors in an integrated process, product success is increased. Peter Koen argues that in the FFE for incremental, platform and radical projects, three separate strategies and processes are typically involved. The traditional Stage Gate (TM) process was designed for incremental product development, namely for a single product. The FFE for developing a new platform must start out with a strategic vision of where the company wants to develop products and this will lead to a family of products. Projects for breakthrough products start out with a similar strategic vision, but are associated with technologies which require new discoveries.

Incremental, platform and breakthrough products include :

- *Incremental products* are considered to be cost reductions, improvements to existing product lines, additions to existing platforms and repositioning of existing products introduced in markets.
- *Breakthrough products* are new to the company or new to the world and offer a 5-10 times or greater improvement in performance combined with a 30–50% or greater reduction in costs.
- *Platform products* establish a basic architecture for a next generation product or process and are substantially larger in scope and resources than incremental projects.

CONCURRENT ENGINEERING

Concurrent engineering is a work methodology based on the parallelization of tasks (*i.e.* performing tasks concurrently). It refers to an approach used in product development in which functions of design engineering, manufacturing engineering and other functions are integrated to reduce the elapsed time required to bring a new product to the market.

Introduction

A publication in 2008 described the concurrent engineering method as a relatively new design management system that has had the opportunity to mature in recent years to become a well-defined systems approach towards optimizing engineering design cycles. Because of this, concurrent engineering has been implemented in a number of companies, organizations and universities, most notably in the aerospace industry.

The basic premise for concurrent engineering revolves around two concepts. The first is the idea that all elements of a product's life-cycle, from functionality, producibility, assembly, testability, maintenance issues, environmental impact and finally disposal and recycling, should be taken into careful consideration in the early design phases.

The second concept is that the preceding design activities should all be occurring at the same time, *i.e.*, concurrently. The idea is that the concurrent nature of these processes significantly increases productivity and product quality. This way, errors and redesigns can be discovered early in the design process when the project is still flexible. By locating and fixing these issues early, the design team can avoid what often become costly errors as the project moves to more complicated computational models and eventually into the actual manufacturing of hardware.

As mentioned above, part of the design process is to ensure that the entire product's life cycle is taken into consideration. This includes establishing user requirements, propagating early conceptual designs, running computational models, creating physical prototypes and eventually manufacturing the product. Included in the process is taking into full account funding, work force capability and time. A study in 2006 claimed that a correct implementation of the concurrent design process can save a significant amount of money, and that organizations have been moving to concurrent design for this reason.

Concurrent engineering replaces the more traditional sequential design flow, or 'Waterfall Model'. In concurrent engineering an iterative or integrated development method is used instead. The difference between these two methods is that the 'Waterfall' method moves in a linear fashion by starting with user requirements and sequentially moving forward to design, implementation and additional steps until you have a finished product. In this design system, a design team would not look backwards or forwards from the step it is on to fix possible problems. In the case that something does go wrong, the design usually must be scrapped or heavily altered. On the other hand, the iterative design process is more cyclic in that, all aspects of the life cycle of the product are taken into account, allowing for a more evolutionary approach to design. The difference between the two design processes can be seen graphically in Figure 1.

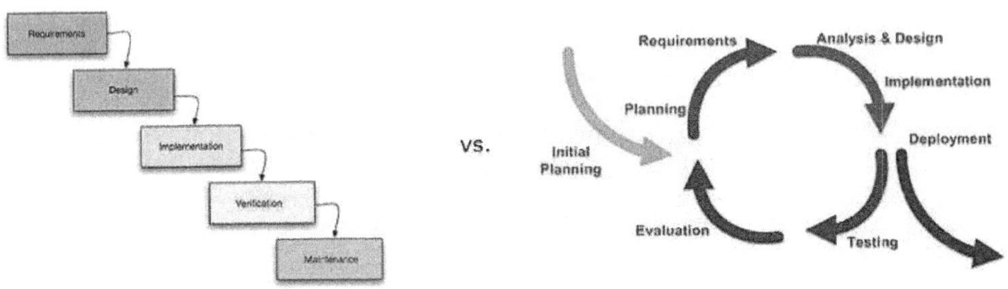

Fig.: Traditional "Waterfall" or Sequential Development Method vs. Iterative Development Method in concurrent engineering.

A significant part of the concurrent design method is that the individual engineer is given much more say in the overall design process due to the collabourative nature of concurrent engineering. Giving the designer ownership is claimed to improve the productivity of the employee and quality of the product that is being produced, based on the assumption that people who are given a sense of gratification and ownership over their work tend to work harder and design a more robust product, as opposed to an employee that is assigned a task with little say in the general process.

Disadvantages of Concurrent Design

Concurrent design creates its own issues, such as the implementation of early design reviews, the dependency on efficient communication between engineers and teams, software compatibility, and opening up the design process. A concurrent design process usually requires that computer models (computer aided design, finite element analysis) are exchanged efficiently, something that can be difficult in practice. If such issues are not addressed properly, concurrent design may not work effectively.

Service providers exist that specialize in this field. Not only training people how to perform concurrent design effectively, but also providing the tools to enhance the communication between the team members. Organizations such as *European Space Agency's Concurrent Design Facility make use of concurrent design to perform feasibility studies for future missions.

Concurrent Engineering Elements

Cross-functional Teams

Include members from various disciplines involved in the process, including manufacturing, hardware and software design, marketing, and so forth.

Concurrent Product Realization

Process activities are at the heart of concurrent engineering. Doing several things at once, such as designing various subsystems simultaneously, is critical to reducing design time.

Incremental Information Sharing

It helps minimize the chance that concurrent product realization will lead to surprises. As soon as new information becomes available, it is shared and integrated into the design. Cross functional teams are important to the effective sharing of information in a timely fashion.

Integrated Project Management

It ensures that someone is responsible for the entire project, and that responsibility is not abdicated once one aspect of the work is done.

Definition

Several definitions of concurrent engineering are in use.

The first one is used by the Concurrent Design Facility (ESA) :

- "Concurrent Engineering (CE) is a systematic approach to integrated product development that emphasizes the response to customer expectations. It embodies team values of co-operation, trust and sharing in such a manner that decision-making is by consensus, involving all perspectives in parallel, from the beginning of the product life cycle."

The second one is by Winner, *et. al.*, 1988 :

- "Concurrent Engineering is a systematic approach to the integrated, concurrent design of products and their related processes, including, manufacturing and support. This approach is intended to cause the developers from the very outset to consider all elements of the product life cycle, from conception to disposal, including quality, cost, schedule, and user requirements."

FLEXIBLE PRODUCT DEVELOPMENT

Flexible product development is the ability to make changes in the *product being developed* or in how it is developed, even relatively late in development, without being too disruptive. Consequently, the later one can make changes, the more flexible the process is, the less disruptive the change is, the greater the flexibility.

Flexibility is important because the development of a new product naturally involves change from what came before it. Change can be expected in what the customer wants and how the customer might use the product, in how competitors might respond, and in the new *technologies* being applied in the product or in its *manufacturing* process. The more innovative a new product is, the more likely it is that the development team will have to make changes during development.

Flexible development counteracts the tendencies of many contemporary management approaches to plan a project completely at its outset and discourage change thereafter. These include *Six Sigma*, which aims to drive variation out of a process; lean, which acts to drive out waste; and traditional project management and phased development systems (including the popular *Phase–gate model*), which encourage upfront planning and following the plan. Although these methodologies have strengths, their side effect is encouraging rigidity in a process that needs flexibility to be effective, especially for truly innovative products.

For more mature product categories, flexibility techniques are not only overly expensive but often unwise. Consequently, flexibility techniques must be used with discretion, for instance, only in the portions of a product likely to undergo change.

When applied to the development of *software* products, these methods are commonly known as *agile software development*. However, agile software methods generally rely on special characteristics of the software medium, especially object technologies, which are not available to non-software products. Consequently, flexible product development draws from some of the roots of agile software development but tends to use other tools and approaches that apply beyond the software medium.

Flexible development uses several techniques to keep the cost of change low and to make decisions at the last responsible moment. These techniques include modular architectures to encapsulate change, experimentation and iteration to sample results and check them out with the customer frequently, set-based design to build and maintain options, and emergent processes that develop during a project in response to its needs.

QUALITY FUNCTION DEPLOYMENT

Quality function deployment (QFD) is a "method to transform user demands into design quality, to deploy the functions forming quality, and to deploy methods for achieving the design quality into sub-systems and component parts, and ultimately to specific elements of the manufacturing process.", as described by Dr. Yoji Akao, who originally developed QFD in Japan in 1966, when the author combined his work in quality assurance and quality control points with function deployment used in value engineering.

QFD is designed to help planners focus on characteristics of a new or existing product or service from the viewpoints of market segments, company, or technology-development needs. The technique yields charts and matrices.

QFD helps transform customer needs (the voice of the customer [VOC]) into engineering characteristics (and appropriate test methods) for a product or service, prioritizing each product or service characteristic while simultaneously setting development targets for product or service.

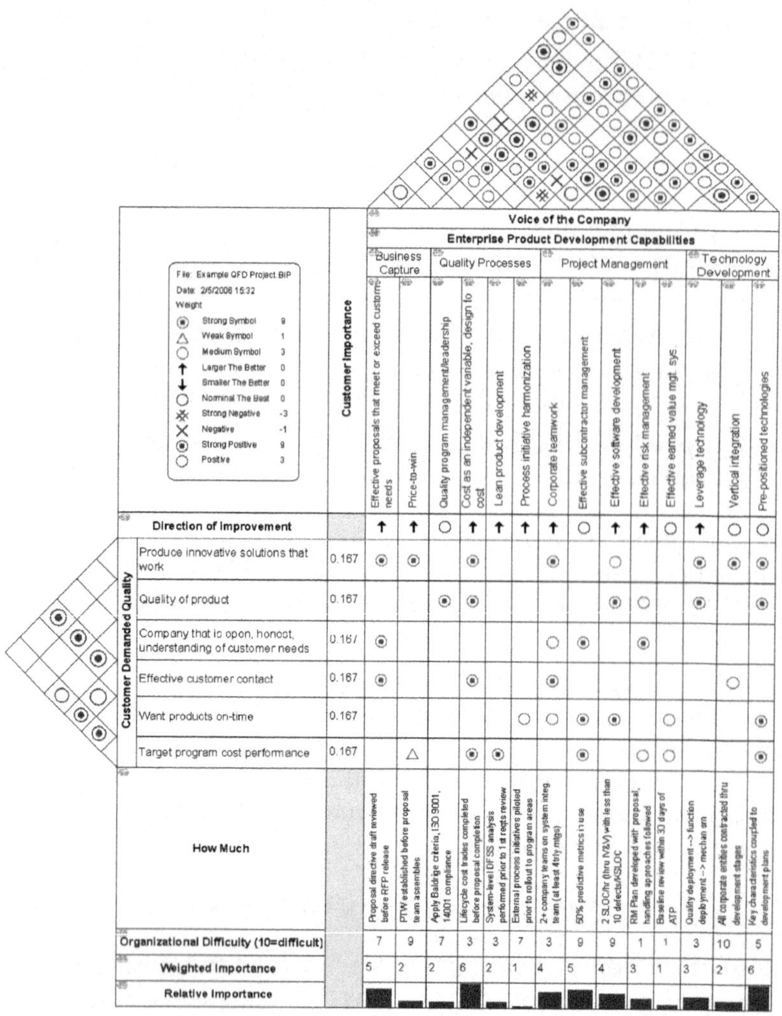

Fig. : **QFD House of Quality for Enterprise Product Development Processes.**

Areas of Application

QFD is applied in a wide variety of services, consumer products, military needs, and emerging technology products. The technique is also included in the new ISO 9000 : 2000 standard which focuses on customer satisfaction.

While many books and articles on "how to do QFD" are available, there is a relative paucity of example matrices available. QFD matrices become highly proprietary due to the high density of product or service information found therein.

Techniques and Tools Based on QFD

House of Quality

House of Quality appeared in 1972 in the design of an oil tanker by Mitsubishi Heavy Industries. Akao has reiterated numerous times that a House of Quality is not QFD, it is just an example of one tool.

A Flash tutorial exists showing the build process of the traditional QFD "House of Quality" (HOQ). (Although this example may violate QFD principles, the basic sequence of HOQ building are illustrative.) There are also free QFD templates available that walk users through the process of creating a House of Quality.

Other tools extend the analysis beyond quality to cost, technology, reliability, function, parts, technology, manufacturing, and service deployments.

In addition, the same technique can extend the method into the constituent product sub-systems, configuration items, assemblies, and parts. From these detail level components, fabrication and assembly process QFD charts can be developed to support statistical process control techniques.

Pugh Concept Selection

Pugh Concept Selection can be used in co-ordination with QFD to select a promising product or service configuration from among listed alternatives.

Modular Function Deployment

Modular Function Deployment uses QFD to establish customer requirements and to identify important design requirements with a special emphasis on modularity. There are three main differences to QFD as applied in Modular Function Deployment compared to House of Quality :

- The benchmarking data is mostly gone.
- The checkboxes and crosses have been replaced with circles.
- The triangular "roof" is missing.

There are also other minor differences between the application of QFD in Modular Function Deployment as compared to House of Quality, for example the term "Customer Attribute" is replaced by "Customer Value", and the term "Engineering Characteristics" is replaced by "Product Properties". But the terms have similar meanings in the two applications.

Relationship to Other Techniques

The QFD-associated Hoshin Kanri process somewhat resembles Management by objectives (MBO), but adds a significant element in the goal setting process, called "catchball". Use of these Hoshin techniques by U.S. companies such as Hewlett Packard have been successful in focusing and aligning company resources to follow stated strategic goals throughout an organizational hierarchy.

Since the early introduction of QFD, the technique has been developed to shorten the time span and reduce the required group efforts.

PHASE-GATE MODEL

A **Phase-gate model**, also referred to as a **phase-gate process**, is a project management technique in which an initiative or project (*e.g.*, new product development, process improvement, business change) is divided into *stages* or *phases*, separated by *gates*. At each gate, the continuation of the process is decided by (typically) a manager or a steering committee. The decision is based on the information available at the time, including the business case, risk analysis, and availability of necessary resources (*e.g.*, money, people with correct competencies). The Phase-gate model may also be known as *stage-limited commitment* or *creeping commitment*.

History

Phased scope development and investment decision is a fundamental concept of chemical engineering and engineering economics, particularly since the 1940s as chemical complexity and scale of chemical processes grew. One source describes 8 phases in the development of a chemical product starting with research, economic study, scale-up (pilot plant) and design leading to full funds authorization. In 1958, the American Association of Cost Engineers created four standard cost estimate type classifications to match these development and approval phases. Other industries with complex products and projects picked up on the process. For example, NASA practiced the concept of phased development in the 1960s with its phased project planning or what is often called *phased review process*. The phased review process was intended to break up the development of any project into a series of phases that could be individually reviewed in sequence. Review points at the end of each phase required that a number of criteria be met before the project could progress to the next phase. The phased review process consisted of five phases (Preliminary Analysis, Definition, Design, Development, Operations) with periodic development reviews between phases. NASA's phased review process is considered a first generation process because it did not take into consideration the analysis of external markets in new product development.

The Phase-gate model refers to the use of funnel tools in decision-making when dealing with new product development. "Gates" or decision points are placed at places in the product development process that are most beneficial to making decisions regarding continuance of product development. These production areas between the gates are idea generation, establishment of feasibility, development of capability, testing and validation and product launch. At the conclusion of each of these areas of development of a new product, it is the responsibility of senior management to make a decision as to whether or not the product should continue to be developed. The passing of gate to gate can be accomplished either formally, with some sort of documentation, or informally, decided upon based on the preferences and culture of the organization.

In the chemical engineering world, and engineering and construction industry, phase-gate processes are often called front-end loading. Phase-gate is the particular version and terminology elabourated in the following discussion.

Phases

A common model is composed of the following phases : ideation, preliminary analysis, business case, development, testing, launch.

A Phase-gate model is a conceptual and operational road map for moving a new project from idea to launch-a blueprint for managing the new-product process to improve effectiveness and efficiency.

The traditional phase-gate process has five phases and five gates. The phases are :

1. Scoping
2. Build business case
3. Development
4. Testing and validation
5. Launch.

Ahead of this process there is often a preliminary or ideation phase called *discovery*, and after the 5th phase the process ends with the post-launch review. Major new product projects go through the full five-phase process. Moderate risk projects, including extensions, modification and improvements, use the short XPress version. Very minor changes (*e.g.* sales force and marketing requests) may be executed using a lighter process (*Phase-gate lite*). Each phase consists of a set of prescribed, cross-functional, and parallel activities undertaken by a team of people from different functional areas. Phases have a common structure and consist of three main elements : *Activities, Integrated Analysis,* and *Deliverables.* Activities consist mainly in information gathering by the project team to reduce key project uncertainties and risks. An integrated analysis of the results of the activities is undertaken by the project team. Deliverables of phases are the results of integrated analysis that are used as input for the next Gate.

Gates

Gates provide various points during the process where an assessment of the quality of an idea is undertaken. It includes three main issues :

- *Quality of execution* : Checks whether the previous step is executed in a quality fashion.
- *Business rationale* : Does the project continue to look like an attractive idea from an economic and business perspective.
- *Action plan* : The proposed action plan and the requested resources reasonable and sound.

A gate meeting can lead to four results : *go, kill, hold, recycle,* or "Conditional go".

Gates have a common structure and consist of three main elements :

- *Deliverables* : What the project manager and team deliver to the decision point. These deliverables are decided at the output of the previous gate, and are based on a standard menu of deliverables for each gate.
- *Criteria* : Questions or metrics on which the project is judged in order to determine a result (*go/kill/hold/recycle*) and make a prioritization decision.
- *Outputs* : Results of the gate review—a decision (*go/kill/hold/recycle*), along with an approved action plan for the next gate, and a list of deliverables and date for the next gate.

Phases in Detail

Phase 0 : Discovery

Deciding what projects the company wants and is capable to pursue. During this phase it is common for companies to take part in idea generation activities such as brainstorming or other group thinking exercises. Once the idea generation team has selected a project that they would like to go forward with, it must be passed on to the first gate and therefore screened by the organization's decision-makers.

When searching for new product ideas it is beneficial for an organization to look to the outside world to suggest business opportunities. Using methods such as those found in empathic design can be quite helpful. Communicating with customers to understand how and why they use products can produce great strides in idea generation. Specifically, communicating with lead users can provide great feedback to the developers, as these customers are most likely to feel most passionately about the product. In addition to communication with lead users, it may be helpful for developers to communicate with suppliers. By understanding all of the types of business that their materials are being used for, developers may be able to act upon previously untapped possibilities.

Phase 1 : Scoping

The second phase of the product development process is scoping. During this step the main goal is to evaluate the product and its corresponding market. The researchers must recognize the strengths and weaknesses of the product and what it is going to offer to the potential consumer. The competition must also be evaluated during this phase. It is important for the researchers to understand who and what is already in the market as well as what can potentially be developed. By determining the relative level of threat from competitors, the management team will be able to recognize whether or not they should go forward with the production of the product.

Phase 2 : Building the Business Case and Plan

Once the new product passes through the gate at the end of the scoping phase, the next phase in the phase-gate model is building the business case and plan. This is the last phase of concept development where it is crucial for companies to perform a solid analysis before they begin developing the product. This phase is generally difficult, complex, and resource-intensive. However, companies must put forth a strong effort in this phase for it is directly related to the success and development of a new product. There are four main steps that comprise this phase : product definition and analysis, building the business case, building the project plan, and feasibility review.

Product Definition and Analysis

The first step, product definition and analysis, is composed of a series of activities that will provide the information to define and justify the development of a new product. One of the first of these activities is the user needs and wants study where customer value is determined. This addresses questions about the product such as what benefits does the product provide and what features should the product have. During this time the company should conduct surveys and interviews with existing and potential customers, along with staff members. Next, the company must conduct a market analysis. They must determine the market size and segmentation, rate of growth, customer trends and behaviour, and what channels reach these customers. Once the market analysis is complete the company must then conduct a competitive analysis. It is important to know how your competitors operate in addition to their strengths and weaknesses. This will not only help you build a great product, but will also help in determining how and where to launch your new product. Together these activities will help define the product and provide a foundation for the marketing strategy. Next, the company must build a technically feasible product concept, which includes the substance and methods needed to produce the new product. Once this is completed the company can then produce a production and operations cost analysis along with a market and launch costs analysis. Next, the company can begin to test the concept they have developed. This is when early prototypes are developed and presented to staff and consumers to gain feedback and gauge customer reaction. From this the company can make the necessary changes and see the sales potential of the product. This feedback will also help the company build a solid product definition. Lastly, the company will then conduct the business analysis, risk analysis, and financial analysis of the new product.

Building the Business Case

The business case is a document that defines the product and provides the rationale for developing it. This document will vary in format amongst companies, but the primary components are the following : results of the activities of product definition and analysis; legal and regulatory requirements; safety, health, and environmental considerations; assumptions needed to draw the conclusions made, and why it is believed they are valid and reasonable; and out-of-bounds criteria that indicate certain changes/events which will mandate an emergency business case review. This document will be referred to throughout the development process and edited when necessary.

Building the Project Plan

The project plan includes : A scheduled list of tasks and events along with timelines for milestones throughout the development process; the personnel, time, and financial resources needed to complete the project; and an expected launch date for the release of the new product.

Feasibility Review

The last step of building the business case and plan is the feasibility review. This is when management, along with other departments of the company, reviews the rationale for pursuing the product. They analyze the information provided by the previous steps in this process to decide whether or not the product should move forward. If it is decided to be pursued then it passes through gate two and moves on to the product development phase.

Phase 3 : Development

During the development phase, plans from previous steps are actually executed. The product's design and development is carried out, including some early, simple tests of the product and perhaps some early customer testing. The product's marketing and production plans are also developed. It is important that the company adheres to their overall goal of the project, which is reflected in these production and marketing plans. Doing this will allow them to definitively decide who they will market their product to and how they will get the product to that target audience. The development team maps out a realistic timeline with specific milestones that are described as SMART : *specific, measurable, actionable, realistic,* and *time-bound.* The timeline is frequently reviewed and updated, helping the team stay on task and giving management information about the product's progress. In the development phase, the product builds momentum as the company commits more resources to the project and makes full use of cross-functional teamwork as the marketing, technical, manufacturing, and sales departments all come together to offer their expert opinions. Having a diversified and parallel development phase ensures that the product continues to meet the company's technical and financial goals. A diverse team allows specific roles and leadership positions to develop as team members make contributions using their strongest attributes. With members having clearly defined roles, tasks can be performed concurrently ensuring a much more efficient development process. The ultimate deliverable of the development phase is the prototype, which will undergo extensive testing and evaluation in the next phase of the process.

Phase 4 : Testing and Validation

This phase provides validation for the entire project. Areas that will be evaluated include : the product itself, the production/manufacturing process, customer acceptance, and the financial merit of the project. This phase includes three types of testing : *near testing, field testing,* and *market testing.*

Near Testing

The main objective of *near testing* is to find any bugs or issues with a product. A key point to remember here is that the product is no longer a prototype and that it has almost all the features of the commercial model. Testing will be done initially by in-house staff, and customers and partners who are close to the firm. It is important to ensure that those testing have an understanding of how the product should perform, so they know what it should or shouldn't be doing. Members of the research and development team are usually present to observe the participants using the product and take any notes or data that may be useful.

Field Testing

Field testing, or beta testing, is done by those who can provide valuable feedback on the product. This usually lasts a long period of time and the participants can include customers, partners, or anyone who is not familiar with the producing company. At this juncture the product fully resembles its planned launch model in all aspects; therefore the participants' interaction rate will be higher because they know all the features and benefits. During this phase there are three primary objectives to be achieved. The first objective is to see how much the participant is interested. It is also worthwhile to note which individual attribute they prefer and if they would buy the product. Next, determine how the customer uses the product and evaluate its durability. Confirm the environment in which the customers will be using the product. Recording and analyzing customer feedback is the final step in the field testing phase. This feedback may be used to help inform any minor design improvements that need to be made. The sales and marketing team will also be a beneficiary of field testing feedback; they can use this information to help focus their sales presentation.

Market Testing

Unlike the other two test activities, market testing is considered optional. A solid marketing and launch plan along with confidence in the product's ability to sell helps to inform the key decision-makers at the test and validation gate. If there is any uncertainty in the marketing or launch plans there are two options to consider. First, a simulated market test may be run, in which customers will be exposed to new products in an advertising and purchasing situation. The goal of this test is to obtain an early forecast of sales and make any necessary adjustments to the marketing plan. The second test involves trial sales, and is done through specific channels, regions, or consumer demographics.

Phase 5 : Product Launch

The product launch is the culmination of the product having passed all previous gates. The producer must come up with a marketing strategy to generate customer demand for the product. The producer must also decide how large they anticipate the market for a new product to be and thus determine the size of their starting volume production. Part of the launch phase is training sales and support personnel to be familiar with the product so that they can assist in sales of this product. Setting a product price is an aspect of the product launch that the producer must consider. They should avoid either undershooting or overpricing the potential market. Finally, distribution is a major decision-making part of the launch process. Selecting a distributor or value-added reseller for a product must be done with careful thought and potential sales in mind.

Having a smooth launch process that includes effective marketing and a knowledgeable and prepared sales force may result in faster time to profit due to early customer acceptance.

Effective Gating

Most firms suffer from having far too many projects in their product development pipelines, for the limited resources available. "Gates with teeth" help to prune the development portfolio of weak projects and deal with a gridlocked pipeline. Also, a robust innovation strategy, coupled with strategic buckets, refocuses resources on high value development initiatives.

Note that gates are not merely project review points, status reports or information updates. Rather, they are tough decision meetings, where the critical go/kill and prioritization decisions are made on projects. Thus the gates become the quality control check points in the process ensuring that you do the right projects and also do the projects right.

Gates must have clear and visible criteria so that senior managers can make go/kill and prioritization decisions objectively. Most importantly, these criteria must be effective—that is, they must be operational (easy to use), realistic (make use of available information) and discriminating (differentiate the good projects from the mediocre ones). These criteria can be :

- *Must meet* : Knock-out questions in a check list, designed to kill poor projects outright.
- *Should meet* : Highly desirable characteristics which are rated and added in a point-count scheme.

A sample list of criteria is shown below, from which a scorecard can be developed that can then be used to score projects at a gate meeting.

- *Must meet* (checklist-yes/no)
  - Strategic alignment (fits business unit strategy)
  - Reasonable likelihood of technical feasibility
  - Meet EH&S policies
  - Positive return versus risk.
- *Should meet* (scored on 0–10 scale)
  - Strategic
    - Degree to which projects aligns with business unit strategy
    - Strategic importance
  - Product advantage
    - Unique benefits
    - Meets customer needs better than existing or competing product
    - Value for money
  - Market attractiveness
    - Market size
    - Market growth
    - Competitive situation
  - Synergies (leverages core competencies)
    - Marketing synergies
    - Technological synergies
    - Manufacturing/processing synergies
  - Technical feasibility
    - Technical gap
    - Complexity
    - Technical uncertainty
  - Risk versus return
    - Expected profitability (*e.g.*, net present value)
    - Return (*e.g.*, internal rate of return)
    - Payback period
    - Certainty of return

If the answers were "no" or "low" to any of these questions, the decision certainly would not be to kill the project-hence they're poor go/kill criteria.

Advantages and Disadvantages

There are a number of advantages to using the phase-gate model for product development, which typically result from its ability to identify problems and assess progress before the project's conclusion. Poor projects can be quickly rejected by disciplined use of the model. When using the phase-gate model on a large project, the model can help reduce complexity of what could be a large and limiting innovation process into a straightforward rule-based approach. When a phase-gate model incorporates cost and fiscal analysis tools such as net present value, the organization can potentially be provided with quantitative information regarding the feasibility of developing potential product ideas. Finally, the model is an opportunity to validate the updated business case by a project's executive sponsors.

One problem with the phase-gate process is the potential for structural organization to interfere with creativity and innovation, as overly structured processes may cause creativity to be reduced in importance and to hinder the largely iterative process of innovation.

Opportunity Management

The opportunity management funnel is a visual representation of phase-gate decision-making. Opportunity management has been defined as "a process to identify business and community development opportunities that could be implemented to sustain or improve a local economy." The components of opportunity management are :

1. Identifying opportunities.
2. *Evaluating and prioritizing these opportunities :* This may involve developing criteria, deliberating, and ranking the alternatives.
3. *Driving opportunities :* Involves assigning leads, accountability, action plans, and project management
4. *Constant monitoring :* May require one of the following actions :
   - *Advance :* Commit additional resources to move the idea forward
   - *Rework :* More investigation/rethinking
   - *Kill :* Stop working on the idea and move on.

The goal of the opportunity management funnel is to eliminate weak or bad ideas before money or resources are contributed to realize these opportunities. The benefit if the opportunity management funnel when utilizing phase-gate decision-making is that the funnel generates efficiencies where weak ideas are efficiently eliminated leaving a strong set of viable alternatives. To fulfill its mandate, the opportunity management funnel filters the broadest range of opportunities and ensures that all priority sectors are represented. When selecting which opportunities to filter through the process, economic developers should be aware that initially, there are no bad ideas or limits. The unviable alternatives will be filtered out throughout the process using phase-gate decision-making process

USER-CENTERED DESIGN

User-centered design (UCD) is a process (not restricted to interfaces or technologies) in which the needs, wants, and limitations of end users of a product, service or process are given extensive attention at each stage of the design process. User-centered design can be characterized as a multi-stage problem solving process that not only requires designers to analyse and foresee how users are likely to use a product, but also to test the validity of their assumptions with regard to user behaviour in real world tests with actual users. Such testing is necessary as it is often very difficult for the designers of a product to understand intuitively what a first-time user of their design experiences, and what each user's learning curve may look like.

The chief difference from other product design philosophies is that user-centered design tries to optimize the product around how users can, want, or need to use the product, rather than forcing the users to change their behaviour to accommodate the product.

UCD Models and Approaches

For example, the user-centered design process can help software designers to fulfill the goal of a product engineered for their users. User requirements are considered right from the beginning and included into the whole product cycle. These requirements are noted and refined through investigative methods including : ethnographic study, contextual inquiry, prototype testing, usability testing and other methods. Generative methods may also be used including : card sorting, affinity diagraming and participatory design sessions. In addition, user requirements can be inferred by careful analysis of usable products similar to the product being designed.

- *Co-operative design* : Involving designers and users on an equal footing. This is the Scandinavian tradition of design of IT artifacts and it has been evolving since 1970.
- Participatory design (PD), a North American term for the same concept, inspired by Co-operative Design, focusing on the participation of users. Since 1990, there has been a bi-annual Participatory Design Conference.
- Contextual design, "customer-centered design" in the actual context, including some ideas from Participatory design

All these approaches follow the ISO standard Human-centred design for interactive systems (ISO 9241-210, 2010).

The ISO standard describes 6 key principles that will ensure a design is user centered :

1. The design is based upon an explicit understanding of users, tasks and environments.
2. Users are involved throughout design and development.
3. The design is driven and refined by user-centered evaluation.
4. The process is iterative.
5. The design addresses the whole user experience.
6. The design team includes multidisciplinary skills and perspectives.

Purpose

UCD answers questions about users and their tasks and goals, then uses the findings to make decisions about development and design. UCD of a web site, for instance, seeks to answer the following questions :

- Who are the users of the document?
- What are the users' tasks and goals?
- What are the users' experience levels with the document, and documents like it?
- What functions do the users need from the document?
- What information might the users need, and in what form do they need it?
- How do users think the document should work?
- What are the extreme environments?
- Is the user multi-tasking?
- Does the interface utilize different inputs modes such as touching, spoken, gestures, or orientation?

Elements

As examples of UCD viewpoints, the essential elements of UCD of a web site are considerations of visibility, accessibility, legibility and language.

Visibility

Visibility helps the user construct a mental model of the document. Models help the user predict the effect(s) of their actions while using the document. Important elements (such as those that aid navigation) should be emphatic. Users should be able to tell from a glance what they can and cannot do with the document.

Accessibility

Users should be able to find information quickly and easily throughout the document, regardless of its length. Users should be offered various ways to find information (such as navigational elements, search functions, table of contents, clearly labelled sections, page numbers, colour-coding, etc.). Navigational elements should be consistent with the genre of the document. 'Chunking' is a useful strategy that involves breaking information into small pieces that can be organized into some type meaningful order or hierarchy. The ability to skim the document allows users to find their piece of information by scanning rather than reading. Bold and italic words are often used.

Legibility

Text should be easy to read : Through analysis of the rhetorical situation, the designer should be able to determine a useful font style. Ornamental fonts and text in all capital letters are hard to read, but italics and bolding can be helpful when used correctly. Large or small body text is also hard to read. (Screen size of 10-12 pixel sans serif and 12-16 pixel serif is recommended.) High figure-ground contrast between text and background increases legibility. Dark text against a light background is most legible.

Language

Depending on the rhetorical situation, certain types of language are needed. Short sentences are helpful, as are well-written texts used in explanations and similar bulk-text situations. Unless the situation calls for it, jargon or technical terms should not be used. Many writers will choose to use active voice, verbs (instead of noun strings or nominals), and simple sentence structure.

Rhetorical Situation

A user-centered design is focused around the rhetorical situation. The rhetorical situation shapes the design of an information medium. There are three elements to consider in a rhetorical situation : Audience, Purpose, and Context.

Audience

The audience is the people who will be using the document. The designer must consider their age, geographical location, ethnicity, gender, education, etc.

Purpose

The purpose is what the document targets or what problem the document is trying to address.

Context

The context is the circumstances surrounding the situation. The context often answers the question : What situation has prompted the need for this document? Context also includes any social or cultural issues that may surround the situation.

Analysis Tools Used in User-centered Design

There are a number of tools that are used in the analysis of user-centered design, mainly : persona, scenarios, and essential use cases.

Persona

During the UCD process, a Persona of the user's need may be created. It is a fictional character with all the characteristics of the user. Personas are created after the field research process, which typically consists of members of the primary stakeholder (user) group being observed on their behaviour, and additionally answering questionnaires or participating in interviews, or a mixture of both.

After results are gathered from the field research, they are used to create personas of the primary stakeholder group. Often, there may be several personas concerning the same group of individuals, since it is almost impossible to apply all the characteristics of the stakeholder group onto one character. The character depicts a "typical" stakeholder, not an "average" individual in the primary stakeholder group, and is referred to throughout the entire design process. There are also what's called a secondary persona, where the character is not a member of the primary stakeholder group and is not the main target of the design, but their needs should be met and problems solved if possible. They exist to help account for further possible problems and difficulties that may occur even though the primary stakeholder group is satisfied with their solution. There is also an anti-persona, which is the character which the design process is not made for. Personas usually include a name and picture, demographics, roles and responsibilities, goals and tasks, motivations and needs, environment and context, and a quote that can represent the character's personality. Personas are useful in the sense that they create a common shared understanding of the user group for which the design process is built around. Also, they help to prioritize the design considerations by providing a context of what the user needs and what functions are simply nice to add and have. They can also provide a human face and existence to a diversified and scattered user group, and can also create some empathy and add emotions when referring to the users. However, since personas are a generalized perception of the primary stakeholder group from collected data, the characteristics may be too broad and typical, or too much of an "average joe". Sometimes, personas can have stereotypical properties also, which may hurt the entire design process. Overall, personas are a useful tool that can be used since designers in the design process can have an actual person to make design measure around other than referring to a set of data or a wide range of individuals.

Scenario

A scenario created in the UCD process is a fictional story about the "daily life of" or a sequence of events with the primary stakeholder group as the main character. Typically, a persona that was created earlier is used as the main character of this story. The story should be specific of the events happening that relate to the problems of the primary stakeholder group, and normally the main research questions the design process is built upon. These may turn out to be a simple story about the daily life of an individual, but small details from the events should imply details about the users, and may include emotional or physical characteristics. There can be the "best case scenario", where everything works out best for the main character, the "worst case scenario", where the main character experiences everything going wrong around him or her, and an "average case scenario", which is the typical life of the individual, where nothing really special or really depressing occurs, and the day just moves on. Scenarios create a social context to which the personas exist in, and also create an actual physical world, instead of imagining a character with internal characteristics from gathered data an nothing else; there is more action involved in the persona's existence. A scenario is also more easily understood by people, since it is in the form of a story, and is easier to follow. Yet, like the personas, these scenarios are assumptions made by the researcher and designer, and is also created from a set of organized data. Some even say such scenarios are unrealistic to real life occurrences. Also, it is difficult to explain and inform low level tasks that occur, like the thought process of the persona before acting.

Use Case

In short, a use case describes the interaction between an individual and the rest of the world. Each use case describes an event that may occur for a short period of time in real life, but may consist of intricate details and interactions between the actor and the world. It is represented as a series of simple steps for the character to achieve his or her goal, in the form of a cause-and effect scheme. Use cases are normally written in the form of a chart with two columns : first column labelled actor, second column labelled world, and the actions performed by each side written in order in the respective columns. The following is an example of a use case for performing a song on a guitar in front of an audience.

Actor	*World*
Choose music to play	
Pick up guitar	
	Display sheet music
Perform each note on sheet music using guitar	
	Convey note to audience using sound
	Audience provides feedback to performer
Assess performance and adjust as needed based on audience feedback.	
Complete song with required adjustments	
	Audience applause

The interaction between actor and the world is an act that can be seen in everyday life, and we take them as granted and don't think too much about the small detail that needs to happen in order for an act like performing a piece of music to exist. It is similar to the fact that when speaking our mother tongue, we don't think too much about grammar and how to phrase words; they just come out since we are so used to saying them. The actions between an actor and the world, notably, the primary stakeholder (user) and the world in this case, should be thought about in detail, and hence use cases are created to understand how these tiny interactions occur.

An essential use case is a special kind of use case, also called an "abstract use case". Essential use cases describe the essence of the problem, and deals with the nature of the problem itself. While writing use cases, no assumptions about unrelated details should be made. In additions, the goals of the subject should be separated from the process and implementation to reach that particular goal. Below is an example of an essential use case with the same goal as the former example.

Actor	*World*
Choose sheet music to perform	
Gathers necessary resources	
	Provides access to resources
Performs piece sequentially	
	Convey and interprets performance
	Provides feedback
Completes performance	

Use cases are useful because they help identify useful levels of design work. They allow the designers to see the actual low level processes that are involved for a certain problem, which makes the problem easier to handle, since certain minor steps and details the user makes are exposed. The designers' job should take into consideration of these small problems in order to arrive at a final solution that works. Another way to say this is that use cases breaks a complicated task into smaller bits, where these bits are useful units. Each bit completes a small task, which then builds up to the final bigger task. Like writing code on a computer, it is easier to write the basic smaller parts and make them work first, and then put them together to finish the larger more complicated code, instead to tackling the entire code from the very beginning. The first solution is less risky because if something goes wrong with the code, it is easier to look for the problem in the smaller bits, since the segment with the problem will be the one that does not work, while in the latter solution, the programmer may have to look through the entire code to search for a single error, which proves time consuming. The same reasoning goes for writing use cases in UCD. Lastly, use cases convey useful and important tasks where the designer can see which one are of higher importance than others. Some drawbacks of writing use cases include the fact that each actions, by the actor or the world, consist of little detail, and is simply a small action. This may possibly lead to further imagination and different interpretation of action from different designers. Also, during the process, it is really easy to oversimplify a task, since a small task from a larger task may consist of even smaller tasks. Picking up a guitar may involve thinking of which guitar to pick up, which pick to use, and think about where the guitar is located first. These tasks may then be divided into smaller tasks, such as first thinking of what colour of guitar fits the place to perform the piece, and other related details. Tasks may be split further down into even tinier tasks, and it is up to the designer to determine what is a suitable place to stop splitting up the tasks. Tasks may not only be oversimplified, they may also be omitted in whole, thus the designer should be aware of all the detail and all the key steps that are involved in an event or action when writing use cases.

User-centered Design, Needs and Emotions

The book "The Design of Everyday Things" (originally called "The Psychology of Everyday Things") was first published in 1986. In this book, Donald A. Norman describes the psychology behind what he deems 'good' and 'bad' design through examples and offers principles of 'good' design. He exalts the importance of design in our everyday lives, and the consequences of errors caused by bad designs.

In his book, Norman uses the term "user-centered design" to describe design based on the needs of the user, leaving aside what he considers secondary issues like aesthetics. User-centered design involves simplifying the structure of tasks, making things visible, getting the mapping right, exploiting the powers of constraint, and designing for error. Norman's overly reductive approach in this text was readdressed by him later in his own publication "Emotional Design."

Other books in a similar vein include "Designing Pleasurable Products" by Patrick W. Jordan, in which the author suggests that different forms of pleasure should be included in a user-centered approach in addition to traditional definitions of usability.

User-centered Design in Product Lifecycle Management Systems

Software applications (or often suites of applications) used in product lifecycle management (typically including CAD, CAM and CAx processes) can be typically characterized by the need for these solutions to serve the needs of a broad range of users, with each user having a particular job role and skill level. For example, a CAD digital mockup might be utilized by a novice analyst, design engineer of moderate skills, or a manufacturing planner of advanced skills.

Chapter 12

SIMILARITY METRICS

JACCARD INDEX

The **Jaccard index**, also known as the **Jaccard similarity coefficient** (originally coined *coefficient de communauté* by Paul Jaccard), is a statistic used for comparing the similarity and diversity of sample sets. The Jaccard coefficient measures similarity between finite sample sets, and is defined as the size of the intersection divided by the size of the union of the sample sets :

$$J(A, B) = \frac{|A \cap B|}{|A \cup B|}.$$

(If A and B are both empty, we define $J(A,B)=1$.) Clearly,

$0 \leq J(A, B) \leq 1.$

The MinHash min-wise independent permutations locality sensitive hashing scheme may be used to efficiently compute an accurate estimate of the Jaccard similarity coefficient of pairs of sets, where each set is represented by a constant-sized signature derived from the minimum values of a hash function.

The **Jaccard distance**, which measures *dis*similarity between sample sets, is complementary to the Jaccard coefficient and is obtained by subtracting the Jaccard coefficient from 1, or, equivalently, by dividing the difference of the sizes of the union and the intersection of two sets by the size of the union :

$$d_J(A, B) = 1 - J(A, B) = \frac{|A \cup B| - |A \cap B|}{|A \cup B|}.$$

An alternate interpretation of the Jaccard distance is as the ratio of the size of the symmetric difference $A \Delta B = (A \cup B) - (A \cap B)$ to the union.

This distance is a proper metric on sets of sets.

There is also a version of the Jaccard distance for measures, including probability measures. If μ is a measure on a measurable space X, then we define the Jaccard coefficient by $J_\mu(A, B) = \frac{\mu(A \cap B)}{\mu(A \cup B)}$, and the Jaccard distance by $d_\mu(A, B) = 1 - J_\mu(A, B) = \frac{\mu(A \Delta B)}{\mu(A \Delta B)}$. Care must be taken if $\mu(A \cup B) = 0$ or ∞, since these formulas are not well defined in that case.

Similarity of Asymmetric Binary Attributes

Given two objects, A and B, each with n binary attributes, the Jaccard coefficient is a useful measure of the overlap that A and B share with their attributes. Each attribute of A and B can either be 0 or 1. The total number of each combination of attributes for both A and B are specified as follows :

M_{11} represents the total number of attributes where A and B both have a value of 1.

M_{01} represents the total number of attributes where the attribute of A is 0 and the attribute of B is 1.

M_{10} represents the total number of attributes where the attribute of A is 1 and the attribute of B is 0.

M_{00} represents the total number of attributes where A and B both have a value of 0.

Each attribute must fall into one of these four categories, meaning that

$M_{11} + M_{01} + M_{10} + M_{00} = n.$

The Jaccard similarity coefficient, J, is given as

$$J = \frac{M_{11}}{M_{01} + M_{10} + M_{11}}.$$

The Jaccard distance, d_J, is given as

$$d_J = \frac{M_{01} + M_{10}}{M_{01} + M_{10} + M_{11}}.$$

Tanimoto Similarity and Distance

Various forms of functions described as Tanimoto Similarity and Tanimoto Distance occur in the literature and on the Internet. Most of these are synonyms for Jaccard Similarity and Jaccard Distance, but some are mathematically different. Many sources cite an unavailable IBM Technical Report as the seminal reference.

In "A Computer Program for Classifying Plants", published in October 1960, a method of classification based on a similarity ratio, and a derived distance function, is given. It seems that this is the most authoritative source for the meaning of the terms "Tanimoto Similarity" and "Tanimoto Distance". The similarity ratio is equivalent to Jaccard similarity, but the distance function is *not* the same as Jaccard Distance.

Tanimoto's Definitions of Similarity and Distance

In that paper, a "similarity ratio" is given over bitmaps, where each bit of a fixed-size array represents the presence or absence of a characteristic in the plant being modelled. The definition of the ratio is the number of common bits, divided by the number of bits set in either sample.

Presented in mathematical terms, if samples X and Y are bitmaps, X_i is the ith bit of X, and $\wedge$, $\vee$ are bitwise *and, or* operators respectively, then the similarity ratio T_s is

$$T_s(X, Y) = \frac{\sum_i (X_i \wedge Y_i)}{\sum_i (X_i \vee Y_i)}$$

If each sample is modelled instead as a set of attributes, this value is equal to the Jaccard Coefficient of the two sets. Jaccard is not cited in the paper, and it seems likely that the authors were not aware of it.

Tanimoto goes on to define a distance coefficient based on this ratio, defined over values with non-zero similarity :

$$T_d(X, Y) = -\log_2(T_s(X, Y))$$

This coefficient is, deliberately, not a distance metric. It is chosen to allow the possibility of two specimens, which are quite different from each other, to both be similar to a third. It is easy to construct an example which disproves the property of triangle inequality.

Other Definitions of Tanimoto Distance

Tanimoto Distance is often referred to, erroneously, as a synonym for Jaccard Distance ($1 - T_s$). This function is a proper distance metric. "Tanimoto Distance" is often stated as being a proper distance metric, probably because of its confusion with Jaccard Distance.

If Jaccard or Tanimoto Similarity is expressed over a bit vector, then it can be written as

$$f(A, B) = \frac{A \cdot B}{|A|^2 + |B|^2 - A \cdot B}$$

where the same calculation is expressed in terms of vector scalar product and magnitude. This representation relies on the fact that, for a bit vector (where the value of each dimension is either 0 or 1) then $A \cdot B = \sum_i A_i B_i = \sum_i (A_i \wedge B_i)$ and

$$|A|^2 = \sum_i A_i^2 = \sum_i A_i.$$

This is a potentially confusing representation, because the function as expressed over vectors is more general, unless its domain is explicitly restricted. Properties of T_s do not necessarily extend to f. In particular, the difference function $(1 - f)$ does not preserve triangle inequality, and is not therefore a proper distance metric, whereas $(1 - T_s)$ is.

There is a real danger that the combination of "Tanimoto Distance" being defined using this formula, along with the statement "Tanimoto Distance is a proper distance metric" will lead to the false conclusion that the function $(1 - f)$ is in fact a distance metric over vectors or multi-sets in general, whereas its use in similarity search or clustering algorithms may fail to produce correct results.

Lipkus uses a definition of Tanimoto similarity which is equivalent to f, and refers to Tanimoto distance as the function $(1 - f)$. It is however made clear within the paper that the context is restricted by the use of a (positive) weighting vector W such that, for any vector A being considered, $A_i \in \{0, W_i\}$. Under these circumstances, the function is a proper distance metric, and so a set of vectors governed by such a weighting vector forms a metric space under this function.

GROUP TECHNOLOGY

We study a technique called Group Technology, including the central idea, its range of applications and advantages, and its practical application. In particular, we shall see how we can apply it to (a) optimize planning of manufacturing processes; and (b) how this is related to the design of manufacturing cells.

Since manufacturing cells are popular in several modern industries, we shall also put some effort into evaluating their performance, in terms of scheduling of operations etc.

The foundation of Group Technology arise out of the following observation :

Most tasks performed in any system can be categorized into a smaller number of sets, where each set has tasks that are very similar to each other.

For each of these sets, if we have the detailed plan for one task, we can use this as a starting point (or template), and make minor modifications to obtain the exact plan for any other task of that set.

This technique is quite general, and has applications in many areas of manufacturing systems :

- Design
- Process planning
- Layout planning
- Scheduling and routing.

Group Technology in Design

We know that generating the CAD model of a part design requires specification of its shape (geometry). So it makes sense to categorize, or classify a set of parts based on their shape.

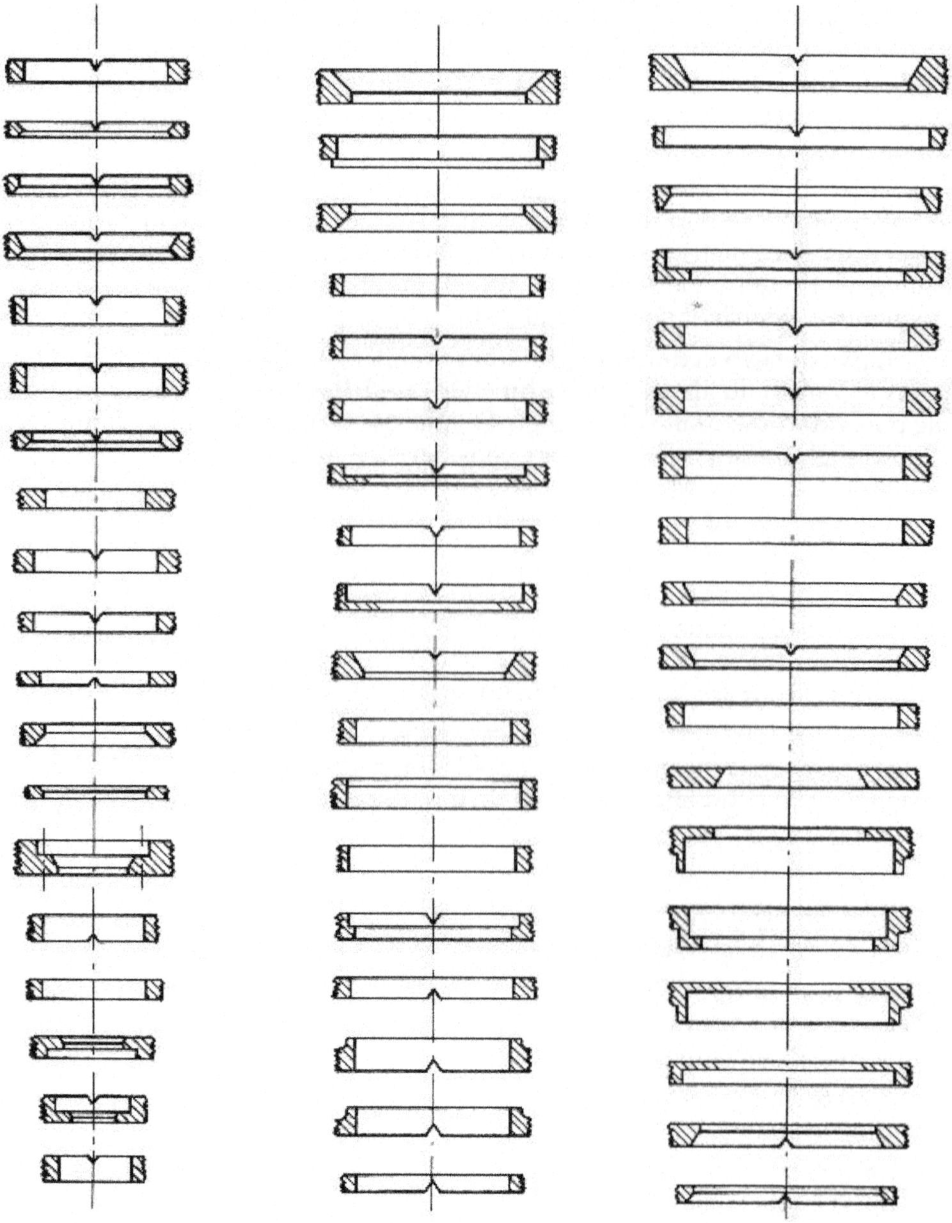

Fig. : A Design Family [source : Chang, Wang and Wysk, CAM, pub. : prentice-hall].

Problem

How can we decide whether two parts/products are similar?

It turns out that there is no good solution to this problem. However, some classification techniques have been successfully utilized in practice. We shall restrict our attention to individual parts. Similar techniques can (and have) been applied to products/assemblies also.

Coding and Classification

Coding : assignment of a symbol (or a set of symbols) to represent information.

Classification : A protocol that is used to separate a large group of objects into separate sub-groups.

Therefore, we first estimate (or survey) all possible different shapes we may be dealing with. Next, we identify some "features" that define something distinguishing about the shape of a part.

Examples

This process is largely intuitive (it would be nice to have mathematical representations of such qualifiers).

Then, for a given application where we wish to apply GT, we select a small set of "representative features". Each of these are given one (or a few) code digits.

For any given feature, we can further classify the variations—and therefore assign different values to the code digit (say 1, 2, 3,..., 9) depending on the size of the actual feature.

Based upon this idea, several schemes of coding and classification have been developed :

(a) Hierarchical Structures (Monocodes)

Here, each digit (or position) in the code represents a feature/sub-group. The first digit represents an entire group. The next digit represents sub-groups of the feature, and so on.

In this sense, each subsequent digit is qualified by the preceding digits (or, in an object-oriented sense, each subsequent digit inherits the properties of the previous digits).

Example

Consider all parts to be classified in terms of a feature : rotational symmetry.

1 == Non-rotational (prismatic) parts

2 == Rotational parts.

Within these groups, we can further classify by feature : presence of hole(s).

0 == No holes

1 == Has holes

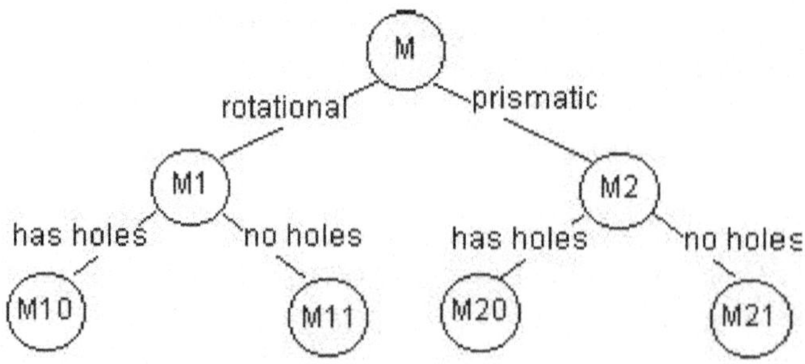

Advantages of monocodes :

1. With just a few digits, a very large amount of information can be stored.
2. The hierarchical structure allows parts of the code to be used for information at different levels of abstraction.

Disadvantages :

1. Impossible to get a good hierarchical structure for most features/groups.
2. Different sub-groups may have different levels of sub-groups, thereby leading to blank codes in some positions.

(b) Chain Codes (Polycodes)

In this method, the code digit represents one feature. Thus, the value of any given digit (or position) within the code has no relation to the other digits.

Example : A feature to be represented may be the longest linear dimension;

The value of this digit may be represented by an integer that is obtained by rounding off the length (longest dimension) [in inches] of the part being coded.

What if we are dealing with parts longer than 9.5 inches ?

We may need to reserve two (or more) digits for this feature in that case.

Advantages :

1. Easy to formulate.

Disadvantages :

1. Less information is stored per digit; therefore to get a meaningful comparison of, say, shape, very long codes will be required.
2. Comparison of coded parts (to check for similarity) requires more work.

(c) Hybrid Structure

In this case, the code for a part is a mixture of polycodes and monocodes. Such coding methods use monocodes where they can, and use polycodes for the other digits—in such a way as to obtain a code structure that captures the essential information about a part shape. This is the most commonly used method of coding and classification.

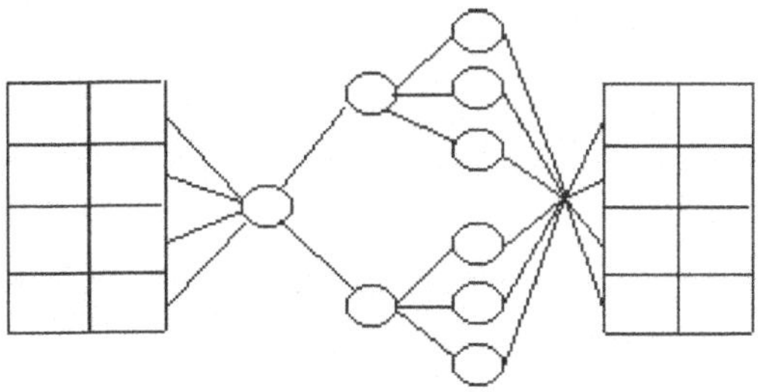

We shall study two popular coding methods to gain better understanding of coding and classification : ***Opitz and KK3.***

Properties :

1. Hybrid codes
2. Useful in shape classification for design and machining applications.

Opitz Code

The original Opitz code was a 5-digit code.

Digit/Position	*Type*	*Feature/Group*
1	Integer	Part class (rotational/non-rotational)
2	Integer	External shape
3	Integer	Internal shape
4	Integer	Plane surface features and machining
5	Integer	Auxiliary features (off-axis holes, gear teeth etc.)

Later, 4 more digits were added to the coding scheme, in order to increase the manufacturing information. These last four digits are also called *supplementary digits*. All four are integers, and respectively represent :

Dimensions, Material, Original shape of raw stock, and Accuracy.

Example

Bevel Gear, Length = 1.75″, Diameter = 3.5″

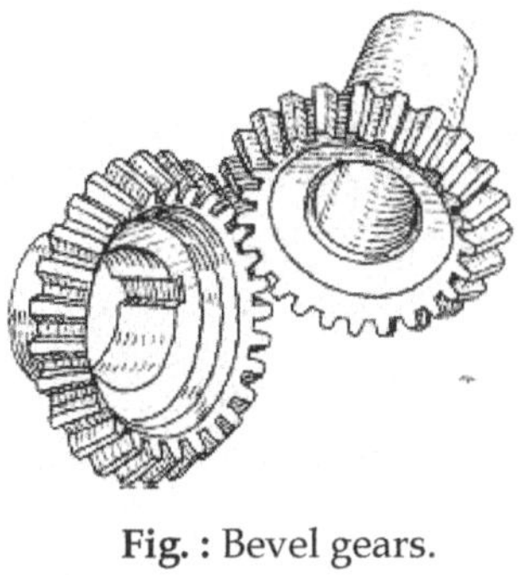

Fig. : Bevel gears.

[*Source* : French and Vierck, Engineering drawing & Graphic Technology, pub. : McGraw-Hill].

The KK-3 Code

It was originally developed by the Japan Society for Promotion of Machining Industry. The domain is machining and grinding parts. The code is 21-digits, each of which is an integer.

Digit	Items		(Rotational Components)
1	Parts name		General classification
2			Detailed classification
3	Materials		General classification
4			Detailed classification
5	Chief Dimensions		Length
6			Diameter
7	Primary Shapes and ratio of major dimensions		
8	Shape details and kinds of processes	External surface	External primary shape
9			Concentric screw threaded parts
10			Functional cut-off parts
11			Extraordinary shaped parts
12			Forming
13			Cylindrical surfaces
14		Internal surface	Internal primary shape
15			Internal curved surface
16			Internal flat / cylindrical surface
17		End surface	
18		Non-concentric holes	Regularly located holes
19			Special holes
20		Non-cutting processes	
21	Accuracy		

Structure of the KK-3 coding system.

[*Source* : Chang, Wang, Wysk, CAM, pub. : Prentice-Hall] [Original source : Japan Society for the Promotion of Machine Industry].

GT Applications : Design

The common process of design involves a designer making many decisions that are non-critical to the ultimate functionality.

[Example : Gear sizes for idling gears in a gear-box]

The process of product manufacturing involves :

1. Making such high level design decision,
2. Calculation of the exact dimensions of the components (usually from the sketch),
3. Creating the detailed drafting/CAD model of the part,

4. Generating the process plan for the part, including :
   - Determination of machine tool, tools, fixtures, machining conditions etc.

Much time is spent, in this process, in the last two steps. In this sequence of tasks, we introduce another one :

(2a) Perform coding and classification for the part.

Then, by comparing the code of the new part with those already existing, if the designer can discover a part very similar to the one we need, then :

1. He may use the existing part in the design. [What are the benefits ?]
2. He may be able to perform tasks (3) and (4) by minor modifications to the design and process plan of the existing part.

Applications : Machine Selection

This is a big issue in job-shops and small batch size factories.

Characteristics of such factories :

- Many different part shapes and specifications;
- Shop floor has many general purpose machine tools (m/c centers etc.).

Problems :

1. A large per cent of the total in-process time of a part (= average difference between the time when a raw stock enters the shop floor and when the final part leaves the factory) is setup time [up to 85% of total machining time].

   During much of the setup time, the machine tool is idle.
2. Many parts that are assigned to a machine may under-utilize its capabilities.

How does GT Help?

1. By scheduling parts on machine tools based on similarity, the machine capabilities are stressed rather than the usual scheduling heuristics (such as : first come first served (FCFS)).
2. By scheduling parts of the same "family" on the same (or similar) machine tool, much of the setup and fixturing time can be cut.

Applications : Factory Layout and Work-flow Planning

A conventional job-shop has a layout where similar machine tools are stationed together. The advantage of this layout is optimized tooling and fixtures allocation.

Disadvantages :

A part requiring several operations on different machines must be moved between these regions according to its operations sequence.

Usually, in such a layout, there are specialized operators for each machine. In many cases, this leads to non-optimal manpower allocation.

For job-shops that handle a variety of parts which can be classified into a small number of part families, a *cellular layout* is more efficient in terms of :

- Lower waiting time for each part (=> higher throughput)
- Lower manpower requirements for the entire shop-floor
- Better job conditions and training for operators (human advantage)
- Large reductions in materials handling requirements.

In order to understand the basic concept of cellular manufacturing, we begin with an understanding of process plans. What is the information needed for processing the raw stock ?

- Machining operations (and their sequence);
- Tools needed for the operation;
- Machine tool on which operation can be carried out;
- Fixtures needed for the operations;

We shall concentrate our attention to the first three of these.

The plan for processing each component is coded into OP codes (operation codes). Each OP code represents a set of machining operations. Thus the complete plan of a part is represented by means of a sequence of OP codes.

Operation code		Operation plan
01	Saw 01	Cut to size
02	Lathe 02	Face end Center drill Drill Ream Bore Turn straight Turn groove Chamfer Cutoff Face Chamfer
03	Grind 05	Grind
04	Insp 06	Inspect dimension Inspect finish

(a) Operation plan code (OP Code) and Operation plan

01	Saw 01
02	Lathe 02
03	Grind 05
04	Insp 06

(b) OP Code sequence

Operation plan and OP Code sequence.

Thus, for each part, we can represent the entire process plan in terms of a simple sequence of a few OP codes.

If we have a group of parts which have a very similar OP code sequence, then we can say that all parts in this group belong to the same family. Also, since each of these parts can be machined on the same set of machines, if we locate these machines close together on the shop floor, then we achieve a cellular layout.

Problem : how do we decide which parts have "similar OP code sequence" ?

Solution : By a process called ***clustering***.

Clustering Algorithms : We shall study a few popular clustering methods : Rank-Order clustering, Direct Clustering, and Similarity coefficients.

In the frist two methods, clustering is achieved by manipulation of matrices depicting the process plans of all the parts that have been produced in the shop.

Component	Processes (OP Code sequences)			
A-112	Saw01	Lathe 02	Grind 05	Insp 06
A-115	Mill 02	Drill 01	Insp 03	
A-120	Saw 01	Lathe 02	Insp 06	
A-123	Saw 01	Lathe 01	Insp 06	
A-131	Saw 01	Lathe 02	Insp 06	
A-212	Mill 05	Insp 03		
A-230	Mill 05	Insp 03		
A-432	Saw 01	Lathe 02	Insp 06	
A-451	Saw 01	Lathe 02	Insp 06	
A-510	Mill 05	Drill 01	Grind 06	Insp 06

Example Data.

Rank-Order Clustering

	A-112	A-115	A-120	A-123	A-131	A-212	A-230	A-432	A-451	A-510
SAW01	1		1	1	1			1	1	
LATHE01				1						
LATHE02	1		1		1			1	1	
DRL01		1								1
MILL02		1								
MILL05						1	1			1
GRIND05	1									
GRIND06	1		1							1

Step 1 : Calculate total weight of each column, $w_j = \sum_j 2^i M_{ij}$

Step 2 : Arrange columns by ascending weight

Step 3 : Calculate total weight of each row, $w_j = \sum_j 2^i M_{ij}$

Step 4 : If rows are in ascending order by weight, STOP.

else Arrange rows by ascending weight, GOTO Step 1.

The following figures show how this algorithm works :

	A-112	A-115	A-120	A-123	A-131	A-212	A-230	A-432	A-451	A-510	2^i
SAW01	1		1	1	1			1	1		2
LATHE01				1							4
LATHE02	1		1		1			1	1		8
DRL01		1								1	16
MILL02		1									32
MILL05						1	1			1	64
GRIND05	1										128
GRIND06										1	256
w_j	138	48	10	6	10	64	64	10	10	326	
Rank	9	6	2	1	3	7	8	4	5	10	

	A-123	A-120	A-131	A-432	A-451	A-115	A-212	A-230	A-112	A-510	w_j	Rank
SAW01	1	1	1	1	1				1		574	4
LATHE01	1										2	1
LATHE02		1	1	1	1				1		602	5
DRL01						1				1	1088	7
MILL02						1					64	2
MILL05							1	1		1	1408	8
GRIND05									1		512	3
GRIND06										1	1024	6
2^j	2	4	8	16	32	64	128	256	512	1024		

	A-123	A-120	A-131	A-432	A-451	A-115	A-212	A-230	A-112	A-510	2^i
LATHE01	1										2
MILL02						1					4
GRIND05									1		8
SAW01	1	1	1	1	1				1		16
LATHE02		1	1	1	1				1		32
GRIND06										1	64
DRL01						1				1	128
MILL05							1	1		1	256
w_j	18	48	48	48	48	132	256	256	56	448	
Rank	1	2	3	4	5	7	8	9	6	10	

	A-123	A-120	A-131	A-432	A-451	A-112	A-115	A-212	A-230	A-510	w_i	Rank
LATHE01	1										2	1
MILL02							1				128	5
GRIND05						1					64	2
SAW01	1	1	1	1	1	1					126	4
LATHE02		1	1	1	1	1					124	3
GRIND06										1	1024	6
DRL01							1			1	1152	7
MILL05								1	1	1	1792	8
2^j	2	4	8	16	32	64	128	256	512	1024		

	A-123	A-120	A-131	A-432	A-451	A-112	A-115	A-212	A-230	A-510	
LATHE01	1										Family 1
GRIND05						1					Family 1
LATHE02		1	1	1	1	1					Family 1
SAW01	1	1	1	1	1	1					Family 1
MILL02							1				Family 2
GRIND06										1	Family 2
DRL01							1			1	Family 2
MILL05								1	1	1	Family 2
w_j	18	24	24	24	24	28	160	256	256	510	

The Direct Clustering Algorithm

A problem with the rank-order clustering is that computation of weights can become problematic when the number of parts is large. For instance, if a shop has data for 2000 parts, then the weight factor for the rightmost column will be 2**2000, which is too large to compute directly !! Even if we use indirect representations, we need 2000 bits to represent a weight of each row. (WHY ?)

To avoid this problem, King and Nakornchai proposed the direct clustering algorithm, which is given as :

Step 1 : Calculate weight of each row, $w_j = \sum_j M_y$

Step 2 : Sort rows in descending order

Step 3 : Calculate weight of each column, $w_j = \sum_j M_y$

Step 4 : Sort columns in ascending order

Step 5 : For $i = 1$ to n, move all columns j where $M_{ij} = 1$, to the right while maintaining the order of the previous rows.

Step 6 : For $j = m$ to 1, move all rows i, where $M_{ij} = 1$, to the top, maintaining the order of the previous columns.

Step 7 : If current matrix = previous matrix, STOP;

Else go to Step 5.

The figures below show how this algorithm works :

0. Data	*a112*	*a115*	*a120*	*a123*	*a131*	*a212*	*a230*	*a432*	*a451*	*a510*	Σr
Saw01	1		1	1	1			1	1		6
Lathe01				1							1
Lathe02	1		1		1			1	1		5
Drill01		1								1	2
Mill02		1									1
Mill05						1	1			1	3
Grind05	1										1
Grind06										1	1
Sc	3	2	2	2	2	1	1	2	2	3	

1. Sort	*a212*	*a230*	*a115*	*a120*	*a123*	*a131*	*a432*	*a451*	*a112*	*a510*	Σr
Saw01				1	1	1	1	1	1		6
Lathe02				1		1	1	1	1		5
Mill05	1	1								1	3
Drill01			1							1	2
Lathe01					1						1

(Contd...)

(Contd...)

1. Sort	*a212*	*a230*	*a115*	*a120*	*a123*	*a131*	*a432*	*a451*	*a112*	*a510*	Σ*r*
Mill02			1								1
Grind05									1		1
Grind06										1	1
Sc	1	1	2	2	2	2	2	2	3	3	

2. Saw01	*a212*	*a230*	*a115*	*a510*	*a120*	*a123*	*a131*	*a432*	*a451*	*a112*
Saw01					1	1	1	1	1	1
Lathe02					1		1	1	1	1
Mill05	1	1		1						
Drill01			1	1						
Lathe01						1				
Mill02			1							
Grind05										1
Grind06				1						

3.All rows	*a115*	*a212*	*a230*	*a510*	*a123*	*a120*	*a131*	*a432*	*a451*	*a112*
Saw01					1	1	1	1	1	1
Lathe02						1	1	1	1	1
Mill05		1	1	1						
Drill01	1			1						
Lathe01					1					
Mill02	1									
Grind05										1
Grind06				1						

4. a112	*a115*	*a212*	*a230*	*a510*	*a123*	*a120*	*a131*	*a432*	*a451*	*a112*
Saw01					1	1	1	1	1	1
Lathe02						1	1	1	1	1
Grind05										1
Mill05		1	1	1						
Drill01	1			1						
Lathe01					1					
Mill02	1									
Grind06				1						

5. All cols	*a115*	*a212*	*a230*	*a510*	*a123*	*a120*	*a131*	*a432*	*a451*	*a112*
Saw01					1	1	1	1	1	1
Lathe02						1	1	1	1	1
Grind05										1
Lathe01					1					
Mill05		1	1	1						
Drill01	1			1						
Grind06				1						
Mill02	1									

Now we can form the part families in a manner similar to Rank Order Clustering.

Questions

1. What are the families ?
2. What are the cells ?
3. How do we use the part family information for layout planning?

Basically, the machine tools that are associated with a given family are put close to each other in the same physical area on the shop floor. This area then makes up one cell. For each family, there is usually one cell, or area, on the shop floor.

The Method of Similarity Coefficients

Why do we need the concept of similarity?

Mainly because a quantitative measure of how 'similar' two objects (*e.g.* parts), or actors (*e.g.* machines) are, allows us to use algorithms to process information about them.

What is similarity?

It depends on what you are studying. If you want to categorize parts, you want to develop a quantitative measure of how similar two parts are in shape, size, or other physical/chemical properties (such as hardness, surface finish, colour etc.)

If you are studying manufacturing systems design, you may be interested to measure how similar two machines are. You could base this measure on type of shapes the machines can produce, size of parts they can handle etc., a good practical measure is to look at what kind of parts are being processed at each machine. This is the concept we shall use.

How do we assign a number value to similarity?

Several methods can be used. Consider two machines, which process the following jobs :

Machine\Part	*P1*	*P2*	*P3*	*P4*	*P5*	*P6*
A			1		1	
B		1	1			

One way to define similarity of Machines A and B would be to look at how many parts visit both the machines in relation to the total number of parts that go to either machine. In the case above, the is one part, P3, which visits both machines, out of three parts that visit either.

So our measure will give a similarity = 1/3, or 0.333.

Note :

There are other, arguably equally good, measures of similarity. For example, one could define :

Similarity(A, B) = (no. of parts visiting A and B)/(no. of parts visiting A) = 1/2 = 0.5

Similarity(B, A) = (no. of parts visiting B and A)/(no. of parts visiting B) = 1/2 = 0.5

And then define common-similarity(A,B) = max(Similarity(A,B), Similarity(B,A)).

Or, if we are conservative, common-similarity(A,B) = min(Similarity(A,B), Similarity(B,A)).

How do we use similarity to apply GT ?

One way to use similarity in forming groups is by making a graph, called a dendogram. I will describe the process by using an example :

Machine/Part	*P1*	*P2*	*P3*	*P4*	*P5*	*P6*
A			1		1	
B		1	1			
C	1			1		
D		1	1		1	
E	1			1		1

For which we can easily find the similarity matrix :

M/c/M/c	*A*	*B*	*C*	*D*	*E*
A	1	0.33	0	0.67	0
B	0.33	1	0	0.67	0
C	0	0	1	0	0.67
D	0.67	0.67	0	1	0.67
E	0	0	0.67	0	1

If one is asked now to group machines together such that they are completely similar (*i.e.*, similarity = 1), we can look at the table and see that we would get five groups : each group has a single machine in it !

But if we say that we are allowed some "fuzziness" in our vision, and machines that are 2/3 similar look as if they are same : then machines C and E, *e.g.*, would 'look' the same to us.

Thus we may say that for a person with filter strength of 0.67, C and E belong to the same group.

So do Machines A and D (since their similarity = 0.67 also).

And of course, so are machines B and D.

Problem :

Thus, may we say that Since A, D are in the same 0.67-group, and B is in the same group as D, then all A, B, and D belong to the same group?

The answer is : It depends. You (the design engineer) have to decide what to do with such ambiguous data.

If you feel that Machine D is an expensive machine, but is used only for small periods by the parts that need it, you may decide to share this machine between groups.

In that case, you can define Similarity of machine B with Group (A, D) as the maximum of the similarity of B with A or D :

Similarity(B, (A, D)) = max(Similarity(B, A), Similarity(B, D)) = max(1/3, 2/3) = 2/3.

Under this assumption, we can say that the 0.67-group contains machines A, B, D.

(Of course, C and E form another 0.67-group).

Looking at the similarity table, it is then clear that if we want to further merge the machines into groups, we need to have vision so "fuzzy", that we cannot distinguish machines with similarity of 0 !

This yields the following dendogram :

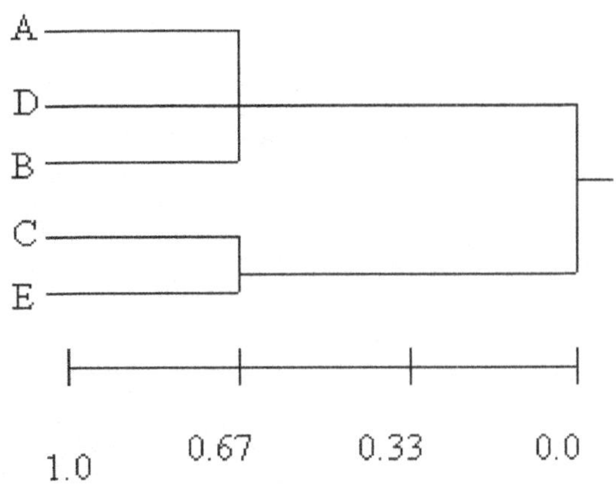

Of course, we might have decided, if machine D is cheap and heavily used, that it is easy to purchase multiple copies of D. If that is the case, we can afford to be more conservative in our definition of similarity of a machine with a group :

We can say that Similarity(X, Y, Z) = minimum of pair-wise similarities of (X,Y), (Y,Z), and (X,Y).

In this case, at a similarity level of 0.67, we are unable to group A, B, and D. So we make a copy of machine D, and get separate groups : (A,D) and (B, D).

Then, at a similarity level of 0.33, all of A, B and D start to look similar, by our new definition of similarity. This results in a new dendogram as follows :

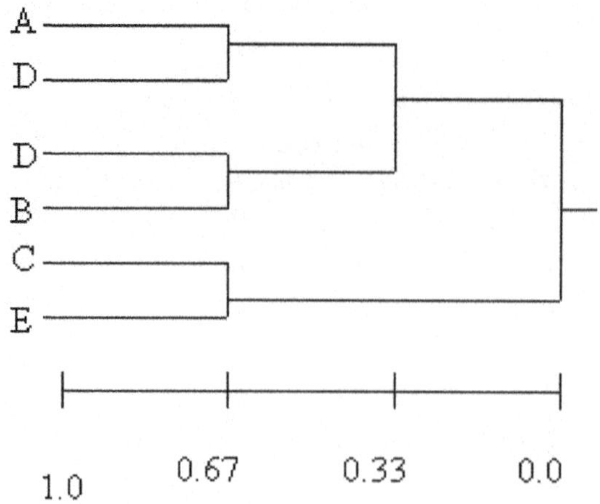

Note, at "fuzzy" vision which can distinguish higher than 0.67 similarity, this has three groups : AD, BD, and CE. If vision is so fuzzy that that we cannot distinguish machines which are better than 0.33 similar, then we see two groups: ABD and CE. And finally, if our vision is so bad that we cannot distinguish between objects with 0 similarity, then we see only 1 group !

Postscript

If you look-up references on the use of similarity coefficients, you may discover many different formulas to define such coefficients. It is difficult to say which is better. In general, all measures of similarity are mappings from a complex, multi-dimensional domain (*e.g.* design description) to real numbers.

Once the machine families have been identified, we can isolate them to form cells within the manufacturing system. The advantages of reduced material handling through cells are easy to visualize.

How (and when) do cells help in efficient manpower allocation ?

In a small cell (4–5 machine tools or less), sometimes a single operator can run parts on all the machines. The first part entering the system is loaded on the machine, and while it is being machined, the next part is set up on the fixtures needed. When the first part is completed on the first machine, it is fixtured on the next machine tool, and while it machines, the operator can load and begin running the second part on the first machine. Thus the same operator can be operating more than one machines at the same time, utilizing the machining time (when he would be idle) to process other parts. Due to the physical closeness of the machine tools, very little time/resources are wasted in material handling or moving the part between machines. Also, since each part is sequentially moved along the machines in the cell, the average time in queue (waiting for the machine to become idle) is eliminated for a well planned cell schedule.

Chapter 13

Decision-Making with Consideration of Association in Supply Chains

ABC ANALYSIS

The **ABC Analysis** is a business term used to define an *inventory* categorization technique often used in *materials management*. It is also known as *Selective Inventory Control*. Policies based on ABC analysis :

- *A ITEMS :* very tight control and accurate records.
- *B ITEMS :* less tightly controlled and good records.
- *C ITEMS :* simplest controls possible and minimal records.

The ABC analysis provides a mechanism for identifying items that will have a significant impact on overall inventory cost, while also providing a mechanism for identifying different categories of stock that will require different management and controls.

The ABC analysis suggests that inventories of an organization are not of equal value. Thus, the inventory is grouped into three categories (**A, B, and C**) in order of their estimated importance.

'**A'** items are very important for an organization. Because of the high value of these 'A' items, frequent value analysis is required. In addition to that, an organization needs to choose an appropriate order pattern (*e.g.* 'Just-in-time') to avoid excess capacity. '**B'** items are important, but of course less important than 'A' items and more important than 'C' items. Therefore 'B' items are intergroup items. '**C'** items are marginally important.

ABC Analysis Categories

There are no fixed threshold for each class, different proportion can be applied based on objective and criteria. ABC Analysis is similar to the Pareto principle in that the 'A' items will typically account for a large proportion of the overall value but a small percentage of number of items. Example of ABC class are

- **'A'** items—20% of the items accounts for 70% of the annual consumption value of the items.
- **'B'** items—30% of the items accounts for 25% of the annual consumption value of the items.
- **'C'** items—50% of the items accounts for 5% of the annual consumption value of the items.

Another recommended breakdown of ABC classes :

1. "A" approximately 10% of items or 66.6% of value
2. "B" approximately 20% of items or 23.3% of value
3. "C" approximately 70% of items or 10.1% of value.

ABC Analysis in ERP Packages

Major ERP packages (SAP, Oracle, Microsoft, etc.) have built in function of ABC analysis. User can execute ABC analysis based on user defined criteria and system apply ABC code to items (parts).

Example of the Application of Weighed Operation Based on ABC Class

Actual distribution of ABC class in the electronics manufacturing company with 4051 active parts.

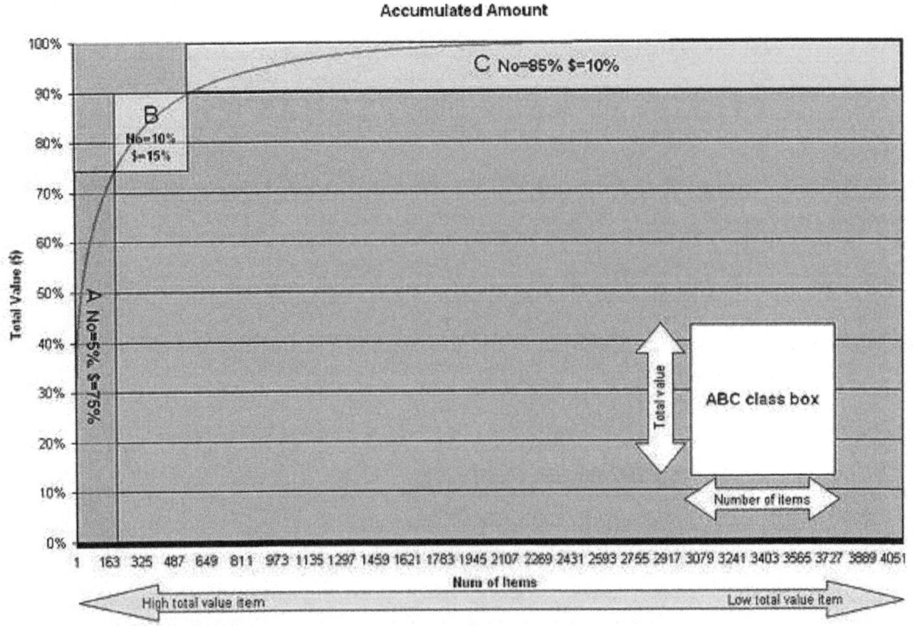

Distribution of ABC class

ABC class	*Number of items*	*Total amount required*
A	10%	70%
B	20%	20%
C	70%	10%
Total	100%	100%

Using this distribution of ABC class and change total number of the parts to 14213.

- *Uniform Purchase :* When you apply equal purchasing policy to all 14213 components, example weekly delivery and re-order point (safety stock) of 2 week supply assuming that there are no lot size constraints, the factory will have 16000 delivery in 4 weeks and average inventory will be 2.5 week supply.

Application of Weighed Purchasing condition

Uniform condition		*Weighed condition*	
Items	*Conditions*	*Items*	*Conditions*
All items 14213	Re-order point = 2 week supply Delivery frequency = weekly	A-class items 200	Re-order point = 1 week supply Delivery frequency = weekly
		B-class items 400	Re-order point = 2 week supply Delivery frequency = bi-weekly
		C-class items 3400	Re-order point = 3 week supply Delivery frequency = every 4 weeks

Weighed Purchase

In comparison, when weighed purchasing policy applied based on ABC class, example C class monthly (every 4 week) delivery with re-order point of 3 week supply, B class Bi-weekly delivery with re-order point of 2 weeks supply, A class weekly delivery with re-order point of 1 week supply, total number of delivery in 4 weeks will be (A 200 × 4 = 800) + (B 400 × 2 = 800) + (C 3400 × 1 = 3400) = 5000 and average inventory will be (A 75% × 1.5 weeks) + (B 15% × 3 weeks) + (C 10% × 3.5 weeks) = 1.925 week supply.

- *Result :* By applying weighed control based on ABC classification, required man-hours and inventory level are drastically reduced.

Comparison of "Equal" and "Weighed" Purchase (4 weeks span)

ABC class	*No. of items*	*% of total value*	*Equal purchase*		*Weighed purchase*		*Note*
			No. of delivery in 4 weeks	*Average supply level*	*No. of delivery in 4 weeks*	*Average supply level*	
A	200	75%	800	2.5 weeks	800	1.5 weeks[a]	same delivery frequency, safety stock reduced from 2.5 to 1.5 weeks[a], require tighter control with more man · hours.
B	400	15%	1600	2.5 weeks	800	3 weeks	increased safety stock level by 20%, delivery frequency reduced to half. Less man · hour required.
C	3400	10%	13600	2.5 weeks	3400	3.5 weeks	increased safety stock from 2.5 to 3.5 week supply, delivery frequency is one quarter. Drastically reduced man · hour requirement.
Total	4000	100%	16000	2.5 weeks	5000	1.925 weeks	average inventory value reduced by 23%, delivery frequency reduced by 69%. Over-all reduction of man · hour requirement.

a) A class item can be applied much tighter control like JIT daily delivery. If daily delivery with one day stock is applied, delivery frequency will be 4000 and average inventory level of A class item will be 1.5 days supply and total inventory level will be 1.025 week supply. reduction of inventory by 59%. Total delivery frequency also reduced to half from 16000 to 8200.

SEQUENTIAL PATTERN MINING

Sequential Pattern mining is a topic of data-mining concerned with finding statistically relevant patterns between data examples where the values are delivered in a sequence. It is usually presumed that the values are discrete, and thus time series mining is closely related, but usually considered a different activity. Sequential pattern mining is a special case of structured data-mining.

There are several key traditional computational problems addressed within this field. These include building efficient databases and indexes for sequence information, extracting the frequently occurring patterns, comparing sequences for similarity, and recovering missing sequence members. In general, sequence mining problems can be classified as *string mining* which is typically based on string processing algorithms and *item-set mining* which is typically based on association rule learning.

String Mining

String mining typically deals with a limited alphabet for items that appear in a sequence, but the sequence itself may be typically very long. Examples of an alphabet can be those in the ASCII character set used in natural language text, nucleotide bases 'A', 'G', 'C' and 'T' in DNA sequences, or amino acids for protein sequences. In biology applications analysis of the arrangement of the alphabet in strings can be used to examine gene and protein sequences to determine their properties. Knowing the sequence of letters of a DNA a protein is not an ultimate goal in itself. Rather, the major task is to understand the sequence, in terms of its structure and biological function. This is typically achieved first by identifying individual regions or structural units within each sequence and then assigning a function to each structural unit. In many cases this requires comparing a given sequence with previously studied ones. The comparison between the strings becomes complicated when insertions, deletions and mutations occur in a string.

A survey and taxonomy of the key algorithms for sequence comparison for bioinformatics is presented by Abouelhoda & Ghanem, which include :

- *Repeat-related problems :* That deal with operations on single sequences and can be based on exact string matching or approximate string matching methods for finding dispersed fixed length and maximal length repeats, finding tandem repeats, and finding unique subsequences and missing (un-spelled) subsequences.
- *Alignment problems :* That deal with comparison between strings by first aligning one or more sequences; examples of popular methods include BLAST for comparing a single sequence with multiple sequences in a database, and Clustal W for multiple alignments. Alignment algorithms can be based on either exact or approximate methods, and can also be classified as global alignments, semi-global alignments and local alignment.

Itemset Mining

Some problems in sequence mining lend themselves discovering frequent itemsets and the order they appear, for example, one is seeking rules of the form "if a {customer buys a car}, he or she is likely to {buy insurance} within 1 week", or in the context of stock prices, "if {Nokia up and Ericsson Up}, it is likely that {Motorola up and Samsung up} within 2 days". Traditionally, itemset mining is used in marketing applications for discovering regularities between frequently co-occurring items in large transactions. For example, by analysing transactions of customer shopping baskets in a supermarket, one can produce a rule which reads "if a customer buys onions and potatoes together, he or she is likely to also buy hamburger meat in the same transaction".

A survey and taxonomy of the key algorithms for item set mining is presented by Han *et. al.*.

The two common techniques that are applied to sequence databases for frequent itemset mining are the influential apriori algorithm and the more-recent FP-Growth technique.

Application

With a great variation of products and user buying behaviours, shelf on which products are being displayed is one of the most important resources in retail environment. Retailers can not only increase their profit but, also decrease cost by proper management of shelf space allocation and products display. To solve this problem, George and Binu have proposed an approach to mine user buying patterns using Prefix Span algorithm and place the products on shelves based on the order of mined purchasing patterns.

PERFORMANCE MEASUREMENTS AND METRICS : AN ANALYSIS OF SUPPLIER EVALUATION

Based on the more current partnership approach, several techniques are used by companies to evaluate suppliers and measure performance. The first step in implementing any of the techniques being discussed is to determine the attributes that should be considered. A firm should focus on the attributes that it finds most important. Some attributes are easy to measure while others are not. A rule of thumb is to consider the total costs associated with a purchased product/service, not solely the purchase price. Some of the metrics that can be considered are :

- Quality level
- Service level
- Correct quantity
- On-time delivery
- Price/cost of product
- Use of electronic data interchange (EDI)
- Willingness to share sensitive information
- Presence of certification or other documentation
- Flexibility to respond to unexpected demand changes
- Communication skills/systems (phone, fax, e-mail, Internet)
- Quick response time in case of emergency, problem, or special request
- Willingness to change their products and services to meet your changing needs
- Willingness to participate in your firm's new product development and value analysis.

Some models are proficient in considering total costs, but they are usually very difficult to implement and time consuming. Thus, the resources available to the firm's purchasing function will drive the firm's model choice.

The three most common approaches are the categorical system, weighted-point average system and the cost-based system. The categorical system is the most subjective technique since it does not differentiate between the weights of the attributes considered. The weighted-point average system overcomes this drawback by assigning weights to each attribute. The cost-based system is the most objective of the three methods because it also considers non-performance costs.

Categorical System

In this method, the buyer chooses attributes that are most important to its particular situation. The buyer assigns either a preferred (+), unsatisfactory (-), or neutral (O) rating for each of the selected attributes to every vendor. Then the ratings are totaled for each vendor. For example, ratings resulting in scores of two preferred (++), one unsatisfactory (-), and one neutral (O) would total one positive (+). A comparison of total scores reveals the highest rated vendor. The representatives of the involved divisions agree upon the ratings.

Categorical supplier measurement is the easiest method to implement but suffers from subjectivity. It does not provide a detailed insight into the supplier's true performance because the attributes being measured are weighted equally. Also, because the representatives of the involved divisions assign the ratings, this method relies on an individual's perception about performance and not on quantitative data.

Weighted Point Method

In the weighted-point method, the relevant attributes are chosen and each are assigned a weight depending on the importance to the overall performance. The firm reaches a consensus on weight assignments to prevent or minimize subjectivity. The weight for each performance category is then multiplied by the performance score that is assigned to it. Finally, these products are totaled to determine a final rating for each supplier.

Firms often use the weighted point system because it is highly reliable and its implementation costs are moderate. In addition, it combines qualitative and quantitative performance factors into a common system. Because users can change the weights assigned to each performance category, or change the performance categories themselves depending on the strategic priorities of the firm, the system is flexible. The weighted-point method overcomes the subjectivity of the categorical system, but it has some drawbacks. It requires the buyer to specify the value of one performance measure relative to another, which is often difficult in practice.

Cost-based System

Using the cost-based system, a buyer is able to quantify the additional costs incurred if a supplier fails to perform as expected. The total cost of doing business with the supplier can be calculated by the supplier performance index (SPI). This index is calculated for each item or commodity provided by the supplier and has a base value of 1. It is represented by the following formula : SPI = (Purchase Price + Non-performance Cost)/ (Purchase Price). As derived from the equation, the closer SPI is to 1, the better the supplier. Non-costs should include qualitative factors.

Benefits a buyer can achieve by using this approach include :

- The ability to source requirements based on total cost consideration
- A methodology to increase supplier accountability and control
- An equitable and consistent evaluation tool
- Definition of supplier performance expectation
- Communication of the firm's buying priorities to suppliers
- The ability to perform sourcing risk assessment
- Enhancement of internal communication for reporting critical supplier sourcing information
- The ability to provide positive supplier reinforcement
- A basis for a supplier award program.

This system is the least subjective of the three because it quantifies the total cost of doing business by considering non-performance costs. The main difficulty in the use of the system is its complexity and its requirement that users have a developed cost accounting system. Although this sounds like an ideal way of dealing with costs, it is difficult to identify the costs of supplier non-performance.

Against their subjectivity and drawbacks, the categorical method, the weighted-point method and the cost-ratio method are the most widely used techniques in supplier evaluation due to their ease of use and implementation. Several other models and techniques have been proposed for supplier measurement and evaluation. These include : total cost of ownership approach, analytical hierarchy process, weighted linear model approaches, mixed integer programming, matrix method, human judgment models, interpretive structural modelling, statistical analysis, linear programming models, and neural networks, multi-objective programming. Of these methods, Total Cost of Ownership (TCO) method stands out as a commonly endorsed approach.

Total Cost of Ownership Approach

The Total Cost of Ownership (TCO) approach is very similar to the cost ratio method. However, it is more comprehensive because it considers all the costs associated with quality, delivery and service. These costs include a number of non-value added activities such as service costs, receiving costs, quality costs (inspection, rework, reject costs), failure costs, and administrative costs including management time, maintenance, disposition and life-cycle costs. Lifecycle costs are costs incurred throughout the life of a product or service. These costs may include maintenance, downtime, repair, overhead, and idle time. In a study by Ellram, it was concluded that the price of a piece of production equipment for a firm's operations was only 35 per cent of the total cost of that piece of equipment over its life cycle. Despite the high percentage of the non-value added costs, firms tend to either underestimate or completely ignore them.

TCO is a proactive and comprehensive system. Rather than focusing only on price, it examines issues outside of the supplier's cost structure. But using TCO presents the firm with new challenges. The method is complex to implement and maintain. Thus, it consumes a great deal of time.

Frequency of Assessment

Supplier evaluation methods are usually time consuming, so they are not performed frequently. In practice, many firms report supplier performance on a monthly or quarterly basis. In fact, some firms evaluate suppliers only once a year. However, frequent meetings with suppliers facilitate the prevention of inefficient practices at an early stage and encourage continuous improvement of suppliers. These assessments, however, are mutually beneficial only if both parties are willing to cooperate and provide the necessary inputs.

How Measurements are Used

The information gained from supplier evaluation is valuable only when both buyer and supplier use it to improve their partnership. Supplier measurement systems are most commonly used for the following purposes : (1) Track performance of supplier; (2) identify supplier improvement opportunities; (3) develop supplier; (4) benchmark suppliers against best practices.

Chapter 14

DATA WAREHOUSE

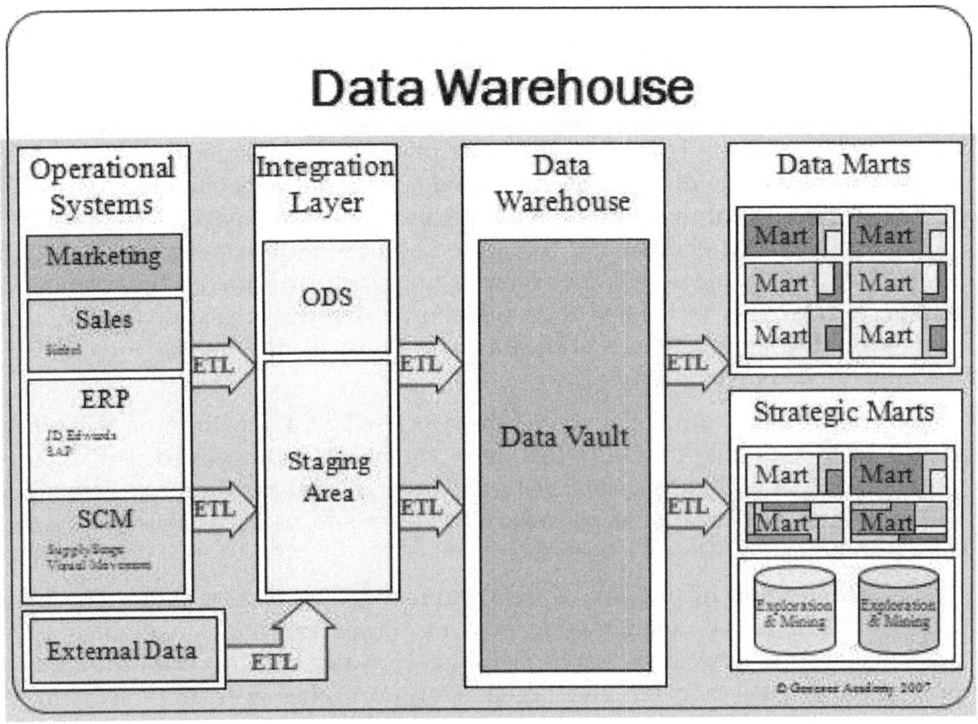

Fig. : Data Warehouse Overview.

In computing, a **data warehouse (DW, DWH)**, or an **enterprise data warehouse (EDW)**, is a database used for reporting and data analysis. Integrating data from one or more disparate sources creates a central repository of data, a data warehouse (DW). Data warehouses store current and historical data and are used for creating trending reports for senior management reporting such as annual and quarterly comparisons.

The data stored in the warehouse is uploaded from the operational systems (such as marketing, sales, etc., shown in the figure to the right). The data may pass through an operational data store for additional operations before it is used in the DW for reporting.

The typical extract-transform-load (ETL)-based data warehouse uses staging, data integration, and access layers to house its key functions. The staging layer or staging database stores raw data extracted from each of the disparate source data systems. The integration layer integrates the disparate data-sets by transforming the data from the staging layer often storing this transformed data in an operational data store (ODS) database. The integrated data are then moved to yet another database, often called the data warehouse database, where the data is arranged into hierarchical groups often called dimensions and into facts and aggregate facts. The combination of facts and dimensions is sometimes called a star schema. The access layer helps users retrieve data.

A data warehouse constructed from integrated data source systems does not require ETL, staging databases, or operational data store databases. The integrated data source systems may be considered to be a part of a distributed operational data store layer. Data federation methods or data virtualization methods may be used to access the distributed integrated source data systems to consolidate and aggregate data directly into the data warehouse database tables. Unlike the ETL-based data warehouse, the integrated source data systems and the data warehouse are all integrated since there is no transformation of dimensional or reference data. This integrated data warehouse architecture supports the drill down from the aggregate data of the data warehouse to the transactional data of the integrated source data systems.

A data mart is a small data warehouse focused on a specific area of interest. Data warehouses can be sub-divided into data marts for improved performance and ease of use within that area. Alternatively, an organization can create one or more data marts as first steps towards a larger and more complex enterprise data warehouse.

This definition of the data warehouse focuses on data storage. The main source of the data is cleaned, transformed, cataloged and made available for use by managers and other business professionals for data mining, online analytical processing, market research and decision support. However, the means to retrieve and analyze data, to extract, transform and load data, and to manage the data dictionary are also considered essential components of a data warehousing system. Many references to data warehousing use this broader context. Thus, an expanded definition for data warehousing includes business intelligence tools, tools to extract, transform and load data into the repository, and tools to manage and retrieve metadata.

BENEFITS OF A DATA WAREHOUSE

A data warehouse maintains a copy of information from the source transaction systems. This architectural complexity provides the opportunity to :

- Congregate data from multiple sources into a single database so a single query engine can be used to present data.
- Mitigate the problem of database isolation level lock contention in transaction processing systems caused by attempts to run large, long running, analysis queries in transaction processing databases.
- Maintain data history, even if the source transaction systems do not.
- Integrate data from multiple source systems, enabling a central view across the enterprise. This benefit is always valuable, but particularly so when the organization has grown by merger.
- Improve *data quality*, by providing consistent codes and descriptions, flagging or even fixing bad data.
- Present the organization's information consistently.
- Provide a single common data model for all data of interest regardless of the data's source.
- Restructure the data so that it makes sense to the business users.
- Restructure the data so that it delivers excellent query performance, even for complex analytic queries, without impacting the *operational systems*.
- Add value to operational business applications, notably *customer relationship management* (CRM) systems.
- Making decision-support queries easier to write.

GENERIC DATA WAREHOUSE ENVIRONMENT

The environment for data warehouses and marts includes the following :

- Source systems that provide data to the warehouse or mart;
- Data integration technology and processes that are needed to prepare the data for use;
- Different architectures for storing data in an organization's data warehouse or data marts;
- Different tools and applications for the variety of users;
- Metadata, data quality, and governance processes must be in place to ensure that the warehouse or mart meets its purposes.

In regards to source systems listed above, Rainer states, "A common source for the data in data warehouses is the company's operational databases, which can be relational databases".

Regarding data integration, Rainer states, "It is necessary to extract data from source systems, transform them, and load them into a data mart or warehouse".

Rainer discusses storing data in an organization's data warehouse or data marts."

Metadata are data about data. "IT personnel need information about data sources; database, table, and column names; refresh schedules; and data usage measures."

Today, the most successful companies are those that can respond quickly and flexibly to market changes and opportunities. A key to this response is the effective and efficient use of data and information by analysts and managers. A "data warehouse" is a repository of historical data that are organized by subject to support decision-makers in the organization. Once data are stored in a data mart or warehouse, they can be accessed.

HISTORY

The concept of data warehousing dates back to the late 1980s when IBM researchers Barry Devlin and Paul Murphy developed the "business data warehouse". In essence, the data warehousing concept was intended to provide an architectural model for the flow of data from operational systems to *decision support environments*. The concept attempted to address the various problems associated with this flow, mainly the high costs associated with it. In the absence of a data warehousing architecture, an enormous amount of redundancy was required to support multiple decision support environments. In larger corporations it was typical for multiple decision support environments to operate independently. Though each environment served different users, they often required much of the same stored data. The process of gathering, cleaning and integrating data from various sources, usually from long-term existing operational systems (usually referred to as *legacy systems*), was typically in part replicated for each environment. Moreover, the operational systems were frequently re-examined as new decision support requirements emerged. Often new requirements necessitated gathering, cleaning and integrating new data from "*data marts*" that were tailored for ready access by users.

Key developments in early years of data warehousing were :

- *1960s : General Mills* and *Dartmouth College*, in a joint research project, develop the terms *dimensions* and *facts*.
- *1970s : ACNielsen* and IRI provide dimensional data marts for retail sales.
- *1970s : Bill Inmon* begins to define and discuss the term : Data Warehouse
- *1975 : Sperry Univac* introduces *MAPPER* (MAintain, Prepare, and Produce Executive Reports) is a database management and reporting system that includes the world's first *4GL*. It was the first platform designed for building Information Centers (a forerunner of contemporary Enterprise Data Warehousing platforms)
- *1983 : Teradata* introduces a database management system specifically designed for decision support.

- *1983 : Sperry Corporation* Martyn Richard Jones defines the Sperry Information Center approach, which while not being a true DW in the Inmon sense, did contain many of the characteristics of DW structures and process as defined previously by Inmon, and later by Devlin. First used at the *TSB England & Wales*
- *1984 : Metaphor Computer Systems,* founded by *David Liddle* and Don Massaro, releases Data Interpretation System (DIS). DIS was a hardware/software package and GUI for business users to create a database management and analytic system.
- *1988 :* Barry Devlin and Paul Murphy publish the article *An architecture for a business and information system* in *IBM Systems Journal* where they introduce the term "business data warehouse."
- *1990 :* Red Brick Systems, founded by *Ralph Kimball,* introduces Red Brick Warehouse, a database management system specifically for data warehousing.
- *1991 :* Prism Solutions, founded by *Bill Inmon,* introduces Prism Warehouse Manager, software for developing a data warehouse.
- *1992 : Bill Inmon* publishes the book *Building the Data Warehouse.*
- *1995 :* The Data Warehousing Institute, a for-profit organization that promotes data warehousing, is founded.
- *1996 : Ralph Kimball* publishes the book *The Data Warehouse Toolkit.*
- *2000 : Daniel Linstedt* releases the *Data Vault,* enabling real time auditable Data Warehouses warehouse.

INFORMATION STORAGE

Facts

A fact is a value or measurement, which represents a fact about the managed entity or system.

Facts as reported by the reporting entity are said to be at raw level.

E.g. if a BTS received 1,000 requests for traffic channel allocation, it allocates for 820 and rejects the remaining then it would report 3 **facts** or measurements to a management system :

- tch_req_total = 1000
- tch_req_success = 820
- tch_req_fail = 180

Facts at raw level are further aggregated to higher levels in various *dimensions* to extract more service or business-relevant information out of it. These are called aggregates or summaries or aggregated facts.

E.g. if there are 3 BTSs in a city, then facts above can be aggregated from BTS to city level in network dimension. *E.g.*

- tch_req_success_city = tch_req_success_bts1 + tch_req_success_bts2 + tch_req_sucess_bts3
- avg_tch_req_success_city = (tch_req_success_bts1 + tch-req_success_bts2 + tch_req_success_bts3)/3

Dimensional *vs.* Normalized Approach for Storage of Data

There are three or more leading approaches to storing data in a data warehouse – the most important approaches are the dimensional approach and the normalized approach.

The dimensional approach, whose supporters are referred to as "Kimballites", believe in *Ralph Kimball*'s approach in which it is stated that the data warehouse should be modelled using a Dimensional Model/ *star schema*. The normalized approach, also called the *3NF* model (Third Normal Form), whose supporters are referred to as "Inmonites", believe in Bill Inmon's approach in which it is stated that the data warehouse should be modelled using an E-R model/normalized model.

In a *dimensional approach, transaction data* are partitioned into "facts", which are generally numeric transaction data, and "*dimensions*", which are the reference information that gives context to the facts. For example, a sales transaction can be broken up into facts such as the number of products ordered and the price paid for the products, and into dimensions such as order date, customer name, product number, order ship-to and bill-to locations, and salesperson responsible for receiving the order.

A key advantage of a dimensional approach is that the data warehouse is easier for the user to understand and to use. Also, the retrieval of data from the data warehouse tends to operate very quickly. Dimensional structures are easy to understand for business users, because the structure is divided into measurements/facts and context/dimensions. Facts are related to the organization's business processes and operational system whereas the dimensions surrounding them contain context about the measurement.

The main disadvantages of the dimensional approach are the following :

1. In order to maintain the integrity of facts and dimensions, loading the data warehouse with data from different operational systems is complicated.
2. It is difficult to modify the data warehouse structure if the organization adopting the dimensional approach changes the way in which it does business.

In the normalized approach, the data in the data warehouse are stored following, to a degree, *database normalization* rules. Tables are grouped together by *subject areas* that reflect general data categories (*e.g.*, data on customers, products, finance, etc.). The normalized structure divides data into entities, which creates several tables in a relational database. When applied in large enterprises the result is dozens of tables that are linked together by a web of joins. Furthermore, each of the created entities is converted into separate physical tables when the database is implemented. The main advantage of this approach is that it is straightforward to add information into the database. Some disadvantages of this approach are that, because of the number of tables involved, it can be difficult for users to join data from different sources into meaningful information and to access the information without a precise understanding of the sources of data and of the *data structure* of the data warehouse.

It should be noted that both normalized and dimensional models can be represented in entity-relationship diagrams as both contain joined relational tables. The difference between the two models is the degree of normalization (also known as *Normal Forms*).

These approaches are not mutually exclusive, and there are other approaches. Dimensional approaches can involve normalizing data to a degree.

In Information-Driven Business, Robert Hillard proposes an approach to comparing the two approaches based on the information needs of the business problem. The technique shows that normalized models hold far more information than their dimensional equivalents (even when the same fields are used in both models) but this extra information comes at the cost of usability. The technique measures information quantity in terms of *Information Entropy* and usability in terms of the Small Worlds data transformation measure.

TOP-DOWN VERSUS BOTTOM-UP DESIGN METHODOLOGIES

Bottom-up Design

Ralph Kimball, designed an approach to data warehouse design known as *bottom-up*.

In the *bottom-up* approach, data marts are first created to provide reporting and analytical capabilities for specific business processes.

Data marts contain, primarily, dimensions and facts. Facts can contain either atomic data and, if necessary, summarized data. The single data mart often models a specific business area such as "Sales" or "Production." These data marts can eventually be integrated to create a comprehensive data warehouse. The data warehouse bus architecture is primarily an implementation of "the bus", a collection of conformed dimensions and conformed facts, which are dimensions that are shared (in a specific way) between facts in two or more data marts.

The integration of the data marts in the data warehouse is centered on the conformed dimensions (residing in "the bus") that define the possible integration "points" between data marts. The actual integration of two or more data marts is then done by a process known as "Drill across". A drill-across works by grouping (summarizing) the data along the keys of the (shared) conformed dimensions of each fact participating in the "drill across" followed by a join on the keys of these grouped (summarized) facts.

Maintaining tight management over the data warehouse bus architecture is fundamental to maintaining the integrity of the data warehouse. The most important management task is making sure dimensions among data marts are consistent.

Business value can be returned as quickly as the first data marts can be created, and the method lends itself well to an exploratory and iterative approach to building data warehouses. For example, the data warehousing effort might start in the "Sales" department, by building a Sales-data mart. Upon completion of the Sales-data mart, the business might then decide to expand the warehousing activities into the, say, "Production department" resulting in a Production data mart. The requirement for the Sales data mart and the Production data mart to be integrable, is that they share the same "Bus", that will be, that the data warehousing team has made the effort to identify and implement the conformed dimensions in the bus, and that the individual data marts links that information from the bus. The Sales-data mart is good as it is (assuming that the bus is complete) and the Production-data mart can be constructed virtually independent of the Sales-data mart (but not independent of the Bus).

If integration via the bus is achieved, the data warehouse, through its two data marts, will not only be able to deliver the specific information that the individual data marts are designed to do, in this example either "Sales" or "Production" information, but can deliver integrated Sales-Production information, which, often, is of critical business value.

Top-down Design

Bill Inmon, has defined a data warehouse as a centralized repository for the entire enterprise. The *top-down* approach is designed using a normalized enterprise data model. "Atomic" data, that is, data at the lowest level of detail, are stored in the data warehouse. Dimensional data marts containing data needed for specific business processes or specific departments are created from the data warehouse. In the Inmon vision, the data warehouse is at the center of the "Corporate Information Factory" (CIF), which provides a logical framework for delivering business intelligence (BI) and business management capabilities. Gartner released a research note confirming Inmon's definition in 2005 with additional clarity plus they added one additional attribute.

The data warehouse is :

Subject-oriented : The data in the data warehouse is organized so that all the data elements relating to the same real-world event or object are linked together.

Non-volatile : Data in the data warehouse are never over-written or deleted – once committed, the data are static, read-only, and retained for future reporting.

Integrated : The data warehouse contains data from most or all of an organization's operational systems and these data are made consistent.

Time-variant : For an **operational system,** the stored data contains the current value. The data warehouse, however, contains the history of data values.

No virtualization : A data warehouse is a physical repository.

The top-down design methodology generates highly consistent dimensional views of data across data marts since all data marts are loaded from the centralized repository. Top-down design has also proven to be robust against business changes. Generating new dimensional data marts against the data stored in the data warehouse is a relatively simple task. The main disadvantage to the top-down methodology is that it represents a very large project with a very broad scope. The up-front cost for implementing a data warehouse using the top-down methodology is significant, and the duration of time from the start of project to the point that end users experience initial benefits can be substantial. In addition, the top-down methodology can be inflexible and unresponsive to changing departmental needs during the implementation phases.

Hybrid Design

Data warehouse (DW) solutions often resemble the hub and spokes architecture. Legacy systems feeding the DW/BI solution often include customer relationship management (CRM) and enterprise resource planning solutions (ERP), generating large amounts of data. To consolidate these various data models, and facilitate the extract transform load (ETL) process, DW solutions often make use of an operational data store (ODS). The information from the ODS is then parsed into the actual DW. To reduce data redundancy, larger systems will often store the data in a normalized way. Data marts for specific reports can then be built on top of the DW solution.

It is important to note that the DW database in a hybrid solution is kept on third normal form to eliminate data redundancy. A normal relational database however, is not efficient for business intelligence reports where dimensional modelling is prevalent. Small data marts can shop for data from the consolidated warehouse and use the filtered, specific data for the fact tables and dimensions required. The DW effectively provides a single source of information from which the data marts can read, creating a highly flexible solution from a BI point of view. The hybrid architecture allows a DW to be replaced with a master data management solution where operational, not static information could reside.

The Data Vault Modelling components follow hub and spokes architecture. This modelling style is a hybrid design, consisting of the best practices from both 3rd normal form and star schema. The Data Vault model is not a true 3rd normal form, and breaks some of the rules that 3NF dictates be followed. It is however, a top-down architecture with a bottom up design. The Data Vault model is geared to be strictly a data warehouse. It is not geared to be end-user accessible, which when built, still requires the use of a data mart or star schema based release area for business purposes.

DATA WAREHOUSES *VERSUS* OPERATIONAL SYSTEMS

Operational systems are optimized for preservation of data integrity and speed of recording of business transactions through use of database normalization and an entity-relationship model. Operational system designers generally follow the Codd rules of database normalization in order to ensure data integrity. Codd defined five increasingly stringent rules of normalization. Fully normalized database designs (that is, those satisfying all five Codd rules) often result in information from a business transaction being stored in dozens to hundreds of tables. Relational databases are efficient at managing the relationships between these tables. The databases have very fast insert/update performance because only a small amount of data in those tables is affected each time a transaction is processed. Finally, in order to improve performance, older data are usually periodically purged from operational systems.

Evolution in Organization Use

These terms refer to the level of sophistication of a data warehouse :

Offline operational data warehouse : Data warehouses in this stage of evolution are updated on a regular time cycle (usually daily, weekly or monthly) from the operational systems and the data is stored in an integrated reporting-oriented data.

Offline data warehouse : Data warehouses at this stage are updated from data in the operational systems on a regular basis and the data warehouse data are stored in a data structure designed to facilitate reporting.

On time data warehouse : Online Integrated Data Warehousing represent the real time Data warehouses stage data in the warehouse is updated for every transaction performed on the source data.

Integrated data warehouse : These data warehouses assemble data from different areas of business, so users can look up the information they need across other systems.

Chapter 15

Concept of Data Warehouse

DATABASE

A **database** is an organized collection of data. The data are typically organized to model relevant aspects of reality in a way that supports processes requiring this information. For example, modelling the availability of rooms in hotels in a way that supports finding a hotel with vacancies.

Database management systems (**DBMSs**) are specially designed software applications that interact with the user, other applications, and the database itself to capture and analyze data. A general-purpose DBMS is a software system designed to allow the definition, creation, querying, update, and administration of databases. Well-known DBMSs include MySQL, MariaDB, PostgreSQL, SQLite, Microsoft SQL Server, Oracle, SAP HANA, dBASE, FoxPro, IBM DB2, LibreOffice Base and FileMaker Pro. A database is not generally portable across different DBMSs, but different DBMSs can interoperate by using standards such as SQL and ODBC or JDBC to allow a single application to work with more than one database.

Terminology and Overview

Formally, "database" refers to the data themselves and supporting data structures. Databases are created to operate large quantities of information by inputting, storing, retrieving and managing that information. Databases are set up so that one set of software programs provides all users with access to all the data.

A "database management system" (DBMS) is a suite of computer software providing the interface between users and a database or databases. Because they are so closely related, the term "database" when used casually often refers to both a DBMS and the data it manipulates.

Outside the world of professional information technology, the term *database* is sometimes used casually to refer to any collection of data (perhaps a spreadsheet, maybe even a card index). This article is concerned only with databases where the size and usage requirements necessitate use of a database management system.

The interactions catered for by most existing DBMSs fall into four main groups :

- *Data definition :* Defining new data structures for a database, removing data structures from the database, modifying the structure of existing data.
- *Update :* Inserting, modifying, and deleting data.
- *Retrieval :* Obtaining information either for end-user queries and reports or for processing by applications.
- *Administration :* Registering and monitoring users, enforcing data security, monitoring performance, maintaining data integrity, dealing with concurrency control, and recovering information if the system fails.

A DBMS is responsible for maintaining the integrity and security of stored data, and for recovering information if the system fails.

Both a database and its DBMS conform to the principles of a particular database model. "Database system" refers collectively to the database model, database management system, and database.

Physically, database servers are dedicated computers that hold the actual databases and run only the DBMS and related software. Database servers are usually multi-processor computers, with generous memory and RAID disk arrays used for stable storage. RAID is used for recovery of data if any of the disks fail. Hardware database accelerators, connected to one or more servers via a high-speed channel, are also used in large volume transaction processing environments. DBMSs are found at the heart of most database applications. DBMSs may be built around a custom multi-tasking kernel with built-in networking support, but modern DBMSs typically rely on a standard operating system to provide these functions. Since DBMSs comprise a significant economical market, computer and storage vendors often take into account DBMS requirements in their own development plans.

Databases and DBMSs can be categorized according to the database model(s) that they support (such as relational or XML), the type(s) of computer they run on (from a server cluster to a mobile phone), the query language(s) used to access the database (such as SQL or XQuery), and their internal engineering, which affects performance, scalability, resilience, and security.

Applications and Roles

Most organizations in developed countries today depend on databases for their business operations. Increasingly, databases are not only used to support the internal operations of the organization, but also to underpin its online interactions with customers and suppliers. Databases are not used only to hold administrative information, but are often embedded within applications to hold more specialized data : for example engineering data or economic models. Examples of database applications include computerized library systems, flight reservation systems, and computerized parts inventory systems.

Client-server or transactional DBMSs are often complex to maintain high performance, availability and security when many users are querying and updating the database at the same time. Personal, desktop-based database systems tend to be less complex. For example, File Maker and Microsoft Access come with built-in graphical user interfaces.

General-Purpose and Special-Purpose DBMSs

A DBMS has evolved into a complex software system and its development typically requires thousands of person-years of development effort. Some general-purpose DBMSs such as Adabas, Oracle and DB2 have been undergoing upgrades since the 1970s. General-purpose DBMSs aim to meet the needs of as many applications as possible, which adds to the complexity. However, the fact that their development cost can be spread over a large number of users means that they are often the most cost-effective approach. However, a general-purpose DBMS is not always the optimal solution : in some cases a general-purpose DBMS may introduce unnecessary overhead. Therefore, there are many examples of systems that use special-purpose databases. A common example is an email system : e-mail systems are designed to optimize the handling of email messages, and do not need significant portions of a general-purpose DBMS functionality.

Many databases have application software that accesses the database on behalf of end-users, without exposing the DBMS interface directly. Application programmers may use a wire protocol directly, or more likely through an application programming interface. Database designers and database administrators interact with the DBMS through dedicated interfaces to build and maintain the applications' databases, and thus need some more knowledge and understanding about how DBMSs operate and the DBMSs' external interfaces and tuning parameters.

General-purpose databases are usually developed by one organization or community of programmers, while a different group builds the applications that use it. In many companies, specialized database administrators maintain databases, run reports, and may work on code that runs on the databases themselves (rather than in the client application).

History

Following the technology progress in the areas of processors, computer memory, computer storage and computer networks, the sizes, capabilities, and performance of databases and their respective DBMSs have grown in orders of magnitude. The development of database technology can be divided into three eras based on data model or structure : navigational, SQL/relational, and post-relational.

The two main early navigational data models were the hierarchical model, epitomized by IBM's IMS system, and the CODASYL model (network model), implemented in a number of products such as IDMS.

The relational model, first proposed in 1970 by Edgar F. Codd, departed from this tradition by insisting that applications should search for data by content, rather than by following links. The relational model employs sets of ledger-style tables, each used for a different type of entity. Only in the mid-1980s did computing hardware became powerful enough to allow the wide deployment of relational systems (DBMSs plus applications). By the early 1990s, however, relational systems dominated in all large-scale data processing applications, and as of 2014 they remain dominant except in niche areas. The dominant database language, standardised SQL for the relational model, has influenced database languages for other data models.

Object databases developed in the 1980s to overcome the inconvenience of object-relational impedance mismatch, which led to the coining of the term "post-relational" and also the development of hybrid object-relational databases.

The next generation of post-relational databases in the late 2000s became known as NoSQL databases, introducing fast key-value stores and document-oriented databases. A competing "next generation" known as NewSQL databases attempted new implementations that retained the relational/SQL model while aiming to match the high performance of NoSQL compared to commercially available relational DBMSs.

1960s, Navigational DBMS

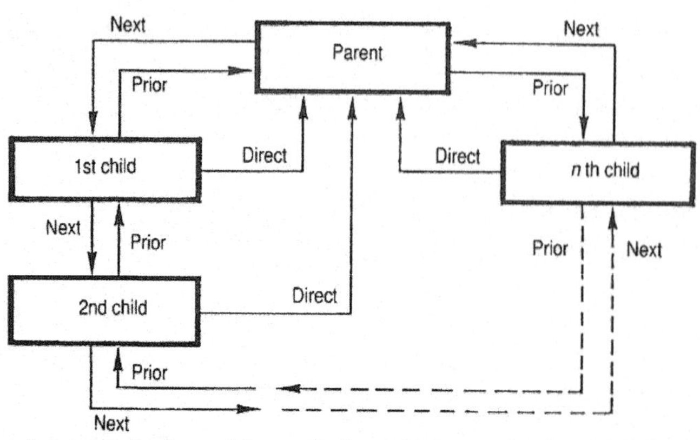

A closed chain of records in a navigational database model (e.g. CODASYL), with ***next pointers***, ***prior pointers*** and ***direct pointers*** provided by keys in the various records.

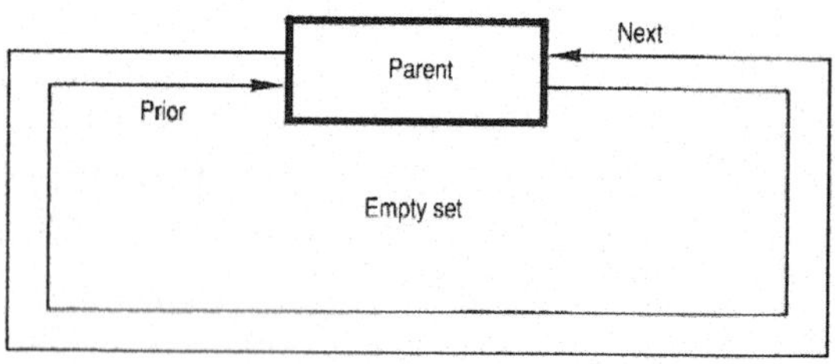

Illustration of an ***empty set***

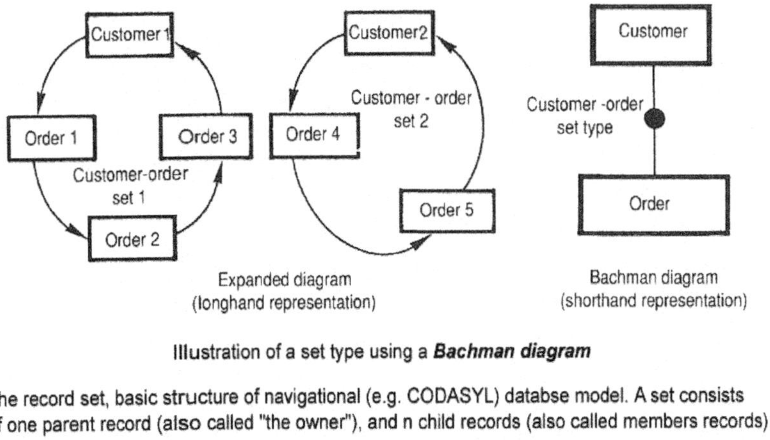

Fig. : Basic structure of navigational CODASYL database model.

Further information : Navigational database

The introduction of the term *database* coincided with the availability of direct-access storage (disks and drums) from the mid-1960s onwards. The term represented a contrast with the tape-based systems of the past, allowing shared interactive use rather than daily batch processing. The Oxford English dictionary cites a 1962 report by the System Development Corporation of California as the first to use the term "data-base" in a specific technical sense.

As computers grew in speed and capability, a number of general-purpose database systems emerged; by the mid-1960s a number of such systems had come into commercial use. Interest in a standard began to grow, and Charles Bachman, author of one such product, the Integrated Data Store (IDS), founded the "Database Task Group" within CODASYL, the group responsible for the creation and standardization of COBOL. In 1971 the Database Task Group delivered their standard, which generally became known as the "CODASYL approach", and soon a number of commercial products based on this approach entered the market.

The CODASYL approach relied on the "manual" navigation of a linked data set which was formed into a large network. Applications could find records by one of three methods :

- Use of a primary key (known as a CALC key, typically implemented by hashing)
- Navigating relationships (called sets) from one record to another
- Scanning all the records in a sequential order.

Later systems added B-Trees to provide alternate access paths. Many CODASYL databases also added a very straightforward query language. However, in the final tally, CODASYL was very complex and required significant training and effort to produce useful applications.

IBM also had their own DBMS system in 1968, known as *IMS*. IMS was a development of software written for the Apollo program on the System/360. IMS was generally similar in concept to CODASYL, but used a strict hierarchy for its model of data navigation instead of CODASYL's network model. Both concepts later became known as navigational databases due to the way data was accessed, and Bachman's 1973 Turing Award presentation was *The Programmer as Navigator*. IMS is classified as a hierarchical database. IDMS and Cincom Systems' TOTAL database are classified as network databases. IMS remains in use as of 2014.

1970s, Relational DBMS

Edgar Codd worked at IBM in San Jose, California, in one of their offshoot offices that was primarily involved in the development of hard disk systems. He was unhappy with the navigational model of the CODASYL approach, notably the lack of a "search" facility. In 1970, he wrote a number of papers that outlined a new approach to database construction that eventually culminated in the ground-breaking *A Relational Model of Data for Large Shared Data Banks*.

In this paper, he described a new system for storing and working with large databases. Instead of records being stored in some sort of linked list of free-form records as in CODASYL, Codd's idea was to use a "table" of fixed-length records, with each table used for a different type of entity. A linked-list system would be very inefficient when storing "sparse" databases where some of the data for any one record could be left empty. The relational model solved this by splitting the data into a series of normalized tables (or *relations*), with optional elements being moved out of the main table to where they would take up room only if needed. Data may be freely inserted, deleted and edited in these tables, with the DBMS doing whatever maintenance needed to present a table view to the application/user.

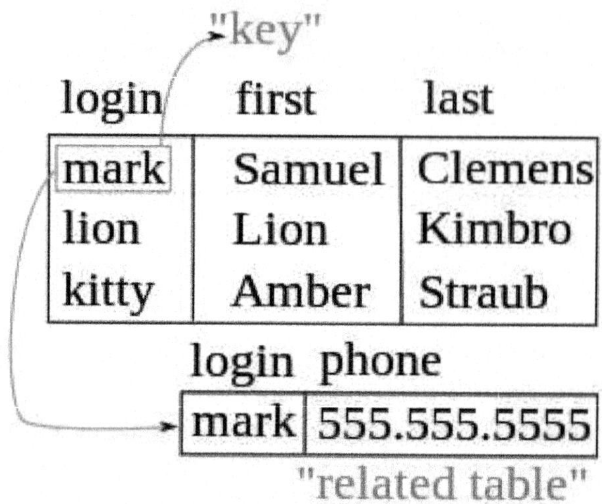

Fig.: In the relational model, related records are linked together with a "key".

The relational model also allowed the content of the database to evolve without constant rewriting of links and pointers. The relational part comes from entities referencing other entities in what is known as one-to-many relationship, like a traditional hierarchical model, and many-to-many relationship, like a navigational (network) model. Thus, a relational model can express both hierarchical and navigational models, as well as its native tabular model, allowing for pure or combined modelling in terms of these three models, as the application requires.

For instance, a common use of a database system is to track information about users, their name, login information, various addresses and phone numbers. In the navigational approach all of these data would be placed in a single record, and unused items would simply not be placed in the database. In the relational approach, the data would be *normalized* into a user table, an address table and a phone number table (for instance). Records would be created in these optional tables only if the address or phone numbers were actually provided.

Linking the information back together is the key to this system. In the relational model, some bit of information was used as a "key", uniquely defining a particular record. When information was being collected about a user, information stored in the optional tables would be found by searching for this key. For instance, if the login name of a user is unique, addresses and phone numbers for that user would be recorded with the login name as its key. This simple "re-linking" of related data back into a single collection is something that traditional computer languages are not designed for.

Just as the navigational approach would require programs to loop in order to collect records, the relational approach would require loops to collect information about any *one* record. Codd's solution to the necessary looping was a set-oriented language, a suggestion that would later spawn the ubiquitous SQL. Using a branch of mathematics known as tuple calculus, he demonstrated that such a system could support all the operations of normal databases (inserting, updating etc.) as well as providing a simple system for finding and returning *sets* of data in a single operation.

Codd's paper was picked up by two people at Berkeley, Eugene Wong and Michael Stonebraker. They started a project known as INGRES using funding that had already been allocated for a geographical database project and student programmers to produce code. Beginning in 1973, INGRES delivered its first test products which were generally ready for widespread use in 1979. INGRES was similar to System R in a number of ways, including the use of a "language" for data access, known as QUEL. Over time, INGRES moved to the emerging SQL standard.

IBM itself did one test implementation of the relational model, PRTV, and a production one, Business System 12, both now discontinued. Honeywell wrote MRDS for Multics, and now there are two new implementations : Alphora Dataphor and Rel. Most other DBMS implementations usually called *relational* are actually SQL DBMSs.

In 1970, the University of Michigan began development of the MICRO Information Management System based on D.L. Childs' Set-Theoretic Data model. Micro was used to manage very large data-sets by the US Department of Labour, the U.S. Environmental Protection Agency, and researchers from the University of Alberta, the University of Michigan, and Wayne State University. It ran on IBM mainframe computers using the Michigan Terminal System. The system remained in production until 1998.

Integrated Approach

In the 1970s and 1980s attempts were made to build database systems with integrated hardware and software. The underlying philosophy was that such integration would provide higher performance at lower cost. Examples were IBM System/38, the early offering of Teradata, and the Britton Lee, Inc. database machine.

Another approach to hardware support for database management was ICL's CAFS accelerator, a hardware disk controller with programmable search capabilities. In the long-term, these efforts were generally unsuccessful because specialized database machines could not keep pace with the rapid development and progress of general-purpose computers. Thus most database systems nowadays are software systems running on general-purpose hardware, using general-purpose computer data storage. However this idea is still pursued for certain applications by some companies like Netezza and Oracle (Exadata).

Late 1970s, SQL DBMS

IBM started working on a prototype system loosely based on Codd's concepts as *System R* in the early 1970s. The first version was ready in 1974/5, and work then started on multi-table systems in which the data could be split so that all of the data for a record (some of which is optional) did not have to be stored in a single large "chunk". Subsequent multi-user versions were tested by customers in 1978 and 1979, by which time a standardized query language-SQL-had been added. Codd's ideas were establishing themselves as both workable and superior to CODASYL, pushing IBM to develop a true production version of System R, known as *SQL/DS*, and, later, *Database 2* (DB2).

Larry Ellison's Oracle started from a different chain, based on IBM's papers on System R, and beat IBM to market when the first version was released in 1978.

Stonebraker went on to apply the lessons from INGRES to develop a new database, Postgres, which is now known as PostgreSQL. PostgreSQL is often used for global mission critical applications (the.org and.info domain name registries use it as their primary data store, as do many large companies and financial institutions).

In Sweden, Codd's paper was also read and Mimer SQL was developed from the mid-1970s at Uppsala University. In 1984, this project was consolidated into an independent enterprise. In the early 1980s, Mimer introduced transaction handling for high robustness in applications, an idea that was subsequently implemented on most other DBMSs.

Another data model, the entity-relationship model, emerged in 1976 and gained popularity for database design as it emphasized a more familiar description than the earlier relational model. Later on, entity-relationship constructs were retrofitted as a data modelling construct for the relational model, and the difference between the two have become irrelevant.

1980s, on the Desktop

The 1980s ushered in the age of desktop computing. The new computers empowered their users with spreadsheets like Lotus 1,2,3 and database software like dBASE. The dBASE product was lightweight and easy for any computer user to understand out of the box. C. Wayne Ratliff the creator of dBASE stated : "dBASE was different from programs like BASIC, C, FORTRAN, and COBOL in that a lot of the dirty work had already been done. The data manipulation is done by dBASE instead of by the user, so the user can concentrate on what he is doing, rather than having to mess with the dirty details of opening, reading, and closing files, and managing space allocation." dBASE was one of the top selling software titles in the 1980s and early 1990s.

1980s, object-oriented

The 1980s, along with a rise in object-oriented programming, saw a growth in how data in various databases were handled. Programmers and designers began to treat the data in their databases as objects. That is to say that if a person's data were in a database, that person's attributes, such as their address, phone number, and age, were now considered to belong to that person instead of being extraneous data. This allows for relations between data to be relations to objects and their attributes and not to individual fields. The term "object-relational impedance mismatch" described the inconvenience of translating between programmed objects and database tables. Object databases and object-relational databases attempt to solve this problem by providing an object-oriented language (sometimes as extensions to SQL) that programmers can use as alternative to purely relational SQL. On the programming side, libraries known as object-relational mappings (ORMs) attempt to solve the same problem.

2000s, NoSQL and NewSQL

The next generation of post-relational databases in the 2000s became known as NoSQL databases, including fast key-value stores and document-oriented databases. XML databases are a type of structured document-oriented database that allows querying based on XML document attributes. XML databases are mostly used in enterprise database management, where XML is being used as the machine-to-machine data interoperability standard. XML databases are mostly commercial software systems, that include Clusterpoint, MarkLogic and Oracle XML DB.

NoSQL databases are often very fast, do not require fixed table schemas, avoid join operations by storing denormalized data, and are designed to scale horizontally. The most popular NoSQL systems include MongoDB, Couchbase, Riak, memcached, Redis, CouchDB, Hazelcast, Apache Cassandra and HBase, which are all open-source software products.

In recent years there was a high demand for massively distributed databases with high partition tolerance but according to the CAP theorem it is impossible for a distributed system to simultaneously provide consistency, availability and partition tolerance guarantees. A distributed system can satisfy any two of these guarantees at the same time, but not all three. For that reason many NoSQL databases are using what is called eventual consistency to provide both availability and partition tolerance guarantees with a maximum level of data consistency.

NewSQL is a class of modern relational databases that aims to provide the same scalable performance of NoSQL systems for online transaction processing (read-write) workloads while still using SQL and maintaining the ACID guarantees of a traditional database system. Such databases include Clustrix, EnterpriseDB, NuoDB and VoltDB.

Research

Database technology has been an active research topic since the 1960s, both in academia and in the research and development groups of companies (for example IBM Research). Research activity includes theory and development of prototypes. Notable research topics have included models, the atomic transaction concept and related concurrency control techniques, query languages and query optimization methods, RAID, and more.

The database research area has several dedicated academic journals (for example, ACM Transactions on Database Systems—TODS, Data and Knowledge Engineering—DKE) and annual conferences (*e.g.*, ACM SIGMOD, ACM PODS, VLDB, IEEE ICDE).

Examples

One way to classify databases involves the type of their contents, for example : bibliographic, document-text, statistical, or multimedia objects. Another way is by their application area, for example, accounting, music compositions, movies, banking, manufacturing, or insurance. A third way is by some technical aspect, such as the database structure or interface type. This section lists a few of the adjectives used to characterize different kinds of databases.

- An in-memory database is a database that primarily resides in main memory, but is typically backed-up by non-volatile computer data storage. Main memory databases are faster than disk databases, and so are often used where response time is critical, such as in telecommunications network equipment. SAP HANA platform is a very hot topic for in-memory database. By May 2012, HANA was able to run on servers with 100TB main memory powered by IBM. The co founder of the company claimed that the system was big enough to run the 8 largest SAP customers.

- An active database includes an event-driven architecture which can respond to conditions both inside and outside the database. Possible uses include security monitoring, alerting, statistics gathering and authorization. Many databases provide active database features in the form of database triggers.
- A cloud database relies on cloud technology. Both the database and most of its DBMS reside remotely, "in the cloud", while its applications are both developed by programmers and later maintained and utilized by (application's) end-users through a web browser and Open APIs.
- Data warehouses archive data from operational databases and often from external sources such as market research firms. The warehouse becomes the central source of data for use by managers and other end-users who may not have access to operational data. For example, sales data might be aggregated to weekly totals and converted from internal product codes to use UPCs so that they can be compared with ACNielsen data. Some basic and essential components of data warehousing include retrieving, analyzing, and mining data, transforming, loading and managing data so as to make them available for further use.
- A deductive database combines logic programming with a relational database, for example by using the Datalog language.
- A distributed database is one in which both the data and the DBMS span multiple computers.
- A document-oriented database is designed for storing, retrieving, and managing document-oriented, or semi-structured data, information. Document-oriented databases are one of the main categories of NoSQL databases.
- An embedded database system is a DBMS which is tightly integrated with an application software that requires access to stored data in such a way that the DBMS is hidden from the application's end-users and requires little or no ongoing maintenance.
- **End-user databases** consist of data developed by individual end-users. Examples of these are collections of documents, spreadsheets, presentations, multimedia, and other files. Several products exist to support such databases. Some of them are much simpler than full fledged DBMSs, with more elementary DBMS functionality.
- A federated database system comprises several distinct databases, each with its own DBMS. It is handled as a single database by a federated database management system (FDBMS), which transparently integrates multiple autonomous DBMSs, possibly of different types (in which case it would also be a heterogeneous database system), and provides them with an integrated conceptual view.
- Sometimes the term *multi-database* is used as a synonym to federated database, though it may refer to a less integrated (*e.g.*, without an FDBMS and a managed integrated schema) group of databases that co-operate in a single application. In this case typically middleware is used for distribution, which

typically includes an atomic commit protocol (ACP), *e.g.*, the two-phase commit protocol, to allow distributed (global) transactions across the participating databases.

- A graph database is a kind of NoSQL database that uses graph structures with nodes, edges, and properties to represent and store information. General graph databases that can store any graph are distinct from specialized graph databases such as triplestores and network databases.
- In a hypertext or hypermedia database, any word or a piece of text representing an object, *e.g.*, another piece of text, an article, a picture, or a film, can be hyperlinked to that object. Hypertext databases are particularly useful for organizing large amounts of disparate information. For example, they are useful for organizing online encyclopeadias, where users can conveniently jump around the text. The World Wide Web is thus a large distributed hypertext database.
- A knowledge base (abbreviated **KB**, **kb** or Δ) is a special kind of database for knowledge management, providing the means for the computerized collection, organization, and retrieval of knowledge. Also a collection of data representing problems with their solutions and related experiences.
- A mobile database can be carried on or synchronized from a mobile computing device.
- Operational databases store detailed data about the operations of an organization. They typically process relatively high volumes of updates using transactions. Examples include customer databases that record contact, credit, and demographic information about a business' customers, personnel databases that hold information such as salary, benefits, skills data about employees, enterprise resource planning systems that record details about product components, parts inventory, and financial databases that keep track of the organization's money, accounting and financial dealings.
- A parallel database seeks to improve performance through parallelization for tasks such as loading data, building indexes and evaluating queries.

The major parallel DBMS architectures which are induced by the underlying hardware architecture are :

- **Shared memory architecture**, where multiple processors share the main memory space, as well as other data storage.
- **Shared disk architecture**, where each processing unit (typically consisting of multiple processors) has its own main memory, but all units share the other storage.
- **Shared nothing architecture**, where each processing unit has its own main memory and other storage.
- Probabilistic databases employ fuzzy logic to draw inferences from imprecise data.
- Real-time databases process transactions fast enough for the result to come back and be acted on right away.

- A spatial database can store the data with multi-dimensional features. The queries on such data include location based queries, like "Where is the closest hotel in my area?"
- A temporal database has built-in time aspects, for example a temporal data model and a temporal version of SQL. More specifically the temporal aspects usually include valid-time and transaction-time.
- A terminology-oriented database builds upon an object-oriented database, often customized for a specific field.
- An unstructured data database is intended to store in a manageable and protected way diverse objects that do not fit naturally and conveniently in common databases. It may include email messages, documents, journals, multimedia objects, etc. The name may be misleading since some objects can be highly structured. However, the entire possible object collection does not fit into a predefined structured framework. Most established DBMSs now support unstructured data in various ways, and new dedicated DBMSs are emerging.

Design and Modelling

The first task of a database designer is to produce a conceptual data model that reflects the structure of the information to be held in the database. A common approach to this is to develop an entity-relationship model, often with the aid of drawing tools. Another popular approach is the Unified Modelling Language. A successful data model will accurately reflect the possible state of the external world being modeled : for example, if people can have more than one phone number, it will allow this information to be captured. Designing a good conceptual data model requires a good understanding of the application domain; it typically involves asking deep questions about the things of interest to an organisation, like "can a customer also be a supplier?", or "if a product is sold with two different forms of packaging, are those the same product or different products?", or "if a plane flies from New York to Dubai via Frankfurt, is that one flight or two (or maybe even three)?". The answers to these questions establish definitions of the terminology used for entities (customers, products, flights, flight segments) and their relationships and attributes.

Producing the conceptual data model sometimes involves input from business processes, or the analysis of workflow in the organization. This can help to establish what information is needed in the database, and what can be left out. For example, it can help when deciding whether the database needs to hold historic data as well as current data.

Having produced a conceptual data model that users are happy with, the next stage is to translate this into a schema that implements the relevant data structures within the database. This process is often called logical database design, and the output is a logical data model expressed in the form of a schema. Whereas the conceptual data model is (in theory at least) independent of the choice of database technology, the logical data model will be expressed in terms of a particular database model supported by the chosen DBMS. (The terms *data model* and *database model* are often used interchangeably, but in this article we use *data model* for the design of a specific database, and *database model* for the modelling notation used to express that design.)

The most popular database model for general-purpose databases is the relational model, or more precisely, the relational model as represented by the SQL language. The process of creating a logical database design using this model uses a methodical approach known as normalization. The goal of normalization is to ensure that each elementary "fact" is only recorded in one place, so that insertions, updates, and deletions automatically maintain consistency.

The final stage of database design is to make the decisions that affect performance, scalability, recovery, security, and the like. This is often called *physical database design*. A key goal during this stage is data independence, meaning that the decisions made for performance optimization purposes should be invisible to end-users and applications. Physical design is driven mainly by performance requirements, and requires a good knowledge of the expected workload and access patterns, and a deep understanding of the features offered by the chosen DBMS.

Another aspect of physical database design is security. It involves both defining access control to database objects as well as defining security levels and methods for the data itself.

Models

A database model is a type of data model that determines the logical structure of a database and fundamentally determines in which manner data can be stored, organized, and manipulated. The most popular example of a database model is the relational model (or the SQL approximation of relational), which uses a table-based format.

Common logical data models for databases include :

- Hierarchical database model
- Network model
- Relational model
- Entity–relationship model
    - Enhanced entity–relationship model
- Object model
- Document model

- Entity–attribute–value model
- Star schema.

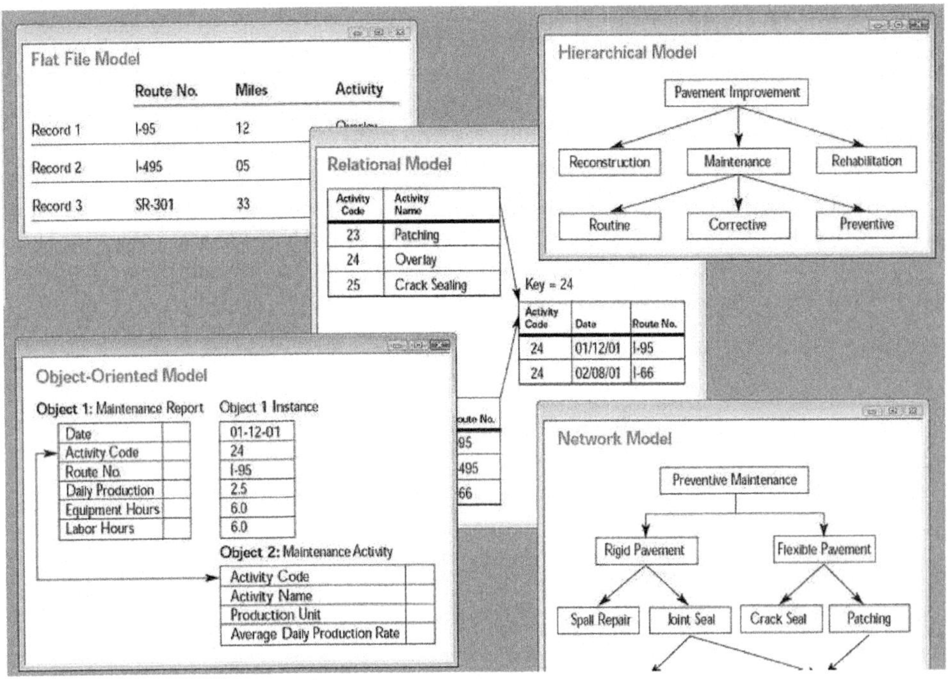

Fig. : Collage of five types of database models.

An object-relational database combines the two related structures.

Physical data models include :

- Inverted index
- Flat file.

Other models include :

- Associative model
- Multi-dimensional model
- Multi-value model
- Semantic model
- XML database
- Named graph.

External, Conceptual, and Internal Views

A database management system provides three views of the database data :

- The **external level** defines how each group of end-users sees the organization of data in the database. A single database can have any number of views at the external level.

- The **conceptual level** unifies the various external views into a compatible global view. It provides the synthesis of all the external views. It is out of the scope of the various database end-users, and is rather of interest to database application developers and database administrators.
- The **internal level** (or *physical level*) is the internal organization of data inside a DBMS. It is concerned with cost, performance, scalability and other operational matters. It deals with storage layout of the data, using storage structures such as indexes to enhance performance. Occasionally it stores data of individual views (materialized views), computed from generic data, if performance justification exists for such redundancy. It balances all the external views' performance requirements, possibly conflicting, in an attempt to optimize overall performance across all activities.

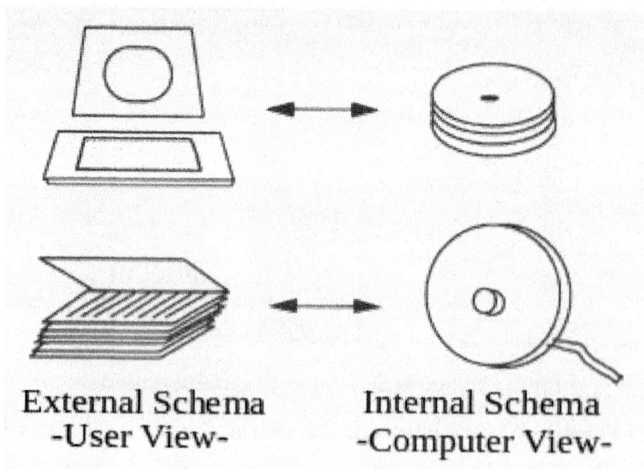

Fig. : Traditional view of data.

While there is typically only one conceptual (or logical) and physical (or internal) view of the data, there can be any number of different external views. This allows users to see database information in a more business-related way rather than from a technical, processing viewpoint. For example, a financial department of a company needs the payment details of all employees as part of the company's expenses, but does not need details about employees that are the interest of the human resources department. Thus different departments need different *views* of the company's database.

The three-level database architecture relates to the concept of *data independence* which was one of the major initial driving forces of the relational model. The idea is that changes made at a certain level do not affect the view at a higher level. For example, changes in the internal level do not affect application programs written using conceptual level interfaces, which reduces the impact of making physical changes to improve performance.

The conceptual view provides a level of indirection between internal and external. On one hand it provides a common view of the database, independent of different external view structures, and on the other hand it abstracts away details of how the data is stored or managed (internal level). In principle every level, and even every external view, can be presented by a different data model. In practice usually a given DBMS uses the same data model for both the external and the conceptual levels (*e.g.*, relational model). The internal level, which is hidden inside the DBMS and depends on its implementation, requires a different level of detail and uses its own types of data structure types.

Separating the *external*, *conceptual* and *internal* levels was a major feature of the relational database model implementations that dominate 21st century databases.

Languages

Database languages are special—purpose languages, which do one or more of the following :

- Data definition language—defines data types and the relationships among them
- Data manipulation language—performs tasks such as inserting, updating, or deleting data occurrences
- Query language—allows searching for information and computing derived information.

Database languages are specific to a particular data model. Notable examples include :

- SQL combines the roles of data definition, data manipulation, and query in a single language. It was one of the first commercial languages for the relational model, although it departs in some respects from the relational model as described by Codd (for example, the rows and columns of a table can be ordered). SQL became a standard of the American National Standards Institute (ANSI) in 1986, and of the International Organization for Standardization (ISO) in 1987. The standards have been regularly enhanced since and is supported (with varying degrees of conformance) by all mainstream commercial relational DBMSs.
- OQL is an object model language standard (from the Object Data Management Group). It has influenced the design of some of the newer query languages like JDOQL and EJB QL.
- XQuery is a standard XML query language implemented by XML database systems such as MarkLogic and eXist, by relational databases with XML capability such as Oracle and DB2, and also by in-memory XML processors such as Saxon.
- SQL/XML combines XQuery with SQL.

A database language may also incorporate features like :

- DBMS-specific Configuration and storage engine management.
- Computations to modify query results, like counting, summing, averaging, sorting, grouping, and cross-referencing.

- Constraint enforcement (*e.g.* in an automotive database, only allowing one engine type per car)
- Application programming interface version of the query language, for programmer convenience.

Performance, Security, and Availability

Because of the critical importance of database technology to the smooth running of an enterprise, database systems include complex mechanisms to deliver the required performance, security, and availability, and allow database administrators to control the use of these features.

Storage

Database storage is the container of the physical materialization of a database. It comprises the *internal* (physical) *level* in the database architecture. It also contains all the information needed (*e.g.*, metadata, "data about the data", and internal data structures) to reconstruct the *conceptual level* and *external level* from the internal level when needed. Putting data into permanent storage is generally the responsibility of the database engine a.k.a. "storage engine". Though typically accessed by a DBMS through the underlying operating system (and often utilizing the operating systems' file systems as intermediates for storage layout), storage properties and configuration setting are extremely important for the efficient operation of the DBMS, and thus are closely maintained by database administrators. A DBMS, while in operation, always has its database residing in several types of storage (*e.g.*, memory and external storage). The database data and the additional needed information, possibly in very large amounts, are coded into bits. Data typically reside in the storage in structures that look completely different from the way the data look in the conceptual and external levels, but in ways that attempt to optimize (the best possible) these levels' reconstruction when needed by users and programs, as well as for computing additional types of needed information from the data (*e.g.*, when querying the database).

Some DBMSs support specifying which character encoding was used to store data, so multiple encodings can be used in the same database.

Various low-level database storage structures are used by the storage engine to serialize the data model so it can be written to the medium of choice. Techniques such as indexing may be used to improve performance. Conventional storage is row-oriented, but there are also column-oriented and correlation databases.

Materialized Views

Often storage redundancy is employed to increase performance. A common example is storing *materialized views*, which consist of frequently needed *external views* or query results. Storing such views saves the expensive computing of them each time they are needed. The downsides of materialized views are the overhead incurred when updating them to keep them synchronized with their original updated database data, and the cost of storage redundancy.

Replication

Occasionally a database employs storage redundancy by database objects replication (with one or more copies) to increase data availability (both to improve performance of simultaneous multiple end-user accesses to a same database object, and to provide resiliency in a case of partial failure of a distributed database). Updates of a replicated object need to be synchronized across the object copies. In many cases the entire database is replicated.

Security

Database security deals with all various aspects of protecting the database content, its owners, and its users. It ranges from protection from intentional unauthorized database uses to unintentional database accesses by unauthorized entities (*e.g.*, a person or a computer program).

Database access control deals with controlling who (a person or a certain computer program) is allowed to access what information in the database. The information may comprise specific database objects (*e.g.*, record types, specific records, data structures), certain computations over certain objects (*e.g.*, query types, or specific queries), or utilizing specific access paths to the former (*e.g.*, using specific indexes or other data structures to access information). Database access controls are set by special authorized (by the database owner) personnel that uses dedicated protected security DBMS interfaces.

This may be managed directly on an individual basis, or by the assignment of individuals and privileges to groups, or (in the most elaborate models) through the assignment of individuals and groups to roles which are then granted entitlements. Data security prevents unauthorized users from viewing or updating the database. Using passwords, users are allowed access to the entire database or sub-sets of it called "sub-schemas". For example, an employee database can contain all the data about an individual employee, but one group of users may be authorized to view only payroll data, while others are allowed access to only work history and medical data. If the DBMS provides a way to interactively enter and update the database, as well as interrogate it, this capability allows for managing personal databases.

Data security in general deals with protecting specific chunks of data, both physically (*i.e.*, from corruption, or destruction, or removal; *e.g.*, see physical security), or the interpretation of them, or parts of them to meaningful information.

Change and access logging records who accessed which attributes, what was changed, and when it was changed. Logging services allow for a forensic database audit later by keeping a record of access occurrences and changes. Sometimes application-level code is used to record changes rather than leaving this to the database. Monitoring can be set up to attempt to detect security breaches.

Transactions and Concurrency

Database transactions can be used to introduce some level of fault tolerance and data integrity after recovery from a crash. A database transaction is a unit of work, typically encapsulating a number of operations over a database (*e.g.*, reading a database object, writing, acquiring lock, etc.), an abstraction supported in database and also other systems. Each transaction has well defined boundaries in terms of which program/code executions are included in that transaction (determined by the transaction's programmer via special transaction commands).

The acronym ACID describes some ideal properties of a database transaction : Atomicity, Consistency, Isolation, and Durability.

Migration

A database built with one DBMS is not portable to another DBMS (*i.e.*, the other DBMS cannot run it). However, in some situations it is desirable to move, migrate a database from one DBMS to another. The reasons are primarily economical (different DBMSs may have different total costs of ownership or TCOs), functional, and operational (different DBMSs may have different capabilities). The migration involves the database's transformation from one DBMS type to another. The transformation should maintain (if possible) the database related application (*i.e.*, all related application programs) intact. Thus, the database's conceptual and external architectural levels should be maintained in the transformation. It may be desired that also some aspects of the architecture internal level are maintained. A complex or large database migration may be a complicated and costly (one-time) project by itself, which should be factored into the decision to migrate. This in spite of the fact that tools may exist to help migration between specific DBMSs. Typically a DBMS vendor provides tools to help importing databases from other popular DBMSs.

Building, Maintaining, and Tuning

After designing a database for an application, the next stage is building the database. Typically an appropriate general-purpose DBMS can be selected to be utilized for this purpose. A DBMS provides the needed user interfaces to be utilized by database administrators to define the needed application's data structures within the DBMS's respective data model. Other user interfaces are used to select needed DBMS parameters (like security related, storage allocation parameters, etc.).

When the database is ready (all its data structures and other needed components are defined) it is typically populated with initial application's data (database initialization, which is typically a distinct project; in many cases using specialized DBMS interfaces that support bulk insertion) before making it operational. In some cases the database becomes operational while empty of application data, and data is accumulated during its operation.

After the database is created, initialised and populated it needs to be maintained. Various database parameters may need changing and the database may need to be tuned (tuning) for better performance; application's data structures may be changed or added, new related application programs may be written to add to the application's functionality, etc. Databases are often confused with spreadsheets such as Microsoft Excel (Microsoft Access is a database management system, Excel is a spreadsheet program). Both can be used to store information, however a database is more efficient and flexible at storing large amounts of data. Below is a simple comparison of spreadsheets and databases.

Spreadsheet strengths	*Spreadsheet Weaknesses*
Very simple data storage	Data integrity problems, including inaccurate, inconsistent and out of date data and formulas.
Relatively easy to use	Difficult to validate data *e.g.* an incorrect formula
Require less planning	

Database strengths	*Database Weaknesses*
Methods for keeping data up to date and consistent	Require more planning and designing
Data is of higher quality than data stored in spreadsheets	Harder to change structure once database is built
Good for storing and organizing information.	Requires more technical knowledge to administrate

Backup and Restore

Sometimes it is desired to bring a database back to a previous state (for many reasons, *e.g.*, cases when the database is found corrupted due to a software error, or if it has been updated with erroneous data). To achieve this a **backup** operation is done occasionally or continuously, where each desired database state (*i.e.*, the values of its data and their embedding in database's data structures) is kept within dedicated backup files (many techniques exist to do this effectively). When this state is needed, *i.e.*, when it is decided by a database administrator to bring the database back to this state (*e.g.*, by specifying this state by a desired point in time when the database was in this state), these files are utilized to **restore** that state.

Other

Other DBMS features might include :

- Database logs
- Graphics component for producing graphs and charts, especially in a data warehouse system
- **Query optimizer** – Performs query optimization on every query to choose for it the most efficient *query plan* (a partial order (tree) of operations) to be executed to compute the query result. May be specific to a particular storage engine.
- Tools or hooks for database design, application programming, application program maintenance, database performance analysis and monitoring, database configuration monitoring, DBMS hardware configuration (a DBMS

and related database may span computers, networks, and storage units) and related database mapping (especially for a distributed DBMS), storage allocation and database layout monitoring, storage migration, etc.

DATABASE TESTING

Databases, the collection of interconnected files on a server, storing information, may not deal with the same *type* of data, *i.e.* databases may be heterogeneous. As a result, many kinds of implementation and integration errors may occur in large database systems, which negatively affect the system's performance, reliability, consistency and security. Thus, it is important to **test** in order to obtain a database system which satisfies the ACID properties (Atomicity, Consistency, Isolation, and Durability) of a database management system.

Purposes

Database testing usually consists of a layered process, including the user interface (UI) layer, the business layer, the data access layer and the database itself. The UI layer deals with the interface design of the database, while the business layer includes databases supporting business strategies. The most critical layer is the data access layer, which deals with databases directly during the communication process. Database testing mainly takes place at this layer and involves testing strategies such as quality control and quality assurance of the product databases. Testing at these different layers is frequently used to maintain consistency of database systems, most commonly seen in the following examples :

- *Data is critical from a business point of view :* Companies such as Google or Symantec, who are associated with data storage, need to have a durable and consistent database system. If database operations such as insert, delete, and update are performed without testing the database for consistency first, the company risks a crash of the entire system.
- Some companies have different types of databases, and also different goals and missions. In order to achieve a level of functionality to meet said goals, they need to test their database system.
- The current approach of testing may not be sufficient in which developers formally test the databases. However, this approach is not sufficiently effective since database developers are likely to slow down the testing process due to communication gaps. A separate database testing team seems advisable.
- Database testing mainly deals with finding errors in the databases so as to eliminate them. This will improve the quality of the database or web-based system.
- Database testing should be distinguished from strategies to deal with other problems such as database crashes, broken insertions, deletions or updates. Here, database refactoring is an evolutionary technique that may apply.

Types of testings and processes

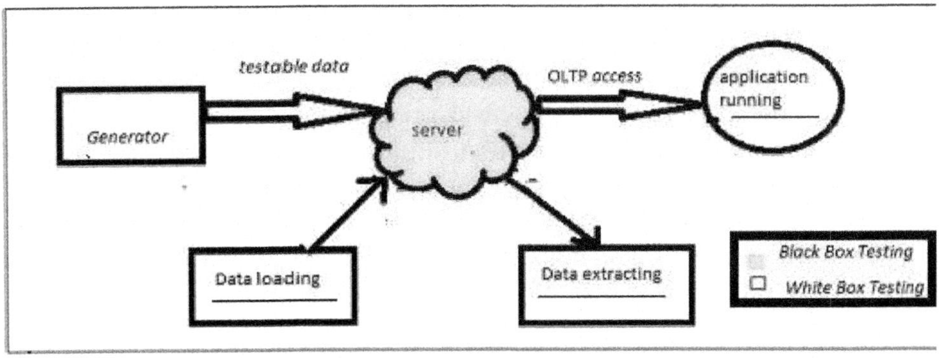

Fig. : Black box and white box testing in database test.

The figure indicates the areas of testing involved during different database testing methods, such as black-box testing and white-box testing.

Black Box Testing in Database Testing

Black box testing involves testing interfaces and the integration of the database, which includes :

1. Mapping of data (including metadata)
2. Verifying incoming data
3. Verifying outgoing data from query functions
4. Various techniques such as Cause effect graphing technique, equivalence partitioning and boundary-value analysis.

With the help of these techniques, the functionality of the database can be tested thoroughly.

Pros and Cons of black box testing include : Test case generation in black box testing is fairly simple. Their generation is completely independent of software development and can be done in an early stage of development. As a consequence, the programmer has better knowledge of how to design the database application and uses less time for debugging. Cost for development of black box test cases is lower than development of white box test cases. The major drawback of black box testing is that it is unknown how much of the program is being tested. Also, certain errors cannot be detected.

White Box Testing in Database Testing

White box testing mainly deals with the internal structure of the database. The specification details are hidden from the user.

1. It involves the testing of database triggers and logical views which are going to support database refactoring.

2. It performs module testing of database functions, triggers, views, SQL queries etc.
3. It validates database tables, data models, database schema etc.
4. It checks rules of Referential integrity.
5. It selects default table values to check on database consistency.
6. The techniques used in white box testing are condition coverage, decision coverage, statement coverage, cyclomatic complexity.

The main advantage of white box testing in database testing is that coding error are detected, so internal bugs in the database can be eliminated. The limitation of white box testing is that SQL statements are not covered.

The WHODATE Approach for Database Testing

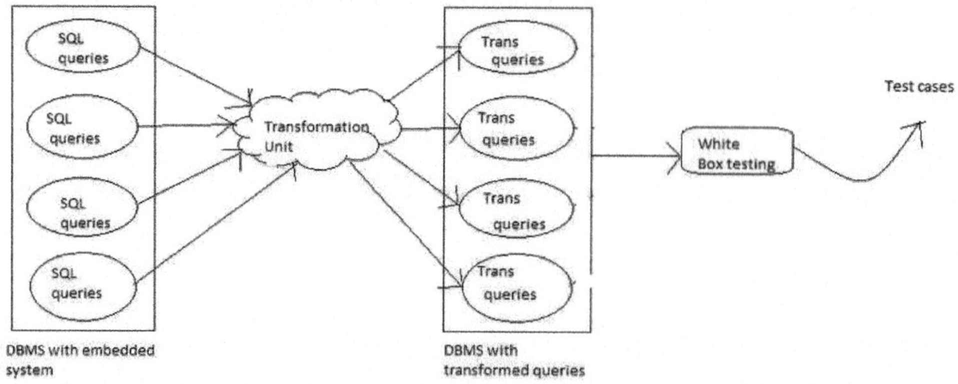

Fig. : WHODATE approach for SQL statement transformation.

While generating test cases for database testing, the semantics of SQL statement need to be reflected in the test cases. For that purpose, a technique called White box Database Application TEchnique "(WHODATE)" is used. As shown in the figure, SQL statements are independently converted into GPL statements, followed by traditional white box testing to generate test cases which include SQL semantics.

Four Stages of Database Testing

- Set Fixture
- Test run
- Outcome verification
- Tear down.

A set fixture describes the initial state of the database before entering the testing. After setting fixtures, database behaviour is tested for defined test cases. Depending on the outcome, test cases are either modified or kept as is. The "tear down" stage either results in terminating testing or continuing with other test cases.

For successful database testing the following workflow executed by each single test is commonly executed :

1. *Clean up the database :* If the testable data is already present in the database, the database needs to be emptied.
2. *Set up Fixture :* A tool like PHP Unit will then iterate over fixtures and do insertions into the database.
3. *Run test, Verify outcome and then Tear down :* After resetting the database to empty and listing the fixtures, the test is run and the output is verified. If the output is as expected, the tear down process follows, otherwise testing is repeated.

Some Problems in Database Testing

1. The setup for database testing is costly and complex to maintain because database systems are constantly changing with expected insert, delete and update operations.
2. Extra overhead is involved in order to determine the state of the database transactions.
3. After cleaning up the database, new test cases have to be designed.
4. An SQL generator is needed to transform SQL statements in order to include the SQL semantic into database test cases.

Basic Techniques

- SQL Query Analyzer is a helpful tool when using Microsoft SQL Server.
- One commonly used function, create_input_dialog ["label"], is used to validate the output with user inputs.
- The design of forms for automated database testing, form front-end and back-end, is helpful to database maintenance workers.
- *Data load testing :*
  - For data load testing, knowledge about source database and destination database is required.
  - Workers check the compatibility between source database and destination database using the DTS package.
  - When updating the source database, workers make sure to compare it with the target database.
  - Database load testing measures the capacity of the database server to handle queries as well as the response time of database server and client.
- In database testing, issues such as atomicity, consistency, isolation, durability, integrity, execution of triggers, and recovery are often considered.

DIMENSION

In a data warehouse, **Dimensions** provide structured labelling information to otherwise unordered numeric measures. The dimension is a data set composed of individual, non-overlapping data elements. The primary functions of dimensions are threefold : to provide filtering, grouping and labelling.

These functions are often described as "slice and dice". Slicing refers to filtering data. Dicing refers to grouping data. A common data warehouse example involves sales as the measure, with customer and product as dimensions. In each sale a customer buys a product. The data can be sliced by removing all customers except for a group under study, and then diced by grouping by product.

A dimensional data element is similar to a categorical variable in statistics.

Typically dimensions in a data warehouse are organized internally into one or more hierarchies. "Date" is a common dimension, with several possible hierarchies :

- "Days (are grouped into) Months (which are grouped into) Years",
- "Days (are grouped into) Weeks (which are grouped into) Years"
- "Days (are grouped into) Months (which are grouped into) Quarters (which are grouped into) Years"
- etc.

Types

Conformed Dimension

A conformed dimension is a set of data attributes that have been physically referenced in multiple database tables using the same key value to refer to the same structure, attributes, domain values, definitions and concepts. A conformed dimension cuts across many facts.

Dimensions are conformed when they are either exactly the same (including keys) or one is a perfect subset of the other. Most important, the row headers produced in two different answer sets from the same conformed dimension(s) must be able to match perfectly.

Conformed dimensions are either identical or strict mathematical sub-sets of the most granular, detailed dimension. Dimension tables are not conformed if the attributes are labelled differently or contain different values. Conformed dimensions come in several different flavours. At the most basic level, conformed dimensions mean exactly the same thing with every possible fact table to which they are joined. The date dimension table connected to the sales facts is identical to the date dimension connected to the inventory facts.

Junk Dimension

A junk dimension is a convenient grouping of typically low-cardinality flags and indicators. By creating an abstract dimension, these flags and indicators are removed from the fact table while placing them into a useful dimensional framework. A Junk Dimension is a dimension table consisting of attributes that do not belong in the fact table or in any of the existing dimension tables. The nature of these attributes is usually text or various flags, *e.g.* non-generic comments or just simple yes/no or true/false indicators. These kinds of attributes are typically remaining when all the obvious dimensions in the business process have been identified and thus the designer is faced with the challenge of where to put these attributes that do not belong in the other dimensions.

One solution is to create a new dimension for each of the remaining attributes, but due to their nature, it could be necessary to create a vast number of new dimensions resulting in a fact table with a very large number of foreign keys. The designer could also decide to leave the remaining attributes in the fact table but this could make the row length of the table unnecessarily large if, for example, the attributes is a long text string.

The solution to this challenge is to identify all the attributes and then put them into one or several Junk Dimensions. One Junk Dimension can hold several true/false or yes/no indicators that have no correlation with each other, so it would be convenient to convert the indicators into a more describing attribute. An example would be an indicator about whether a package had arrived, instead of indicating this as "yes" or "no", it would be converted into "arrived" or "pending" in the junk dimension. The designer can choose to build the dimension table so it ends up holding all the indicators occurring with every other indicator so that all combinations are covered. This sets up a fixed size for the table itself which would be 2^x rows, where x is the number of indicators. This solution is appropriate in situations where the designer would expect to encounter a lot of different combinations and where the possible combinations are limited to an acceptable level. In a situation where the number of indicators are large, thus creating a very big table or where the designer only expect to encounter a few of the possible combinations, it would be more appropriate to build each row in the junk dimension as new combinations are encountered. To limit the size of the tables, multiple junk dimensions might be appropriate in other situations depending on the correlation between various indicators.

Junk dimensions are also appropriate for placing attributes like non-generic comments from the fact table. Such attributes might consist of data from an optional comment field when a customer places an order and as a result will probably be blank in many cases. Therefore the junk dimension should contain a single row representing the blanks as a surrogate key that will be used in the fact table for every row returned with a blank comment field

Degenerate Dimension

A degenerate dimension is a key, such as a transaction number, invoice number, ticket number, or bill-of-lading number, that has no attributes and hence does not join to an actual dimension table. Degenerate dimensions are very common when the grain of a fact table represents a single transaction item or line item because the degenerate dimension represents the unique identifier of the parent. Degenerate dimensions often play an integral role in the fact table's primary key.

Role-playing Dimension

Dimensions are often recycled for multiple applications within the same database. For instance, a "Date" dimension can be used for "Date of Sale", as well as "Date of Delivery", or "Date of Hire". This is often referred to as a "role-playing dimension".

Use of ISO Representation Terms

When referencing data from a metadata registry such as ISO/IEC 11179, representation terms such as **Indicator** (a boolean true/false value), **Code** (a set of non-overlapping enumerated values) are typically used as dimensions. For example, using the National Information Exchange Model (NIEM) the data element name would be **Person Gender Code** and the enumerated values would be **male, female** and **unknown**.

Common Patterns

Date and Time

Since many fact tables in a data warehouse are time series of observations, one or more date dimensions are often needed. One of the reasons to have date dimensions is to place calendar knowledge in the data warehouse instead of hard coded in an application. While a simple SQL date/timestamp is useful for providing accurate information about the time a fact was recorded, it can not give information about holidays, fiscal periods, etc. An SQL date/timestamp can still be useful to store in the fact table, as it allows for precise calculations.

Having both the date and time of day in the same dimension, may easily result in a huge dimension with millions of rows. If a high amount of detail is needed it is usually a good idea to split date and time into two or more separate dimensions. A time dimension with a grain of seconds in a day will only have 86400 rows. A more or less detailed grain for date/time dimensions can be chosen depending on needs. As examples, date dimensions can be accurate to year, quarter, month or day and time dimensions can be accurate to hours, minutes or seconds.

As a rule of thumb, time of day dimension should only be created if hierarchical groupings are needed or if there are meaningful textual descriptions for periods of time within the day (ex. "evening rush" or "first shift").

If the rows in a fact table are coming from several timezones, it might be useful to store date and time in both local time and a standard time. This can be done by having two dimensions for each date/time dimension needed-one for local time, and one for standard time. Storing date/time in both local and standard time, will allow for analysis on when facts are created in a local setting and in a global setting as well. The standard time chosen can be a global standard time (ex. UTC), it can be the local time of the business' headquarter, or any other time zone that would make sense to use.

DIMENSIONAL MODELLING

Dimensional modelling (DM) is the name of a set of techniques and concepts used in data warehouse design. It is considered to be different from entity-relationship modelling (ER). Dimensional Modelling does not necessarily involve a relational database. The same modelling approach, at the logical level, can be used for any physical form, such as multi-dimensional database or even flat files. According to data warehousing consultant Ralph Kimball, DM is a design technique for databases intended to support end-user queries in a data warehouse. It is oriented around understandability and performance. According to him, although transaction-oriented ER is very useful for the transaction capture, it should be avoided for end-user delivery.

Dimensional modelling always uses the concepts of facts (measures), and dimensions (context). Facts are typically (but not always) numeric values that can be aggregated, and dimensions are groups of hierarchies and descriptors that define the facts. For example, sales amount is a fact; timestamp, product, register#, store#, etc. are elements of dimensions. Dimensional models are built by business process area, *e.g.* store sales, inventory, claims, etc. Because the different business process areas share some but not all dimensions, efficiency in design, operation, and consistency, is achieved using conformed dimensions, *i.e.* using one copy of the shared dimension across subject areas. The term "conformed dimensions" was originated by Ralph Kimball.

Dimensional Modelling Process

The dimensional model is built on a star-like schema, with dimensions surrounding the fact table. To build the schema, the following design model is used :

1. Choose the business process
2. Declare the grain
3. Identify the dimensions
4. Identify the fact.

Choose the Business Process

The process of dimensional modelling builds on a 4-step design method that helps to ensure the usability of the dimensional model and the use of the data warehouse. The basics in the design build on the actual business process which the data warehouse should cover. Therefore the first step in the model is to describe the business process which the model builds on. This could for instance be a sales situation in a retail store. To describe the business process, one can choose to do this in plain text or use basic Business Process Modelling Notation (BPMN) or other design guides like the Unified Modelling Language (UML).

Declare the Grain

After describing the Business Process, the next step in the design is to declare the grain of the model. The grain of the model is the exact description of what the dimensional model should be focusing on. This could for instance be "An individual line item on a customer slip from a retail store". To clarify what the grain means, you should pick the central process and describe it with one sentence. Furthermore the grain (sentence) is what you are going to build your dimensions and fact table from. You might find it necessary to go back to this step to alter the grain due to new information gained on what your model is supposed to be able to deliver.

Identify the Dimensions

The third step in the design process is to define the dimensions of the model. The dimensions must be defined within the grain from the second step of the 4-step process. Dimensions are the foundation of the fact table, and is where the data for the fact table is collected. Typically dimensions are nouns like date, store, inventory etc. These dimensions are where all the data is stored. For example, the date dimension could contain data such as year, month and weekday.

Identify the Facts

After defining the dimensions, the next step in the process is to make keys for the fact table. This step is to identify the numeric facts that will populate each fact table row. This step is closely related to the business users of the system, since this is where they get access to data stored in the data warehouse. Therefore, most of the fact table rows are numerical, additive figures such as quantity or cost per unit, etc.

Dimension Normalization

Dimensional normalization or snowflaking removes redundant attributes, which are known in the normal flatten de-normalized dimensions. Dimensions are strictly joined together in sub dimensions.

Snowflaking has an influence on the data structure that differs from many philosophies of data warehouses. Single data (fact) table surrounded by multiple descriptive (dimension) tables.

Developers often don't normalize dimensions due to several reasons :

1. Normalization makes the data structure more complex
2. Performance can be slower, due to the many joins between tables
3. The space savings are minimal
4. Bitmap indexes can't be used
5. Query Performance, 3NF databases suffer from performance problems when aggregating or retrieving many dimensional values that may require analysis. If you are only going to do operational reports then you may be able to get by with 3NF because your operational user will be looking for very fine grain data.

There are some arguments on why normalization can be useful. It can be an advantage when part of hierarchy is common to more than one dimension. For example, a geographic dimension may be reusable because both the customer and supplier dimensions use it.

Benefits of Dimensional Modelling

Benefits of the dimensional modelling are following :

- *Understandability* : Compared to the normalized model, the dimensional model is easier to understand and more intuitive. In dimensional models, information is grouped into coherent business categories or dimensions, making it easier to read and interpret. Simplicity also allows software to navigate databases efficiently. In normalized models, data is divided into many discrete entities and even a simple business process might result in dozens of tables joined together in a complex way.
- *Query performance* : Dimensional models are more denormalized and optimized for data querying, while normalized models seek to eliminate data redundancies and are optimized for transaction loading and updating. The predictable framework of a dimensional model allows the database to make strong assumptions about the data that aid in performance. Each dimension is an equivalent entry point into the fact table, and this symmetrical structure allows effective handling of complex queries. Query optimization for star join databases is simple, predictable, and controllable.
- *Extensibility* : Dimensional models are extensible and easily accommodate unexpected new data. Existing tables can be changed in place either by simply adding new data rows into the table or executing SQL alter table commands. No queries or other applications that sit on top of the Warehouse need to be reprogrammed to accommodate changes. Old queries and applications continue to run without yielding different results. But in normalized models each modification should be considered carefully, because of the complex dependencies between database tables.

ONLINE ANALYTICAL PROCESSING

In computing, **online analytical processing**, or **OLAP**, is an approach to answering multi-dimensional analytical (MDA) queries swiftly. OLAP is part of the broader category of business intelligence, which also encompasses relational database, report writing and data-mining. Typical applications of OLAP include business reporting for sales, marketing, management reporting, business process management (BPM), budgeting and forecasting, financial reporting and similar areas, with new applications coming up, such as agriculture. The term *OLAP* was created as a slight modification of the traditional database term OLTP (Online Transaction Processing).

OLAP tools enable users to analyze multi-dimensional data interactively from multiple perspectives. OLAP consists of three basic analytical operations :

consolidation (roll-up), drill-down, and slicing and dicing. Consolidation involves the aggregation of data that can be accumulated and computed in one or more dimensions. For example, all sales offices are rolled up to the sales department or sales division to anticipate sales trends. By contrast, the drill-down is a technique that allows users to navigate through the details. For instance, users can view the sales by individual products that make up a region's sales. Slicing and dicing is a feature whereby users can take out (slicing) a specific set of data of the OLAP cube and view (dicing) the slices from different viewpoints.

Databases configured for OLAP use a multi-dimensional data model, allowing for complex analytical and *ad hoc* queries with a rapid execution time. They borrow aspects of navigational databases, hierarchical databases and relational databases.

Overview of OLAP Systems

The core of any OLAP system is an OLAP cube (also called a 'multi-dimensional cube' or a hypercube). It consists of numeric facts called *measures* which are categorized by *dimensions*. The measures are placed at the intersections of the hypercube, which is spanned by the dimensions as a Vector space. The usual interface to manipulate an OLAP cube is a matrix interface like Pivot tables in a spreadsheet program, which performs projection operations along the dimensions, such as aggregation or averaging.

The cube metadata is typically created from a star schema or snowflake schema or fact constellation of tables in a relational database. Measures are derived from the records in the fact table and dimensions are derived from the dimension tables.

Each *measure* can be thought of as having a set of *labels*, or meta-data associated with it. A *dimension* is what describes these *labels*; it provides information about the *measure*.

A simple example would be a cube that contains a store's sales as a *measure*, and Date/Time as a *dimension*. Each Sale has a Date/Time *label* that describes more about that sale.

Any number of *dimensions* can be added to the structure such as Store, Cashier, or Customer by adding a foreign key column to the fact table. This allows an analyst to view the *measures* along any combination of the *dimensions*.

For example :

```
  Sales Fact Table
+-------------+----------+
| sale_amount | time_id  |
+-------------+----------+                Time Dimension
|      2008.10|     1234 |---+     +---------+-------------------+
+-------------+----------+   |     | time_id | timestamp         |
                             |     +---------+-------------------+
                             +---->|   1234  | 20080902 12:35:43 |
                                   +---------+-------------------+
```

Multi-dimensional Databases

Multi-dimensional structure is defined as "a variation of the relational model that uses multi-dimensional structures to organize data and express the relationships between data". The structure is broken into cubes and the cubes are able to store and access data within the confines of each cube. "Each cell within a multi-dimensional structure contains aggregated data related to elements along each of its dimensions". Even when data is manipulated it remains easy to access and continues to constitute a compact database format. The data still remains interrelated. Multi-dimensional structure is quite popular for analytical databases that use online analytical processing (OLAP) applications. Analytical databases use these databases because of their ability to deliver answers to complex business queries swiftly. Data can be viewed from different angles, which gives a broader perspective of a problem unlike other models.

Aggregations

It has been claimed that for complex queries OLAP cubes can produce an answer in around 0.1% of the time required for the same query on OLTP relational data. The most important mechanism in OLAP which allows it to achieve such performance is the use of *aggregations*. Aggregations are built from the fact table by changing the granularity on specific dimensions and aggregating up data along these dimensions. The number of possible aggregations is determined by every possible combination of dimension granularities.

The combination of all possible aggregations and the base data contains the answers to every query which can be answered from the data.

Because usually there are many aggregations that can be calculated, often only a predetermined number are fully calculated; the remainder are solved on demand. The problem of deciding which aggregations (views) to calculate is known as the view selection problem. View selection can be constrained by the total size of the selected set of aggregations, the time to update them from changes in the base data, or both. The objective of view selection is typically to minimize the average time to answer OLAP queries, although some studies also minimize the update time. View selection is NP-Complete. Many approaches to the problem have been explored, including greedy algorithms, randomized search, genetic algorithms and A* search algorithm.

Types

OLAP systems have been traditionally categorized using the following taxonomy.

Multi-dimensional

MOLAP is a "multi-dimensional online analytical processing". 'MOLAP' is the 'classic' form of OLAP and is sometimes referred to as just OLAP. MOLAP stores this data in an optimized multi-dimensional array storage, rather than in a relational database. Therefore it requires the pre-computation and storage of information in the cube—the operation known as processing. MOLAP tools generally utilize a pre—calculated data set referred to as a data cube. The data cube contains all the possible answers to a given range of questions. MOLAP tools have a very fast response time and the ability to quickly write back data into the data-set.

Advantages of MOLAP

- Fast query performance due to optimized storage, multi-dimensional indexing and caching.
- Smaller on-disk size of data compared to data stored in relational database due to compression techniques.
- Automated computation of higher level aggregates of the data.
- It is very compact for low dimension data-sets.
- Array models provide natural indexing.
- Effective data extraction achieved through the pre-structuring of aggregated data.

Disadvantages of MOLAP

- Within some MOLAP Solutions the processing step (data load) can be quite lengthy, especially on large data volumes. This is usually remedied by doing only incremental processing, *i.e.*, processing only the data which have changed (usually new data) instead of reprocessing the entire data-set.
- MOLAP tools traditionally have difficulty querying models with dimensions with very high cardinality (*i.e.*, millions of members).
- Some MOLAP products have difficulty updating and querying models with more than ten dimensions. This limit differs depending on the complexity and cardinality of the dimensions in question. It also depends on the number of facts or measures stored. Other MOLAP products can handle hundreds of dimensions.
- Some MOLAP methodologies introduce data redundancy.

Relational

ROLAP works directly with relational databases. The base data and the dimension tables are stored as relational tables and new tables are created to hold the aggregated information. Depends on a specialized schema design. This methodology relies on manipulating the data stored in the relational database to give the appearance of traditional OLAP's slicing and dicing functionality. In essence, each action of slicing and dicing is equivalent to adding a "WHERE" clause in the SQL statement. ROLAP tools do not use pre-calculated data cubes but instead pose the query to the standard relational database and its tables in order to bring back the data required to answer the question. ROLAP tools feature the ability to ask any question because the methodology does not limit to the contents of a cube. ROLAP also has the ability to drill down to the lowest level of detail in the database.

Hybrid

There is no clear agreement across the industry as to what constitutes "Hybrid OLAP", except that a database will divide data between relational and specialized storage. For example, for some vendors, a HOLAP database will use relational tables to hold the larger quantities of detailed data, and use specialized storage for at least some aspects of the smaller quantities of more-aggregate or less-detailed data. HOLAP addresses the shortcomings of MOLAP and ROLAP by combining the capabilities of both approaches. HOLAP tools can utilize both pre-calculated cubes and relational data sources.

Comparison

Each type has certain benefits, although there is disagreement about the specifics of the benefits between providers.

- Some MOLAP implementations are prone to database explosion, a phenomenon causing vast amounts of storage space to be used by MOLAP databases when certain common conditions are met : high number of dimensions, pre-calculated results and sparse multi-dimensional data.
- MOLAP generally delivers better performance due to specialized indexing and storage optimizations. MOLAP also needs less storage space compared to ROLAP because the specialized storage typically includes compression techniques.
- ROLAP is generally more scalable. However, large volume pre-processing is difficult to implement efficiently so it is frequently skipped. ROLAP query performance can therefore suffer tremendously.
- Since ROLAP relies more on the database to perform calculations, it has more limitations in the specialized functions it can use.
- HOLAP encompasses a range of solutions that attempt to mix the best of ROLAP and MOLAP. It can generally pre-process swiftly, scale well, and offer good function support.

Other Types

The following acronyms are also sometimes used, although they are not as widespread as the ones above :

- **WOLAP** – Web-based OLAP
- **DOLAP** – **Desktop** OLAP
- **RTOLAP** – Real-Time OLAP.

APIs and Query Languages

Unlike relational databases, which had SQL as the standard query language, and widespread APIs such as ODBC, JDBC and OLEDB, there was no such unification in the OLAP world for a long time. The first real standard API was OLE DB for

OLAP specification from Microsoft which appeared in 1997 and introduced the MDX query language. Several OLAP vendors—both server and client—adopted it. In 2001 Microsoft and Hyperion announced the XML for Analysis specification, which was endorsed by most of the OLAP vendors. Since this also used MDX as a query language, MDX became the *de facto* standard. Since September-2011 LINQ can be used to query SSAS OLAP cubes from Microsoft.NET.

Products

History

The first product that performed OLAP queries was *Express,* which was released in 1970 (and acquired by Oracle in 1995 from Information Resources). However, the term did not appear until 1993 when it was coined by Edgar F. Codd, who has been described as "the father of the relational database". Codd's paper resulted from a short consulting assignment which Codd undertook for former Arbor Software (later Hyperion Solutions, and in 2007 acquired by Oracle), as a sort of marketing coup. The company had released its own OLAP product, *Essbase,* a year earlier. As a result Codd's "twelve laws of online analytical processing" were explicit in their reference to Essbase. There was some ensuing controversy and when Computerworld learned that Codd was paid by Arbor, it retracted the article. OLAP market experienced strong growth in late 90s with dozens of commercial products going into market. In 1998, Microsoft released its first OLAP Server—Microsoft Analysis Services, which drove wide adoption of OLAP technology and moved it into mainstream.

Market Structure

Below is a list of top OLAP vendors in 2006, with figures in millions of US Dollars.

Vendor	*Global Revenue*	*Consolidated company*
Microsoft Corporation	1,806	Microsoft
Hyperion Solutions Corporation	1,077	Oracle
Cognos	735	IBM
Business Objects	416	SAP
MicroStrategy	416	Micro-Strategy
SAP AG	330	SAP
Cartesis (SAP)	210	SAP
Applix	205	IBM
Infor	199	Infor
Oracle Corporation	159	Oracle
Others	152	Others
Total	**5,700**	

STAR SCHEMA

In computing, the **Star Schema** (also called **star-join schema**) is the simplest style of data mart schema. The star schema consists of one or more fact tables referencing any number of dimension tables. The star schema is an important special case of the snowflake schema, and is more effective for handling simpler queries.

The star schema gets its name from the physical model's resemblance to a star with a fact table at its center and the dimension tables surrounding it representing the star's points.

Model

The star schema separates business process data into facts, which hold the measurable, quantitative data about a business, and dimensions which are descriptive attributes related to fact data. Examples of fact data include sales price, sale quantity, and time, distance, speed, and weight measurements. Related dimension attribute examples include product models, product colours, product sizes, geographic locations, and salesperson names.

A star schema that has many dimensions is sometimes called a *centipede schema*. Having dimensions of only a few attributes, while simpler to maintain, results in queries with many table joins and makes the star schema less easy to use.

Fact Tables

Fact tables record measurements or metrics for a specific event. Fact tables generally consist of numeric values, and foreign keys to dimensional data where descriptive information is kept. Fact tables are designed to a low level of uniform detail (referred to as "granularity" or "grain"), meaning facts can record events at a very atomic level. This can result in the accumulation of a large number of records in a fact table over time. Fact tables are defined as one of three types :

- Transaction fact tables record facts about a specific event (*e.g.*, sales events)
- Snapshot fact tables record facts at a given point in time (*e.g.*, account details at month end)
- Accumulating snapshot tables record aggregate facts at a given point in time (*e.g.*, total month-to-date sales for a product)

Fact tables are generally assigned a surrogate key to ensure each row can be uniquely identified.

Dimension Tables

Dimension tables usually have a relatively small number of records compared to fact tables, but each record may have a very large number of attributes to describe the fact data. Dimensions can define a wide variety of characteristics, but some of the most common attributes defined by dimension tables include :

- Time dimension tables describe time at the lowest level of time granularity for which events are recorded in the star schema
- Geography dimension tables describe location data, such as country, state, or city
- Product dimension tables describe products
- Employee dimension tables describe employees, such as sales people
- Range dimension tables describe ranges of time, dollar values, or other measurable quantities to simplify reporting.

Dimension tables are generally assigned a surrogate primary key, usually a single-column integer data type, mapped to the combination of dimension attributes that form the natural key.

Benefits

Star schemas are denormalized, meaning the normal rules of normalization applied to transactional relational databases are relaxed during star schema design and implementation. The benefits of star schema denormalization are :

- Simpler queries—star schema join logic is generally simpler than the join logic required to retrieve data from a highly normalized transactional schemas.
- Simplified business reporting logic—when compared to highly normalized schemas, the star schema simplifies common business reporting logic, such as period-over-period and as-of reporting.
- Query performance gains—star schemas can provide performance enhancements for read-only reporting applications when compared to highly normalized schemas.
- Fast aggregations—the simpler queries against a star schema can result in improved performance for aggregation operations.
- Feeding cubes—star schemas are used by all OLAP systems to build proprietary OLAP cubes efficiently; in fact, most major OLAP systems provide a ROLAP mode of operation which can use a star schema directly as a source without building a proprietary cube structure.

Disadvantages

The main disadvantage of the star schema is that data integrity is not enforced as well as it is in a highly normalized database. One-off inserts and updates can result in data anomalies which normalized schemas are designed to avoid. Generally speaking, star schemas are loaded in a highly controlled fashion via batch processing or near-real time "trickle feeds", to compensate for the lack of protection afforded by normalization.

Example

Consider a database of sales, perhaps from a store chain, classified by date, store and product. The image of the schema to the right is a star schema version of the sample schema provided in the snowflake schema article.

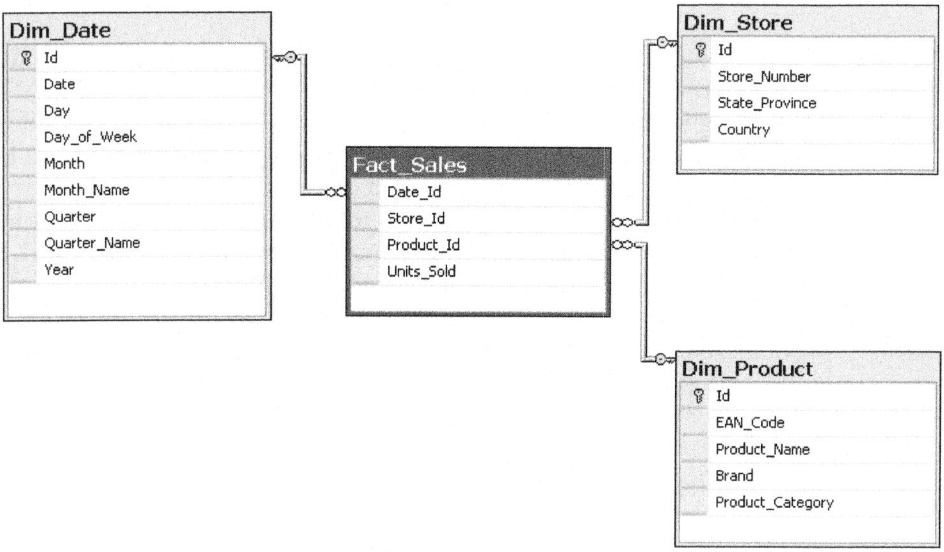

Star schema used by example query.

`Fact_Sales` is the fact table and there are three dimension tables `Dim_Date`, `Dim_Store` and `Dim_Product`.

Each dimension table has a primary key on its Id column, relating to one of the columns (viewed as rows in the example schema) of the `Fact_Sales` table's three-column (compound) primary key (`Date_Id`, `Store_Id`, `Product_Id`). The non-primary key `Units_Sold` column of the fact table in this example represents a measure or metric that can be used in calculations and analysis. The non-primary key columns of the dimension tables represent additional attributes of the dimensions (such as the Year of the `Dim_Date` dimension).

For example, the following query answers how many TV sets have been sold, for each brand and country, in 1997 :

```
SELECT
        P.Brand,
        S.Country AS Countries,
        SUM(F.Units_Sold)
FROM Fact_Sales F
INNERJOIN Dim_Date D     ON F.Date_Id = D.Id
INNERJOIN Dim_Store S    ON F.Store_Id = S.Id
INNERJOIN Dim_Product P ON F.Product_Id = P.Id
WHERE
        D.YEAR=1997
AND P.Product_Category ='tv'
GROUP BY
        P.Brand,
        S.Country
```

AGGREGATE

Aggregates are used in dimensional models of the data warehouse to produce dramatic positive effects on the time it takes to query large sets of data. At the simplest form an **aggregate** is a simple summary table that can be derived by performing a *Group by* SQL query. A more common use of aggregates is to take a dimension and change the granularity of this dimension. When changing the granularity of the dimension the fact table has to be partially summarized to fit the new grain of the new dimension, thus creating new dimensional and fact tables, fitting this new level of grain. Aggregates are sometimes referred to as pre-calculated summary data, since aggregations are usually pre-computed, partially summarized data, that are stored in new aggregated tables. When facts are aggregated, it is either done by eliminating dimensionality or by associating the facts with a rolled up dimension. Rolled up dimensions should be shrunken versions of the dimensions associated with the granular base facts. This way, the aggregated dimension tables should conform to the base dimension tables. So the reason why aggregates can make such a dramatic increase in the performance of the data warehouse is the reduction of the number of rows to be accessed when responding to a query.

Ralph Kimball, who is widely regarded as one of the original architects of data warehousing, says :

The single most dramatic way to affect performance in a large data warehouse is to provide a proper set of aggregate (summary) records that coexist with the primary base records. Aggregates can have a very significant effect on performance, in some cases speeding queries by a factor of one hundred or even one thousand. No other means exist to harvest such spectacular gains.

Having aggregates and atomic data increases the complexity of the dimensional model. This complexity should be transparent to the users of the data warehouse, thus when a request is made, the data warehouse should return data from the table with the correct grain. So when requests to the data warehouse are made, aggregate navigator functionality should be implemented, to help determine the correct table with the correct grain. The number of possible aggregations is determined by every possible combination of dimension granularities. Since it would produce a lot of overhead to build all possible aggregations, it is a good idea to choose a sub-set of tables on which to make aggregations. The best way to choose this subset and decide which aggregations to build is to monitor queries and design aggregations to match query patterns.

Aggregate Navigator

Having aggregate data in the dimensional model makes the environment more complex. To make this extra-complexity transparent to the user, functionality known as aggregate navigation is used to query the dimensional and fact tables with the correct grain level.

Implementations of aggregate navigators can be found in a range of technologies :

- OLAP engines
- Materialized views
- Relational OLAP (ROLAP) services
- BI application servers or query tools.

It is generally recommended to use either of the first three technologies, since the benefits in the latter case is restricted to a single front end BI tool.

Problems/Challenges

- Since dimensional models only gains from aggregates on large data sets, at what size of the data sets should one start considering using aggregates?
- Similarly, is a data warehouses always handling data sets that are too large for direct queries, or is it sometimes a good idea to omit the aggregate tables, when starting a new data warehouse project. Thus will, omitting aggregates in the first iteration of building a new data warehouse, make the structure of the dimensional model simpler?

Chapter 16

VARIANTS OF DATA WAREHOUSE

ANCHOR MODELLING

Anchor Modelling is an agile database modelling technique suited for information that change over time both in structure and content. It provides a graphical notation used for conceptual modelling similar to that of entity—relationship modelling, with extensions for working with temporal data. The modelling technique is based around four modelling constructs : the anchor, attribute, tie and knot, each capturing different aspects of the domain being modelled. The resulting models can be translated to physical database designs using formalized rules. When such a translation is done the tables in the relational database will mostly be in the sixth normal form.

Philosophy and History

Anchor Modelling was created in order to take advantage of the benefits from a high degree of normalization while avoiding its drawbacks. Advantages such as being able to non-destructively evolve the model, avoid null values, and keep the information free from redundancies are gained. Performance issues due to extra joins are largely avoided thanks to a feature in modern database engines called 'table elimination'. In order to handle changes in the information content Anchor Modelling emulates aspects of a temporal database in the resulting relational database schema.

The earliest installations using Anchor Modelling were made in Sweden with the first dating back to 2004, when a data warehouse for an insurance company was built using the technique. In 2007 the technique was being used in a few data warehouses and one OLTP system, and it was presented internationally by Lars Rönnbäck at the TDWI (The Data Warehousing Institute) conference in Amsterdam. This stirred enough interest for the technique to warrant a more formal description. Since then research concerning Anchor Modelling is being done in a collaboration between the creators Olle Regardt and Lars Rönnbäck and a team at the Department of Computer and Systems Sciences, Stockholm University. The first paper, in which Anchor Modelling is formalized, was presented at the 28th International Conference on Conceptual Modelling and won the best paper award.

Basic Notions

Anchor Modelling has four basic modelling concepts, anchors, attributes, ties, and knots. Anchors are used to model entities and events, attributes are used to model properties of anchors, ties model the relationships between anchors, and knots are used to model shared properties, such as states. Attributes and ties can be historized when changes in the information they model need to be kept.

An example model showing the different graphical symbols for all the concepts can be seen below. The symbols resemble those used in Entity-Relationship modelling, with a couple of extensions. A double outline on an attribute or tie indicates that a history of changes is kept and the knot symbol (an outlined square with rounded edges) is also available.

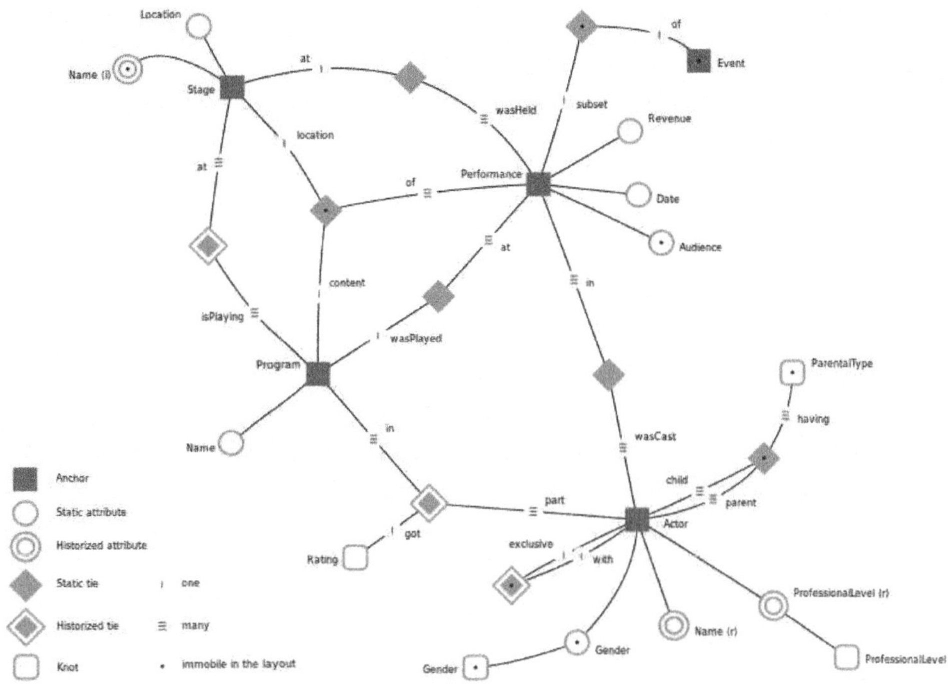

Temporal Aspects

Anchor Modelling handles two types of informational evolution, structural changes and content changes. Changes to the structure of information is represented through extensions. The high degree of normalization makes it possible to non-destructively add the necessary modelling concepts needed to capture a change, in such a way that every previous schema always remains as a sub-set of the current schema. Since the existing schema is not touched, this gives the benefit of being able to evolve the database in a highly iterative manner and without causing any downtime.

Changes in the content of information is done by emulating similar features of a temporal database in a relational database. In Anchor Modelling, pieces of information can be tied to points in time or to intervals of time (both open and closed). The time points when events occur are modelled using attributes, *e. g.* the birth dates of persons or the time of a purchase. The intervals of time in which a value is valid are captured through the historization of attributes and ties, *e. g.* the changes of hair colour of a person or the period of time during which a person was married. In a relational database this is achieved by adding a single column, with a data type granular enough to capture the speed of the changes, to the table corresponding to the historized attribute or tie. This adds a slight complexity as more than one row in the table have to be examined in order to know if an interval is closed or not.

Points or intervals of time not directly related to the domain being modelled, such as the points of time information entered the database, are handled through the use of metadata in Anchor Modelling, rather than any of the above mentioned constructs. If information about such changes to the database needs to be kept Bitemporal Anchor Modelling can be used, where in addition to updates, also delete statements become non-destructive.

Relational Representation

In Anchor Modelling there is a one-to-one mapping between the symbols used in the conceptual model and tables in the relational database. Every anchor, attribute, tie, and knot have a corresponding table in the database with an unambiguously defined structure. A conceptual model can thereby be translated to a relational database schema using simple automated rules, and *vice versa.* This is different from many other modelling techniques in which there are complex and sometimes subjective translation steps between the conceptual, logical, and physical levels.

Anchor tables contain a single column in which identities are stored. An identity is assumed to be the only property of an entity that is always present and immutable. As identities are rarely available from the domain being modelled, they are instead technically generated, *e. g.* from an incrementing number sequence.

An example of an anchor for the identities of the nephews of Donald Duck is a set of 1-tuples :

```
{<#42>, <#43>, <#44>}
```

Knots can be thought of as the combination of an anchor and a single attribute. Knot tables contain two columns, one for an identity and one for a value. Due to storing identities and values together, knots cannot be historized. Their usefulness comes from being able to reduce storage requirements and improve performance, since tables referencing knots can store a short value rather than a long string.

An example of a knot for genders is a set of 2-tuples :

```
{<#1, 'Male'>, <#2, 'Female'>}
```

Static attribute tables contain two columns, one for the identity of the entity to which the value belongs and one for the actual property value. Historized attribute tables have an extra column for storing the starting point of a time interval. In a knotted attribute table, the value column is an identity that references a knot table.

An example of a static attribute for their names is a set of 2-tuples :

```
{<#42, 'Huey'>, <#43, 'Dewey'>, <#44, 'Louie'>}
```

An example of a knotted static attribute for their genders is a set of 2-tuples :

```
{<#42, #1>, <#43, #1>, <#44, #1>}
```

An example of a historized attribute for the (changing) colours of their outfits is a set of 3-tuples :

```
{<#44, 'Orange', 1938-04-15>, <#44, 'Green', 1939-04-28>,
<#44, 'Blue', 1940-12-13>}
```

Static tie tables relate two or more anchors to each other, and contain two or more columns for storing the identities. Historized tie tables have an extra column for storing the starting point of a time interval. Knotted tie tables have an additional column for each referenced knot.

An example of a static tie for the sibling relationship is a set of 2-tuples :

```
{<#42, #43>, <#42, #44>, <#43, #42>, <#43, #44>, <#44,
#42>, <#44, #43>}
```

The resulting tables will all be in sixth normal form except for ties in which not all columns are part of the primary key.

COLUMN-ORIENTED DBMS

A **column-oriented DBMS** is a database management system (DBMS) that stores data tables as sections of columns of data rather than as rows of data. In comparison, most relational DBMSs store data in rows. This column-oriented DBMS has advantages for data warehouses, customer relationship management (CRM) systems, and library card catalogs, and other *ad hoc* inquiry systems where aggregates are computed over large numbers of similar data items.

It is possible to achieve some of the benefits of column-oriented and row-oriented organization with any DBMSs. Denoting one as column-oriented refers to both the ease of expression of a column-oriented structure and the focus on optimizations for column-oriented workloads. This approach is in contrast to **row-oriented** or **row store** databases and with correlation databases, which use a value-based storage structure.

Description

Background

A relational database management system provides data that represents a two-dimensional table, of columns and rows. For example, a database might have this table :

EmpId	*Lastname*	*Firstname*	*Salary*
10	Smith	Joe	40000
12	Jones	Mary	50000
11	Johnson	Cathy	44000
22	Jones	Bob	55000

This simple table includes an employee identifier (EmpId), name fields (Lastname and Firstname) and a salary (Salary). This two-dimensional format exists only in theory, in practice, storage hardware requires the data to be serialized into one form or another.

The most expensive operations involving hard drives are seeks. In order to improve overall performance, related data should be stored in a fashion to minimize the number of seeks. This is known as locality of reference, and the basic concept appears in a number of different contexts. Hard drives are organized into a series of blocks of a fixed size, typically enough to store several rows of the table. By organizing the data so rows fit within the blocks, and related rows are grouped together, the number of blocks that need to be read or sought is minimized.

Row-oriented systems

The common solution to the storage problem is to serialize each row of data, like this;

```
001 : 10,Smith,Joe,40000;002 : 12,Jones,Mary,50000;003: 11,
Johnson, Cathy,44000;004 :22,Jones,Bob,55000;
```

As data is inserted into the table, it is assigned an internal ID, the rowid that is used internally in the system to refer to data. In this case the records have sequential rowids independent of the user-assigned empid. In this example, the DBMS uses short integers to store rowids, in practice larger numbers, 64-bit or 128-bit, are normally used.

Row-based systems are designed to efficiently return data for an entire row, or record, in as few operations as possible. This matches the common use-case where the system is attempting to retrieve information about a particular object, say the contact information for a user in a rolodex system, or product information for an online shopping system. By storing the record's data in a single block on the disk, along with related records, the system can quickly retrieve records with a minimum of disk operations.

Row-based systems are not efficient at performing operations that apply to the entire data-set, as opposed to a specific record. For instance, in order to find all the records in the example table that have salaries between 40,000 and 50,000, the DBMS would have to seek through the entire data-set looking for matching records. While the example table shown above will likely fit in a single disk block, a table with even a few hundred rows would not, and multiple disk operations would be needed to retrieve the data and examine it.

To improve the performance of these sorts of operations, most DBMS's support the use of database indexes, which store all the values from a set of columns along with pointers back into the original rowid. An index on the salary column would look something like this :

```
001:40000;002:50000;003:44000;004:55000;
```

As they store only single pieces of data, rather than entire rows, indexes are generally much smaller than the main table stores. By scanning smaller sets of data the number of disk operations is reduced. If the index is heavily used, it can provide dramatic time savings for common operations. However, maintaining indexes adds overhead to the system, especially when new data is written to the database. In this case not only is the record stored in the main table, but any attached indexes have to be updated as well.

Database indexes on one or more columns are typically sorted by value, which makes operations like range queries fast.

There are a number of row-oriented databases that are designed to fit entirely in RAM, an in-memory database. These systems do not depend on disk operations, and have equal-time access to the entire dataset. This reduces the need for indexes, as it is required the same amount of operations to full scan the original data as a complete index for typical aggregation purposes. Such systems may be therefore simpler and smaller, but can only manage databases that will fit in memory.

Column-oriented Systems

A column-oriented database serializes all of the values of a column together, then the values of the next column, and so on. For our example table, the data would be stored in this fashion;

```
10:001,12:002,11:003,22:004;Smith:001,Jones:002,Johnson:00,
Jones:004;Joe:001,Mary:002,Cathy: 003,Bob:004;40000:001,500
00:002,44000:003,55000:004;
```

In this layout, any one of the columns more closely matches the structure of an index in a row-based system. This causes confusion about how a column-oriented store "is really just" a row-store with an index on every column. However, it is the mapping of the data that differs dramatically. In a row-oriented indexed system, the primary key is the rowid that is mapped to indexed data. In the column-oriented system primary key is the data, mapping back to rowids. This may seem subtle, but the difference can be seen in this common modification to the same store :

```
...;Smith:001,Jones:002,004,Johnson:003;...
```

As two of the records store the same value, "Jones", it is possible to store this only once in the column store, along with pointers to all of the rows that match it. For many common searches, like "find all the people with the last name Jones", the answer is retrieved in a single operation. Other operations, like counting the number of matching records or performing math over a set of data, can be greatly improved through this organization.

Whether or not a column-oriented system will be more efficient in operation depends heavily on the workload being automated. It would appear that operations that retrieve data for objects would be slower, requiring numerous disk operations to collect data from multiple columns to build up the record. However, these whole-row operations are generally rare. In the majority of cases, only a limited sub-set of data is retrieved. In a rolodex application, for instance, operations collecting the first and last names from many rows in order to build a list of contacts is far more common than operations reading the data for any single address. This is even more true for writing data into the database, especially if the data tends to be "sparse" with many optional columns. For this reason, column stores have demonstrated excellent real-world performance in spite of any theoretical disadvantages.

This is a simplification. Moreover, partitioning, indexing, caching, views, OLAP cubes, and transactional systems such as write ahead logging or multiversion concurrency control all dramatically affect the physical organization of either system. That said, online transaction processing (OLTP)-focused RDBMS systems are more row-oriented, while online analytical processing (OLAP)-focused systems are a balance of row-oriented and column-oriented.

Benefits

Comparisons between row-oriented and column-oriented data layouts are typically concerned with the efficiency of hard-disk access for a given workload, as seek time is incredibly long compared to the other delays in computers. Sometimes, reading a megabyte of sequentially stored data takes no more time than one random access. Further, because seek time is improving much more slowly than CPU power, this focus will likely continue on systems that rely on hard disks for storage. Following is a set of oversimplified observations which attempt to paint a picture of the trade-offs between column-and row-oriented organizations. Unless, of course, the application can be reasonably assured to fit most/all data into memory, in which case huge optimizations are available from in-memory database systems.

1. Column-oriented organizations are more efficient when an aggregate needs to be computed over many rows but only for a notably smaller sub-set of all columns of data, because reading that smaller sub-set of data can be faster than reading all data.
2. Column-oriented organizations are more efficient when new values of a column are supplied for all rows at once, because that column data can be

written efficiently and replace old column data without touching any other columns for the rows.

3. Row-oriented organizations are more efficient when many columns of a single row are required at the same time, and when row-size is relatively small, as the entire row can be retrieved with a single disk seek.
4. Row-oriented organizations are more efficient when writing a new row if all of the row data is supplied at the same time, as the entire row can be written with a single disk seek.

In practice, row-oriented storage layouts are well-suited for OLTP-like workloads which are more heavily loaded with interactive transactions. Column-oriented storage layouts are well-suited for OLAP-like workloads (*e.g.*, data warehouses) which typically involve a smaller number of highly complex queries over all data (possibly terabytes).

Compression

Column data is of uniform type; therefore, there are some opportunities for storage size optimizations available in column-oriented data that are not available in row-oriented data. For example, many popular modern compression schemes, such as LZW or run-length encoding, make use of the similarity of adjacent data to compress. While the same techniques may be used on row-oriented data, a typical implementation will achieve less effective results.

To improve compression, sorting rows can also help. For example, using bitmap indexes, sorting can improve compression by an order of magnitude. To maximize the compression benefits of the lexicographical order with respect to run-length encoding, it is best to use low-cardinality columns as the first sort keys. For example, given a table with columns sex, age, name, it would be best to sort first on the value sex (cardinality of two), then age (cardinality of <150), then name.

Columnar compression achieves a reduction in disk space at the expense of efficiency of retrieval. Retrieving all data from a single row is more efficient when that data is located in a single location, such as in a row-oriented architecture. Further, the greater adjacent compression achieved, the more difficult random-access may become, as data might need to be uncompressed to be read. Therefore, column-oriented architectures are sometimes enriched by additional mechanisms aimed at minimizing the need for access to compressed data.

History

Column stores or transposed files have been implemented from the early days of DBMS development. TAXIR was the first application of a column-oriented database storage system with focus on information-retrieval in biology in 1969. Statistics Canada implemented the RAPID system in 1976 and used it for processing and retrieval of the Canadian Census of Population and Housing as well as several other statistical applications. RAPID was shared with other statistical organizations throughout the world and used widely in the 1980s. It continued to be used by Statistics Canada until the 1990s.

KDB was the first commercially available column oriented database developed in 1993 followed in 1995 by Sybase IQ. However, that has changed rapidly since about 2005 with many open source and commercial implementations.

Implementations

While even a traditional row-oriented RDBMS system can achieve some benefits of column-oriented layout, specialisation of the storage layer and of the query-execution engine provide further benefits. While nothing precludes providing both row-and column-optimized capabilities in a single DBMS, typically products specialize in one of these directions.

DATA VAULT MODELLING

Data Vault Modelling is a database modelling method that is designed to provide long-term historical storage of data coming in from multiple operational systems. It is also a method of looking at historical data that, apart from the modelling aspect, deals with issues such as auditing, tracing of data, loading speed and resilience to change.

Data Vault Modelling focuses on several things. First, it emphasizes the need to trace of where all the data in the database came from. This means that every row in a Data Vault must be accompanied by record source and load date attributes, enabling an auditor to trace values back to the source.

Second, it makes no distinction between good and bad data ("bad" meaning not conforming to business rules). This is summarized in the statement that a Data Vault stores "a single version of the facts" (also expressed by Dan Linstedt as "all the data, all of the time") as opposed to the practice in other data warehouse methods of storing "a single version of the truth" where data that does not conform to the definitions is removed or "cleansed."

Third, the modelling method is designed to be resilient to change in the business environment where the data being stored is coming from, by explicitly separating structural information from descriptive attributes.

Finally, Data Vault is designed to enable parallel loading as much as possible, so that very large implementations can scale out without the need for major redesign.

History and Philosophy

In data warehouse modelling there are two well-known competing options for modelling the layer where the data is stored. Either you model according to Ralph Kimball, with conformed dimensions and an enterprise data bus, or you model according to Bill Inmon with the database normalized. Both techniques have issues when dealing with changes in the systems feeding the data warehouse. For conformed dimensions you also have to cleanse data (to conform it) and this is undesirable in a number of cases since this inevitably will lose information. Data Vault is designed to avoid or minimize the impact of those issues, by moving them to areas of the data warehouse that are outside the historical storage area (cleansing is done in the data marts) and by separating the structural items (business keys and the associations between the business keys) from the descriptive attributes.

Dan Linstedt, the creator of the method, describes the resulting database as follows :

The Data Vault is a detail oriented, historical tracking and uniquely linked set of normalized tables that support one or more functional areas of business. It is a hybrid approach encompassing the best of breed between 3rd normal form (3NF) and star schema. The design is flexible, scalable, consistent and adaptable to the needs of the enterprise

Data Vault's philosophy is that all data is relevant data, even if it is not in line with established definitions and business rules. If data is not conforming to these definitions and rules then that is a problem for the business, not the data warehouse. The determination of data being "wrong" is an interpretation of the data that stems from a particular point of view that may not be valid for everyone, or at every point in time. Therefore the Data Vault must capture all data and only when reporting or extracting data from the Data Vault is the data being interpreted.

Another issue to which Data Vault is a response is that more and more there is a need for complete auditability and traceability of all the data in the datawarehouse. Due to Sarbanes-Oxley in the USA and similar measures in Europe this is a relevant topic for many business intelligence implementations, hence the focus of any Data Vault implementation on complete traceability and auditability of all information.

Data Vault 2.0 is the new specification, but is not an open standard. The new specification contains components which define the implementation best practices, the methodology (SEI/CMMI, Six Sigma, SDLC, etc..), the Architecture, and the model. Data Vault 2.0 has a focus on including new components such as Big Data, NoSQL-and also focuses on performance of the existing model. The old specification (documented here for the most part) is highly focused on Data Vault Modelling.

It is necessary to evolve the specification to include the new components, along with the best practices in order to keep the EDW and BI systems current with the needs and desires of today's businesses.

History

Data Vault Modelling was originally conceived by Dan Linstedt in 1990 and was released in 2000 as a public domain modelling method. In a series of five articles on The Data Administration Newsletter the basic rules of the Data Vault method are expanded and explained. These contain a general overview, an overview of the components, a discussion about end dates and joins, link tables, and an article on loading practices.

An alternative (and seldom used) name for the method is "Common Foundational Integration Modelling Architecture."

Data Vault 1.0 Specification and certification (CDVDM) is now outdated. It focused highly on *just* Data Vault modelling.

Data Vault 2.0 has arrived on the scene as of 2013 and brings to the table : Big Data, NoSQL, Unstructured, Semi-Structured seamless integration. Along with methodology, architecture, and implementation best practices.

Alternative Interpretations

According to Dan Linstedt, the Data Model is inspired by (or patterned off) a simplistic view of neurons, dendrites, and synapses-where neurons are associated with Hubs and Hub Satellites, Links are dendrites (vectors of information), and other Links are synapses (vectors in the opposite direction). By utilizing a data-mining set of algorithms, links can be scored with confidence and strength ratings. They can be created and dropped on the fly in accordance with learning about relationships that currently don't exist. The model can be automatically morphed, adapted, and adjusted as it is used and fed new structures.

Another view is that a Data Vault model provides an ontology of the Enterprise in the sense that it describes the terms in the domain of the enterprise (Hubs) and the relationships among them (Links), adding descriptive attributes (Satellites) where necessary.

Another way to think of a Data Vault Model is as a graph model. The Data Vault model actually provides a "graph based" model with hubs and relationships in a relational database world. In this manner, the developer can use SQL to get at graph based relationships with sub-second responses.

Basic Notions

Data Vault attempts to solve the problem of dealing with change in the environment by separating the business keys (that do not mutate as often, because they uniquely identify a business entity) and the associations between those business keys, from the descriptive attributes of those keys.

The business keys and their associations are structural attributes, forming the skeleton of the data model. The Data Vault method has as one of its main axioms that real business keys only change when the business changes and are therefore the most stable elements from which to derive the structure of a historical database. If you use these keys as the backbone of a Data Warehouse, you can organize the rest of the data around them. This means that choosing the correct keys for the Hubs is of prime importance for the stability of your model. The keys are stored in tables with a few constraints on the structure. These key-tables are called Hubs.

Hubs

Hubs contain a list of unique business keys with low propensity to change. Hubs also contain a surrogate key for each Hub item and metadata describing the origin of the business key. The descriptive attributes for the information on the Hub (such as the description for the key, possibly in multiple languages) are stored in structures called Satellite tables which will be discussed below :

The Hub contains at least the following fields :

- A surrogate key, used to connect the other structures to this table.
- A business key, the driver for this hub. The business key can consist of multiple fields.
- The record source, can be used to see where the business keys come from and if the primary loading system has all of the keys available in other systems as well.
- Optionally, you can also have metadata fields with information about manual updates (user/time) and the extraction date.

A Hub is not allowed to contain multiple business keys, except when two systems deliver the same business key but with collisions that have different meanings.

Hubs should normally have at least one satellite.

Hub Example

This is an example for a Hub-table containing Cars, surprisingly called "Car" (H_CAR). The driving key is Vehicle Identification Number.

Fieldname	*Description*	*Mandatory?*	*Comment*
H_CAR_ID	Sequence ID and surrogate key for the Hub	No	Recommended but optional
VEHICLE_ID_NR	The business key that drives this hub. Can be more than one field for a composite business key	Yes	
H_RSRC	The record source of this key when first loaded	Yes	
LOAD_AUDIT_ID	An ID into a table with audit information, such as load time, duration of load, number of lines, etc.	No	

Links

Associations or transactions between business keys (relating for instance the Hubs for Customer and Product with each other through the Purchase transaction) are modelled using Link tables. These tables are basically many-to-many join tables, with some metadata.

Links can link to other Links, to deal with changes in granularity (for instance, adding a new key to a database table would change the grain of the database table). For instance, if you have an association between Customer and Address, you could add a reference to a Link between the Hubs for Product and Transport Company. This could be a Link called Delivery. Referencing a Link in another Link is considered a bad practice, because it introduces dependencies between Links that make parallel loading more difficult. Since a Link to another Link is the same as a new Link with the Hubs from the other Link, in these cases a new Link is the preferred solution.

Links sometimes link Hubs to information that is not by itself enough to construct a Hub. This occurs when one of the business keys associated by the Link is not a real business key. As an example, take an order form with "order number" as key, and order lines that are keyed with a semi-random number to make them unique. Let's say, "unique number". The latter key is not a real business key, so it is no Hub. However, we do need to use it in order to guarantee the correct granularity for the Link. In this case, we do not use a Hub with surrogate key, but add the business key "unique number" itself to the Link. This is done only when there is no possibility of ever using the business key for another Link or as key for attributes in a Satellite. This construct has been called a 'peg-legged Link' by Dan Linstedt on his (now defunct) forum.

Links contain the surrogate keys for the Hubs that are linked, their own surrogate key for the Link and metadata describing the origin of the association. The descriptive attributes for the information on the association (such as the time, price or amount) are stored in structures called Satellite tables which are discussed below.

Link Example

This is an example for a Link-table between two Hubs for Cars (H_CAR) and Persons (H_PERSON). The Link is called "Driver" (L_DRIVER).

Fieldname	*Description*	*Mandatory?*	*Comment*
L_DRIVER_ID	Sequence ID and surrogate key for the Link	No	Recommended but optional
H_CAR_ID	Surrogate key for the Car Hub, the first anchor of the link	Yes	
H_PERSON_ID	Surrogate key for the Person Hub, the second anchor of the link	Yes	
L_RSRC	The record-source of this association when first loaded	Yes	
LOAD_AUDIT_ID	An ID into a table with audit information, such as load time, duration of load, number of lines, etc.	No	

Satellites

The Hubs and Links form the structure of the model, but have no temporal attributes and hold no descriptive attributes. These are stored in separate tables called Satellites. These consist of metadata linking them to their parent Hub or Link, metadata describing the origin of the association and attributes, as well as a timeline with start and end dates for the attribute. Where the Hubs and Links provide the structure of the model, the Satellites provide the "meat" of the model, the context for the business processes that are captured in Hubs and Links. These attributes are stored both with regards to the details of the matter as well as the timeline and can range from quite complex (all of the fields describing a clients complete profile) to quite simple (a satellite on a Link with only a Valid-indicator and a timeline).

Usually the attributes are grouped in Satellites by source system. However, descriptive attributes such as size, cost, speed, amount or colour can change at different rates, so you can also split these attributes up in different Satellites based on their rate of change.

All the tables contain metadata, minimally describing at least the source system and the date on which this entry became valid, giving a complete historical view of the data as it enters the data warehouse.

Satellite Example

This is an example for a satellite on the Drivers-link between the Hubs for Cars and Persons, called Driver Insurance. This Satellite contains attributes that are specific to the insurance of the relationship between the Car and the Person driving it, for instance an indicator whether this is the primary driver, the name of the insurance company for this car and person (could also be a separate Hub) and a summary of the number of accidents involving this combination of vehicle and driver. Also included is a reference to a lookup-or reference table called R_RISK_CATEGORY containing the codes for the risk category in which this relationship is deemed to fall.

Fieldname	*Description*	*Mandatory?*	*Comment*
S_DRIVER_INSURANCE_ID	Sequence ID and surrogate key for the Satellite on the Link	No	Recommended but optional
L_DRIVER_ID	(Surrogate) primary key for the Driver Link, the parent of the Satellite	Yes	
S_SEQ_NR	Ordering or sequence number, to enforce uniqueness if there are several valid satellites for one parent key	No(**)	This can happen if, for instance, you have a hub COURSE and the name of the course is an attribute but in several different languages.

(Contd...)

(Contd...)

Fieldname	*Description*	*Mandatory?*	*Comment*
S_LDTS	Load Date (start-date) for the validity of this combination of attribute values for parent key L_DRIVER_ID	Yes	
S_LEDTS	Load End Date (end-date) for the validity of this combination of attribute values for parent key L_DRIVER_ID	Yes	
IND_PRIMARY_DRIVER	Indicator whether the driver is the primary driver for this car	No (*)	
INSURANCE_COMPANY	The name of the insurance company for this vehicle and this driver	No (*)	
NR_OF_ACCIDENTS	The number of accidents by this driver in this vehicle	No (*)	
R_RISK_CATEGORY_CD	The risk category for the driver. This is a reference to R_RISK_CATEGORY	No (*)	
S_RSRC	The record-source of the information in this satellite when first loaded	Yes	
LOAD_AUDIT_ID	An ID into a table with audit information, such as load time, duration of load, number of lines, etc.	No	

(*) at least one attribute is mandatory. (**) sequence number becomes mandatory if it is needed to enforce uniqueness for multiple valid satellites on the same hub or link.

Reference Tables

Reference tables are a normal part of a healthy Data Vault model. They are there to prevent redundant storage of simple reference data that is referenced a lot. More formally, Dan Linstedt defines reference data as follows :

Any information deemed necessary to resolve descriptions from codes, or to translate keys in to (sic) a consistent manner. Many of these fields are "descriptive" in nature and ***describe*** *a specific state of the other more important information. As such, reference data lives in separate tables from the raw Data Vault tables.*

Reference tables are referenced from Satellites, but never bound with physical foreign keys. There is no prescribed structure for reference tables : use what works best in your specific case, ranging from simple lookup tables to small Data Vaults or even Stars. They can be historical or have no history, but it is recommended that you stick to the natural keys and not create surrogate keys in that case. Normally, Data Vaults have a lot of reference tables, just like any other Data Warehouse.

Reference Example

This is an example of a reference table with risk categories for drivers of vehicles. It can be referenced from any Satellite in the Data Vault. For now we reference it from Satellite S_DRIVER_INSURANCE. The reference table is R_RISK_CATEGORY.

Fieldname	*Description*	*Mandatory?*
R_RISK_CATEGORY_CD	The code for the risk category	Yes
RISK_CATEGORY_DESC	A description of the risk category	No (*)

(*) at least one attribute is mandatory.

Loading Practices

The ETL for updating a Data Vault model is fairly straightforward. First you have to load all the Hubs, creating surrogate ID's for any new business keys. Having done that, you can now resolve all business keys to surrogate ID's if you query the Hub. The second step is to resolve the Links between Hubs and create surrogate ID's for any new associations. At the same time, you can also create all satellites that are attached to Hubs, since you can resolve the key to a surrogate ID. Once you have created all the new Links with their surrogate keys, you can add the Satellites to all the Links.

Since the Hubs are not joined to each other except through Links, you can load all the Hubs in parallel. Since Links are not attached directly to each other, you can load all the Links in parallel as well. Since Satellites can be attached only to Hubs and Links, you can also load these in parallel.

The ETL is quite straightforward and lends itself to easy automation or templating. Problems occur only with Links relating to other Links, because resolving the business keys in the Link only leads to another Link that has to be resolved as well. Due to the equivalence of this situation with a Link to multiple Hubs, this difficulty can be avoided by remodelling such cases and this is in fact the recommended practice.

Data is never deleted from the Data Vault, unless you have a technical error while loading data.

Data Vault and Dimensional Modelling

The Data Vault modelled layer is normally used to store data. It is not optimized for query performance, nor is it easy to query by the well-known query-tools such as Cognos, SAP Business Objects, Pentaho *et. al.* Since these end-user computing tools expect or prefer their data to be contained in a dimensional model, a conversion is usually necessary.

For this purpose, the Hubs and related Satellites on those Hubs can be considered as Dimensions and the Links and related Satellites on those Links can be viewed as Fact tables in a dimensional model. This enables you to quickly prototype a dimensional model out of a Data Vault model using views. For performance reasons the dimensional model will usually be implemented in relational tables, after approval.

Note that while it is relatively straightforward to move data from a Data Vault model to a (cleansed) dimensional model, the reverse is not as easy.

Data Vault Methodology

The Data Vault methodology is based on SEI/CMMI Level 5 best practices. It includes multiple components of CMMI Level 5, and combines them with best practices from Six Sigma, TQM, and SDLC. Particularly, it is focused on Scott Ambler's Agile Methodology for build out and deployment. Data Vault projects have a short, scope controlled release cycle and should consist of a production release every 2 to 3 weeks.

Teams using the Data Vault methodology will automatically adopt to the repeatable, consistent, and measurable projects that are expected at CMMI Level 5. Data that flows through the EDW Data Vault system will begin to follow the TQM (total quality management) life-cycle that has long been missing from BI (business intelligence) projects.

HOLAP

HOLAP (**H**ybrid **O**nline **A**nalytical **P**rocessing) is a combination of ROLAP (Relational OLAP) and MOLAP (Multi-dimensional OLAP) which are other possible implementations of OLAP. HOLAP allows storing part of the data in a MOLAP store and another part of the data in a ROLAP store, allowing a tradeoff of the advantages of each. The degree of control that the cube designer has over this partitioning varies from product to product.

Vertical Partitioning

In this mode HOLAP stores *aggregations* in MOLAP for fast query performance, and detailed data in ROLAP to optimize time of cube *processing*.

Horizontal Partitioning

In this mode HOLAP stores some slice of data, usually the more recent one (*i.e.* sliced by Time dimension) in MOLAP for fast query performance, and older data in ROLAP. Moreover, we can store some dices in MOLAP and others in ROLAP, leveraging the fact that in a large cuboid, there will be dense and sparse sub-regions.

Products

The first product to provide HOLAP storage was Holos, but the technology also became available in other commercial products such as Microsoft Analysis Services, Oracle Database OLAP Option, Micro-Strategy and SAP AG BI Accelerator. The hybrid OLAP approach combines ROLAP and MOLAP technology, benefiting from the greater scalability of ROLAP and the faster computation of MOLAP. For example, a HOLAP server may allow large volumes of detail data to be stored in a relational database, while aggregations are kept in a separate MOLAP store. The Microsoft SQL Server 7.0 OLAP Services supports a hybrid OLAP server

MOLAP

MOLAP stands for Multi-dimensional Online Analytical Processing.

MOLAP is an alternative to the ROLAP (Relational OLAP) technology. While both ROLAP and MOLAP analytic tools are designed to allow analysis of data through the use of a multi-dimensional data model, MOLAP differs significantly in that (in some software) it requires the pre-computation and storage of information in the cube — the operation known as *processing*. Most MOLAP solutions store these data in an optimized multi-dimensional array storage, rather than in a relational database (*i.e.* in ROLAP).

There are many methodologies and algorithms for efficient data storage, aggregation and *implementation specific* business logic with a MOLAP Solution. As a result there are many misconceptions to what the term specifically implies.

MOLAP *vs.* ROLAP

Advantages of MOLAP

- Fast query performance due to optimized storage, multi-dimensional indexing and caching.
- Smaller on-disk size of data compared to data stored in relational database due to compression techniques.
- Automated computation of higher level aggregates of the data.
- It is very compact for low dimension data-sets.
- Array models provide natural indexing.
- Effective data extraction achieved through the pre-structuring of aggregated data.

Disadvantages of MOLAP

- Within some MOLAP Solutions the processing step (data load) can be quite lengthy, especially on large data volumes. This is usually remedied by doing only incremental processing, *i.e.*, processing only the data which have changed (usually new data) instead of reprocessing the entire data-set.

- MOLAP tools traditionally have difficulty querying models with dimensions with very high cardinality (*i.e.*, millions of members).
- Some MOLAP products have difficulty updating and querying models with more than ten dimensions. This limit differs depending on the complexity and cardinality of the dimensions in question. It also depends on the number of facts or measures stored. Other MOLAP products can handle hundreds of dimensions.
- Some MOLAP methodologies introduce data redundancy.

Trends

Many commercial OLAP tools now use a "Hybrid OLAP" (HOLAP) approach, which allows the model designer to decide which portion of the data will be stored in MOLAP.

Products

Examples of commercial products that use MOLAP are Cognos Powerplay, Oracle Database OLAP Option, Micro-Strategy, Microsoft Analysis Services, Essbase, TM1, Jedox, Board Toolkit, Lilith Hicare and Daptech Keystone. There is also an open source MOLAP server Palo.

Not all MOLAP are generic databases. Some commercial products focus on a specific business function or type, such as Oracle© RPAS (Retail Predictive Application Server).

ROLAP

ROLAP stands for **R**elational **O**nline **A**nalytical **P**rocessing.

ROLAP is an alternative to the MOLAP (Multi-dimensional OLAP) technology. While both ROLAP and MOLAP analytic tools are designed to allow analysis of data through the use of a multi-dimensional data model, ROLAP differs significantly in that it does not require the pre-computation and storage of information. Instead, ROLAP tools access the data in a relational database and generate SQL queries to calculate information at the appropriate level when an end user requests it. With ROLAP, it is possible to create additional database tables (*summary tables* or *aggregations*) which summarize the data at any desired combination of dimensions.

While ROLAP uses a relational database source, generally the database must be carefully designed for ROLAP use. A database which was designed for OLTP will not function well as a ROLAP database. Therefore, ROLAP still involves creating an additional copy of the data. However, since it is a database, a variety of technologies can be used to populate the database.

ROLAP *vs.* MOLAP

The discussion of the advantages and disadvantages of ROLAP below, focus on those things that are true of the most widely used ROLAP and MOLAP tools available today. In some cases there will be tools which are exceptions to any generalization made.

Advantages of ROLAP

- ROLAP is considered to be more scalable in handling large data volumes, especially models with dimensions with very high cardinality (*i.e.*, millions of members).
- With a variety of data loading tools available, and the ability to fine tune the ETL code to the particular data model, load times are generally much shorter than with the automated MOLAP loads.
- The data are stored in a standard relational database and can be accessed by any SQL reporting tool (the tool does not have to be an OLAP tool).
- ROLAP tools are better at handling *non-aggregatable facts* (*e.g.*, textual descriptions). MOLAP tools tend to suffer from slow performance when querying these elements.
- By decoupling the data storage from the multi-dimensional model, it is possible to successfully model data that would not otherwise fit into a strict dimensional model.
- The ROLAP approach can leverage database authorization controls such as row-level security, whereby the query results are filtered depending on preset criteria applied, for example, to a given user or group of users (SQL WHERE clause).

Disadvantages of ROLAP

- There is a consensus in the industry that ROLAP tools have slower performance than MOLAP tools. However.
- The loading of *aggregate tables* must be managed by custom ETL code. The ROLAP tools do not help with this task. This means additional development time and more code to support.
- When the step of creating aggregate tables is skipped, the query performance then suffers because the larger detailed tables must be queried. This can be partially remedied by adding additional aggregate tables, however it is still not practical to create aggregate tables for all combinations of dimensions/ attributes.
- ROLAP relies on the general purpose database for querying and caching, and therefore several special techniques employed by MOLAP tools are not available (such as special hierarchical indexing). However, modern ROLAP tools take advantage of latest improvements in SQL language such as CUBE and ROLLUP operators, DB2 Cube Views, as well as other SQL OLAP extensions. These SQL improvements can mitigate the benefits of the MOLAP tools.
- Since ROLAP tools rely on SQL for all of the computations, they are not suitable when the model is heavy on calculations which don't translate well

into SQL. Examples of such models include budgeting, allocations, financial reporting and other scenarios.

Performance of ROLAP

OLAP Survey

In the OLAP industry ROLAP is usually perceived as being able to scale for large data volumes, but suffering from slower query performance as opposed to MOLAP. The OLAP Survey, the largest independent survey across all major OLAP products, being conducted for 6 years have consistently found that companies using ROLAP report slower performance than those using MOLAP even when data volumes were taken into consideration.

However, as with any survey there are a number of subtle issues that must be taken into account when interpreting the results.

- The survey shows that ROLAP tools have 7 times more users than MOLAP tools within each company. Systems with more users will tend to suffer more performance problems at peak usage times.
- There is also a question about complexity of the model, measured both in number of dimensions and richness of calculations. The survey does not offer a good way to control for these variations in the data being analyzed.

Downside of Flexibility

Some companies select ROLAP because they intend to re-use existing relational database tables — these tables will frequently not be optimally designed for OLAP use. The superior flexibility of ROLAP tools allows this less than optimal design to work, but performance suffers. MOLAP tools in contrast would force the data to be re-loaded into an optimal OLAP design.

Trends

The undesirable trade-off between additional ETL cost and slow query performance has ensured that most commercial OLAP tools now use a "Hybrid OLAP" (HOLAP) approach, which allows the model designer to decide which portion of the data will be stored in MOLAP and which portion in ROLAP.

Products

Examples of commercial products using ROLAP include Microsoft Analysis Services, Micro-Strategy, SAP Business Objects, Oracle Business Intelligence Suite Enterprise Edition (the former Siebel Analytics) and Tableau Software. There is also an open source ROLAP server, Mondrian.

Chapter 17

ELEMENTS OF DATA WAREHOUSE

DATA DICTIONARY

A **data dictionary**, or metadata repository, as defined in the *IBM Dictionary of Computing*, is a "centralized repository of information about data such as meaning, relationships to other data, origin, usage, and format." The term may have one of several closely related meanings pertaining to databases and database management systems (DBMS) :

- A document describing a database or collection of databases
- An integral component of a DBMS that is required to determine its structure
- A piece of middleware that extends or supplants the native data dictionary of a DBMS.

Documentation

The term **data dictionary** and **data repository** are used to indicate a more general software utility than a catalogue. A **catalogue** is closely coupled with the DBMS software. It provides the information stored in it to the user and the DBA, but it is mainly accessed by the various software modules of the DBMS itself, such as DDL and DML compilers, the query optimiser, the transaction processor, report generators, and the constraint enforcer. On the other hand, a **data dictionary** is a data structure that stores metadata, *i.e.*, (structured) data about data. The software package for a stand-alone data dictionary or data repository may interact with the software modules of the DBMS, but it is mainly used by the designers, users and administrators of a computer system for information resource management. These systems are used to maintain information on system hardware and software configuration, documentation, application and users as well as other information relevant to system administration.

If a data dictionary system is used only by the designers, users, and administrators and not by the DBMS Software, it is called a **passive data dictionary.**

Otherwise, it is called an **active data dictionary** or **data dictionary.** When a passive data dictionary is updated, it is done so manually and independently from any changes to a DBMS (database) structure. With an active data dictionary, the dictionary is updated first and changes occur in the DBMS automatically as a result.

Database users and application developers can benefit from an authoritative data dictionary document that catalogs the organization, contents, and conventions of one or more databases. This typically includes the names and descriptions of various tables (records or Entities) and their contents (fields) plus additional details, like the type and length of each data element. Another important piece of information that a data dictionary can provide is the relationship between Tables. This is sometimes referred to in Entity-Relationship diagrams, or if using Set descriptors, identifying in which Sets database Tables participate.

In an active data dictionary constraints may be placed upon the underlying data. For instance, a Range may be imposed on the value of numeric data in a data element (field), or a Record in a Table may be FORCED to participate in a set relationship with another Record-Type. Additionally, a distributed DBMS may have certain location specifics described within its active data dictionary (*e.g.* where Tables are physically located).

The data dictionary consists of record types (tables) created in the database by systems generated command files, tailored for each supported back-end DBMS. Command files contain SQL Statements for CREATE TABLE, CREATE UNIQUE INDEX, ALTER TABLE (for referential integrity), etc., using the specific statement required by that type of database.

There is no universal standard as to the level of detail in such a document.

Middleware

In the construction of database applications, it can be useful to introduce an additional layer of data dictionary software, *i.e.* middleware, which communicates with the underlying DBMS data dictionary. Such a "high-level" data dictionary may offer additional features and a degree of flexibility that goes beyond the limitations of the native "low-level" data dictionary, whose primary purpose is to support the basic functions of the DBMS, not the requirements of a typical application. For example, a high-level data dictionary can provide alternative entity-relationship models tailored to suit different applications that share a common database. Extensions to the data dictionary also can assist in query optimization against distributed databases. Additionally, DBA functions are often automated using restructuring tools that are tightly coupled to an active data dictionary.

Software frameworks aimed at rapid application development sometimes include high-level data dictionary facilities, which can substantially reduce the amount of programming required to build menus, forms, reports, and other components of a database application, including the database itself. For example, PHPLens includes a PHP class library to automate the creation of tables, indexes, and foreign key constraints portably for multiple databases. Another PHP-based data dictionary, part of the RADICORE toolkit, automatically generates program objects, scripts, and SQL code for menus and forms with data validation and complex joins. For the ASP.NET environment, Base One's data dictionary provides cross-DBMS facilities for automated database creation, data validation, performance enhancement (caching and index utilization), application security, and extended data types. Visual DataFlex features provides the ability to use Data Dictionaries as class files to form middle layer between the user interface and the underlying database. The intent is to create standardized rules to maintain data integrity and enforce business rules throughout one or more related applications.

Platform-specific Examples

Data description specifications (DDS) allow the developer to describe data attributes in file descriptions that are external to the application program that processes the data, in the context of an IBM System i.

METADATA

Metadata is "data about data". The term is ambiguous, as it is used for two fundamentally different concepts (types). **Structural metadata** is about the design and specification of data structures and is more properly called "data about the containers of data"; **descriptive metadata**, on the other hand, is about individual instances of application data, the data content.

Metadata are traditionally found in the card catalogs of libraries. As information has become increasingly digital, metadata are also used to describe digital data using metadata standards specific to a particular discipline. By describing the contents and context of data files, the quality of the original data/files is greatly increased. For example, a webpage may include metadata specifying what language it is written in, what tools were used to create it, and where to go for more on the subject, allowing browsers to automatically improve the experience of users.

The main purpose of metadata is to facilitate in the discovery of relevant information, more often classified as resource discovery. Metadata also helps organize electronic resources, provide digital identification, and helps support archiving and preservation of the resource. Metadata assists in resource discovery by "allowing resources to be found by relevant criteria, identifying resources, bringing similar resources together, distinguishing dissimilar resources, and giving location information."

Definition

Metadata (metacontent) are defined as the data providing information about one or more aspects of the data, such as :

- Means of creation of the data
- Purpose of the data
- Time and date of creation

- Creator or author of the data
- Location on a computer network where the data were created
- Standards used.

For example, a digital image may include metadata that describe how large the picture is, the colour depth, the image resolution, when the image was created, and other data. A text document's metadata may contain information about how long the document is, who the author is, when the document was written, and a short summary of the document.

Metadata are data. As such, metadata can be stored and managed in a database, often called a metadata registry or metadata repository. However, without context and a point of reference, it might be impossible to identify metadata just by looking at them. For example, by itself, a database containing several numbers, all 13 digits long could be the results of calculations or a list of numbers to plug into an equation—without any other context, the numbers themselves can be perceived as the data. But if given the context that this database is a log of a book collection, those 13-digit numbers may now be identified as ISBNs-information that refers to the book, but is not itself the information within the book.

The term "metadata" was coined in 1968 by Philip Bagley, in his book "Extension of programming language concepts" where it is clear that he uses the term in the ISO 11179 "traditional" sense, which is "structural metadata" *i.e.* "data about the containers of data"; rather than the alternate sense "content about individual instances of data content" or metacontent, the type of data usually found in library catalogues. Since then the fields of information management, information science, information technology, librarianship and GIS have widely adopted the term. In these fields the word *metadata* is defined as "data about data". While this is the generally accepted definition, various disciplines have adopted their own more specific explanation and uses of the term.

Libraries

Metadata have been used in various forms as a means of cataloging archived information. The Dewey Decimal System employed by libraries for the classification of library materials is an early example of metadata usage. Library catalogues used 3x5 inch cards to display a book's title, author, subject matter, and a brief plot synopsis along with an abbreviated alpha-numeric identification system which indicated the physical location of the book within the library's shelves. Such data help classify, aggregate, identify, and locate a particular book. Another form of older metadata collection is the use by US Census Bureau of what is known as the "Long Form." The Long Form asks questions that are used to create demographic data to find patterns of distribution.

Photographs

Metadata may be written into a digital photo file that will identify who owns it, copyright and contact information, what camera created the file, along with exposure information and descriptive information such as keywords about the photo, making the file searchable on the computer and/or the Internet. Some metadata are written by the camera and some is input by the photographer and/or software after downloading to a computer. However, not all digital cameras enable you to edit metadata; this functionality has been available on most Nikon DSLRs since the Nikon D3 and on most new Canon cameras since the Canon EOS 7D.

Photographic Metadata Standards are governed by organizations that develop the following standards. They include, but are not limited to :

- IPTC Information Interchange Model IIM (International Press Telecommunications Council),
- IPTC Core Schema for XMP
- XMP—Extensible Metadata Platform (an ISO standard)
- Exif—Exchangeable image file format, Maintained by CIPA (Camera & Imaging Products Association) and published by JEITA (Japan Electronics and Information Technology Industries Association)
- Dublin Core (Dublin Core Metadata Initiative—DCMI)
- PLUS (Picture Licensing Universal System).

Video

Metadata are particularly useful in video, where information about its contents (such as transcripts of conversations and text descriptions of its scenes) are not directly understandable by a computer, but where efficient search is desirable.

Web Pages

Web pages often include metadata in the form of meta tags. Description and keywords meta tags are commonly used to describe the Web page's content. Most search engines use these data when adding pages to their search index.

Creation of Metadata

Metadata can be created either by automated information processing or by manual work. Elementary metadata captured by computers can include information about when an object was created, who created it, when it was last updated, file size and file extension.

For the purposes of this article, an "object" refers to any of the following :

- A physical item such as a book, CD, DVD, map, chair, table, flower pot, etc.
- An electronic file such as a digital image, digital photo, document, program file, database table, etc.

Metadata Types

The metadata application is manyfold covering a large variety of fields of application there are nothing but specialised and well accepted models to specify types of metadata. Bretheron & Singley distinguish between two distinct classes: structural/control metadata and guide metadata. *Structural metadata* are used to describe the structure of database objects such as tables, columns, keys and indexes. *Guide metadata* are used to help humans find specific items and are usually expressed as a set of keywords in a natural language. According to Ralph Kimball metadata can be divided into 2 similar categories : technical metadata and business metadata. *Technical metadata* correspond to internal metadata, and *business metadata* correspond to external metadata. Kimball adds a third category named *process metadata*. On the other hand, NISO distinguishes among three types of metadata : descriptive, structural and administrative.

Descriptive metadata are typically used for discovery and identification, as information used to search and locate an object such as title, author, subjects, keywords, publisher. *Structural metadata* give a description of how the components of an object are organized. An example of structural metadata would be how pages are ordered to form chapters of a book. Finally, *administrative metadata* give information to help manage the source. They refer to the technical information including file type or when and how the file was created. Two sub-types of administrative metadata are rights management metadata and preservation metadata. *Rights management metadata* explain intellectual property rights, while *preservation metadata* contain information that is needed to preserve and save a resource.

Metadata Structures

Metadata (metacontent), or more correctly, the vocabularies used to assemble metadata (metacontent) statements, are typically structured according to a standardized concept using a well-defined metadata scheme, including : metadata standards and metadata models. Tools such as controlled vocabularies, taxonomies, thesauri, data dictionaries and metadata registries can be used to apply further standardization to the metadata. Structural metadata commonality is also of paramount importance in data model development and in database design.

Metadata Syntax

Metadata (metacontent) syntax refers to the rules created to structure the fields or elements of metadata (metacontent). A single metadata scheme may be expressed in a number of different markup or programming languages, each of which requires a different syntax. For example, Dublin Core may be expressed in plain text, HTML, XML and RDF.

A common example of (guide) metacontent is the bibliographic classification, the subject, the Dewey Decimal class number. There is always an implied statement in any "classification" of some object. To classify an object as, for example, Dewey class number 514 (Topology) (*i.e.* books having the number 514 on their spine) the implied statement is : "<book><subject heading><514>. This is a subject-predicate-object triple, or more importantly, a class-attribute-value triple. The first two elements of the triple (class, attribute) are pieces of some structural metadata having a defined semantic. The third element is a value, preferably from some controlled vocabulary, some reference (master) data. The combination of the metadata and master data elements results in a statement which is a metacontent statement *i.e.* "metacontent = metadata + master data". All these elements can be thought of as "vocabulary". Both metadata and master data are vocabularies which can be assembled into metacontent statements. There are many sources of these vocabularies, both meta and master data : UML, EDIFACT, XSD, Dewey/UDC/LoC, SKOS, ISO-25964, Pantone, Linnaean Binomial Nomenclature etc. Using controlled vocabularies for the components of metacontent statements, whether for indexing or finding, is endorsed by ISO-25964 : "If both the indexer and the searcher are guided to choose the same term for the same concept, then relevant documents will be retrieved." This is particularly relevant when considering the behemoth of the internet, Google. It simply indexes pages then matches text strings using its complex algorithm, there is no intelligence or "inferencing" occurring. Just the illusion thereof.

Hierarchical, Linear and Planar Schemata

Metadata schema can be hierarchical in nature where relationships exist between metadata elements and elements are nested so that parent-child relationships exist between the elements. An example of a hierarchical metadata schema is the IEEE LOM schema where metadata elements may belong to a parent metadata element. Metadata schema can also be one-dimensional, or linear, where each element is completely discrete from other elements and classified according to one dimension only. An example of a linear metadata schema is Dublin Core schema which is one dimensional. Metadata schema are often two dimensional, or planar, where each element is completely discrete from other elements but classified according to two orthogonal dimensions.

Metadata Hypermapping

In all cases where the metadata schemata exceed the planar depiction, some type of hypermapping is required to enable display and view of metadata according to chosen aspect and to serve special views. Hypermapping frequently applies to layering of geographical and geological information overlays.

Granularity

The degree to which the data or metadata are structured is referred to as their granularity. Metadata with a high granularity allow for deeper structured information and enable greater levels of technical manipulation however, a lower level of granularity means that metadata can be created for considerably lower costs but will not provide as detailed information. The major impact of granularity is not only on creation and capture, but moreover on maintenance. As soon as the metadata structures get outdated, the access to the referred data will get outdated. Hence granularity shall take into account the effort to create as well as the effort to maintain.

Metadata Standards

International standards apply to metadata. Much work is being accomplished in the national and international standards communities, especially ANSI (American National Standards Institute) and ISO (International Organization for Standardization) to reach consensus on standardizing metadata and registries.

The core standard is ISO/IEC 11179-1 : 2004 and subsequent standards. All yet published registrations according to this standard cover just the definition of metadata and do not serve the structuring of metadata storage or retrieval neither any administrative standardisation. It is important to note that this standard refers to metadata as the data about containers of the data and not to metadata (metacontent) as the data about the data contents. It should also be noted that this standard describes itself originally as a "data element" registry, describing disembodied data elements, and explicitly disavows the capability of containing complex structures. Thus the original term "data element" is more applicable than the later applied buzzword "metadata".

The Dublin Core metadata terms are a set of vocabulary terms which can be used to describe resources for the purposes of discovery. The original set of 15 classic metadata terms, known as the Dublin Core Metadata Element Set are endorsed in the following standards documents :

- IETF RFC 5013
- ISO Standard 15836-2009
- NISO Standard Z39.85.

Although not a standard, Microformat (also mentioned in the section metadata on the internet below) is a web-based approach to semantic markup which seeks to re-use existing HTML/XHTML tags to convey metadata. Microformat follows XHTML and HTML standards but is not a standard in itself. One advocate of microformats, Tantek Çelik, characterized a problem with alternative approaches :

"Here's a new language we want you to learn, and now you need to output these additional files on your server. It's a hassle. (Microformats) lower the barrier to entry."

Metadata Usage

Data Virtualization

Data virtualization has emerged as the new software technology to complete the virtualization stack in the enterprise. Metadata are used in data virtualization servers which are enterprise infrastructure components, alongside database and application servers. Metadata in these servers are saved as persistent repository and describe business objects in various enterprise systems and applications. Structural metadata commonality is also important to support data virtualization.

Statistics and Census Services

Standardization work has had a large impact on efforts to build metadata systems in the statistical community. Several metadata standards are described, and their importance to statistical agencies is discussed. Applications of the standards at the Census Bureau, Environmental Protection Agency, Bureau of Labour Statistics, Statistics Canada, and many others are described. Emphasis is on the impact a metadata registry can have in a statistical agency.

Library and Information Science

Libraries employ metadata in library catalogues, most commonly as part of an Integrated Library Management System. Metadata are obtained by cataloguing resources such as books, periodicals, DVDs, web pages or digital images. These data are stored in the integrated library management system, ILMS, using the MARC metadata standard. The purpose is to direct patrons to the physical or electronic location of items or areas they seek as well as to provide a description of the item/s in question.

More recent and specialized instances of library metadata include the establishment of digital libraries including e-print repositories and digital image libraries. While often based on library principles, the focus on non-librarian use, especially in providing metadata, means they do not follow traditional or common cataloging approaches. Given the custom nature of included materials, metadata fields are often specially created *e.g.* taxonomic classification fields, location fields, keywords or copyright statement. Standard file information such as file size and format are usually automatically included.

Standardization for library operation has been a key topic in international standardization (ISO) for decades. Standards for metadata in digital libraries include Dublin Core, METS, MODS, DDI, ISO standard Digital Object Identifier (DOI), ISO standard Uniform Resource Name (URN), PREMIS schema, Ecological Metadata Language, and OAI-PMH. Leading libraries in the world give hints on their metadata standards strategies.

Metadata and the Law

United States

Problems involving metadata in litigation in the United States are becoming widespread. Courts have looked at various questions involving metadata, including the discoverability of metadata by parties. Although the Federal Rules of Civil Procedure have only specified rules about electronic documents, subsequent case law has elaborated on the requirement of parties to reveal metadata. In October 2009, the Arizona Supreme Court has ruled that metadata records are public record.

Document metadata have proven particularly important in legal environments in which litigation has requested metadata, which can include sensitive information detrimental to a party in court.

Using metadata removal tools to "clean" documents can mitigate the risks of unwittingly sending sensitive data. This process partially protects law firms from potentially damaging leaking of sensitive data through electronic discovery.

Metadata in Healthcare

Australian researches in medicine started a lot of metadata definition for applications in health care. That approach offers the first recognized attempt to adhere to international standards in medical sciences instead of defining a proprietary standard under the WHO umbrella first.

The medical community yet did not approve the need to follow metadata standards despite respective research.

Metadata and Data Warehousing

Data warehouse (DW) is a repository of an organization's electronically stored data. Data warehouses are designed to manage and store the data whereas the business intelligence (BI) focuses on the usage of the data to facilitate reporting and analysis.

The purpose of a data warehouse is to house standardized, structured, consistent, integrated, correct, cleansed and timely data, extracted from various operational systems in an organization. The extracted data are integrated in the data warehouse environment in order to provide an enterprise wide perspective, one version of the truth. Data are structured in a way to specifically address the reporting and analytic requirements. The design of structural metadata commonality using a data modelling method such as entity relationship model diagramming is very important in any data warehouse development effort.

An essential component of a data warehouse/business intelligence system is the metadata and tools to manage and retrieve the metadata. Ralph Kimball describes metadata as the DNA of the data warehouse as metadata defines the elements of the data warehouse and how they work together.

Kimball *et. al.* refers to three main categories of metadata : Technical metadata, business metadata and process metadata. Technical metadata are primarily definitional, while business metadata and process metadata are primarily descriptive. Keep in mind that the categories sometimes overlap.

- **Technical metadata** define the objects and processes in a DW/BI system, as seen from a technical point of view. The technical metadata include the system metadata which define the data structures such as : tables, fields, data types, indexes and partitions in the relational engine, and databases, dimensions, measures, and data-mining models. Technical metadata define the data model and the way it is displayed for the users, with the reports, schedules, distribution lists and user security rights.
- **Business metadata** are a content from the data warehouse described in more user-friendly terms. The business metadata tell you what data you have, where

they come from, what they mean and what their relationship is to other data in the data warehouse. Business metadata may also serve as a documentation for the DW/BI system. Users who browse the data warehouse are primarily viewing the business metadata.

- **Process metadata** are used to describe the results of various operations in the data warehouse. Within the ETL process, all key data from tasks are logged on execution. This includes start time, end time, CPU seconds used, disk reads, disk writes and rows processed. When troubleshooting the ETL or query process, this sort of data becomes valuable. Process metadata are the fact measurement when building and using a DW/BI system. Some organizations make a living out of collecting and selling this sort of data to companies – in that case the process metadata becomes the business metadata for the fact and dimension tables. Collecting process metadata is in the interest of business people who can use the data to identify the users of their products, which products they are using and what level of service they are receiving.

Metadata on the Internet

The HTML format used to define web pages allows for the inclusion of a variety of types of metadata, from basic descriptive text, dates and keywords to further advanced metadata schemes such as the Dublin Core, e-GMS, and AGLS standards. Pages can also be geotagged with co-ordinates. Metadata may be included in the page's header or in a separate file. Micro-formats allow metadata to be added to on-page data in a way that users do not see, but computers can readily access.

Interestingly, many search engines are cautious about using metadata in their ranking algorithms due to exploitation of metadata and the practice of search engine optimization, SEO, to improve rankings. Studies show that search engines respond to web pages with metadata implementations.

Metadata in the Broadcast Industry

In broadcast industry, metadata are linked to audio and video Broadcast media to :

- *Identify the media :* Clip or playlist names, duration, timecode, etc.
- *Describe the content :* Notes regarding the quality of video content, rating, description (for example, during a sport event, keywords like *goal, red card* will be associated to some clips)
- *Classify media :* Metadata allow to sort the media or to easily and quickly find a video content (a TV news could urgently need some archive content for a subject). For example, the BBC have a large subject classification system, Lonclass, a customized version of the more general-purpose Universal Decimal Classification.

These metadata can be linked to the video media thanks to the video servers. All latest broadcasted sport events like FIFA World Cup or Olympic Games use these metadata to distribute their video content to TV stations through keywords. It's often the host broadcaster who is in charge of organizing metadata through its

International Broadcast Centre and its video servers. Those metadata are recorded with the images and are entered by metadata operators (*loggers*) who associate in live metadata available in *metadata grids* through software (such as Multicam(LSM) or IPDirector used during FIFA World Cup or Olympic Games).

Geospatial Metadata

Metadata that describe geographic objects (such as datasets, maps, features, or simply documents with a geospatial component) have a history dating back to at least 1994 (refer MIT Library page on FGDC Metadata). This class of metadata is described more fully on the Geospatial metadata page.

Ecological and Environmental Metadata

Ecological and environmental metadata are intended to document the who, what, when, where, why, and how of data collection for a particular study. Metadata should be generated in a format commonly used by the most relevant science community, such as Darwin Core, Ecological Metadata Language, or Dublin Core. Metadata editing tools exist to facilitate metadata generation (*e.g.* Metavist, Mercury : Metadata Search System, Morpho). Metadata should describe provenance of the data (where they originated, as well as any transformations the data underwent) and how to give credit for (cite) the data products.

Digital Music

Metadata is "information about information" and it is one of the really useful features of digital audio files. When audio went from analogue to digital, it became possible to label or encode audio files with more information than could be contained in just the file name. That descriptive information is called "metadata."

Metadata can be used to name, describe, catalogue and indicate ownership or copyright for a digital audio file, and its presence makes it much easier to locate a specific audio file within a group—through use of a search engine that accesses the metadata. As different digital audio formats were developed, it was agreed that a standardized and specific location would be set aside within the digital files where this information could be stored.

As a result, almost all digital audio formats, including mp3, broadcast wav and AIFF files, have similar standardized locations that can be populated with metadata.

CDs such as recordings of music will carry a layer of metadata about the recordings such as dates, artist, genre, copyright owner, etc. The metadata, not normally displayed by CD players, can be accessed and displayed by specialized music playback and/or editing applications.

The metadata for compressed and uncompressed digital music is often encoded in the ID3 tag. Common editors such as TagLib support MP3, Ogg Vorbis, FLAC, MPC, Speex, WavPack TrueAudio, WAV, AIFF, MP4 and ASF file formats.

Cloud Applications

With the availability of Cloud applications, which include those to add metadata to content, metadata is increasingly available over the Internet.

Metadata Administration and Management

Metadata Storage

Metadata can be stored either *internally*, in the same file or structure as the data (this is also called *embedded metadata*), or *externally*, in a separate file or field from the described data. A data repository typically stores the metadata *detached* from the data, but can be designed to support embedded metadata approaches. Each option has advantages and disadvantages :

- Internal storage means metadata always travel as part of the data they describe; thus, metadata are always available with the data, and can be manipulated locally. This method creates redundancy (precluding normalization), and does not allow managing all of a system's metadata in one place. It arguably increases consistency, since the metadata is readily changed whenever the data is changed.
- External storage allows collocating metadata for all the contents, for example in a database, for more efficient searching and management. Redundancy can be avoided by normalizing the metadata's organization. In this approach, metadata can be united with the content when information is transferred, for example in Streaming media; or can be referenced (for example, as a web link) from the transferred content. On the down side, the division of the metadata from the data content, especially in standalone files that refer to their source metadata elsewhere, increases the opportunity for misalignments between the two, as changes to either may not be reflected in the other.

Metadata can be stored in either human-readable or binary form. Storing metadata in a human-readable format such as XML can be useful because users can understand and edit it without specialized tools. On the other hand, these formats are rarely optimized for storage capacity, communication time, and processing speed. A binary metadata format enables efficiency in all these respects, but requires special libraries to convert the binary information into human-readable content.

Metadata Management

Metadata management is the end-to-end process and governance framework for creating, controlling, enhancing, attributing, defining and managing a metadata schema, model or other structured aggregation, either independently or within a repository and the associated supporting processes (often to enable the management of content). The world Wide Web Consortium (W3C) has identified Governance as a key challenge in the advancement of third generation Web Technologies

(Web 3.0, Semantic Web), and a number of research prototypes, such as S3DB, explore the use of semantic modelling to identify practical solutions.

Database Management

Each relational database system has its own mechanisms for storing metadata. Examples of relational-database metadata include :

- Tables of all tables in a database, their names, sizes and number of rows in each table.
- Tables of columns in each database, what tables they are used in, and the type of data stored in each column.

In database terminology, this set of metadata is referred to as the catalog. The SQL standard specifies a uniform means to access the catalog, called the information schema, but not all databases implement it, even if they implement other aspects of the SQL standard. For an example of database-specific metadata access methods. Programmatic access to metadata is possible using APIs such as JDBC, or SchemaCrawler.

DATA MART

A **data mart** is the access layer of the data warehouse environment that is used to get data out to the users. The data mart is a sub-set of the data warehouse that is usually oriented to a specific business line or team. Data marts are small slices of the data warehouse. Whereas data warehouses have an enterprise-wide depth, the information in data marts pertains to a single department. In some deployments, each department or business unit is considered the *owner* of its data mart including all the *hardware*, *software* and *data*. This enables each department to use, manipulate and develop their data any way they see fit; without altering information inside other data marts or the data warehouse. In other deployments where conformed dimensions are used, this business unit ownership will not hold true for shared dimensions like customer, product, etc.

The related term spreadmart is a derogatory label describing the situation that occurs when one or more business analysts develop a system of linked spreadsheets to perform a business analysis, then grow it to a size and degree of complexity that makes it nearly impossible to maintain.

The primary use for a data mart is business intelligence (BI) applications. BI is used to gather, store, access and analyze data. The data mart can be used by smaller businesses to utilize the data they have accumulated. A data mart can be less expensive than implementing a data warehouse, thus making it more practical for the small business. A data mart can also be set up in much less time than a data warehouse, being able to be set up in less than 90 days. Since most small businesses only have use for a small number of BI applications, the low cost and quick set up of the data mart makes it a suitable method for storing data.

Design Schemas

- Star schema-fairly popular design choice; enables a relational database to emulate the analytical functionality of a multidimensional database.
- Snowflake schema.

Reasons for Creating a Data Mart

- Easy access to frequently needed data
- Creates collective view by a group of users
- Improves end-user response time
- Ease of creation
- Lower cost than implementing a full data warehouse
- Potential users are more clearly defined than in a full data warehouse
- Contains only business essential data and is less cluttered.

Dependent Data Mart

According to the Inmon school of data warehousing, a **dependent data mart** is a logical sub-set (view) or a physical sub-set (extract) of a larger data warehouse, isolated for one of the following reasons :

- *A need refreshment for a special data model or schema : E.g.,* to restructure for OLAP
- *Performance :* To offload the data mart to a separate computer for greater efficiency or to obviate the need to manage that workload on the centralized data warehouse.
- *Security :* To separate an authorized data sub-set selectively
- *Expediency :* To bypass the data governance and authorizations required to incorporate a new application on the Enterprise Data Warehouse
- *Proving Ground :* To demonstrate the viability and ROI (return on investment) potential of an application prior to migrating it to the Enterprise Data Warehouse
- *Politics :* A coping strategy for IT (Information Technology) in situations where a user group has more influence than funding or is not a good citizen on the centralized data warehouse.
- *Politics :* A coping strategy for consumers of data in situations where a data warehouse team is unable to create a usable data warehouse.

According to the Inmon school of data warehousing, tradeoffs inherent with data marts include limited scalability, duplication of data, data inconsistency with other silos of information, and inability to leverage enterprise sources of data.

The alternative school of data warehousing is that of Ralph Kimball. In his view, a data warehouse is nothing more than the union of all the data marts. This view helps to reduce costs and provides fast development, but can create an inconsistent data warehouse, especially in large organizations. Therefore, Kimball's approach is more suitable for small-to-medium corporations.

SIXTH NORMAL FORM

Sixth normal form (6NF) is a term in relational database theory, used in two different ways.

6NF (C. Date's Definition)

A book by Christopher J. Date and others on temporal databases, defined sixth normal form as a normal form for databases based on an extension of the relational algebra.

In this work, the relational operators, such as *join*, are generalized to support a natural treatment of interval data, such as sequences of dates or moments in time. Sixth normal form is then based on this generalized join, as follows :

A relvar R [table] is in **sixth normal form** if and only if it satisfies no nontrivial join dependencies at all—where, as before, a join dependency is trivial if and only if at least one of the projections (possibly U_projections) involved is taken over the set of all attributes of the relvar [table] concerned.[Date *et. al.*]

Any relation in 6NF is also in 5NF : Sixth normal form is intended to decompose relation variables to irreducible components. Though this may be relatively unimportant for non-temporal relation variables, it can be important when dealing with temporal variables or other interval data. For instance, if a relation comprises a supplier's name, status, and city, we may also want to add temporal data, such as the time during which these values are, or were, valid (*e.g.*, for historical data) but the three values may vary independently of each other and at different rates. We may, for instance, wish to trace the history of changes to Status.

DKNF

Some authors use the term **sixth normal form** differently, namely, as a synonym for Domain/key normal form (DKNF). This usage predates Date *et. al.*'s work.

Usage

The sixth normal form is currently being used in some data warehouses where the benefits outweigh the drawbacks, for example using Anchor Modelling. Although using 6NF leads to an explosion of tables, modern databases can prune the tables from select queries (using a process called 'table elimination') where they are not required and thus speed up queries that only access several attributes.

SURROGATE KEY

A **surrogate key** in a database is a unique identifier for either an *entity* in the modelled world or an *object* in the database. The surrogate key is *not* derived from application data.

Definition

There are at least two definitions of a surrogate :

Surrogate (1)-Hall, Owlett and Codd

A surrogate represents an *entity* in the outside world. The surrogate is internally generated by the system but is nevertheless visible to the user or application.

Surrogate (2)-Wieringa and De Jonge (1991)

A surrogate represents an *object* in the database itself. The surrogate is internally generated by the system and is invisible to the user or application.

An important distinction between a surrogate and a primary key depends on whether the database is a current database or a temporal database. Since a *current database* stores only *currently* valid data, there is a one-to-one correspondence between a surrogate in the modelled world and the primary key of the database. In this case the surrogate may be used as a primary key, resulting in the term *surrogate key*. In a temporal database, however, there is a many-to-one relationship between primary keys and the surrogate. Since there may be several objects in the database corresponding to a single surrogate, we cannot use the surrogate as a primary key; another attribute is required, in addition to the surrogate, to uniquely identify each object.

Although Hall et. al. say nothing about this, others have argued that a surrogate should have the following characteristics :

- The value is unique system-wide, hence never reused
- The value is system generated
- The value is not manipulable by the user or application
- The value contains no semantic meaning
- The value is not visible to the user or application
- The value is not composed of several values from different domains.

Surrogates in Practice

In a current database, the surrogate key can be the primary key, generated by the database management system and *not* derived from any application data in the database. The only significance of the surrogate key is to act as the primary key. It is also possible that the surrogate key exists in addition to the database-generated UUID (for example, an HR number for each employee other than the UUID of each employee).

A surrogate key is frequently a sequential number (*e.g.* a Sybase or SQL Server "identity column", a PostgreSQL or Informix serial, an Oracle SEQUENCE or a column defined with AUTO_INCREMENT in MySQL) but doesn't have to be. Having the key independent of all other columns insulates the database relationships from changes in data values or database design (making the database more agile) and guarantees uniqueness.

In a temporal database, it is necessary to distinguish between the surrogate key and the business key. Every row would have both a business key and a surrogate key. The surrogate key identifies one unique row in the database, the business key identifies one unique entity of the modelled world. One table row represents a slice of time holding all the entities attributes for a defined timespan. Those slices depict the whole lifespan of one business entity. For example, a table *EmployeeContracts* may hold temporal information to keep track of contracted working hours. The business key for one contract will be identical (non-unique) in both rows however the surrogate key for each row is unique.

Surrogate Key	*Business Key*	*Employee Name*	*Working Hours Per Week*	*Row Valid From*	*Row Valid To*
1	BOS0120	John Smith	40	2000-01-01	2000-12-31
56	P0000123	Bob Brown	25	1999-01-01	2011-12-31
234	BOS0120	John Smith	35	2001-01-01	2009-12-31

Some database designers use surrogate keys systematically regardless of the suitability of other candidate keys, while others will use a key already present in the data, if there is one.

A *surrogate key* may also be called a synthetic key, an entity identifier, a system-generated key, a database sequence number, a factless key, a technical key, or an arbitrary unique identifier. Some of these terms describe the way of *generating* new surrogate values rather than the *nature* of the surrogate concept.

Approaches to generating surrogates include :

- Universally Unique Identifiers (UUIDs)
- Globally Unique Identifiers (GUIDs)
- Object Identifiers (OIDs)
- Sybase or SQL Server identity column IDENTITY OR IDENTITY(n,n)
- Oracle SEQUENCE
- PostgreSQL or IBM Informix serial
- MySQL AUTO_INCREMENT
- AutoNumber data type in Microsoft Access
- AS IDENTITY GENERATED BY DEFAULT in IBM DB2
- Identity column (implemented in DDL) in Teradata.

Advantages

Immutability

Surrogate keys do not change while the row exists. This has the following advantages :

- Applications cannot lose their reference to a row in the database (since the identifier never changes).

- The primary or natural key data can always be modified, even with databases that do not support cascading updates across related foreign keys.

Requirement Changes

Attributes that uniquely identify an entity might change, which might invalidate the suitability of natural keys. Consider the following example :

An employee's network user name is chosen as a natural key. Upon merging with another company, new employees must be inserted. Some of the new network user names create conflicts because their user names were generated independently (when the companies were separate).

In these cases, generally a new attribute must be added to the natural key (for example, an *original_company* column). With a surrogate key, only the table that defines the surrogate key must be changed. With natural keys, all tables (and possibly other, related software) that use the natural key will have to change.

Some problem domains do not clearly identify a suitable natural key. Surrogate key avoids choosing a natural key that might be incorrect.

Performance

Surrogate keys tend to be a compact data type, such as a four-byte integer. This allows the database to query the single key column faster than it could multiple columns. Furthermore a non-redundant distribution of keys causes the resulting b-tree index to be completely balanced. Surrogate keys are also less expensive to join (fewer columns to compare) than compound keys.

Compatibility

While using several database application development systems, drivers, and object-relational mapping systems, such as Ruby on Rails or Hibernate, it is much easier to use an integer or GUID surrogate keys for every table instead of natural keys in order to support database-system-agnostic operations and object-to-row mapping.

Uniformity

When every table has a uniform surrogate key, some tasks can be easily automated by writing the code in a table-independent way.

Validation

It is possible to design key-values that follow a well-known pattern or structure which can be automatically verified. For instance, the keys that are intended to be used in some column of some table might be designed to "look differently from" those that are intended to be used in another column or table, thereby simplifying the detection of application errors in which the keys have been misplaced. However, this characteristic of the surrogate keys should never be used to drive any of the logic of the applications themselves, as this would violate the principles of Database normalization.

Disadvantages

Disassociation

The values of generated surrogate keys have no relationship to the real-world *meaning* of the data held in a row. When inspecting a row holding a foreign key reference to another table using a surrogate key, the meaning of the surrogate key's row cannot be discerned from the key itself. Every foreign key must be joined to see the related data item. This can also make auditing more difficult, as incorrect data is not obvious.

Surrogate keys are unnatural for data that is exported and shared. A particular difficulty is that tables from two otherwise identical schemas (for example, a test schema and a development schema) can hold records that are equivalent in a business sense, but have different keys. This can be mitigated by not exporting surrogate keys, except as transient data (most obviously, in executing applications that have a "live" connection to the database).

Query Optimization

Relational databases assume a unique index is applied to a table's primary key. The unique index serves two purposes : (i) to enforce entity integrity, since primary key data must be unique across rows and (ii) to quickly search for rows when queried. Since surrogate keys replace a table's identifying attributes—the natural key—and since the identifying attributes are likely to be those queried, then the query optimizer is forced to perform a full table scan when fulfilling likely queries. The remedy to the full table scan is to apply indexes on the identifying attributes, or sets of them. Where such sets are themselves a candidate key, the index can be a unique index.

These additional indexes, however, will take up disk space and slow down inserts and deletes.

Normalization

The presence of a surrogate key can result in the database administrator forgetting to establish, or accidentally removing, a secondary unique index on the natural key of the table. Without a unique index on the natural key, duplicate rows can appear and once present can be difficult to identify.

Business Process Modelling

Because surrogate keys are unnatural, flaws can appear when modelling the business requirements. Business requirements, relying on the natural key, then need to be translated to the surrogate key. A strategy is to draw a clear distinction between the logical model (in which surrogate keys do not appear) and the physical implementation of that model, to ensure that the logical model is correct and reasonably well normalised, and to ensure that the physical model is a correct implementation of the logical model.

Inadvertent Disclosure

Proprietary information can be leaked if sequential key generators are used. By subtracting a previously generated sequential key from a recently generated sequential key, one could learn the number of rows inserted during that time period. This could expose, for example, the number of transactions or new accounts per period. There are a few ways to overcome this problem :

- Increase the sequential number by a random amount.
- Generate a completely random primary key. However, to prevent duplication which would cause an insert rejection, a randomly generated primary key must either be queried (to check that it is not already in use), or the key must contain enough entropy that one can be confident that collisions will not happen.

Inadvertent Assumptions

One might incorrectly infer from sequentially generated surrogate keys that events with a higher primary key value occurred after events with a lower primary key value. The sequential primary key implies nothing of the kind. It is possible for inserts to fail and leave gaps, and for those gaps to be filled at some later time. A sequential key value is not a reliable indicator of chronology. If chronology is important, rely not upon the sequential key but upon a timestamp. A random key would prevent a person from making the assumption that the key has some bearing to real-world chronology only if the person making the assumption is aware that the key is indeed random and has no bearing upon chronology. A randomly generated primary key must be queried before assigned to prevent duplication and cause an insert rejection.

Chapter 18

FACT, DIMENSION AND FILLING

FACT

Fact Table

In data warehousing, a **fact table** consists of the measurements, metrics or facts of a business process. It is located at the center of a star schema or a snowflake schema surrounded by dimension tables. Where multiple fact tables are used, these are arranged as a fact constellation schema. A fact table typically has two types of columns : those that contain facts and those that are foreign keys to dimension tables. The primary key of a fact table is usually a composite key that is made up of all of its foreign keys. Fact tables contain the content of the data warehouse and store different types of measures like additive, non additive, and semi additive measures.

Fact tables provide the (usually) additive values that act as independent variables by which dimensional attributes are analyzed. Fact tables are often defined by their *grain*. The grain of a fact table represents the most atomic level by which the facts may be defined. The grain of a SALES fact table might be stated as "Sales volume by Day by Product by Store". Each record in this fact table is therefore uniquely defined by a day, product and store. Other dimensions might be members of this fact table (such as location/region) but these add nothing to the uniqueness of the fact records. These "affiliate dimensions" allow for additional slices of the independent facts but generally provide insights at a higher level of aggregation (a region contains many stores).

Example

If the business process is SALES, then the corresponding fact table will typically contain columns representing both raw facts and aggregations in rows such as :

- *$12,000,* being "sales for New York store for 15-Jan-2005"

- *$34,000,* being "sales for Los Angeles store for 15-Jan-2005"
- *$22,000,* being "sales for New York store for 16-Jan-2005"
- *$50,000,* being "sales for Los Angeles store for 16-Jan-2005"
- *$21,000,* being "average daily sales for Los Angeles Store for Jan-2005"
- *$65,000,* being "average daily sales for Los Angeles Store for Feb-2005"
- *$33,000,* being "average daily sales for Los Angeles Store for year 2005".

"Average daily sales" is a measurement which is stored in the fact table. The fact table also contains foreign keys from the dimension tables, where time series (*e.g.* dates) and other dimensions (*e.g.* store location, salesperson, product) are stored.

All foreign keys between fact and dimension tables should be surrogate keys, not reused keys from operational data.

Measure Types

- *Additive :* Measures that can be added across any dimension.
- *Non Additive :* Measures that cannot be added across any dimension.
- *Semi Additive :* Measures that can be added across some dimensions.

A fact table might contain either detail level facts or facts that have been aggregated (fact tables that contain aggregated facts are often instead called summary tables).

Special care must be taken when handling ratios and percentage. One good design rule is to never store percentages or ratios in fact tables but only calculate these in the data access tool. Thus only store the numerator and denominator in the fact table, which then can be aggregated and the aggregated stored values can then be used for calculating the ratio or percentage in the data access tool.

In the real world, it is possible to have a fact table that contains no measures or facts. These tables are called "factless fact tables", or "junction tables".

The "Factless fact tables" can for example be used for modelling many-to-many relationships or capture events.

Types of Fact Tables

There are basically four fundamental measurement events, which characterizes all fact tables.

- *Transactional :* A transactional table is the most basic and fundamental. The grain associated with a transactional fact table is usually specified as "one row per line in a transaction", *e.g.*, every line on a receipt. Typically a transactional fact table holds data of the most detailed level, causing it to have a great number of dimensions associated with it.
- *Periodic snapshots :* The periodic snapshot, as the name implies, takes a "picture of the moment", where the moment could be any defined period of time, *e.g.* a performance summary of a salesman over the previous month. A periodic snapshot table is dependent on the transactional table, as it needs the detailed

data held in the transactional fact table in order to deliver the chosen performance output.

- *Accumulating snapshots* : This type of fact table is used to show the activity of a process that has a well-defined beginning and end, *e.g.*, the processing of an order. An order moves through specific steps until it is fully processed. As steps towards fulfilling the order are completed, the associated row in the fact table is updated. An accumulating snapshot table often has multiple date columns, each representing a milestone in the process. Therefore, it's important to have an entry in the associated date dimension that represents an unknown date, as many of the milestone dates are unknown at the time of the creation of the row.
- *Temporal snapshots* : By applying Temporal Database theory and modelling techniques the Temporal Snapshot Fact Table allows to have the equivalent of daily snapshots without really having daily snapshots. It introduces the concept of Time Intervals into a fact table, allowing to save a lot of space, optimizing performances while allowing the end user to have the logical equivalent of the "picture of the moment" he is interested in.

Steps in Designing a Fact Table

- Identify a business process for analysis (like sales).
- Identify measures of facts (sales dollar), by asking questions like 'What number of XX are relevant for the business process?', replacing the XX with various options that make sense within the context of the business.
- Identify dimensions for facts (product dimension, location dimension, time dimension, organization dimension), by asking questions that make sense within the context of the business, like 'Analyse by XX', where XX is replaced with the subject to test.
- List the columns that describe each dimension (region name, branch name, business unit name).
- Determine the lowest level (granularity) of summary in a fact table (*e.g.* sales dollars).

An alternative approach is the four step design process described in Kimball.

Early-arriving Fact

In the Data Warehouse practice of ETL, an **early fact** or **early-arriving fact** denotes the detection of a dimensional natural key during fact table source loading, prior to the assignment of a corresponding primary key or surrogate key in the dimension table. Hence, the fact which cites the dimension arrives early, relative to the definition of the dimension value.

Procedurally, an early fact can be treated as :

- An error, on the presumption that the dimensional attribute values should have been collected before fact source loading.

- A valid fact, whose collection pauses whilst the missing dimensional attribute value itself is collected.
- A valid fact, where a primary key value is generated on the dimension with no attributes (stub/dummy row), the fact completes processing, and the dimension attributes are populated (overwritten) later in the load processing on the new row.

Measure

In a data warehouse, a **measure** is a property on which calculations (*e.g.*, sum, count, average, minimum, maximum) can be made.

Example

For example if a retail store sold a specific product, the quantity and prices of each item sold could be added or averaged to find the total number of items sold and total or average price of the goods.

Use of ISO Representation Terms

When entering data into a metadata registry such as ISO/IEC 11179, representation terms such as **Number**, **Value** and **Measure** are typically used as measures.

DIMENSION

Dimension Table

In data warehousing, a **dimension table** is one of the set of companion tables to a fact table.

The fact table contains business facts (or *measures*), and foreign keys which refer to candidate keys (normally primary keys) in the dimension tables.

Contrary to *fact* tables, *dimension* tables contain descriptive attributes (or fields) that are typically textual fields (or discrete numbers that behave like text). These attributes are designed to serve two critical purposes : query constraining and/or filtering, and query result set labelling.

Dimension attributes should be :

- Verbose (labels consisting of full words)
- Descriptive
- Complete (having no missing values)
- Discretely valued (having only one value per dimension table row)
- Quality assured (having no misspellings or impossible values)

Dimension table rows are uniquely identified by a single key field. It is recommended that the key field be a simple integer because a key value is meaningless, used only for joining fields between the fact and dimension tables.

The use of surrogate dimension keys brings several advantages, including :

- Performance. Join processing is made much more efficient by using a single field (the surrogate key).
- Buffering from operational key management practices. This prevents situations where removed data rows might reappear when their natural keys get reused or reassigned after a long period of dormancy.
- Mapping to integrate disparate sources.
- Handling unknown or not-applicable connections.
- Tracking changes in dimension attribute values.

Although surrogate key use places a burden put on the ETL system, pipeline processing can be improved, and ETL tools have built-in improved surrogate key processing.

The goal of a dimension table is to create standardized, conformed dimensions that can be shared across the enterprise's data warehouse environment, and enable joining to multiple fact tables representing various business processes.

Conformed dimensions are important to the enterprise nature of DW/BI systems because they promote :

- *Consistency :* Every fact table is filtered consistently, so that query answers are labeled consistently.
- *Integration :* Queries can drill into different process fact tables separately for each individual fact table, then join the results on common dimension attributes.
- *Reduced development time to market :* The common dimensions are available without recreating them.

Over time, the attributes of a given row in a dimension table may change. For example, the shipping address for a company may change. Kimball refers to this phenomenon as Slowly Changing Dimensions. Strategies for dealing with this kind of change are divided into three categories :

- *Type One :* Simply overwrite the old value(s).
- *Type Two :* Add a new row containing the new value(s), and distinguish between the rows using Tuple-versioning techniques.
- *Type Three :* Add a new attribute to the existing row.

Degenerate Dimension

The Kimball Definition

According to Ralph Kimball, in a data warehouse, a **degenerate dimension** is a dimension key in the fact table that does not have its own dimension table, because all the interesting attributes have been placed in analytic dimensions. The term "degenerate dimension" was originated by Ralph Kimball.

As Bob Becker says,

Degenerate dimensions commonly occur when the fact table's grain is a single transaction (or transaction line). Transaction control header numbers assigned by the operational business process are typically degenerate dimensions, such as order, ticket, credit card transaction, or check numbers. These degenerate dimensions are natural keys of the "parents" of the line items.

Even though there is no corresponding dimension table of attributes, degenerate dimensions can be quite useful for grouping together related fact tables rows. For example, retail point-of-sale transaction numbers tie all the individual items purchased together into a single market basket. In health care, degenerate dimensions can group the claims items related to a single hospital stay or episode of care.

Other Uses of the Term

Although most writers and practitioners use the term **degenerate dimension** correctly, it is very easy to find misleading definitions in online and printed sources. For example, the Oracle FAQ defines a degenerate dimension as a "data dimension that is stored in the fact table rather than a separate dimension table. This eliminates the need to join to a dimension table. You can use the data in the degenerate dimension to limit or 'slice and dice' your fact table measures."

This common interpretation implies that it is good dimensional modelling practice to place dimension attributes in the fact table, as long as you call them a degenerate dimension. This is not the case; the concept of **degenerate dimension** was developed by Kimball to support a specific, well-defined exception to the otherwise ironclad rule that dimension attributes are always pulled out into dimension tables.

Slowly Changing Dimension

Dimension is a term in data management and data warehousing. It's the logical groupings of data such as geographical location, customer or product information. With **Slowly Changing Dimensions (SCDs)** data changes slowly, rather than changing on a time-based, regular schedule.

For example, you may have a dimension in your database that tracks the sales records of your company's salespeople. Creating sales reports seems simple enough, until a salesperson is transferred from one regional office to another. How do you record such a change in your sales dimension?

You could calculate the sum or average of each salesperson's sales, but if you use that to compare the performance of salespeople, that might give misleading information. If the salesperson was transferred and used to work in a hot market where sales were easy, and now works in a market where sales are infrequent, his/her totals will look much stronger than the other salespeople in their new region. Or you could create a second salesperson record and treat the transferred person as a new sales person, but that creates problems.

Dealing with these issues involves SCD management methodologies referred to as Type 0 through 6. Type 6 SCDs are also sometimes called Hybrid SCDs.

Type 0

The **Type 0** method is passive. It manages dimensional changes and no action is performed. Values remain as they were at the time the dimension record was first inserted. In certain circumstances history is preserved with a Type 0. High order types are employed to guarantee the preservation of history whereas Type 0 provides the least or no control.

The most common types are I, II, and III.

Type 1

This methodology overwrites old with new data, and therefore does not track historical data.

Example of a supplier table :

Supplier_Key	*Supplier_Code*	*Supplier_Name*	*Supplier_State*
123	ABC	Acme Supply Co	CA

In the above example, Supplier_Code is the natural key and Supplier_Key is a surrogate key. Technically, the surrogate key is not necessary, since the row will be unique by the natural key (Supplier_Code). However, to optimize performance on joins use integer rather than character keys.

If the supplier relocates the headquarters to Illinois the record would be overwritten :

Supplier_Key	*Supplier_Code*	*Supplier_Name*	*Supplier_State*
123	ABC	Acme Supply Co	IL

The disadvantage of the Type I method is that there is no history in the data warehouse. It has the advantage however that it's easy to maintain.

If you have calculated an aggregate table summarizing facts by state, it will need to be recalculated when the Supplier_State is changed.

Type 2

This method tracks historical data by creating multiple records for a given natural key in the dimensional tables with separate surrogate keys and/or different version numbers. Unlimited history is preserved for each insert.

For example, if the supplier relocates to Illinois the version numbers will be incremented sequentially :

Supplier_Key	*Supplier_Code*	*Supplier_Name*	*Supplier_State*	*Version.*
123	ABC	Acme Supply Co	CA	0
124	ABC	Acme Supply Co	IL	1

Another method is to add 'effective date' columns.

Supplier_Key	*Supplier_Code*	*Supplier_Name*	*Supplier_State*	*Start_Date*	*End_Date*
123	ABC	Acme Supply Co	CA	01-Jan-2000	21-Dec-2004
124	ABC	Acme Supply Co	IL	22-Dec-2004	

The null End_Date in row two indicates the current tuple version. In some cases, a standardized surrogate high date may be used as an end date, so that the field can be included in an index, and so that null-value substitution is not required when querying.

Transactions that reference a particular surrogate key (Supplier_Key) are then permanently bound to the time slices defined by that row of the slowly changing dimension table. An aggregate table summarizing facts by state continues to reflect the historical state, *i.e.* the state the supplier was in at the time of the transaction; no update is needed.

If there are retrospective changes made to the contents of the dimension, or if new attributes are added to the dimension (for example a Sales_Rep column) which have different effective dates from those already defined, then this can result in the existing transactions needing to be updated to reflect the new situation. This can be an expensive database operation, so Type 2 SCDs are not a good choice if the dimensional model is subject to change.

Type 3

This method tracks changes using separate columns and preserves limited history. The Type 3 preserves limited history as it's limited to the number of columns designated for storing historical data. The original table structure in Type 1 and Type 2 is the same but Type III adds additional columns. In the following example, an additional column has been added to the table to record the supplier's original state-only the previous history is stored.

Supplier_Key	*Supplier_Code*	*Supplier_Name*	*Original_Supplier_State*	*Effective_Date*	*Current_Supplier_State*
123	ABC	Acme Supply Co	CA	22-Dec-2004	IL

This record contains a column for the original state and current state—cannot track the changes if the supplier relocates a second time.

One variation of this is to create the field Previous_Supplier_State instead of Original_Supplier_State which would track only the most recent historical change.

Type 4

The **Type 4** method is usually referred to as using "history tables", where one table keeps the current data, and an additional table is used to keep a record of some or all changes. Both the surrogate keys are referenced in the Fact table to enhance query performance.

For the above example the original table name is **Supplier** and the history table is **Supplier_History.**

Supplier			
Supplier_key	*Supplier_Code*	*Supplier_Name*	*Supplier_State*
123	ABC	Acme Supply Co	IL

Supplier_History				
Supplier_key	*Supplier_Code*	*Supplier_Name*	*Supplier_State*	*Create_Date*
123	ABC	Acme Supply Co	CA	22-Dec-2004

This method resembles how database audit tables and change data capture techniques function.

Type 6/Hybrid

The **Type 6** method combines the approaches of types 1, 2 and 3 (1 + 2 + 3 = 6). One possible explanation of the origin of the term was that it was coined by Ralph Kimball during a conversation with Stephen Pace from Kalido. Ralph Kimball calls this method "Unpredictable Changes with Single-Version Overlay" in *The Data Warehouse Toolkit*.

The Supplier table starts out with one record for our example supplier :

Supplier_ Key	*Supplier_ Code*	*Supplier_ Name*	*Current_ State*	*Histori-cal_State*	*Start_ Date*	*End_ Date*	*Current_ Flag*
123	ABC	Acme Supply Co	CA	CA	01-Jan-2000	31-Dec-9999	Y

The Current_State and the Historical_State are the same. The Current_Flag attribute indicates that this is the current or most recent record for this supplier.

When Acme Supply Company moves to Illinois, we add a new record, as in Type 2 processing :

Supplier_ Key	*Supplier_ Code*	*Supplier_ Name*	*Current_ State*	*Histori-cal_State*	*Start_ Date*	*End_ Date*	*Current_ Flag*
123	ABC	Acme Supply Co	IL	CA	01-Jan-2000	21-Dec-2004	N
124	ABC	Acme Supply Co	IL	IL	22-Dec-2004	31-Dec-9999	Y

We overwrite the Current_State information in the first record (Supplier_Key = 123) with the new information, as in Type 1 processing. We create a new record to track the changes, as in Type 2 processing. And we store the history in a second State column (Historical_State), which incorporates Type 3 processing.

For example if the supplier were to relocate again, we would add another record to the Supplier dimension, and we would overwrite the contents of the Current_State column :

Supplier_ Key	*Supplier_ Code*	*Supplier_ Name*	*Current_ State*	*Histori-cal_State*	*Start_ Date*	*End_ Date*	*Current_ Flag*
123	ABC	Acme Supply Co	NY	CA	01-Jan-2000	21-Dec-2004	N
124	ABC	Acme Supply Co	NY	IL	22-Dec-2004	03-Feb-2008	N
125	ABC	Acme Supply Co	NY	NY	04-Feb-2008	31-Dec-9999	Y

Note that, for the current record (Current_Flag = 'Y'), the Current_State and the Historical_State are always the same.

Type 2/Type 6 Fact Implementation

Type 2 Surrogate Key with Type 3 Attribute

In many Type 2 and Type 6 SCD implementations, the surrogate key from the dimension is put into the fact table in place of the natural key when the fact data is loaded into the data repository. The surrogate key is selected for a given fact record based on its effective date and the Start_Date and End_Date from the dimension table. This allows the fact data to be easily joined to the correct dimension data for the corresponding effective date.

Here is the Supplier table as we created it above using Type 6 Hybrid methodology :

Supplier_ Key	*Supplier_ Code*	*Supplier_ Name*	*Current_ State*	*Histori-cal_State*	*Start_ Date*	*End_ Date*	*Current_ Flag*
123	ABC	Acme Supply Co	NY	CA	01-Jan-2000	21-Dec-2004	N
124	ABC	Acme Supply Co	NY	IL	22-Dec-2004	03-Feb-2008	N
125	ABC	Acme Supply Co	NY	NY	04-Feb-2008	31-Dec-9999	Y

Once the Delivery table contains the correct Supplier_Key, it can easily be joined to the Supplier table using that key. The following SQL retrieves, for each fact record, the current supplier state and the state the supplier was located in at the time of the delivery :

```
SELECT
 delivery.delivery_cost,
```

```
  supplier.supplier_name,
  supplier.historical_state,
  supplier.current_state
FROM delivery
INNER JOIN supplier
  ON delivery.supplier_key = supplier.supplier_key
```

Pure Type 6 Implementation

Having a Type 2 surrogate key for each time slice can cause problems if the dimension is subject to change.

A pure Type 6 implementation does not use this, but uses a Surrogate Key for each master data item (*e.g.* each unique supplier has a single surrogate key).

This avoids any changes in the master data having an impact on the existing transaction data.

It also allows more options when querying the transactions.

Here is the Supplier table using the pure Type 6 methodology :

Supplier_Key	*Supplier_Code*	*Supplier_Name*	*Supplier_State*	*Start_Date*	*End_Date*
456	ABC	Acme Supply Co	CA	01-Jan-2000	21-Dec-2004
456	ABC	Acme Supply Co	IL	22-Dec-2004	03-Feb-2008
456	ABC	Acme Supply Co	NY	04-Feb-2008	31-Dec-9999

The following example shows how the query must be extended to ensure a single supplier record is retrieved for each transaction.

```
SELECT
  supplier.supplier_code,
  supplier.supplier_state
FROM supplier
INNER JOIN delivery
  ON supplier.supplier_key = delivery.supplier_key
  AND delivery.delivery_date >= supplier.start_date
  AND delivery.delivery_date <= supplier.end_date
```

A fact record with an effective date (Delivery_Date) of August 9, 2001 will be linked to Supplier_Code of ABC, with a Supplier_State of 'CA'. A fact record with an effective date of October 11, 2007 will also be linked to the same Supplier_Code ABC, but with a Supplier_State of 'IL'.

Whilst more complex, there are a number of advantages of this approach, including :

1. If there is more than one date on the fact (*e.g.* Order Date, Delivery Date, Invoice Payment Date) you can choose which date to use for a query.
2. You can do "as at now", "as at transaction time" or "as at a point in time" queries by changing the date filter logic.
3. You don't need to reprocess the Fact table if there is a change in the dimension table (*e.g.* adding additional fields retrospectively which change the time slices, or if you make a mistake in the dates on the dimension table you can correct them easily).
4. You can introduce bi-temporal dates in the dimension table.
5. You can join the fact to the multiple versions of the dimension table to allow reporting of the same information with different effective dates, in the same query.

The following example shows how a specific date such as '2012-01-01 00 : 00 : 00' (which could be the current datetime) can be used.

```
SELECT
 supplier.supplier_code,
 supplier.supplier_state
FROM supplier
INNER JOIN delivery
 ON supplier.supplier_key = delivery.supplier_key
 AND '2012-01-01 00 :00 :00' >= supplier.start_date
 AND '2012-01-01 00 :00 :00' <= supplier.end_date
```

Both surrogate and natural key

An alternative implementation is to place *both* the surrogate key and the natural key into the fact table. This allows the user to select the appropriate dimension records based on :

- The primary effective date on the fact record (above),
- The most recent or current information,
- Any other date associated with the fact record.

This method allows more flexible links to the dimension, even if you have used the Type 2 approach instead of Type 6.

Here is the Supplier table as we might have created it using Type 2 methodology :

Supplier_ Key	*Supplier_ Code*	*Supplier_ Name*	*Supplier_ State*	*Start_ Date*	*End_Date*	*Current_ Flag*
123	ABC	Acme Supply Co	CA	01-Jan-2000	21-Dec-2004	N
124	ABC	Acme Supply Co	IL	22-Dec-2004	03-Feb-2008	N
125	ABC	Acme Supply Co	NY	04-Feb-2008	31-Dec-9999	Y

The following SQL retrieves the most current Supplier_Name and Supplier_State for each fact record :

```
SELECT
 delivery.delivery_cost,
 supplier.supplier_name,
 supplier.supplier_state
FROM delivery
INNER JOIN supplier
 ON delivery.supplier_code = supplier.supplier_code
WHERE supplier.current_flag = 'Y'
```

If there are multiple dates on the fact record, the fact can be joined to the dimension using another date instead of the primary effective date. For instance, the Delivery table might have a primary effective date of Delivery_Date, but might also have an Order_Date associated with each record.

The following SQL retrieves the correct Supplier_Name and Supplier_State for each fact record based on the Order_Date :

```
SELECT
 delivery.delivery_cost,
 supplier.supplier_name,
 supplier.supplier_state
FROM delivery
INNER JOIN supplier
 ON delivery.supplier_code = supplier.supplier_code
 AND delivery.order_date >= supplier.start_date
 AND delivery.order_date <= supplier.end_date
```

Some cautions :

- If the join query is not written correctly, it may return duplicate rows and/or give incorrect answers.
- The date comparison might not perform well.
- Some Business Intelligence tools do not handle generating complex joins well.

- The ETL processes needed to create the dimension table needs to be carefully designed to ensure that there are no overlaps in the time periods for each distinct item of reference data.

Combining Types

Different SCD Types can be applied to different columns of a table. For example, we can apply Type 1 to the Supplier_Name column and Type 2 to the Supplier_State column of the same table.

FILLING

Extract, Transform, Load

In computing, **extract, transform, and load (ETL)** refers to a process in database usage and especially in data warehousing that :

- Extracts data from outside sources
- Transforms it to fit operational needs, which can include quality levels
- Loads it into the end target (database, more specifically, operational data store, data mart, or data warehouse)

ETL systems are commonly used to integrate data from multiple applications, typically developed and supported by different vendors or hosted on separate computer hardware. The disparate systems containing the original data are frequently managed and operated by different employees. For example a cost accounting system may combine data from payroll, sales and purchasing.

Extract

The first part of an ETL process involves extracting the data from the source systems. In many cases this is the most challenging aspect of ETL, since extracting data correctly sets the stage for how subsequent processes go further.

Most data warehousing projects consolidate data from different source systems. Each separate system may also use a different data organization and/or format. Common data source formats are relational databases and flat files, but may include non-relational database structures such as Information Management System (IMS) or other data structures such as Virtual Storage Access Method (VSAM) or Indexed Sequential Access Method (ISAM), or even fetching from outside sources such as through web spidering or screen-scraping. The streaming of the extracted data source and load on-the-fly to the destination database is another way of performing ETL when no intermediate data storage is required. In general, the goal of the extraction phase is to convert the data into a single format appropriate for transformation processing.

An intrinsic part of the extraction involves the parsing of extracted data, resulting in a check if the data meets an expected pattern or structure. If not, the data may be rejected entirely or in part.

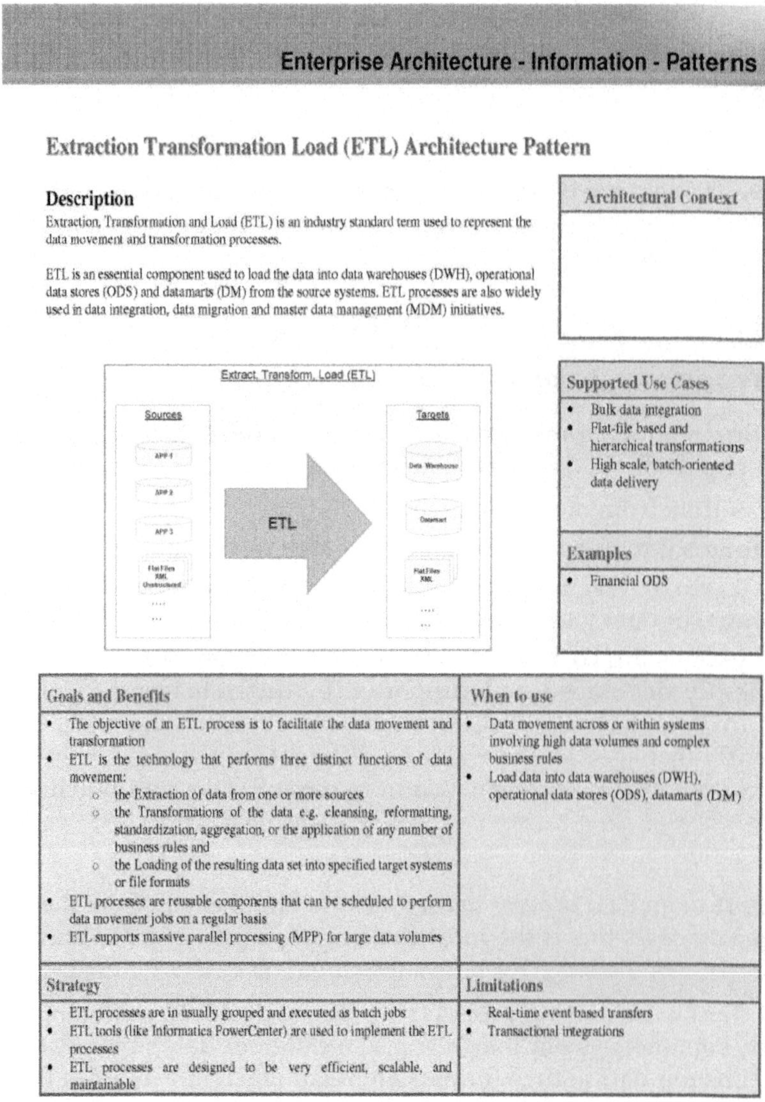

Fig. : ETL Architecture Pattern.

Transform

The transform stage applies a series of rules or functions to the extracted data from the source to derive the data for loading into the end target. Some data sources require very little or even no manipulation of data. In other cases, one or more of the following transformation types may be required to meet the business and technical needs of the target database :

- Selecting only certain columns to load (or selecting null columns not to load). For example, if the source data has three columns (also called attributes), roll_no, age, and salary, then the selection may take only roll_no and salary. Similarly, the selection mechanism may ignore all those records where salary is not present (salary = null).

- Translating coded values (*e.g.*, if the source system stores 1 for male and 2 for female, but the warehouse stores M for male and F for female)
- Encoding free-form values (*e.g.*, mapping "Male" to "M")
- Deriving a new calculated value (*e.g.*, sale_amount = qty * unit_price)
- Sorting
- Joining data from multiple sources (*e.g.*, lookup, merge) and deduplicating the data
- Aggregation (for example, rollup—summarizing multiple rows of data—total sales for each store, and for each region, etc.)
- Generating surrogate-key values
- Transposing or pivoting (turning multiple columns into multiple rows or *vice versa*)
- Splitting a column into multiple columns (*e.g.*, converting a comma-separated list, specified as a string in one column, into individual values in different columns)
- Disaggregation of repeating columns into a separate detail table (*e.g.*, moving a series of addresses in one record into single addresses in a set of records in a linked *address* table)
- Lookup and validate the relevant data from tables or referential files for slowly changing dimensions.
- Applying any form of simple or complex data validation. If validation fails, it may result in a full, partial or no rejection of the data, and thus none, some or all the data is handed over to the next step, depending on the rule design and exception handling. Many of the above transformations may result in exceptions, for example, when a code translation parses an unknown code in the extracted data.

Load

The load phase loads the data into the end target, usually the data warehouse (DW). Depending on the requirements of the organization, this process varies widely. Some data warehouses may overwrite existing information with cumulative information; frequently, updating extracted data is done on a daily, weekly, or monthly basis. Other data warehouses (or even other parts of the same data warehouse) may add new data in a historical form at regular intervals—for example, hourly. To understand this, consider a data warehouse that is required to maintain sales records of the last year. This data warehouse overwrites any data older than a year with newer data. However, the entry of data for any one year window is made in a historical manner. The timing and scope to replace or append are strategic design choices dependent on the time available and the business needs. More complex systems can maintain a history and audit trail of all changes to the data loaded in the data warehouse.

As the load phase interacts with a database, the constraints defined in the database schema—as well as in triggers activated upon data load—apply (for example, uniqueness, referential integrity, mandatory fields), which also contribute to the overall data quality performance of the ETL process.

- For example, a financial institution might have information on a customer in several departments and each department might have that customer's information listed in a different way. The membership department might list the customer by name, whereas the accounting department might list the customer by number. ETL can bundle all this data and consolidate it into a uniform presentation, such as for storing in a database or data warehouse.
- Another way that companies use ETL is to move information to another application permanently. For instance, the new application might use another database vendor and most likely a very different database schema. ETL can be used to transform the data into a format suitable for the new application to use.
- An example of this would be an Expense and Cost Recovery System (ECRS) such as used by accountancies, consultancies and lawyers. The data usually ends up in the time and billing system, although some businesses may also utilize the raw data for employee productivity reports to Human Resources or equipment usage reports to Facilities Management.

Real-life ETL cycle

The typical real-life ETL cycle consists of the following execution steps :

1. Cycle initiation
2. Build reference data
3. Extract (from sources)
4. Validate
5. Transform (clean, apply business rules, check for data integrity, create aggregates or disaggregates)
6. Stage (load into staging tables, if used)
7. Audit reports (for example, on compliance with business rules. Also, in case of failure, helps to diagnose/repair)
8. Publish (to target tables)
9. Archive
10. Clean up.

Challenges

ETL processes can involve considerable complexity, and significant operational problems can occur with improperly designed ETL systems.

The range of data values or data quality in an operational system may exceed the expectations of designers at the time validation and transformation rules are specified. Data profiling of a source during data analysis can identify the data conditions that must be managed by transform rules specifications. This leads to an amendment of validation rules explicitly and implicitly implemented in the ETL process.

Data warehouses are typically assembled from a variety of data sources with different formats and purposes. As such, ETL is a key process to bring all the data together in a standard, homogeneous environment.

Design analysts should establish the scalability of an ETL system across the lifetime of its usage. This includes understanding the volumes of data that must be processed within service level agreements. The time available to extract from source systems may change, which may mean the same amount of data may have to be processed in less time. Some ETL systems have to scale to process terabytes of data to update data warehouses with tens of terabytes of data. Increasing volumes of data may require designs that can scale from daily batch to multiple-day micro batch to integration with message queues or real-time change-data capture for continuous transformation and update.

Performance

ETL vendors benchmark their record-systems at multiple TB (terabytes) per hour (or ~1 GB per second) using powerful servers with multiple CPUs, multiple hard drives, multiple gigabit-network connections, and lots of memory. The fastest ETL record is currently held by Syncsort, Vertica and HP at 5.4 TB in under an hour, which is more than twice as fast as the earlier record held by Microsoft and Unisys.

In real life, the slowest part of an ETL process usually occurs in the database load phase. Databases may perform slowly because they have to take care of concurrency, integrity maintenance, and indices. Thus, for better performance, it may make sense to employ :

- Direct Path Extract method or bulk unload whenever is possible (instead of querying the database) to reduce the load on source system while getting high speed extract.
- Most of the transformation processing outside of the database.
- Bulk load operations whenever possible.

Still, even using bulk operations, database access is usually the bottleneck in the ETL process. Some common methods used to increase performance are :

- Partition tables (and indices). Try to keep partitions similar in size (watch for null values that can skew the partitioning).
- Do all validation in the ETL layer before the load. Disable integrity checking (disable constraint...) in the target database tables during the load.
- Disable triggers (disable trigger...) in the target database tables during the load. Simulate their effect as a separate step.

- Generate IDs in the ETL layer (not in the database).
- Drop the indices (on a table or partition) before the load-and recreate them after the load (SQL : drop index...; create index...).
- Use parallel bulk load when possible—works well when the table is partitioned or there are no indices. Note : attempt to do parallel loads into the same table (partition) usually causes locks—if not on the data rows, then on indices.
- If a requirement exists to do insertions, updates, or deletions, find out which rows should be processed in which way in the ETL layer, and then process these three operations in the database separately. You often can do bulk load for inserts, but updates and deletes commonly go through an API (using SQL).

Whether to do certain operations in the database or outside may involve a trade-off. For example, removing duplicates using distinct may be slow in the database; thus, it makes sense to do it outside. On the other side, if using distinct significantly (x100) decreases the number of rows to be extracted, then it makes sense to remove duplications as early as possible in the database before unloading data.

A common source of problems in ETL is a big number of dependencies among ETL jobs. For example, job "B" cannot start while job "A" is not finished. One can usually achieve better performance by visualizing all processes on a graph, and trying to reduce the graph making maximum use of parallelism, and making "chains" of consecutive processing as short as possible. Again, partitioning of big tables and of their indices can really help.

Another common issue occurs when the data is spread between several databases, and processing is done in those databases sequentially. Sometimes database replication may be involved as a method of copying data between databases—and this can significantly slow down the whole process. The common solution is to reduce the processing graph to only three layers :

- Sources
- Central ETL layer
- Targets.

This allows processing to take maximum advantage of parallel processing. For example, if you need to load data into two databases, you can run the loads in parallel (instead of loading into 1st—and then replicating into the 2nd).

Sometimes processing must take place sequentially. For example, dimensional data is needed before one can get and validate the rows for main "fact" tables.

Parallel Processing

A recent development in ETL software is the implementation of parallel processing. This has enabled a number of methods to improve overall performance of ETL processes when dealing with large volumes of data.

ETL applications implement three main types of parallelism :

- *Data* : By splitting a single sequential file into smaller data files to provide parallel access.
- *Pipeline* : Allowing the simultaneous running of several components on the same data stream. For example : looking up a value on record 1 at the same time as adding two fields on record 2.
- *Component* : The simultaneous running of multiple processes on different data streams in the same job, for example, sorting one input file while removing duplicates on another file.

All three types of parallelism usually operate combined in a single job.

An additional difficulty comes with making sure that the data being uploaded is relatively consistent. Because multiple source databases may have different update cycles (some may be updated every few minutes, while others may take days or weeks), an ETL system may be required to hold back certain data until all sources are synchronized. Likewise, where a warehouse may have to be reconciled to the contents in a source system or with the general ledger, establishing synchronization and reconciliation points becomes necessary.

Rerunnability, Recoverability

Data warehousing procedures usually subdivide a big ETL process into smaller pieces running sequentially or in parallel. To keep track of data flows, it makes sense to tag each data row with "row_id", and tag each piece of the process with "run_id". In case of a failure, having these IDs help to roll back and rerun the failed piece.

Best practice also calls for *checkpoints*, which are states when certain phases of the process are completed. Once at a checkpoint, it is a good idea to write everything to disk, clean out some temporary files, log the state, and so on.

Virtual ETL

As of 2010 data virtualization had begun to advance ETL processing. The application of data virtualization to ETL allowed solving the most common ETL tasks of data migration and application integration for multiple dispersed data sources. So-called Virtual ETL operates with the abstracted representation of the objects or entities gathered from the variety of relational, semi-structured and unstructured data sources. ETL tools can leverage object-oriented modelling and work with entities' representations persistently stored in a centrally located hub-and-spoke architecture. Such a collection that contains representations of the entities or objects gathered from the data sources for ETL processing is called a metadata repository and it can reside in memory or be made persistent. By using a persistent metadata repository, ETL tools can transition from one-time projects to persistent middleware, performing data harmonization and data profiling consistently and in near-real time.

Dealing with Keys

Keys are some of the most important objects in all relational databases, as they tie everything together. A primary key is a column that identifies a given entity, where a foreign key is a column in another table that refers a primary key. These keys can also be made of several columns, in which case they are composite keys. In many cases the primary key is an auto generated integer that has no meaning for the business entity being represented, but solely exists for the purpose of the relational database-commonly referred to as a surrogate key.

As there is usually more than one data source being loaded into the warehouse, the keys are an important concern to be addressed. Your customers might be represented in several data sources, and in one their SSN (Social Security Number) might be the primary key, their phone number in another and a surrogate in the third. All of the customers information needs to be consolidated into one dimension table.

A recommended way to deal with the concern is to add a warehouse surrogate key, which is used as a foreign key from the fact table.

Usually updates occur to a dimension's source data, which obviously must be reflected in the data warehouse. If the primary key of the source data is required for reporting, the dimension already contains that piece of information for each row. If the source data uses a surrogate key, the ware house must keep track of it even though it is never used in queries or reports.

That is done by creating a lookup table that contains the warehouse surrogate key and the originating key. This way the dimension is not polluted with surrogates from various source systems, while the ability to update is preserved.

The lookup table is used in different ways depending on the nature of the source data. There are 5 types to consider, where three selected ones are included here :

Type 1 : The dimension row is simply updated to match the current state of the source system. The warehouse does not capture history. The lookup table is used to identify the dimension row to update or overwrite.

Type 2 : A new dimension row is added with the new state of the source system. A new surrogate key is assigned. Source key is no longer unique in the lookup table.

Fully logged : A new dimension row is added with the new state of the source system, while the previous dimension row is updated to reflect it is no longer active and record time of deactivation.

Tools

Programmers can set up ETL processes using almost any programming language, but building such processes from scratch can become complex. Increasingly, companies are buying ETL tools to help in the creation of ETL processes.

By using an established ETL framework, one may increase one's chances of ending up with better connectivity and scalability. A good ETL tool must be able to communicate with the many different relational databases and read the various file formats used throughout an organization. ETL tools have started to migrate into Enterprise Application Integration, or even Enterprise Service Bus, systems that now cover much more than just the extraction, transformation, and loading of data. Many ETL vendors now have data profiling, data quality, and metadata capabilities. A common use case for ETL tools include converting CSV files to formats readable by relational databases. A typical translation of millions of records is facilitated by ETL tools that enable users to input csv-like data feeds/ files and import it into a database with as little code as possible.

ETL Tools are typically used by a broad range of professionals—from students in computer science looking to quickly import large data sets to database architects in charge of company account management, ETL Tools have become a convenient tool that can be relied on to get maximum performance. ETL tools in most cases contain a GUI that helps users conveniently transform data as opposed to writing large programs to parse files and modify data types—which ETL tools facilitate as much as possible.

Data Extraction

Data extraction is the act or process of retrieving data out of (usually unstructured or poorly structured) data sources for further data processing or data storage (data migration). The import into the intermediate extracting system is thus usually followed by data transformation and possibly the addition of metadata prior to export to another stage in the data workflow.

Usually, the term data extraction is applied when (experimental) data is first imported into a computer from primary sources, like measuring or recording devices. Today's electronic devices will usually present an electrical connector (*e.g.* USB) through which 'raw data' can be streamed into a personal computer.

Typical unstructured data sources include web pages, e-mails, documents, PDFs, scanned text, mainframe reports, spool files etc. Extracting data from these unstructured sources has grown into a considerable technical challenge where as historically data extraction has had to deal with changes in physical hardware formats, the majority of current data extraction deals with extracting data from these unstructured data sources, and from different software formats. This growing process of data extraction from the web is referred to as Web scraping.

The act of adding structure to unstructured data takes a number of forms :

- Using text pattern matching such as regular expressions to identify small or large-scale structure *e.g.* records in a report and their associated data from headers and footers;
- Using a table-based approach to identify common sections within a limited domain *e.g.* in emailed resumes, identifying skills, previous work experience, qualifications etc., using a standard set of commonly used headings (these

would differ from language to language), *e.g.* Education might be found under Education/Qualification/Courses;

- Using text analytics to attempt to understand the text and link it to other information.

Data Transformation

In metadata and data warehouse, a **data transformation** converts a set of data values from the data format of a source data system into the data format of a destination data system.

Data transformation can be divided into two steps :

1. Data mapping maps data elements from the source data system to the destination data system and captures any transformation that must occur
2. Code generation that creates the actual transformation program

Data element to data element mapping is frequently complicated by complex transformations that require one-to-many and many-to-one transformation rules.

The code generation step takes the data element mapping specification and creates an executable program that can be run on a computer system. Code generation can also create transformation in easy-to-maintain computer languages such as Java or XSLT.

A master data recast is another form of data transformation where the entire database of data values is transformed or recast without extracting the data from the database. All data in a well designed database is directly or indirectly related to a limited set of master database tables by a network of foreign key constraints. Each foreign key constraint is dependent upon a unique database index from the parent database table. Therefore, when the proper master database table is recast with a different unique index, the directly and indirectly related data are also recast or restated. The directly and indirectly related data may also still be viewed in the original form since the original unique index still exists with the master data. Also, the database recast must be done in such a way as to not impact the applications architecture software.

When the data mapping is indirect via a mediating data model, the process is also called **data mediation.**

Transformational Languages

There are numerous languages available for performing data transformation. Many transformation languages require a grammar to be provided. In many cases the grammar is structured using something closely resembling Backus–Naur Form (BNF). There are numerous languages available for such purposes varying in their accessibility (cost) and general usefulness. Examples of such languages include :

- AWK—one of the oldest and popular textual data transformation language;
- Perl—a high-level language with both procedural and object-oriented syntax capable of powerful operations on binary or text data.

- Template languages—specialized for transform data into documents;
- TXL—prototyping language-based descriptions, used for source code or data transformation.
- XSLT—the standard XML data transformation language (suitable by XQuery in many applications);

Although transformational languages are typically best suited for transformation, something as simple as regular expressions can be used to achieve useful transformation. A text editor like emacs or Textpad supports the use of regular expressions with arguments. This would allow all instances of a particular pattern to be replaced with another pattern using parts of the original pattern. For example :

```
foo ("some string", 42, gCommon);
bar (someObj, anotherObj);
foo ("another string", 24, gCommon);
bar (myObj, myOtherObj);
```

could both be transformed into a more compact form like :

```
foobar("some string", 42, someObj, anotherObj);
foobar("another string", 24, myObj, myOtherObj);
```

In other words, all instances of a function invocation of foo with three arguments, followed by a function invocation with two invocations would be replaced with a single function invocation using some or all of the original set of arguments.

Another advantage to using regular expressions is that they will not fail the null transform test. That is, using your transformational language of choice, run a sample program through a transformation that doesn't perform any transformations. Many transformational languages will fail this test.

Transforming Source Code

Program synthesis, Automatic programming and other fields use the data transformation strategies to translate, adapt or even generate software source code. Inversely these *source transformation* tools can be used for data transform, typically for transform "document source code" as HTML or another XML dialect.

Generally the different types of transformations fall into one of two categories,

- *Translation* : a transformation from a language X into another language Y.
- *Rephrasing* : a rephrasing involves a transformation within the same language but merely stated a different way.

Example

A difficult problem to address in C++ is "unstructured preprocessor directives". These are preprocessor directives which do not contain blocks of code with simple grammatical descriptions, like in this function definition :

```
void MyFunc ()
     {
      if (x>17)
      { printf("test");
     # ifdef FOO
      } else {
     # endif
      if (gWatch)
      mTest = 42;
      }
     }
```

A really general solution to handling this is very hard because such preprocessor directives can essentially edit the underlying language in arbitrary ways. However, because such directives are not, in practice, used in completely arbitrary ways, one can build practical tools for handling preprocessed languages. The DMS Software Reengineering Toolkit is capable of handling structured macros and preprocessor conditionals. Brabrand and Schwartzbach offer another approach, substituting the C preprocessor by a metamorphic one.

Chapter 19

OPERATIONAL INTELLIGENCE

Operational intelligence (OI) is a category of real-time dynamic, business analytics that delivers visibility and insight into data, streaming events and business operations. Operational Intelligence solutions run queries against streaming data feeds and event data to deliver real-time analytic results as operational instructions. Operational Intelligence provides organizations the ability to make decisions and immediately act on these analytic insights, through manual or automated actions.

PURPOSE

The purpose of OI is to monitor business activities and identify and detect situations relating to inefficiencies, opportunities, and threats and provide operational solutions. Some definitions define operational intelligence an event-centric approach to delivering information that empowers people to make better decisions.

In addition, these metrics act as the starting point for further analysis (drilling down into details, performing root cause analysis—tying anomalies to specific transactions and of the business activity).

Sophisticated OI systems also provide the ability to associate metadata with metrics, process steps, channels, etc. With this, it becomes easy to get related information, *e.g.*, "retrieve the contact information of the person that manages the application that executed the step in the business transaction that took 60% more time than the norm," or "view the acceptance/rejection trend for the customer who was denied approval in this transaction," or "Launch the application that this process step interacted with."

FEATURES

Different operational intelligence solutions may use many different technologies and be implemented in different ways. This section lists the common features of an operational intelligence solution :

- Real-time monitoring
- Real-time situation detection
- Real-time dashboards for different user roles
- Correlation of events
- Industry-specific dashboards
- Multidimensional analysis
  - Root cause analysis
  - Time Series and trending analysis
- *Big Data Analytics* : Operational Intelligence is well suited to address the inherent challenges of Big Data. Operational Intelligence continuously monitors and analyzes the variety of high velocity, high volume Big Data sources. Often performed in memory, OI platforms and solutions then present the incremental calculations and changes, in real-time, to the end-user.

TECHNOLOGY COMPONENTS

Operational intelligence solutions share many features, and therefore many also share technology components. This is a list of some of the commonly found technology components, and the features they enable :

- Business activity monitoring (BAM) – Dashboard customization and personalization.
- Complex event processing (CEP) – Advanced, continuous analysis of real-time information and historical data.
- Business process management (BPM) – To perform model-driven execution of policies and processes defined as Business Process Model and Notation (BPMN) models.
- Metadata framework to model and link events to resources.
- Multi-channel publishing and notification.
- Dimensional database.
- Root cause analysis.
- Multi-protocol event collection.

Operational intelligence is a relatively new market segment (compared to the more mature business intelligence and business process management segments). In addition to companies that produce dedicated and focussed products in this area, there are numerous companies in adjacent areas that provide solutions with some OI components.

Operational intelligence places complete information at one's fingertips, enabling one to make smarter decisions in time to maximize impact. By correlating a wide variety of events and data from both streaming feeds and historical data silos, operational intelligence helps organizations gain real-time visibility of information, in context, through advanced dashboards, real-time insight into business performance, health and status so that immediate action based on business policies and processes can be taken. Operational intelligence applies the benefits of real-time analytics, alerts, and actions to a broad spectrum of use cases across and beyond the enterprise.

One specific technology segment is AIDC (Automatic Identification and Data Capture) represented by barcodes, RFID and voice recognition.

COMPARISON WITH OTHER TECHNOLOGIES OR SOLUTIONS

Business Intelligence

OI is often linked to or compared with business intelligence (BI) or real time business intelligence, in the sense that both help make sense out of large amounts of information. But there are some basic differences : OI is primarily activity-centric, whereas BI is primarily data-centric. As with most technologies, each of these could be sub-optimally coerced to perform the other's task. OI is, by definition, real-time, unlike BI or "On-Demand" BI, which are traditionally after-the-fact and report-based approaches to identifying patterns. Real-time BI (*i.e.*, On-Demand BI) relies on the database as the sole source of events.

OI provides continuous, real-time analytics on data at rest and data in-flight, whereas BI typically looks only at historical data at rest. OI and BI can be complementary. OI is best used for short-term planning, such as deciding on the "next best action," while BI is best used for longer-term planning (over the next days to weeks). BI requires a more reactive approach, often reacting to events that have already taken place.

If all that is needed is a glimpse at historical performance over a very specific period of time, existing BI solutions should meet the requirement. However, historical data needs to be analyzed with events that are happening now, or to reduce the time between when intelligence is received and when action is taken, then Operational Intelligence is the more appropriate approach.

Systems Management

System Management mainly refers to the availability and capability monitoring of IT infrastructure. Availability monitoring refers to monitoring the status of IT infrastructure components such as servers, routers, networks, etc. This usually entails pinging or polling the component and waiting to receive a response. Capability monitoring usually refers to synthetic transactions where user activity is mimicked by a special software program, and the responses received are checked for correctness.

Complex Event Processing

There is a strong relationship between complex event processing companies and operational intelligence, especially since CEP is regarded by many OI companies as a core component of their OI solutions. CEP companies tend to focus solely on development of a CEP framework for other companies to use within their organisations as a pure CEP engine.

Business Activity Monitoring

Business activity monitoring (BAM) is software that aids in monitoring of business processes, as those processes are implemented in computer systems. BAM is an enterprise solution primarily intended to provide a real-time summary of business processes to operations managers and upper management. The main difference between BAM and OI appears to be in the implementation details—real-time situation detection appears in BAM and OI and is often implemented using CEP. Furthermore, BAM focuses on high-level process models whereas OI instead relies on correlation to infer a relationship between different events.

BUSINESS PROCESS MANAGEMENT

Business process management (BPM) has been referred to as a "holistic management" approach to aligning an organization's business processes with the wants and needs of clients. BPM uses a systematic approach in an attempt to continuously improve business effectiveness and efficiency while striving for innovation, flexibility, and integration with technology. It can therefore be described as a "process optimization process." It is argued that BPM enables organizations to be more efficient, more effective and more capable of change than a functionally focused, traditional hierarchical management approach. These processes can impact the cost and revenue generation of an organization. As a managerial approach, BPM sees processes as strategic assets of an organization that must be understood, managed, and improved to deliver value-added products and services to clients. This foundation closely resembles other Total Quality Management or Continuous Improvement Process methodologies or approaches. BPM goes a step further by stating that this approach can be supported, or enabled, through technology to ensure the viability of the managerial approach in times of stress and change. In fact, BPM offers an approach to integrate an organizational "change capability" that is both human and technological. As such, many BPM articles and pundits often discuss BPM from one of two viewpoints : people and/or technology.

BPM or Business Process Management is often referred to as 'Management by Business Processes'. The term "business" can be confusing as it is often linked with a hierarchical view (by function) of a company. It is therefore preferable to define BPM as "corporate management through processes". By adding to BPM the second meaning of 'Business Performance Management' used by August-Wilhelm Scheer in his article "Advanced BPM Assessment", BPM can therefore be defined as "company performance management through processes".

Changes in BPM

Roughly speaking, the idea of business process is as traditional as concepts of tasks, department, production, and outputs. The management and improvement approach as of 2010, with formal definitions and technical modelling, has been around since the early 1990s. Note that the IT community often uses the term "business process" as synonymous with the management of middleware processes; or as synonymous with integrating application software tasks. This viewpoint may be overly restrictive—a limitation to keep in mind when reading software engineering papers that refer to "business processes" or to "business process modelling".

Although BPM initially focused on the automation of business processes with the use of information technology, it has since been extended to integrate human-driven processes in which human interaction takes place in series or parallel with the use of technology. For example (in workflow systems), when individual steps in the business process require deploying human intuition or judgment, these steps are assigned to appropriate members within the organization.

More advanced forms such as "human interaction management" are in the complex interaction between human workers in performing a workgroup task. In this case, many people and systems interact in structured, *ad hoc*, and sometimes completely dynamic ways to complete one to many transactions.

BPM can be used to understand organizations through expanded views that would not otherwise be available to organize and present, such as relationships between processes. When included in a process model, these relationships provide for advanced reporting and analysis. BPM is sometimes regarded as the backbone of enterprise content management.

Because BPM allows organizations to abstract business process from technology infrastructure, it goes far beyond automating business processes (software) or solving business problems (suite). BPM enables business to respond to changing consumer, market, and regulatory demands faster than competitors-creating competitive advantage.

As of 2010 technology has allowed the coupling of BPM to other methodologies, such as Six Sigma. BPM tools allow users to :

- Vision—strategize functions and processes
- Define—define the goal of the change
- Measure—determine the appropriate measure to determine success
- Analyze—compare the various simulations to determine an optimal improvement
- Improve—select and implement the improvement
- Control—deploy this implementation and by use of user-defined dashboards monitor the improvement in real time and feed the performance information back into the simulation model in preparation for the next improvement iteration
- Re-engineer—revamp the processes from scratch for better results

This brings with it the benefit of being able to simulate changes to business processes based on real-life data (not just on assumed knowledge). Also, the coupling of BPM to industry methodologies allows users to continually streamline and optimize the process to ensure that it is tuned to its market need.

As of 2012 research on BPM has paid increasing attention to the compliance of business processes. Although a key aspect of business processes is flexibility, as business processes continuously need to adapt to changes in the environment, compliance with business strategy, policies and government regulations should also be ensured. The compliance aspect in BPM is highly important for governmental organizations. As of 2010 BPM approaches in a governmental context largely focus on operational processes and knowledge representation. Although there have been many technical studies on operational business processes both in the public and in the private sector, researchers have rarely taken legal compliance activities into account, for instance the legal implementation processes in public-administration bodies.

BPM Life-cycle

Business process management activities can be grouped into six categories : vision, design, modelling, execution, monitoring, and optimization.

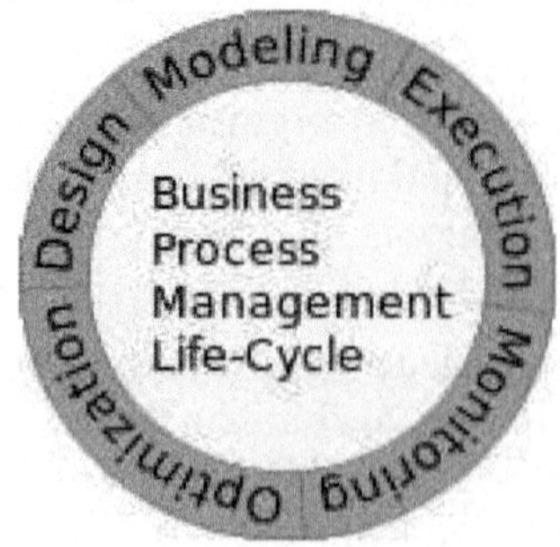

Functions are designed around the strategic vision and goals of an organization. Each function is attached with a list of processes. Each functional head in an organization is responsible for certain sets of processes made up of tasks which are to be executed and reported as planned. Multiple processes are aggregated to function accomplishments and multiple functions are aggregated to achieve organizational goals.

Design

Process design encompasses both the identification of existing processes and the design of "to-be" processes. Areas of focus include representation of the process flow, the factors within it, alerts & notifications, escalations, standard operating procedures, service level agreements, and task hand-over mechanisms.

Good design reduces the number of problems over the lifetime of the process. Whether or not existing processes are considered, the aim of this step is to ensure that a correct and efficient theoretical design is prepared.

The proposed improvement could be in human-to-human, human-to-system or system-to-system workflows, and might target regulatory, market, or competitive challenges faced by the businesses.

The existing process and the design of new process for various applications will have to synchronise and not cause major outage or process interruption.

Modelling

Modelling takes the theoretical design and introduces combinations of variables (*e.g.*, changes in rent or materials costs, which determine how the process might operate under different circumstances).

It also involves running "what-if analysis" on the processes : *"What if I have 75% of resources to do the same task?" "What if I want to do the same job for 80% of the current cost?"*

Execution

One of the ways to automate processes is to develop or purchase an application that executes the required steps of the process; however, in practice, these applications rarely execute all the steps of the process accurately or completely. Another approach is to use a combination of software and human intervention; however this approach is more complex, making the documentation process difficult.

As a response to these problems, software has been developed that enables the full business process (as developed in the process design activity) to be defined in a computer language which can be directly executed by the computer. The system will either use services in connected applications to perform business operations (*e.g.* calculating a repayment plan for a loan) or, when a step is too complex to automate, will ask for human input. Compared to either of the previous approaches, directly executing a process definition can be more straightforward and therefore easier to improve. However, automating a process definition requires flexible and comprehensive infrastructure, which typically rules out implementing these systems in a legacy IT environment.

Business rules have been used by systems to provide definitions for governing behaviour, and a business rule engine can be used to drive process execution and resolution.

Monitoring

Monitoring encompasses the tracking of individual processes, so that information on their state can be easily seen, and statistics on the performance of one or more processes can be provided. An example of the tracking is being able to determine the state of a customer order (*e.g.* order arrived, awaiting delivery, invoice paid) so that problems in its operation can be identified and corrected.

In addition, this information can be used to work with customers and suppliers to improve their connected processes. Examples of the statistics are the generation of measures on how quickly a customer order is processed or how many orders were processed in the last month. These measures tend to fit into three categories : cycle time, defect rate and productivity.

The degree of monitoring depends on what information the business wants to evaluate and analyze and how business wants it to be monitored, in real-time, near real-time or *ad hoc*. Here, business activity monitoring (BAM) extends and expands the monitoring tools generally provided by BPMS.

Process mining is a collection of methods and tools related to process monitoring. The aim of process mining is to analyze event logs extracted through process monitoring and to compare them with an *a priori* process model. Process mining allows process analysts to detect discrepancies between the actual process execution and the *a priori* model as well as to analyze bottlenecks.

Optimization

Process optimization includes retrieving process performance information from modelling or monitoring phase; identifying the potential or actual bottlenecks and the potential opportunities for cost savings or other improvements; and then, applying those enhancements in the design of the process. Overall, this creates greater business value.

Re-engineering

When the process becomes too noisy and optimization is not fetching the desired output, it is recommended to re-engineer the entire process cycle. BPR has become an integral part of organizations to achieve efficiency and productivity at work.

Certification

Currently the certification is being offered by Global Association for Quality Management (GAQM) the Syllabus and Certificate is recognized, approved and managed by the International Accreditation Organization (IAO)

BPM Suites

Forrester Research, Inc recognize the BPM suite space through three different lenses :

- Human-centric BPM
- Integration-centric BPM (Enterprise Service Bus)
- Document-centric BPM (Dynamic Case Management)

However, standalone integration-centric and document-centric offerings have matured into separate, standalone markets.

Gartner's Magic Quadrant identifies 10 core components of an intelligent BPM suite, including predictive analytics and robust rules management.

Practice

Example of Business Process Management (BPM) Service Pattern : This pattern shows how business process management (BPM) tools can be used to implement business processes through the orchestration of activities between people and systems.

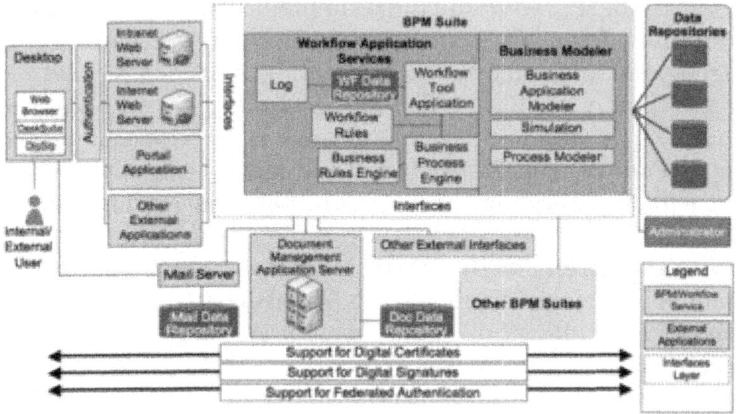

While the steps can be viewed as a cycle, economic or time constraints are likely to limit the process to only a few iterations. This is often the case when an organization uses the approach for short to medium term objectives rather than trying to transform the organizational culture. True iterations are only possible through the collaborative efforts of process participants. In a majority of organizations, complexity will require enabling technology to support the process participants in these daily process management challenges.

To date, many organizations often start a BPM project or program with the objective to optimize an area that has been identified as an area for improvement.

In the financial sector, BPM is critical to make sure the system delivers a quality service while maintaining regulatory compliance.

Currently, the international standards for the task have limited BPM to the application in the IT sector, and ISO/IEC 15944 covers the operational aspects of the business. However, some corporations with the culture of best practices do use standard operating procedures to regulate their operational process. Other standards are currently being worked upon to assist in BPM implementation (BPMN, Enterprise Architecture, Business Motivation Model).

BPM Technology

Some define the BPM system or suite (BPMS) as "the whole of BPM." Others relate the important concept of information moving between enterprise software packages and immediately think of service-oriented architecture. Still others limit the definition to "modelling".

BPM is now considered a critical component of operational intelligence (OI) solutions to deliver real-time, actionable information. This real-time information can be acted upon in a variety of ways—alerts can be sent or executive decisions can be made using real-time dashboards. OI solutions use real-time information to take automated action based on pre-defined rules so that security measures and or exception management processes can be initiated.

These are partial answers and the technological offerings continue to evolve. The BPMS term may not survive. Today it encompasses the concept of supporting the managerial approach through enabling technology. The BPMS should enable all stakeholders to have a firm understanding of an organization and its performance. The BPMS should facilitate business process change throughout the life cycle stated above. This assists in the automation of activities, collaboration, integration with other systems, integrating partners through the value chain, etc. For instance, the size and complexity of daily tasks often requires the use of technology to model efficiently. These models facilitate automation and solutions to business problems. These models can also become executable to assist in monitoring and controlling business processes. As such, some people view BPM as "the bridge between Information Technology (IT) and Business.". In fact, an argument can be made that this "holistic approach" bridges organizational and technological silos.

There are four critical components of a BPM Suite :

- *Process engine :* A robust platform for modelling and executing process-based applications, including business rules.
- *Business analytics :* Enable managers to identify business issues, trends, and opportunities with reports and dashboards and react accordingly.
- *Content management :* Provides a system for storing and securing electronic documents, images, and other files.
- *Collaboration tools :* Remove intra-and interdepartmental communication barriers through discussion forums, dynamic workspaces, and message boards.

BPM also addresses many of the critical IT issues underpinning these business drivers, including :

- Managing end-to-end, customer-facing processes.
- Consolidating data and increasing visibility into and access to associated data and information.
- Increasing the flexibility and functionality of current infrastructure and data
- Integrating with existing systems and leveraging emerging service oriented architecture (SOAs).
- Establishing a common language for business-IT alignment.

Validation of BPMS is another technical issue that vendors and users need to be aware of, if regulatory compliance is mandatory. The validation task could be performed either by an authenticated third party or by the users themselves. Either way, validation documentation will need to be generated. The validation document usually can either be published officially or retained by users.

Cloud Computing BPM

Cloud computing business process management is the use of (BPM) tools that are delivered as software services (SaaS) over a network. Cloud BPM business logic is deployed on an application server and the business data resides in cloud storage.

Market

According to Gartner, by 2016, 20% of all the "shadow business processes" will be supported by BPM cloud platforms. Gartner refers to all the hidden organizational processes that are supported by IT departments as part of legacy business processes such as Excel spreadsheets, routing of e-mails using rules, phone calls routing, etc. These can, of course also be replaced by other technologies such as workflow software.

Benefits

The benefits of using cloud BPM services include removes the need and cost of maintaining specialized technical skill sets in-house and reducing distractions from an enterprise's main focus. It offers controlled IT budgeting and enables geographical mobility.

The details of this are still emerging.

BUSINESS ACTIVITY MONITORING

Business activity monitoring (BAM) is software that aids in monitoring of business activities, as those activities are implemented in computer systems.

The term was originally coined by analysts at Gartner, Inc. and refers to the aggregation, analysis, and presentation of real-time information about activities inside organizations and involving customers and partners. A business activity can either be a business process that is orchestrated by business process management (BPM) software, or a business process that is a series of activities spanning multiple systems and applications. BAM is an enterprise solution primarily intended to provide a real-time summary of business activities to operations managers and upper management.

Goals and Benefits

The goals of business activity monitoring is to provide real time information about the status and results of various operations, processes, and transactions. The main benefits of BAM are to enable an enterprise to make better informed business decisions, quickly address problem areas, and re-position organizations to take full advantage of emerging opportunities.

Key Features

One of the most visible features of BAM solutions is the presentation of information on dashboards that contain key performance indicators (KPIs) used to provide assurance and visibility of activity and performance. This information is used by technical and business operations to provide visibility, measurement, and assurance of key business activities. It is also exploited by event correlation to detect and warn of impending problems.

Although BAM systems usually use a computer dashboard display to present data, BAM is distinct from the dashboards used by business intelligence (BI) insofar as events are processed in real-time or near real-time and pushed to the dashboard in BAM systems, whereas BI dashboards refresh at predetermined intervals by polling or querying databases. Depending on the refresh interval selected, BAM and BI dashboards can be similar or vary widely.

Some BAM solutions additionally provide trouble notification functions, which allows them to interact automatically with the trouble ticket system. For example, whole groups of people can be sent e-mails, voice or text messages, according to the nature of the problem. Automated problem solving, where feasible, can correct and restart failed processes.

Deployment Effort

In nearly all deployments of BAM solutions, extensive tailoring to specific enterprises is required. Many BAM solutions seek to reduce extensive customization and may offer templates that are written to solve common problems in specific lines of business, for example banking, manufacturing, and stock brokering. Due to the high degrees of system integration required for initial deployment, many enterprises use experts that specialize in BAM to implement solutions.

BAM is now considered a critical component of Operational Intelligence (OI) solutions to deliver visibility into business operations. Multiple sources of data can be combined from different organizational silos to provide a common operating picture that uses current information. Wherever real-time insight has the greatest value, OI solutions can be applied to deliver the information and need.

Processing Events

All BAM solutions process events. While most of the first BAM solutions were closely linked to business process management (BPM) solutions and therefore processed events emitted as the process was being orchestrated, this had the disadvantage of requiring enterprises to invest in BPM before being able to acquire and use BAM. Fortunately the newer generation of BAM solutions are based on complex event processing (CEP) technology, and can process high volumes of underlying technical events to derive higher level business events, therefore severing the dependency on BPM, and providing the benefits of BAM to a wider audience of customers.

Examples

A specific brief example follows of how a BAM solution might be used. A bank is interested in minimizing the amount of money it borrows overnight from a central bank. Interbank transfers must be communicated and arranged through automation by a set time each business day. The failure of any vital communication could cost the bank large sums in interest charged by the central bank.

A BAM solution would be programmed to become aware of each message and await confirmation. Failure to receive confirmation within a reasonable amount of time would be grounds for the BAM solution to raise an alarm that would set in motion manual intervention to investigate the cause of the delay and to push the problem toward resolution before it becomes costly.

Another example involves a mobile telecommunications company interested in detecting a situation whereby new customers are not set up promptly and correctly on their network and within the various CRM and billing solutions. Low-level technical events such as messages passing from one application to another over a middleware system, or transactions detected within a database logfile, are processed by a CEP engine. All events relating to an individual customer are correlated in order to detect an anomalous situation whereby an individual customer has not been promptly or correctly provisioned, or set up. An alert can be generated to notify technical operations or to notify business operations, and the failed provisioning step may be restarted automatically.

COMPLEX EVENT PROCESSING

Event processing is a method of tracking and analyzing (processing) streams of information (data) about things that happen (events), and deriving a conclusion from them. **Complex event processing**, or CEP, is event processing that combines data from multiple sources to infer events or patterns that suggest more complicated circumstances. The goal of complex event processing is to identify meaningful events (such as opportunities or threats) and respond to them as quickly as possible.

These events may be happening across the various layers of an organization as sales leads, orders or customer service calls. Or, they may be news items, text messages, social media posts, stock market feeds, traffic reports, weather reports, or other kinds of data. An event may also be defined as a "change of state," when a measurement exceeds a predefined threshold of time, temperature, or other value. Analysts suggest that CEP will give organizations a new way to analyze patterns in real-time, and help the business side communicate better with IT and service departments.

The vast amount of information available about events is sometimes referred to as the event cloud.

Conceptual Description

Among thousands of incoming events, a monitoring system may for instance receive the following three from the same source :

1. Church bells ringing.
2. The appearance of a man in a tuxedo with a woman in a flowing white gown.
3. Rice flying through the air.

From these events the monitoring system may infer a *complex event* : a wedding. CEP as a technique helps discover complex events by analyzing and correlating other events : the bells, the man and woman in wedding attire and the rice flying through the air.

CEP relies on a number of techniques, including :

- Event-pattern detection
- Event abstraction
- Event filtering
- Event aggregation and transformation
- Modelling event hierarchies
- Detecting relationships (such as causality, membership or timing) between events
- Abstracting event-driven processes.

Commercial applications of CEP exist in variety of industries and include algorithmic stock-trading, the detection of credit-card fraud, business activity monitoring, and security monitoring.

History

The CEP area has roots in Discrete event simulation, active database area and some programming languages. The activity in the industry was preceded by a wave of research projects in the 1990s. According to the first project that paved the way to a generic CEP language and execution model was the Rapide project in Stanford University, directed by David Luckham, in parallel there have been three other research projects : Infospheres in California Institute of Technology, directed by K. Mani Chandy, Apama in University of Cambridge directed by John Bates, and Amit in IBM Haifa Research Laboratory, directed by Opher Etzion. The commercial products were dependents of the concepts developed in these and some later research projects. Community efforts started in a series of event processing symposiums organized by the Event Processing Technical Society, and later by the ACM DEBS conference series. One of the community effort was in producing the event processing manifesto.

Related Concepts

CEP is used in Operational Intelligence (OI) solutions to provide insight into business operations by running query analysis against live feeds and event data. OI solutions use real-time data to collect and correlate against historical data to provide insight into and analysis of the current situation. Multiple sources of data can be combined from different organizational silos to provide a common operating picture that uses current information. Wherever real-time insight has the greatest value, OI solutions can be applied to deliver the information and need.

In network management, systems management, application management and service management, people usually refer instead to event correlation. As

CEP engines, event correlation engines (*event correlators*) analyze a mass of events, pinpoint the most significant ones, and trigger actions. However, most of them do not produce new inferred events. Instead, they relate high-level events with low-level events.

Inference engines, *e.g.* rule-based reasoning engines typically produce inferred information in artificial intelligence. However, they do not usually produce new information in the form of complex (*i.e.*, inferred) events.

Example

A more systemic example of CEP involves a car, some sensors and various events and reactions. Imagine that a car has several sensors—one that measures tire pressure, one that measures speed, and one that detects if someone sits on a seat or leaves a seat.

In the first situation, the car is moving and the pressure of one of the tires moves from 45 psi (pound per square inch) to 41 psi over 15 minutes. As the pressure in the tire is decreasing, a series of events containing the tire pressure is generated. In addition, a series of events containing the speed of the car is generated. The car's Event Processor may detect a situation whereby a loss of tire pressure over a relatively long period of time results in the creation of the "loss of Tire Pressure" event. This new event may trigger a reaction process to note the pressure loss into the car's maintenance log, and alert the driver via the car's portal that the tire pressure has reduced.

In the second situation, the car is moving and the pressure of one of the tires drops from 45 psi to 20 psi in 5 seconds. A different situation is detected—perhaps because the loss of pressure occurred over a shorter period of time, or perhaps because the difference in values between each event were larger than a predefined limit. The different situation results in a new event "blow Out Tire" being generated. This new event triggers a different reaction process to immediately alert the driver and to initiate onboard computer routines to assist the driver in bringing the car to a stop without losing control through skidding.

In addition, events that represent detected situations can also be combined with other events in order to detect more complex situations. For example, in the final situation the car was moving normally but suffers a blown tire which results in the car leaving the road and striking a tree and the driver is thrown from the car. A series of different situations are rapidly detected. The combination of "blow Out Tire", "zero Speed" and "driver Left Seat" within a very short space of time results in a new situation being detected : "occupant Thrown Accident". Even though there is no direct measurement that can determine conclusively that the driver was thrown, or that there was an accident, the combination of events allows the situation to be detected and a new event to be created to signify the detected situation. This is the essence of a complex (or composite) event. It is complex because one cannot directly detect the situation; one has to infer or deduce that the situation has occurred from a combination of other events.

Types

Most CEP solutions and concepts can be classified into two main categories :

1. Aggregation-oriented CEP
2. Detection-oriented CEP.

An aggregation-oriented CEP solution is focused on executing on-line algorithms as a response to event data entering the system. A simple example is to continuously calculate an average based in data on the inbound events.

Detection-oriented CEP is focused on detecting combinations of events called events patterns or situations. A simple example of detecting a situation is to look for a specific sequence of events.

Currently many application use hybrid of the two approaches.

Integration with Business Process Management

Of course, rarely does the application of a new technology exist in isolation. A natural fit for CEP has been with business process management, or BPM. BPM very much focuses on end-to-end business processes, in order to continuously optimize and align for its operational environment.

However, the optimization of a business does not rely solely upon its individual, end-to-end processes. Seemingly disparate processes can affect each other significantly. Consider this scenario : In the aerospace industry, it is good practice to monitor breakdowns of vehicles to look for trends (determine potential weaknesses in manufacturing processes, material, etc.). Another separate process monitors current operational vehicles' life cycles and decommissions them when appropriate. Now one use for CEP is to link these separate processes, so that in the case of when the initial process (breakdown monitoring) discovers a malfunction based on metal fatigue (a significant event) an action can be created to exploit the second process (life cycle) to issue a recall on vehicles using the same batch of metal discovered as faulty in the initial process.

The integration of CEP and BPM must exist at two levels, both at the business awareness level (users must understand the potential holistic benefits of their individual processes) and also at the technological level (there needs to be a method by which CEP can interact with BPM implementation).

Computation-oriented CEP's role can arguably be seen to overlap with Business Rule technology.

For example, customer service centers are using CEP for click-stream analysis and customer experience management. CEP software can factor real-time information about millions of events per second into business intelligence and other decision-support applications. These "recommendation applications" help agents provide personalized service based on each customer's experience. The CEP application may collect data about what customers on the phone are currently doing, or how they have recently interacted with the company in other various channels, including in-branch, or on the Web via self-service features, instant messaging and email. The application then analyzes the total customer experience and recommends scripts or next steps that guide the agent on the phone, and hopefully keep the customer happy.

Another example of CEP in practice is in the healthcare industry. The HyReminder system, developed by the Worcester Polytechnic Institute and UMass Medical School, continually tracks healthcare workers for hygiene compliance (*e.g.* sanitizing hands and wearing masks), reminding them to perform hygiene when appropriate to prevent the spread of infectious disease. Each worker wears an RFID badge that displays a green (safe), yellow (warning) or red (violation) light, depending on what behaviour the RFID chip has observed.

In Financial Services

The financial services industry was an early adopter of CEP technology, using complex event processing to structure and contextualize available data so that it could inform trading behaviour, specifically algorithmic trading, by identifying opportunities or threats that indicate traders (or automatic trading systems) should buy or sell. Algorithmic trading is already a practice in stock trading. It is estimated that around 60% of Equity trading in the United States is by way of algo trades. CEP is expected to continue to help financial institutions improve their algorithms and be more efficient.

Recent improvements in CEP technologies have made it more affordable, helping smaller firms to create trading algorithms of their own and compete with larger firms. CEP has evolved from an emerging technology to an essential platform of many capital markets. The technology's most consistent growth has been in banking, serving fraud detection, online banking, and multi-channel marketing initiatives.

Today, a wide variety of financial applications use CEP, including profit, loss, and risk management systems, order and liquidity analysis, quantitative trading and signal generation systems, and others.

Integration with Time Series Databases

A time series database is a software system that is optimized for the handling of data organized by time. Time series are finite or infinite sequences of data items, where each item has an associated timestamp and the sequence of timestamps is non-decreasing. Elements of a time series are often called ticks. The timestamps are not required to be ascending (merely non-decreasing) because in practice the time resolution of some systems such as financial data sources can be quite low (milli-seconds, micro-seconds or even nano-seconds), so consecutive events may carry equal timestamps.

Time series data provides a historical context to the analysis typically associated with complex event processing. This can apply to any vertical industry such as finance and co-operatively with other technologies such as BPM as described elsewhere in this document.

Consider the scenario in finance where there is a need to understand historic price volatility to determine statistical thresholds of future price movements. This is helpful for both trade models and transaction cost analysis.

The ideal case for CEP analysis is to view historical time series and real-time streaming data as a single time continuum. What happened yesterday, last week or last month is simply an extension of what is occurring today and what may occur in the future. An example may involve comparing current market volumes to historic volumes, prices and volatility for trade execution logic. Or the need to act upon live market prices may involve comparisons to benchmarks that include sector and index movements, whose intra-day and historic trends gauge volatility and smooth outliers.

BUSINESS PROCESS MODEL AND NOTATION

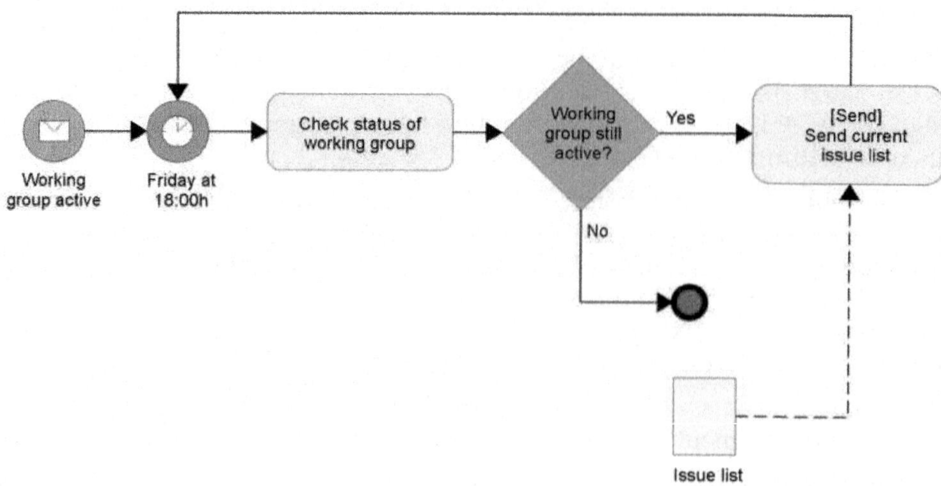

Fig. : Example of a Business Process Model and Notation for a process with a normal flow.

Business Process Model and Notation (BPMN) is a graphical representation for specifying business processes in a business process model. It is also known as *Business Process Modelling Notation*.

Business Process Management Initiative (BPMI) developed BPMN, which has been maintained by the Object Management Group since the two organizations merged in 2005. As of March 2011, the current version of BPMN is 2.0.

Overview

Business Process Model and Notation (BPMN) is a standard for business process modelling that provides a graphical notation for specifying business processes in a *Business Process Diagram* (BPD), based on a flowcharting technique very similar to activity diagrams from Unified Modelling Language (UML). The objective of BPMN is to support business process management, for both technical users and business users, by providing a notation that is intuitive to business users, yet able to represent complex process semantics. The BPMN specification also provides a mapping between the graphics of the notation and the underlying constructs of execution languages, particularly Business Process Execution Language (BPEL).

The primary goal of BPMN is to provide a standard notation readily understandable by all business stakeholders. These include the business analysts who create and refine the processes, the technical developers responsible for implementing them, and the business managers who monitor and manage them. Consequently, BPMN serves as a common language, bridging the communication gap that frequently occurs between business process design and implementation.

Currently there are several competing standards for business process modelling languages used by modelling tools and processes. Widespread adoption of the BPMN will help unify the expression of basic business process concepts (*e.g.*, public and private processes, choreographies), as well as advanced process concepts (*e.g.*, exception handling, transaction compensation).

BPMN Topics

Scope

BPMN is constrained to support only the concepts of modelling applicable to business processes. Other types of modelling done by organizations for non-process purposes are out of scope for BPMN. Examples of modelling excluded from BPMN are :

- Organizational structures
- Functional breakdowns
- Data models.

In addition, while BPMN shows the flow of data (messages), and the association of data artifacts to activities, it is not a data flow diagram.

Elements

BPMN models consist of simple diagrams constructed from a limited set of graphical elements. For both business users and developers, they simplify understanding business activities' flow and process. BPMN's four basic element categories are :

Flow objects

Events, activities, gateways

Connecting objects

Sequence flow, message flow, association

Swim lanes

Pool, lane

Artifacts

Data object, group, annotation.

These four categories enable creation of simple business process diagrams (BPDs). BPDs also permit making new types of flow object or artifact, to make the diagram more understandable.

Flow Objects and Connecting Objects

Event

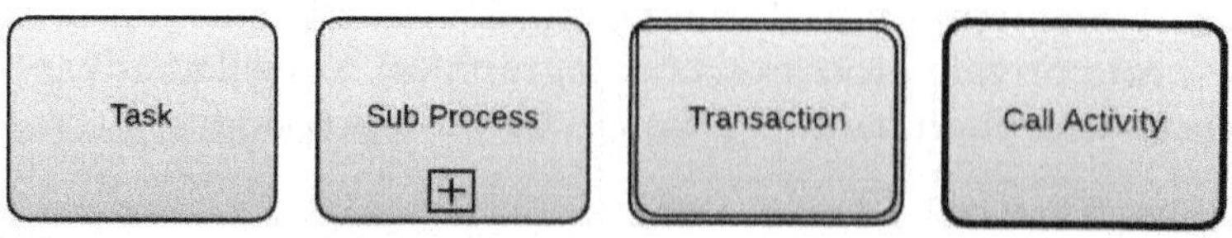

Activity

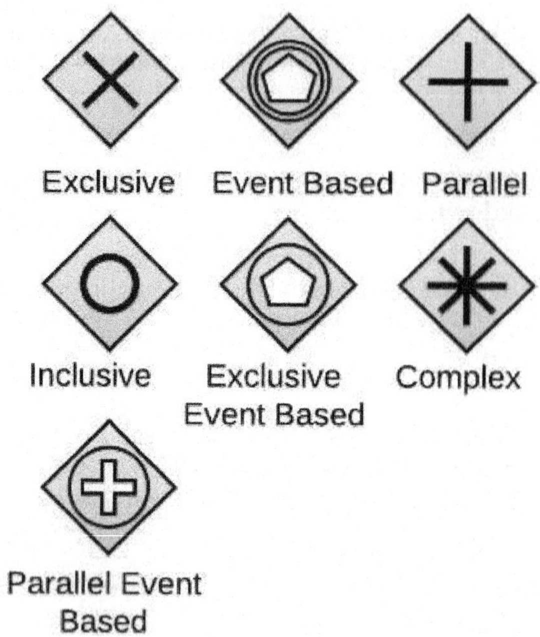

Gateway

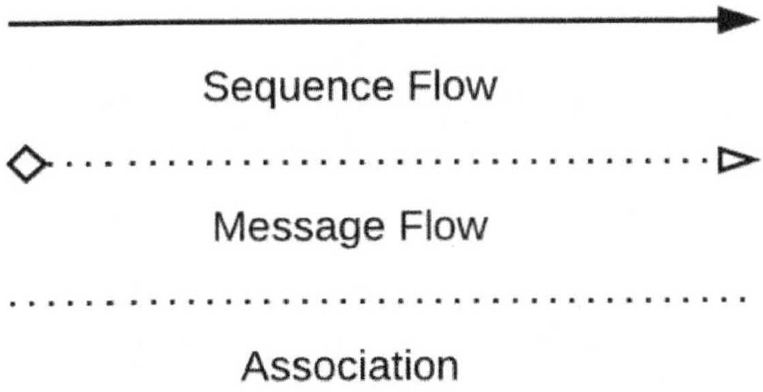

Connections

Flow objects are the main describing elements within BPMN, and consist of three core elements : events, activities, and gateways.

Event

An Event is represented with a circle and denotes something that *happens* (compared with an activity, which is something that is *done*). Icons within the circle denote the type of event (*e.g.*, an envelope representing a message, or a clock representing time). Events are also classified as **Catching** (for example, if catching an incoming message starts a process) or **Throwing** (such as throwing a completion message when a process ends).

Start Event

Acts as a process trigger; indicated by a single narrow border, and can only be *Catch*, so is shown with an open (outline) icon.

Intermediate Event

Represents something that happens between the start and end events; is indicated by a double border, and can *Throw* or *Catch* (using solid or open icons as appropriate). For example, a task could flow to an event that throws a message across to another pool, where a subsequent event waits to catch the response before continuing.

End Event

Represents the result of a process; indicated by a single thick or bold border, and can only *Throw*, so is shown with a solid icon.

Activity

An activity is represented with a rounded-corner rectangle and describes the kind of work which must be done.

Task

A task represents a single unit of work that is not or cannot be broken down to a further level of business process detail without diagramming the steps in a procedure (which is not the purpose of BPMN)

Sub-process

Used to hide or reveal additional levels of business process detail. When collapsed, a sub-process is indicated by a plus sign against the bottom line of the rectangle; when expanded, the rounded rectangle expands to show all flow objects, connecting objects, and artifacts.

Has its own self-contained start and end events; sequence flows from the *parent* process must not cross the boundary.

Transaction

A form of sub-process in which all contained activities must be treated as a whole; *i.e.*, they must all be completed to meet an objective, and if any one of them fails, they must all be compensated (undone). Transactions are differentiated from expanded sub-processes by being surrounded by a double border.

Call Activity

A point in the process where a global process or a global Task is reused. A call activity is differentiated from other activity types by a bolded border around the activity area.

Gateway

A gateway is represented with a diamond shape and determines forking and merging of paths, depending on the conditions expressed.

Exclusive

Used to create alternative flows in a process because only one of the paths can be taken, it is called exclusive

Event Based

The condition determining the path of a process is based on an evaluated event.

Parallel

Used to create parallel paths without evaluating any conditions.

Inclusive

Used to create alternative flows where all paths are evaluated.

Exclusive Event Based

An event is being evaluated to determine which of mutually exclusive paths will be taken.

Complex

Used to model complex synchronization behaviour.

Parallel Event Based

Two parallel process are started based on an event but there is no evaluation of the event.

Connections

Flow objects are connected to each other using **Connecting objects**, which are of three types : sequences, messages, and associations.

Sequence Flow

A Sequence Flow is represented with a solid line and arrowhead, and shows in which order the activities are performed. The *sequence flow* may also have a symbol at its start, a small diamond indicates one of a number of **conditional flows** from an activity, while a diagonal slash indicates the **default flow** from a decision or activity with conditional flows.

Message flow

A Message Flow is represented with a dashed line, an open circle at the start, and an open arrowhead at the end. It tells us what messages flow across organizational boundaries (*i.e.*, between pools). A message flow can never be used to connect activities or events within the same pool.

Association

An Association is represented with a dotted line. It is used to associate an Artifact or text to a Flow Object, and can indicate some directionality using an open arrowhead (toward the artifact to represent a result, from the artifact to represent an input, and both to indicate it is read and updated). No directionality is used when the Artifact or text is associated with a sequence or message flow (as that flow already shows the direction).

Swimlanes and Artifacts

Swimlanes

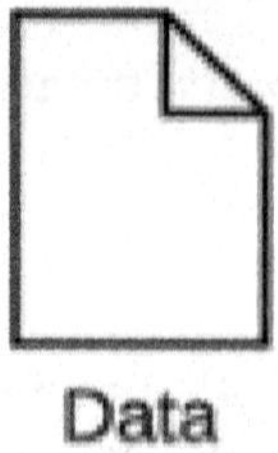

Data objects

Groups

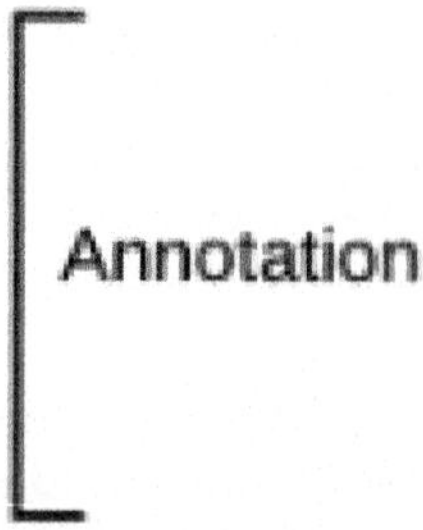

Annotation

Swim lanes are a visual mechanism of organising and categorising activities, based on cross-functional flowcharting, and in BPMN consist of two types :

Pool

Represents major participants in a process, typically separating different organisations. A pool contains one or more lanes (like a real swimming pool). A pool can be open (*i.e.*, showing internal detail) when it is depicted as a large rectangle showing one or more lanes, or collapsed (*i.e.*, hiding internal detail) when it is depicted as an empty rectangle stretching the width or height of the diagram.

Lane

Used to organise and categorise activities within a pool according to function or role, and depicted as a rectangle stretching the width or height of the pool. A lane contains the flow objects, connecting objects and artifacts.

Artifacts allow developers to bring some more information into the model/ diagram. In this way the model/diagram becomes more readable. There are three pre-defined Artifacts and they are :

- *Data objects* : Data objects show the reader which data is required or produced in an activity.
- *Group* : A Group is represented with a rounded-corner rectangle and dashed lines. The group is used to group different activities but does not affect the flow in the diagram.
- *Annotation* : An annotation is used to give the reader of the model/diagram an understandable impression.

Examples of Business Process Diagrams

- Click on small images for full-size version

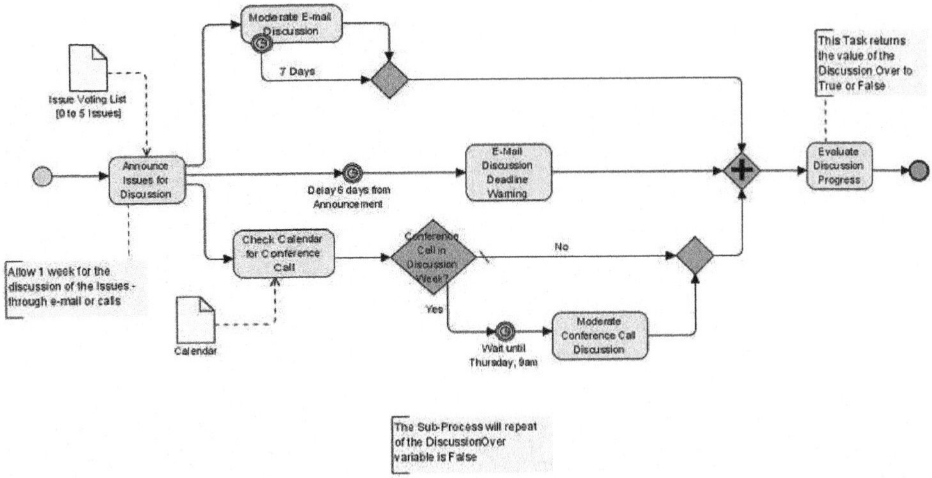

Fig. : Discussion cycle.

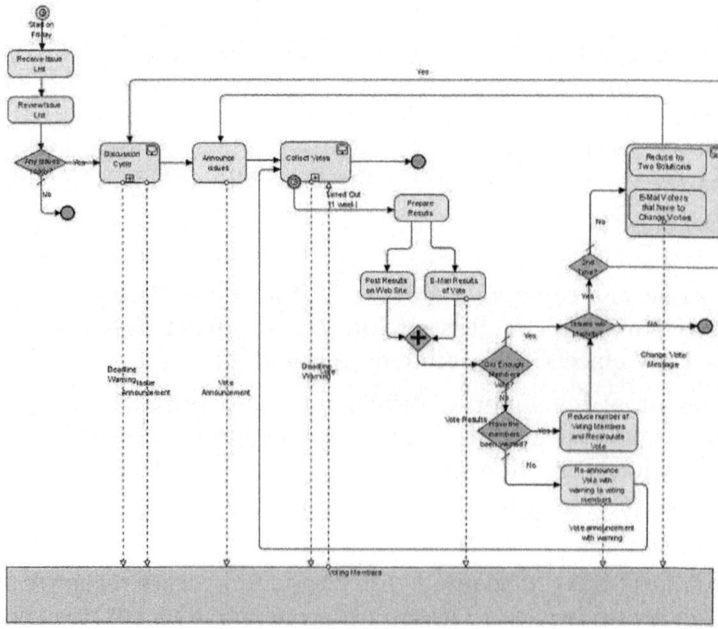

Fig. : E-mail voting process.

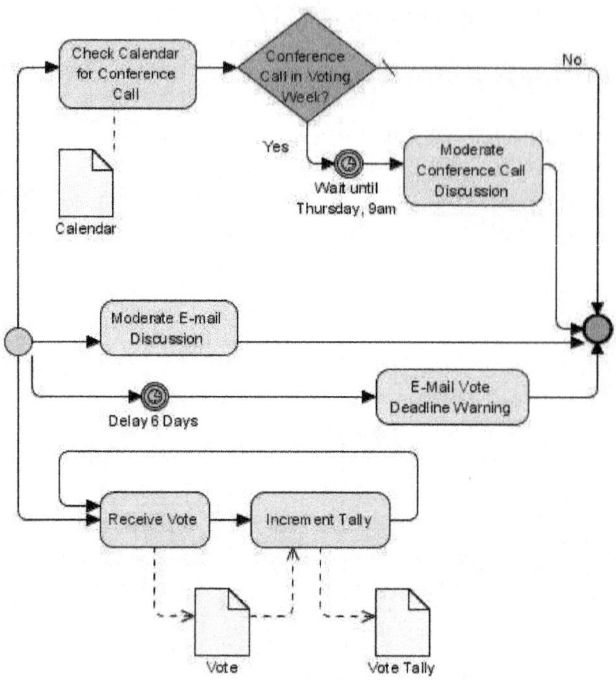

Fig. : Collect votes.

BPMN 2.0

The vision of BPMN 2.0 is to have one single specification for a new Business Process Model and Notation that defines the notation, metamodel and interchange format but with a modified name that still preserves the "BPMN" brand. The features include :

- Aligning BPMN with the business process definition meta model BPDM to form a single consistent language
- Enabling the exchange of business process models and their diagram layouts among process modelling tools to preserve semantic integrity
- Expand BPMN to allow model orchestrations and choreographies as standalone or integrated models
- Support the display and interchange of different perspectives on a model that allow a user to focus on specific concerns
- Serialize BPMN and provide XML schemes for model transformation and to extend BPMN towards business modelling and executive decision support.

The final version of the specification was released in January, 2011.

Comparison of BPMN Versions

Attributes	BPMN 1.0	BPMN 1.1	BPMN 1.2	BPMN 2.0-Beta 1
Consortium	BPMI	OMG	OMG	OMG
Date of release	May 2004	January 2008	January 2009	August 2009
Models	Collaborative (public) B2B processes, internal (private) business processes.			Collaborative (public) B2B processes, internal (private) business processes, a choreography-expected behaviour between two or more business participants, collaborations, which is a collection of participants and their interaction and a conversation—the logical relation of message exchanges.
event	**start** (none, message, timer, rule, link, multiple) **intermediate** (none, message, timer, error, cancel, compensation, rule, link, multiple) **end** (none, message, error, cancel, compensation, link, terminate, multiple)	**start** (none, message, timer, conditional, signal, multiple) **intermediate** (none, message, timer, error, cancel, compensation, conditional, link, signal, multiple) **end** (none, message, error, cancel, compensation, signal, terminate, multiple)		**start** **top-level** (none, message, timer, conditional, signal, multiple, parallel multiple) **event sub-process interrupting** (message, timer, escalation, conditional, error, compensation, signal, multiple, parallel multiple) **event sub-process non-interrupting** (message, timer, escalation, conditional, signal, multiple, parallel multiple)

(Contd...)

(Contd...)

			intermediate **catching** (message, timer, conditional, link, signal, multiple, parallel multiple) **boundary interrupting** (message, timer, escalation, conditional, error, cancel, compensation, signal, multiple, parallel multiple) **boundary non-interrupting** (message, timer, escalation, conditional, signal, multiple, parallel multiple, terminate) **throwing** (none, message, escalation, link, compensation, signal, multiple, parallel multiple) **end** (none, message, escalation, error, cancel, compensation, signal, multiple, terminate)
activity	**task** (atomic) **process/sub-process** (nonatomic) collapsed sub-process expanded sub-process		**task** (atomic) **choreography task** collapsed choreography sub-process expanded choreography sub-process **process/sub-process** (non-atomic) collapsed sub-process expanded sub-process
gateway	**XOR**-exclusive decision and merging. both data-based and event-based. data-based can be shown with or without the "x" marker. **OR**-inclusive decision and merging **complex**-complex conditions and situations **AND**-forking and joining	**exclusive** decision and merging. both data-based and event-based. data-based can be shown with or without the "x" marker. **inclusive** decision and merging. **complex**-complex conditions and situations. **parallel** forking and joining.	**exclusive** decision and merging. both data-based and event-based. exclusive can be shown with or without the "x" marker. **inclusive** gateway decision and merging **complex gateway**-complex conditions and situations **parallel gateway**-forking and joining
sequence flow	**normal flow** uncontrolled flow conditional flow default flow **exception flow**		
message flow	message flow		

(Contd...)

(*Contd...*)

association	association			
pool	pool			
lane	lane			
data objects	data object			data object collection data input data output
groups	group			
annotations	annotations			
message	/	/	/	**message**
other elements	looping activity looping sequence flow looping multiple instances process break transactions nested/embedded sub-process off-page connector compensation association			looping activity looping sequence flow looping multiple instances process break transactions nested/embedded sub-process off-page connector compensation association communication (subcommunication) communication link
Number of all elements	48	55	55	116
Major changes	/	The new specification introduces a categorization of event triggers into "catching" and "throwing" events. *i.e.* there are two kinds of intermediate message events now-one kind responsible for reception of messages ("catching") and one kind responsible	The BPMN 1.2 minor revision changes consist of editorial corrections and implementation bug fixes. Consequently, these minor changes affect modelling tool vendors more than modelers (users).	Choreographies Choreographies-model Conversation-model Complete Metamodel BPMN Core BPMN Execution Semantics BPMN-BPEL Mapping XPDL (BPMN XML Serialization) Diagram Interchange Elements For Abstraction Callable Element Call Activity Global Task Gateways (Updated) Exclusive/Parallel Event-based Gateway (they stand at the beginning of the process) Tasks/SubProcesses (Updated) Event-Subprocess (Used to handle events in the bounding subprocess) BusinessRule task Sequential Multi-Instance Activity Service Task

(*Contd...*)

(*Contd...*)

	for sending messages ("throwing"). In addition to the old types, it introduces a new type, the **signal event**. Start and end link events do not exist any longer in BPMN 1.1. The old "rule events" where renamed to **conditional events**. The semantics and appearance have not changed. The event-based gateway in BPMN 1.1 looks slightly different from what it looked like in 1.0. Instead of the hexagonal star it now has a pentagon in its center. The same shape is also used for the multiple events (start, intermediate, end). There is an additional line separating your lane's description from its content.	Artifacts (Updated) Data Objects (Collection, Data Input, Data Output)

Types of BPMN Sub-model

Business process modelling is used to communicate a wide variety of information to a wide variety of audiences. BPMN is designed to cover this wide range of usage and allows modelling of end-to-end business processes to allow the viewer of the Diagram to be able to easily differentiate between sections of a BPMN Diagram. There are three basic types of sub-models within an end-to-end BPMN model : Private (internal) business processes, Abstract (public) processes, and Collaboration (global) processes :

Private (Internal) Business Processes

Private business processes are those internal to a specific organization and are the type of processes that have been generally called workflow or BPM processes. If swim lanes are used then a private business process will be contained within a single Pool. The Sequence Flow of the Process is therefore contained within the Pool and cannot cross the boundaries of the Pool. Message Flow can cross the Pool boundary to show the interactions that exist between separate private business processes.

Abstract (Public) Processes

This represents the interactions between a private business process and another process or participant. Only those activities that communicate outside the private business process are included in the abstract process. All other "internal" activities of the private business process are not shown in the abstract process. Thus, the abstract process shows to the outside world the sequence of messages that are required to interact with that business process. Abstract processes are contained within a Pool and can be modelled separately or within a larger BPMN Diagram to show the Message Flow between the abstract process activities and other entities. If the abstract process is in the same Diagram as its corresponding private business process, then the activities that are common to both processes can be associated.

Collaboration (Global) Processes

A collaboration process depicts the interactions between two or more business entities. These interactions are defined as a sequence of activities that represent the message exchange patterns between the entities involved. Collaboration processes may be contained within a Pool and the different participant business interactions are shown as Lanes within the Pool. In this situation, each Lane would represent two participants and a direction of travel between them. They may also be shown as two or more Abstract Processes interacting through Message Flow (as described in the previous section). These processes can be modelled separately or within a larger BPMN Diagram to show the Associations between the collaboration process activities and other entities. If the collaboration process is in the same Diagram as one of its corresponding private business process, then the activities that are common to both processes can be associated.

Within and between these three BPMN sub-models, many types of Diagrams can be created. The following are the types of business processes that can be modelled with BPMN (those with asterisks may not map to an executable language) :

- High-level private process activities (not functional breakdown)*
- Detailed private business process
- As-is or old business process*
- To-be or new business process
- Detailed private business process with interactions to one or more external entities (or "Black Box" processes)
- Two or more detailed private business processes interacting
- Detailed private business process relationship to Abstract Process
- Detailed private business process relationship to Collaboration Process
- Two or more Abstract Processes*
- Abstract Process relationship to Collaboration Process*
- Collaboration Process only (*e.g.*, ebXML BPSS or RosettaNet)*
- Two or more detailed private business processes interacting through their Abstract Processes and/or a Collaboration Process.

BPMN is designed to allow all the above types of Diagrams. However, it should be cautioned that if too many types of sub-models are combined, such as three or more private processes with message flow between each of them, then the Diagram may become too hard for someone to understand. Thus, we recommend that the modeler pick a focused purpose for the BPD, such as a private process, or a collaboration process.

Weaknesses of BPMN

The weaknesses of BPMN could relate to :

- Ambiguity and confusion in sharing BPMN models
- Support for routine work
- Support for knowledge work, and
- Converting BPMN models to executable environments.

BPEL and BPMN

The BPMN specification includes an informal and partial mapping from BPMN to BPEL 1.1. A more detailed mapping of BPMN to BPEL has been implemented in a number of tools, including an open-source tool known as BPMN2BPEL. However, the development of these tools has exposed fundamental differences between BPMN and BPEL, which make it very difficult, and in some cases impossible, to generate human-readable BPEL code from BPMN models. Even more difficult is the problem of BPMN-to-BPEL *round-trip* engineering : generating BPEL code from BPMN diagrams and maintaining the original BPMN model and the generated

BPEL code synchronized, in the sense that any modification to one is propagated to the other.

ROOT CAUSE ANALYSIS

Root cause analysis (RCA) is a method of problem solving that tries to identify the root causes of faults or problems.

RCA practice tries to solve problems by attempting to identify and correct the root causes of events, as opposed to simply addressing their symptoms. Focusing correction on root causes has the goal of preventing problem recurrence. RCFA (Root Cause Failure Analysis) recognizes that complete prevention of recurrence by one corrective action is not always possible.

Conversely, there may be several effective measures (methods) that address the root causes of a problem. Thus, RCA is an iterative process and a tool of continuous improvement.

RCA is typically used as a reactive method of identifying event(s) causes, revealing problems and solving them. Analysis is done *after* an event has occurred. Insights in RCA may make it useful as a pre-emptive method. In that event, RCA can be used to *forecast* or predict probable events even *before* they occur. While one follows the other, RCA is a completely separate process to Incident Management.

Root cause analysis is not a single, sharply defined methodology; there are many different tools, processes, and philosophies for performing RCA. However, several very-broadly defined approaches or "schools" can be identified by their basic approach or field of origin : safety-based, production-based, process-based, failure-based, and systems-based.

- Safety-based RCA descends from the fields of accident analysis and occupational safety and health.
- Production-based RCA has its origins in the field of quality control for industrial manufacturing.
- Process-based RCA is basically a follow-on to production-based RCA, but with a scope that has been expanded to include business processes.
- Failure-based RCA is rooted in the practice of failure analysis as employed in engineering and maintenance.
- Systems-based RCA has emerged as an amalgamation of the preceding schools, along with ideas taken from fields such as change management, risk management, and systems analysis.

Despite the different approaches among the various schools of root cause analysis, there are some common principles. It is also possible to define several general processes for performing RCA.

General Principles of Root Cause Analysis

1. The primary aim of root cause analysis is to identify the factors that resulted in the nature, the magnitude, the location, and the timing of the harmful out-

comes (consequences) of one or more past events in order to identify what behaviours, actions, inactions, or conditions need to be changed to prevent recurrence of similar harmful outcomes and to identify the lessons to be learned to promote the achievement of better consequences. ("Success" is defined as the near-certain prevention of recurrence.)

2. To be effective, root cause analysis must be performed systematically, usually as part of an investigation, with conclusions and root causes that are identified backed up by documented evidence. Usually a team effort is required.
3. There may be more than one root cause for an event or a problem, the difficult part is demonstrating the persistence and sustaining the effort required to determine them.
4. The purpose of identifying all solutions to a problem is to prevent recurrence at lowest cost in the simplest way. If there are alternatives that are equally effective, then the simplest or lowest cost approach is preferred.
5. Root causes identified depend on the way in which the problem or event is defined. Effective problem statements and event descriptions (as failures, for example) are helpful, or even required.
6. To be effective, the analysis should establish a sequence of events or timeline to understand the relationships between contributory (causal) factors, root cause(s) and the defined problem or event to prevent in the future.
7. Root cause analysis can help transform a reactive culture (that reacts to problems) into a forward-looking culture that solves problems before they occur or escalate. More importantly, it reduces the frequency of problems occurring over time within the environment where the root cause analysis process is used.
8. Root cause analysis is a threat to many cultures and environments. Threats to cultures often meet with resistance. There may be other forms of management support required to achieve root cause analysis effectiveness and success. For example, a "non-punitive" policy toward problem identifiers may be required.

General Process for Performing and Documenting an RCA-based Corrective Action

RCA forms the most critical part of successful corrective action, because it directs the corrective action at the true root cause of the problem. Knowing the root cause is secondary to the goal of prevention, but without knowing the root cause, it is not possible to determine what an effective corrective action for the defined problem would be :

1. Define the problem or describe the event factually. Include the qualitative and quantitative attributes (properties) of the harmful outcomes. This usually includes specifying the natures, the magnitudes, the locations, and the timing of events.
2. Gather data and evidence, classifying it along a timeline of events to the final failure or crisis. For every behaviour, condition, action, and inaction specify

in the "timeline" what should have been done when it differs from what was done.

3. Ask "why" and identify the causes associated with each step in the sequence towards the defined problem or event. "Why" is taken to mean "What were the factors that directly resulted in the effect?"
4. Classify causes into causal factors that relate to an event in the sequence and root causes, that if eliminated, can be agreed to have interrupted that step of the sequence chain.
5. Identify all other harmful factors that have equal or better claim to be called "root causes." If there are multiple root causes, which is often the case, reveal those clearly for later optimum selection.
6. Identify corrective action(s) that will with certainty prevent recurrence of each harmful effect, including outcomes and factors. Check that each corrective action would, if pre-implemented before the event, have reduced or prevented specific harmful effects.
7. Identify solutions that, when effective, and with consensus agreement of the group, prevent recurrence with reasonable certainty, are within the institution's control, meet its goals and objectives and do not cause or introduce other new, unforeseen problems.
8. Implement the recommended root cause correction(s).
9. Ensure effectiveness by observing the implemented recommendation solutions.
10. Identify other methodologies for problem solving and problem avoidance that may be useful.
11. Identify and address the other instances of each harmful outcome and harmful factor.

Chapter 20

DATA PROCESSING

Data processing is, broadly, "the collection and manipulation of items of data to produce meaningful information." In this sense it can be considered a sub-set of *information processing*, "the change (processing) of information in any manner detectable by an observer."

The term is often used more specifically in the context of a business or other organization to refer to the class of commercial data processing applications.

DATA PROCESSING FUNCTIONS

Data processing may involve various processes, including :

- Validation—Ensuring that supplied data is "clean, correct and useful."
- Sorting—"arranging items in some sequence and/or in different sets."
- Summarization—reducing detail data to its main points.
- Aggregation—combining multiple pieces of data.
- Analysis—the "collection, organization, analysis, interpretation and presentation of data.".
- Reporting—list detail or summary data or computed information.

HISTORY

The United States Census Bureau illustrates the evolution of data processing from manual through electronic procedures.

Manual Data Processing

Although widespread use of the term *data processing* dates only from the nineteen-fifties data processing functions have been performed manually for millennia. For example, book keeping involves functions such as posting transactions and producing reports like the balance sheet and the cash flow statement. Completely manual methods were augmented by the application of mechanical or electronic calculators. A person whose job it was to perform calculations manually or using a calculator was called a "computer."

The 1850 United States Census schedule was the first to gather data by individual rather than household. A number of questions could be answered by making a check in the appropriate box on the form. From 1850 through 1880 the Census Bureau employed "a system of tallying, which, by reason of the increasing number of combinations of classifications required, became increasingly complex. Only a limited number of combinations could be recorded in one tally, so it was necessary to handle the schedules 5 or 6 times, for as many independent tallies." "It took over 7 years to publish the results of the 1880 census" using manual processing methods.

Automatic Data Processing

The term *automatic data processing* was applied to operations performed by means of unit record equipment, such as Herman Hollerith's application of punched card equipment for the 1890 United States Census. "Using Hollerith's punchcard equipment, the Census Office was able to complete tabulating most of the 1890 census data in 2 to 3 years, compared with 7 to 8 years for the 1880 census.... It is also estimated that using Herman Hollerith's system saved some $5 million in processing costs" even with twice as many questions as in 1880.

Electronic Data Processing

Computerized data processing, or Electronic data processing represents the further evolution, with the computer taking the place of several independent pieces of equipment. The Census Bureau first made limited use of electronic computers for the 1950 United States Census, using a UNIVAC I system, delivered in 1952.

Further Evolution

"Data processing (DP)" has also previously been used to refer to the department within an organization responsible for the operation of data processing applications. The term *data processing* has mostly been sub-sumed under the newer and somewhat more general term *information technology* (IT). "Data processing" has acquired a negative connotation, suggesting use of older technologies. As an example, in 1996 the *Data Processing Management Association* (DPMA) changed its name to *Association of Information Technology Professionals*. Nevertheless, the terms are roughly synonymous.

APPLICATIONS

Commercial Data Processing

Commercial data processing involves a large volume of input data, relatively few computational operations, and a large volume of output. For example, an insurance company needs to keep records on tens or hundreds of thousands of policies, print and mail bills, and receive and post payments.

Data Analysis

In a science or engineering field, the terms *data processing* and *information systems* are considered too broad, and the more specialized term data analysis is typically used. Data analysis makes use of specialized and highly accurate algorithms and statistical calculations that are less often observed in the typical general business environment.

One divergence of culture between data processing and data analysis is shown by the numerical representations generally used; In data processing, measurements are typically stored as integers, fixed-point or binary-coded decimal representations of numbers, whereas the majority of measurements in data analysis are stored as floating-point representations of rational numbers.

For data analysis, packages like SPSS or SAS, or their free counterparts such as DAP, gretl or PSPP are often used.

Real Estate Appraisal Data Entry

Real estate appraisal, property valuation or land valuation is the process of valuing real property. In real estate valuations, an appraiser provide all the data he/she needs to input in his/her appraisal reports. Data Entry team does the rest—They searches, consolidates and types the data into reports, such as subject data and comparable grid prior sales history. Most data entry professionals work 24 hours a day, 7 days a week, 365 days a year. Appraiser sends empty reports, and data entry team work all day and night, even while appraiser is sleeping. This process can increase appraiser's efficiency, and free up his/her time.

DATA PROCESSING SYSTEM

A **data processing system** is a combination of machines and people that for a set of inputs produces a defined set of outputs. The inputs and outputs are interpreted as data, facts, information,... depending on the interpreter's relation to the system. A common synonymous term is "information system".

A data processing system may involve some combination of:

- Conversion converting data to another format.
- Validation—Ensuring that supplied data is "clean, correct and useful."
- Sorting—"arranging items in some sequence and/or in different sets."
- Summarization—reducing detail data to its main points.
- Aggregation—combining multiple pieces of data.
- Analysis—the "collection, organization, analysis, interpretation and presentation of data."
- Reporting—list detail or summary data or computed information.

Types of Data Processing Systems

By Application Area

Scientific Data Processing

Scientific data processing "usually involves a great deal of computation (arithmetic and comparison operations) upon a relatively small amount of input data, resulting in a small volume of output."

Commercial Data Processing

Commercial data processing "involves a large volume of input data, relatively few computational operations, and a large volume of output." Accounting programs are the prototypical examples of data processing applications. Information systems (IS) is the field that studies such organizational computer systems.

Data Analysis

"Data analysis is a body of methods that help to describe facts, detect patterns, develop explanations, and test hypotheses." For example, data analysis might be used to look at sales and customer data to "identify connections between products to allow for cross selling campaigns."

By Service Type

- Transaction processing systems
- Information storage and retrieval systems
- Command and control systems
- Computing service systems
- Process control systems
- Message switching systems.

Examples

Simple Example

A very simple example of a data processing system is the process of maintaining a check register. Transactions—checks and deposits—are recorded as they occur and the transactions are summarized to determine a current balance. Monthly the data recorded in the register is reconciled with a hopefully identical list of transactions processed by the bank.

A more sophisticated record keeping system might further identify the transactions—for example, deposits by source or checks by type, such as charitable contributions. This information might be used to obtain information like the total of all contributions for the year.

The important thing about this example is that it is a *system*, in which, all transactions are recorded consistently, and the same method of bank reconciliation is used each time.

Real-world Example

This is a flowchart of a data processing system combining manual and computerized processing to handle accounts receivable, billing, and general ledger

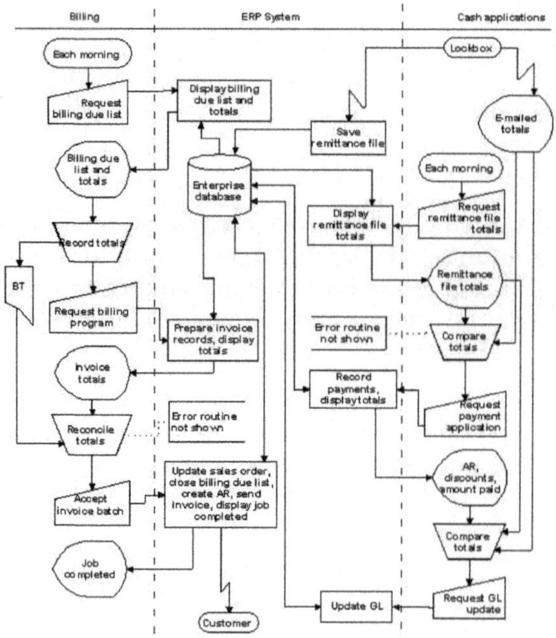

DATA VALIDATION

In computer science, **data validation** is the process of ensuring that a program operates on clean, correct and useful data. It uses routines, often called "validation rules" "validation constraints" or "check routines", that check for correctness, meaningfulness, and security of data that are input to the system. The rules may be implemented through the automated facilities of a data dictionary, or by the inclusion of explicit application program validation logic.

Overview

Data validation is intended to provide certain well-defined guarantees for fitness, accuracy, and consistency for any of various kinds of user input into an application or automated system. Data validation rules can be defined and designed using any of various methodologies, and be deployed in any of various contexts.

Data validation rules may be defined, designed and deployed, for example :

Definition and Design Contexts

- As a part of requirements-gathering phase in a software engineering or designing a software specification.

- As part of an operations modelling phase in business process modelling

Deployment Contexts

- As part of a user-interface
- As a set of programs or business-logic routines in a programming language
- As a set of stored-procedures in a database management system.

For business applications, data validation can be defined through declarative data integrity rules, or procedure-based business rules. Data that does not conform to these rules will negatively affect business process execution. Therefore, data validation should start with business process definition and set of business rules within this process. Rules can be collected through the requirements capture exercise.

Different Kinds of Validation

In evaluating the basics of data validation, generalizations can be made regarding the different types of validation, according to the scope, complexity, and purpose of the various validation operations to be carried out.

For Example

- Data type validation;
- Range and constraint validation;
- Code and Cross-reference validation; and
- Structured validation.

Data Type Validation

Data type validation is customarily carried out on one or more simple data fields.

The simplest kind of data type validation verifies that the individual characters provided through user input are consistent with the expected characters of one or more known primitive data types; as defined in a programming language, data storage and retrieval mechanism, or otherwise.

For example, many database systems allow the specification of the following primitive data types : (1) integer; (2) float (decimal); or (3) string.

Similarly, telephone numbers are routinely expected to include the digits and possibly the characters +, –, (, and) (plus, minus, and parentheses). A more sophisticated data validation routine would check to see the user had entered a valid country code, *i.e.*, that the number of digits entered matched the convention for the country or area specified.

A validation process involves two distinct steps : (a) Validation Check and (b) Post-Check action. The check step uses one or more computational rules to determine if the data is valid. The Post-validation action sends feedback to help enforce validation.

Simple Range and Constraint Validation

Simple range and constraint validation may examine user input for consistency with a minimum/maximum range, or consistency with a test for evaluating a sequence of characters, such as one or more tests against regular expressions.

Code and Cross-reference Validation

Code and cross-reference validation includes tests for data type validation, combined with one or more operations to verify that the user-supplied data is consistent with one or more external rules, requirements, or validity constraints relevant to a particular organization, context or set of underlying assumptions. These additional validity constraints may involve cross-referencing supplied data with a known look-up table or directory information service such as LDAP.

For example, a user may enter a well-formed string that matches the specification for a valid e-mail address, as defined in RFC 5322 but that well-formed string might not actually correspond to a resolvable domain connected to an active e-mail account.

Structured Validation

Structured validation allows for the combination of any of various basic data type validation steps, along with more complex processing. Such complex processing may include the testing of conditional constraints for an entire complex data object or set of process operations within a system.

Validation Methods

Allowed character checks—Checks that ascertain that only expected characters are present in a field. For example a numeric field may only allow the digits 0-9, the decimal point and perhaps a minus sign or commas. A text field such as a personal name might disallow characters such as < and >, as they could be evidence of a markup-based security attack. An e-mail address might require at least one @ sign and various other structural details. Regular expressions are effective ways of implementing such checks.

Batch totals—Checks for missing records. Numerical fields may be added together for all records in a batch. The batch total is entered and the computer checks that the total is correct, *e.g.*, add the 'Total Cost' field of a number of transactions together.

Cardinality check—Checks that record has a valid number of related records. For example if Contact record classified as a Customer it must have at least one associated Order (Cardinality > 0). If order does not exist for a "customer" record then it must be either changed to "seed" or the order must be created. This type of rule can be complicated by additional conditions. For example if contact record in Payroll database is marked as "former employee", then this record must not have any associated salary payments after the date on which employee left organization (Cardinality = 0).

Check digits—Used for numerical data. An extra digit is added to a number which is calculated from the digits. The computer checks this calculation when data are entered. For example the last digit of an ISBN for a book is a check digit calculated modulus 10.[3]

Consistency checks—Checks fields to ensure data in these fields corresponds, *e.g.*, If Title = "Mr.", then Gender = "M".

Control totals—This is a total done on one or more numeric fields which appears in every record. This is a meaningful total, *e.g.*, add the total payment for a number of Customers.

Cross-system consistency checks—Compares data in different systems to ensure it is consistent, *e.g.*, The address for the customer with the same id is the same in both systems. The data may be represented differently in different systems and may need to be transformed to a common format to be compared, *e.g.*, one system may store customer name in a single Name field as 'Doe, John Q', while another in three different fields : First_Name (John), Last_Name (Doe) and Middle_Name (Quality); to compare the two, the validation engine would have to transform data from the second system to match the data from the first, for example, using SQL : Last_Name | | ',' | | First_Name | | substr(Middle_Name, 1, 1) would convert the data from the second system to look like the data from the first 'Doe, John Q'

Data type checks—Checks the data type of the input and give an error message if the input data does not match with the chosen data type, *e.g.*, In an input box accepting numeric data, if the letter 'O' was typed instead of the number zero, an error message would appear.

File existence check—Checks that a file with a specified name exists. This check is essential for programs that use file handling.

Format or picture check—Checks that the data is in a specified format (template), *e.g.*, dates have to be in the format DD/MM/YYYY. Regular expressions should be considered for this type of validation.

Hash totals—This is just a batch total done on one or more numeric fields which appears in every record. This is a meaningless total, *e.g.*, add the Telephone Numbers together for a number of Customers.

Limit check—Unlike range checks, data are checked for one limit only, upper OR lower, *e.g.*, data should not be greater than 2 (<=2).

Logic check—Checks that an input does not yield a logical error, *e.g.*, an input value should not be 0 when it will divide some other number somewhere in a program.

Presence check—Checks that important data is actually present and have not been missed out, *e.g.*, customers may be required to have their telephone numbers listed.

Range check—Checks that the data lie within a specified range of values, *e.g.*, the month of a person's date of birth should lie between 1 and 12.

Referential integrity—In modern Relational database values in two tables can be linked through foreign key and primary key. If values in the primary key field are not constrained by database internal mechanism, then they should be validated. Validation of the foreign key field checks that referencing table must always refer to a valid row in the referenced table.

Spelling and grammar check—Looks for spelling and grammatical errors.

Uniqueness check—Checks that each value is unique. This can be applied to several fields (*i.e.* Address, First Name, Last Name).

Table Look Up Check—A table look up check takes the entered data item and compares it to a valid list of entries that are stored in a database table.

Post-validation Actions

Enforcement Action

Enforcement action typically rejects the data entry request and requires the input actor to make a change that brings the data into compliance. This is most suitable for interactive use, where a real person is sitting on the computer and making entry. It also works well for batch upload, where a file input may be rejected and a set of messages sent back to the input source for why the data is rejected. Another form of enforcement action involves automatically changing the data and saving a conformant version instead of the original version. This is most suitable for cosmetic change. For example, converting an [all-caps] entry to a [Pascal case] entry does not need user input. An inappropriate use of automatic enforcement would be in situations where the enforcement leads to loss of business information. For example, saving a truncated comment if the length is longer than expected. This is not typically a good thing since it may result in loss of significant data.

Advisory Action

Advisory actions typically allow data to be entered unchanged but sends a message to the source actor indicating those validation issues that were encountered. This is most suitable for non-interactive system, for systems where the change is not business critical, for cleansing steps of existing data and for verification steps of an entry process.

Verification Action

Verification actions are special cases of advisory actions. In this case, the source actor is asked to verify that this data is what they would really want to enter, in the light of a suggestion to the contrary. Here, the check step suggests an alternative (*e.g.* : a check of your mailing address returns a different way of formatting that address or suggests a different address altogether). You would want in this case, to give the user the option of accepting the recommendation or keeping their version. This is not a strict validation process, by design and is useful for capturing addresses to a new location or to a location that is not yet supported by the validation databases.

Validation and Security

Failures or omissions in data validation can lead to data corruption or a security vulnerability. Data validation checks that data are valid, sensible, reasonable, and secure before they are processed.

Sorting type

SORTING

Sorting is any process of arranging items according to a certain sequence or in different sets, and therefore, it has two common, yet distinct meanings :

1. *Ordering* : Arranging items of the same kind, class or nature, in some ordered sequence,
2. *Categorizing* : Grouping and labelling items with similar properties together (by sorts).

Sorting Information or Data

In computer science, sorting is one of the most extensively researched subjects because of the need to speed up the operation on thousands or millions of records during a search operation.

The main purpose of sorting information is to optimise its usefulness for specific tasks. In general, there are two ways of grouping information : **by category** *e.g.* a shopping catalogue where items are compiled together under headings such as 'home', 'sport & leisure', 'women's clothes' etc. (nominal scale) and **by the intensity** of some property, such as price, *e.g.* from the cheapest to most expensive (ordinal scale). Richard Saul Wurman, in his book *Information Anxiety*, proposes that the most common sorting purposes are **name**, by **location** and by **time** (these are actually special cases of category and hierarchy). Together these give the acronym LATCH (Location, Alphabetical, Time, Category, Hierarchy) and can be used to describe just about every type of ordered information.

Often information is sorted using different methods at different levels of abstraction : *e.g.* the UK telephone directories which are sorted by location, by category (business or residential) and then alphabetically. New media still subscribe to these basic sorting methods : *e.g.* a Google search returns a list of web pages in a hierarchical list based on its own scoring system for how closely they match the search criteria (from closest match downwards).

The opposite of sorting, rearranging a sequence of items in a random or meaningless order, is called shuffling.

For sorting, either a weak order, "should not come after", can be specified, or a strict weak order, "should come before" (specifying one defines also the other, the two are the complement of the inverse of each other, see operations on binary relations). For the sorting to be unique, these two are restricted to a total order and a strict total order, respectively.

Sorting n-tuples (depending on context also called *e.g.* records consisting of fields) can be done based on one or more of its components. More generally objects can be sorted based on a property. Such a component or property is called a **sort key**.

For example, the items are books, the sort key is the title, subject or author, and the order is alphabetical.

A new sort key can be created from two or more sort keys by lexicographical order. The first is then called the **primary sort key**, the second the **secondary sort key**, etc.

For example, addresses could be sorted using the city as primary sort key, and the street as secondary sort key.

If the sort key values are totally ordered, the sort key defines a weak order of the items : items with the same sort key are equivalent with respect to sorting. If different items have different sort key values then this defines a unique order of the items.

A standard order is often called *ascending* (corresponding to the fact that the standard order of numbers is ascending, *i.e.* A to Z, 0 to 9), the reverse order *descending* (Z to A, 9 to 0). For dates and times, *ascending* means that earlier values precede later ones *e.g.* 1/1/2000 will sort ahead of 1/1/2001.

SUMMARY STATISTIC

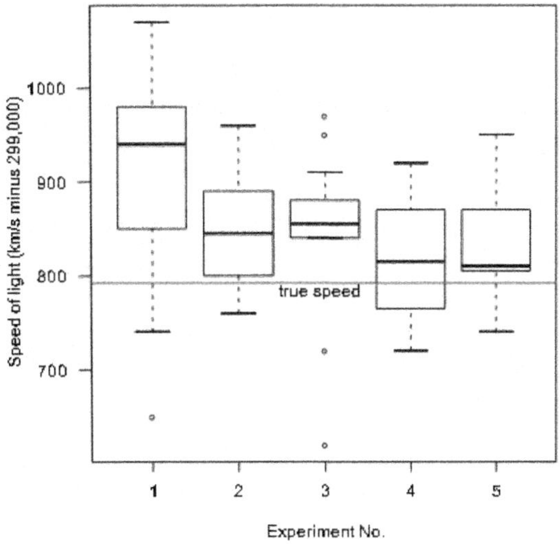

Fig. : Box plot of the Michelson–Morley experiment, showing several summary statistics.

In descriptive statistics, **summary statistics** are used to summarize a set of observations, in order to communicate the largest amount of information as simply as possible. Statisticians commonly try to describe the observations in :

- A measure of location, or central tendency, such as the arithmetic mean
- A measure of statistical dispersion like the standard deviation
- A measure of the shape of the distribution like skewness or kurtosis
- If more than one variable is measured, a measure of statistical dependence such as a correlation coefficient.

A common collection of order statistics used as summary statistics are the five-number summary, sometimes extended to a seven-number summary, and the associated box plot.

Entries in an analysis of variance table can also be regarded as summary statistics.

Examples of Summary Statistics

Location

Common measures of location, or central tendency, are the arithmetic mean, median, mode, and interquartile mean.

Spread

Common measures of statistical dispersion are the standard deviation, variance, range, interquartile range, absolute deviation and the distance standard deviation. Measures that assess spread in comparison to the typical size of data values include the coefficient of variation.

The Gini coefficient was originally developed to measure income inequality and is equivalent to one of the L-moments.

A simple summary of a dataset is sometimes given by quoting particular order statistics as approximations to selected percentiles of a distribution.

Shape

Common measures of the shape of a distribution are skewness or kurtosis, while alternatives can be based on L-moments. A different measure is the distance skewness, for which a value of zero implies central symmetry.

Dependence

The common measure of dependence between paired random variables is the Pearson product-moment correlation coefficient, while a common alternative summary statistic is Spearman's rank correlation coefficient. A value of zero for the distance correlation implies independence.

AGGREGATE DATA

In statistics, **aggregate data** describes data combined from several measurements. When data are aggregated, groups of observations are replaced with summary statistics based on those observations.

In economics, **aggregate data** or **data aggregates** describes high-level data that is composed from a multitude or combination of other more individual data.

- In macro-economics, data such as the overall price level or overall inflation rate
- In micro-economics, data of an entire sector of an economy composed of many firms, or of all households in a city or region.

AUTOMATIC DATA PROCESSING EQUIPMENT

Automatic data processing equipment, legally defined, is any equipment or interconnected system or subsystems of equipment that is used in the :

- Automatic acquisition,
- Storage,
- Manipulation,
- Management,
- Movement,
- Control,
- Display,
- Switching,

- Interchange,
- Transmission, or
- Reception,

  Of data or information
- By a Federal agency, or
- Under a contract with a Federal agency

  Which
- Requires the use of such equipment, or
- Requires the performance of a service or the furnishing of a product which is performed or produced making significant use of such equipment.

  The term includes
- Computer,
- Ancillary equipment,
- Software,
- Firmware, and similar procedures,
- Services, including support services, and
- Related resources as defined by regulations issued by the Administrator for General Services.

BINARY PRIORITIZATION

Binary Prioritization is a sorting algorithm which prioritizes to-do tasks.

Unlike other binary sort methods (*e.g.* binary search) this method assumes that the deferred work will be prioritized in a later process, but their order is not relevant in the first iteration. The faster processing of classified and important tasks is achieved by reducing the cost of sorting by not sorting the sub-set of the less important tasks. In each iteration, the cost is reduced by the sorted elements.

Application Requirements

For the application of the binary prioritization it is assumed that the elements to be prioritized are present in an unsorted heap (abstract : *an unsorted list*).

Algorithm

Step 1

Step 2

Step 3

Step 4

The algorithm of the binary prioritization is as follows :

- One element is taken from the stack (or list).
- The element (the object) is analyzed regarding its priority (importance) relative to the other elements.
- If the item is judged as *important*, it is sorted on a new batch (to a new list) of important elements; if not, it is sorted on a different stack (in a different list) of unimportant elements. This is repeated in the first iteration until all items have been evaluated into two new stacks (or lists).
- The two stacks (lists) are then reunited by the stacking the important elements on the elements that were considered to be unimportant. The last element of the first stack is memorized, *e.g.* by a *separator*.
- The algorithm is then applied in an additional iteration on the united stack up to the separator element.
- If there was only one element in the last step, the algorithm is complete.

ELECTRONIC DATA PROCESSING

Electronic Data Processing (EDP) can refer to the use of automated methods to process commercial data. Typically, this uses relatively simple, repetitive activities to process large volumes of similar information. For example, stock updates applied to an inventory, banking transactions applied to account and customer master files, booking and ticketing transactions to an airline's reservation system, billing for utility services. The modifier "electronic" or "automatic" was used with "data processing" (DP), especially ca. 1960, to distinguish human clerical data processing from that done by computer.

History

The first commercial business computer was developed in the United Kingdom in 1951, by the J. Lyons and Co. catering organization. This was known as the 'Lyons Electronic Office' – or LEO for short. It was developed further and used widely during the 1960s and early 1970s. (Joe Lyons formed a separate company to develop the LEO computers and this subsequently merged to form English Electric Leo Marconi and then International Computers Ltd.) By the end of the 1950s punched card manufacturers, Hollerith, Power-Samas, IBM and others, were also marketing an array of computers [2]. Early commercial systems were installed exclusively by large organizations. These could afford to invest the time and capital necessary to purchase hardware, hire specialist staff to develop bespoke software and work through the consequent (and often unexpected) organizational and cultural changes.

At first, individual organizations developed their own software, including data management utilities, themselves. Different products might also have 'one-off' bespoke software. This fragmented approach led to duplicated effort and the production of management information needed manual effort.

High hardware costs and relatively slow processing speeds forced developers to use resources 'efficiently'. Data storage formats were heavily compacted, for example. A common example is the removal of the century from dates, which eventually led to the 'millennium bug'.

Data input required intermediate processing via punched paper tape or punched card and separate input to a repetitive, labour intensive task, removed from user control and error-prone. Invalid or incorrect data needed correction and re-submission with consequences for data and account reconciliation.

Data storage was strictly serial on paper tape, and then later to magnetic tape : the use of data storage within readily accessible memory was not cost-effective. Significant developments took place in 1959 with IBM announcing the 1401 computer and in 1962 with ICT (International Computers & Tabulators) making delivery of the 1301. Like all machines during this time the core processor together with the peripherals-magnetic tape decks, discs, drums, printers and card and paper tape input and output required considerable space in specially constructed air conditioned accommodation. Often parts of the punched card installation, in particular sorters, were retained to present the card input to the computer in a pre-sort form that reduced the processing time involved in sorting large amounts of data.

Data processing facilities became available to smaller organization in the form of the Computer Services Bureau. These offered processing of specific applications *e.g.* payroll and were often a prelude to the purchase of customers own computer. Organizations used these facilities for testing programs while awaiting the arrival of their own machine.

These initial machines were delivered to customers with limited software. The design staff was divided into two groups. Systems Analysts produced a systems specification and Programmers translated the specification into machine language.

Literature on computers and EDP was sparse through articles appearing in accountancy publications and material supplied by the equipment manufacturers. The first issue of The Computer Journal published by The British Computer Society appeared in mid 1958. The UK Accountancy Body now named The Association of Chartered Certified Accountants formed an Electronic Data Processing Committee in July 1958 with the purpose of informing its members of the opportunities created by the computer. The Committee produced its first booklet in 1959, An Introduction to Electronic Computers. Also in 1958 The Institute of Chartered Accountants in England and Wales produced a paper Accounting by Electronic Methods. The notes indicated what appears capable and the possible implications of using a computer.

Progressive organizations attempted to go beyond the straight systems transfer from punched card equipment and unit accounting machines to the computer, to producing accounts to the trial balance stage and integrated management information systems. New procedures redesigned the way paper flowed, changed organizational structures, called for a rethink of the way information was presented to management and challenged the internal control principles adopted by the designers of accounting systems [3]. But the full realization of these benefits had to await the arrival of the next generation of computers

Today

As with other industrial processes commercial IT has moved in all respects from a custom-order, craft-based industry where the product was tailored to fit the customer; to multi-use components taken off the shelf to find the best-fit in any situation. Mass-production has greatly reduced costs and IT is available to the smallest company.

LEO was hardware tailored for a single client : Today, Intel Pentium and compatible chips are standard and become parts of other components which are combined as needed. One individual change of note was the freeing of computers and removable storage from protected, air-filtered environments. Microsoft and IBM at various times have been influential enough to impose order on IT and the resultant standardizations allowed specialist software to flourish.

Software is available off the shelf: Apart from Microsoft products such as Office, or Lotus, there are also specialist packages for payroll and personnel management, account maintenance and customer management, to name a few. These are highly specialized and intricate components of larger environments, but they rely upon common conventions and interfaces.

Data storage has also standardized : Relational databases are developed by different suppliers to common formats and conventions. Common file formats can be shared by large main-frames and desk-top personal computers, allowing online, real time input and validation.

In parallel, software development has fragmented : There are still specialist technicians, but these increasingly use standardized methodologies where outcomes are predictable and accessible. At the other end of the scale, any office manager can dabble in spreadsheets or databases and obtain acceptable results (but there are risks). Specialized software is software that is written for a specific task rather for a broad application area. These programs provide facilities specifically for the purpose for which they were designed.

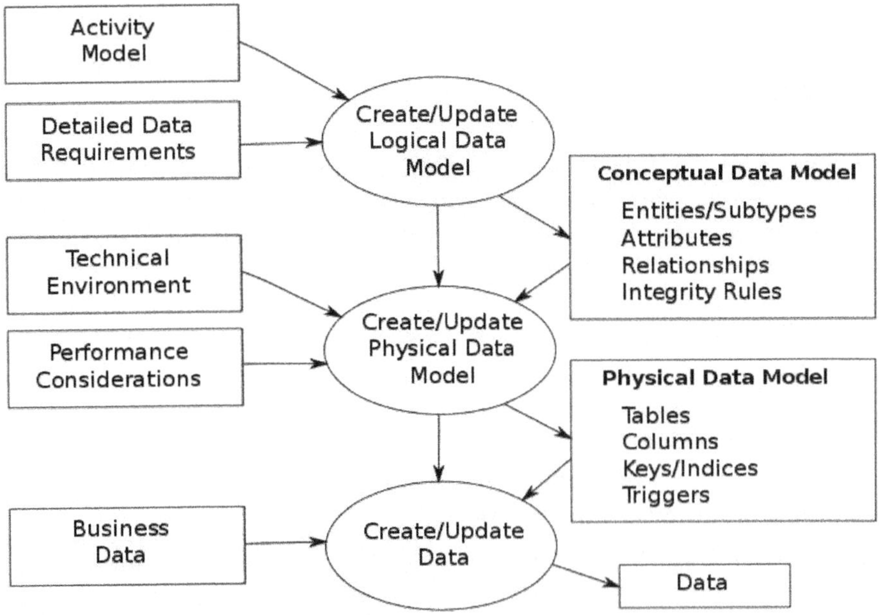

Fig. : The data modelling process. The figure illustrates the way data models are developed and used today. A conceptual data model is developed based on the data requirements for the application that is being developed, perhaps in the context of an activity model. The data model will normally consist of entity types, attributes, relationships, integrity rules, and the definitions of those objects. This is then used as the start point for interface or database design.

DATA MODELLING

Data modelling in software engineering is the process of creating a data model for an information system by applying formal data modelling techniques.

Overview

Data modelling is a process used to define and analyze data requirements needed to support the business processes within the scope of corresponding information systems in organizations. Therefore, the process of data modelling involves professional data modelers working closely with business stakeholders, as well as potential users of the information system.

There are three different types of data models produced while progressing from requirements to the actual database to be used for the information system. The data requirements are initially recorded as a conceptual data model which is essentially a set of technology independent specifications about the data and is used to discuss initial requirements with the business stakeholders. The conceptual model is then translated into a logical data model, which documents structures of the data that can be implemented in databases. Implementation of one conceptual data model may require multiple logical data models. The last step in data modelling is transforming the logical data model to a physical data model that organizes the data into tables, and accounts for access, performance and storage details. Data modelling defines not just data elements, but also their structures and the relationships between them.

Data modelling techniques and methodologies are used to model data in a standard, consistent, predictable manner in order to manage it as a resource. The use of data modelling standards is strongly recommended for all projects requiring a standard means of defining and analyzing data within an organization, *e.g.*, using data modelling :

- To assist business analysts, programmers, testers, manual writers, IT package selectors, engineers, managers, related organizations and clients to understand and use an agreed semi-formal model the concepts of the organization and how they relate to one another
- To manage data as a resource
- For the integration of information systems
- For designing databases/data warehouses (aka data repositories)

Data modelling may be performed during various types of projects and in multiple phases of projects. Data models are progressive; there is no such thing as the final data model for a business or application. Instead a data model should be considered a living document that will change in response to a changing business. The data models should ideally be stored in a repository so that they can be retrieved, expanded, and edited over time. Whitten *et. al.* (2004) determined two types of data modelling :

- *Strategic data modelling :* This is part of the creation of an information systems strategy, which defines an overall vision and architecture for information systems is defined. Information engineering is a methodology that embraces this approach.

- *Data modelling during systems analysis :* In systems analysis logical data models are created as part of the development of new databases.

Data modelling is also used as a technique for detailing business requirements for specific databases. It is sometimes called *database modelling* because a data model is eventually implemented in a database.

Data Modelling Topics

Data Models

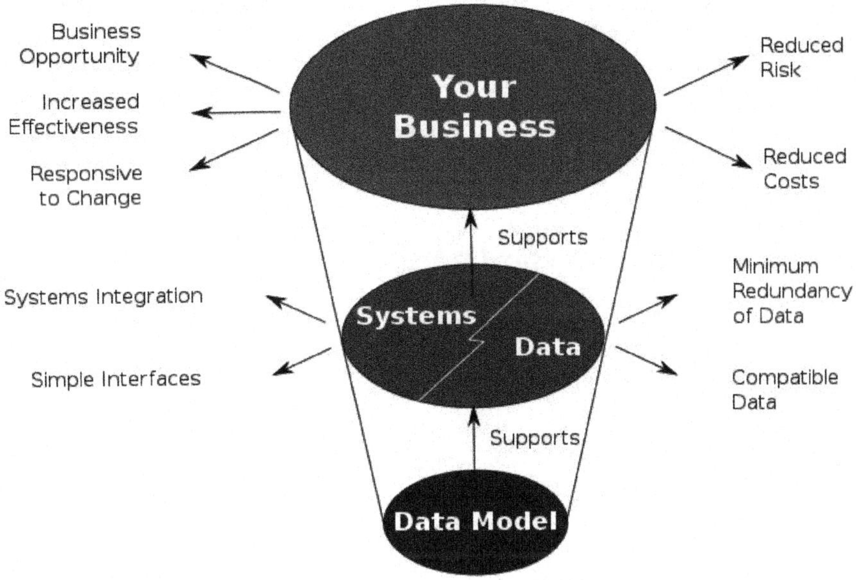

Fig. : How data models deliver benefit.

Data models provide a structure for data used within information systems by providing specific definition and format. If a data model is used consistently across systems then compatibility of data can be achieved. If the same data structures are used to store and access data then different applications can share data seamlessly. The results of this are indicated in the diagram. However, systems and interfaces often cost more than they should, to build, operate, and maintain. They may also constrain the business rather than support it. This may occur when the quality of the data models implemented in systems and interfaces is poor.

- Business rules, specific to how things are done in a particular place, are often fixed in the structure of a data model. This means that small changes in the way business is conducted lead to large changes in computer systems and interfaces. So, business rules need to be implemented in a flexible way that does not result in complicated dependencies, rather the data model should be flexible enough so that changes in the business can be implemented within the data model in a relatively quick and efficient way.

- Entity types are often not identified, or are identified incorrectly. This can lead to replication of data, data structure and functionality, together with the attendant costs of that duplication in development and maintenance.Therefore, data definitions should be made as explicit and easy to understand as possible to minimize misinterpretation and duplication.
- Data models for different systems are arbitrarily different. The result of this is that complex interfaces are required between systems that share data. These interfaces can account for between 25-70% of the cost of current systems. Required interfaces should be considered inherently while designing a data model, as a data model on its own would not be usable without interfaces within different systems.
- Data cannot be shared electronically with customers and suppliers, because the structure and meaning of data has not been standardised. To obtain optimal value from an implemented data model, it is very important to define standards that will ensure that data models will both meet business needs and be consistent.

Conceptual, Logical and Physical Schemas

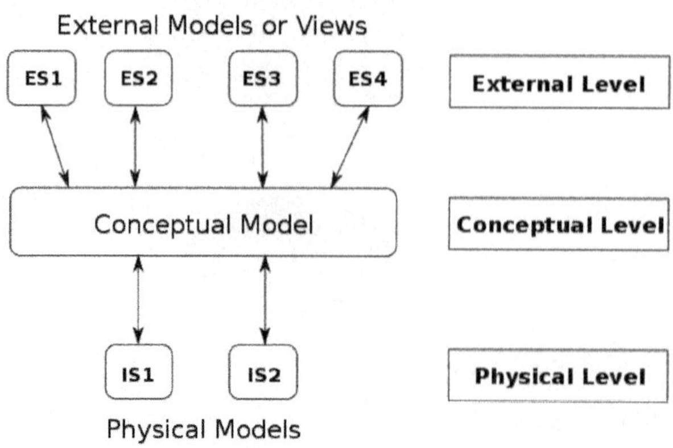

Fig.: *The ANSI/SPARC three level architecture :* This shows that a data model can be an external model (or view), a conceptual model, or a physical model. This is not the only way to look at data models, but it is a useful way, particularly when comparing models.

In 1975 ANSI described three kinds of data-model instance :

- *Conceptual schema :* Describes the semantics of a domain (the scope of the model). For example, it may be a model of the interest area of an organization or of an industry. This consists of entity classes, representing kinds of things of significance in the domain, and relationships assertions about associations between pairs of entity classes. A conceptual schema specifies the kinds of facts or propositions that can be expressed using the model. In that sense, it defines the allowed expressions in an artificial "language" with a scope that is limited by the scope of the model. Simply described, a conceptual schema is the first step in organizing the data requirements.

- *Logical schema* : Describes the structure of some domain of information. This consists of descriptions of (for example) tables, columns, object-oriented classes, and XML tags. The logical schema and conceptual schema are sometimes implemented as one and the same.
- *Physical schema* : Describes the physical means used to store data. This is concerned with partitions, CPUs, tablespaces, and the like.

According to ANSI, this approach allows the three perspectives to be relatively independent of each other. Storage technology can change without affecting either the logical or the conceptual schema. The table/column structure can change without (necessarily) affecting the conceptual schema. In each case, of course, the structures must remain consistent across all schemas of the same data model.

Data Modelling Process

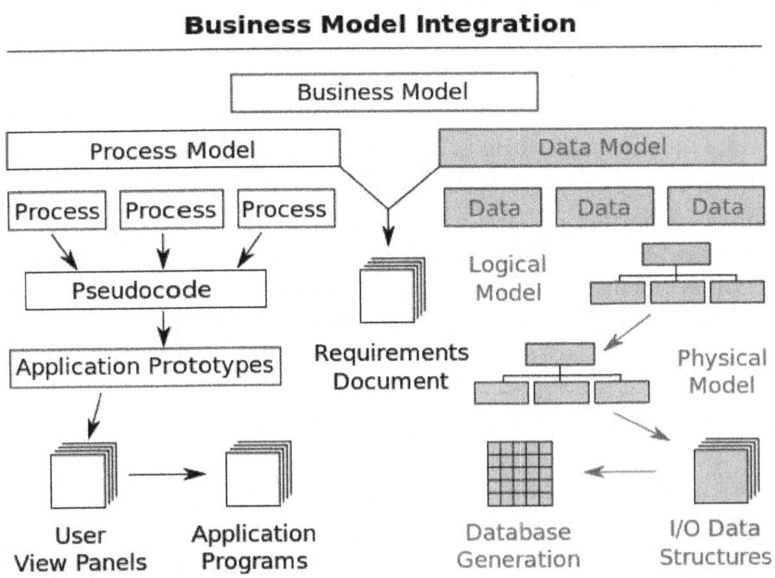

Fig. : Data modelling in the context of Business Process Integration.

In the context of business process integration, data modelling complements business process modelling, and ultimately results in database generation.

The process of designing a database involves producing the previously described three types of schemas-conceptual, logical, and physical. The database design documented in these schemas are converted through a Data Definition Language, which can then be used to generate a database. A fully attributed data model contains detailed attributes (descriptions) for every entity within it. The term "database design" can describe many different parts of the design of an overall database system. Principally, and most correctly, it can be thought of as the logical design of the base data structures used to store the data. In the relational model these are the tables and views. In an object database the entities and relationships map directly to object classes and named relationships. However, the term "database design" could also be used to apply to the overall process of designing, not just the base data structures, but also the forms and queries used as part of the overall database application within the Database Management System or DBMS.

In the process, system interfaces account for 25% to 70% of the development and support costs of current systems. The primary reason for this cost is that these systems do not share a common data model. If data models are developed on a system by system basis, then not only is the same analysis repeated in overlapping areas, but further analysis must be performed to create the interfaces between them. Most systems within an organization contain the same basic data, re-developed for a specific purpose. Therefore, an efficiently designed basic data model can minimize rework with minimal modifications for the purposes of different systems within the organization

Modelling Methodologies

Data models represent information areas of interest. While there are many ways to create data models, according to Len Silverston only two modelling methodologies stand out, top-down and bottom-up :

- Bottom-up models or View Integration models are often the result of a reengineering effort. They usually start with existing data structures forms, fields on application screens, or reports. These models are usually physical, application-specific, and incomplete from an enterprise perspective. They may not promote data sharing, especially if they are built without reference to other parts of the organization.
- Top-down logical data models, on the other hand, are created in an abstract way by getting information from people who know the subject area. A system may not implement all the entities in a logical model, but the model serves as a reference point or template.

Sometimes models are created in a mixture of the two methods : by considering the data needs and structure of an application and by consistently referencing a subject-area model. Unfortunately, in many environments the distinction between a logical data model and a physical data model is blurred. In addition, some CASE tools don't make a distinction between logical and physical data models.

Entity Relationship Diagrams

There are several notations for data modelling. The actual model is frequently called "Entity relationship model", because it depicts data in terms of the entities and relationships described in the data. An entity-relationship model (ERM) is an abstract conceptual representation of structured data. Entity-relationship modelling is a relational schema database modelling method, used in software engineering to produce a type of conceptual data model (or semantic data model) of a system, often a relational database, and its requirements in a top-down fashion.

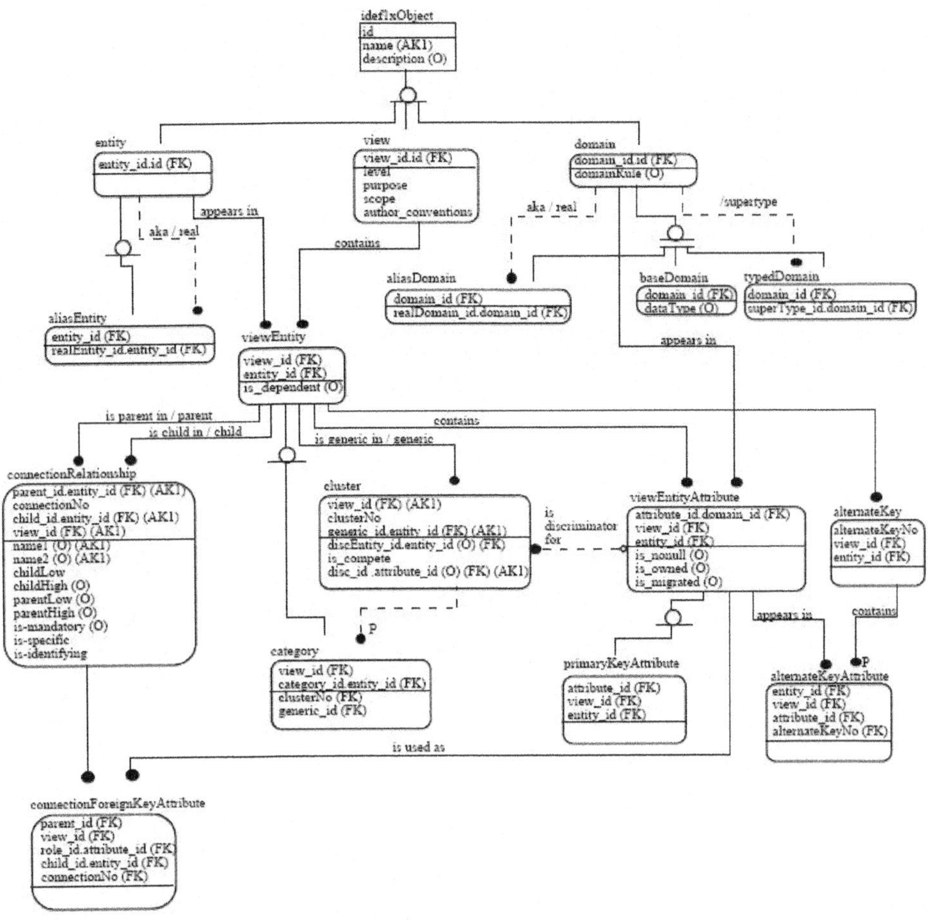

Fig. : Example of an IDEF1X Entity relationship diagrams used to model IDEF1X itself. The name of the view is mm. The domain hierarchy and constraints are also given. The constraints are expressed as sentences in the formal theory of the meta model.

These models are being used in the first stage of information system design during the requirements analysis to describe information needs or the type of information that is to be stored in a database. The data modelling technique can be used to describe any ontology (*i.e.* an overview and classifications of used terms and their relationships) for a certain universe of discourse *i.e.* area of interest.

Several techniques have been developed for the design of data models. While these methodologies guide data modelers in their work, two different people using the same methodology will often come up with very different results. Most notable are :

- Bachman diagrams

- Barker's notation
- Chen's Notation
- Data Vault Modelling
- Extended Backus–Naur form
- IDEF1X
- Object-relational mapping
- Object-Role Modelling
- Relational Model
- Relational Model/Tasmania.

Generic Data Modelling

Figure 1. The backbone of the HL7 Reference Information Model.

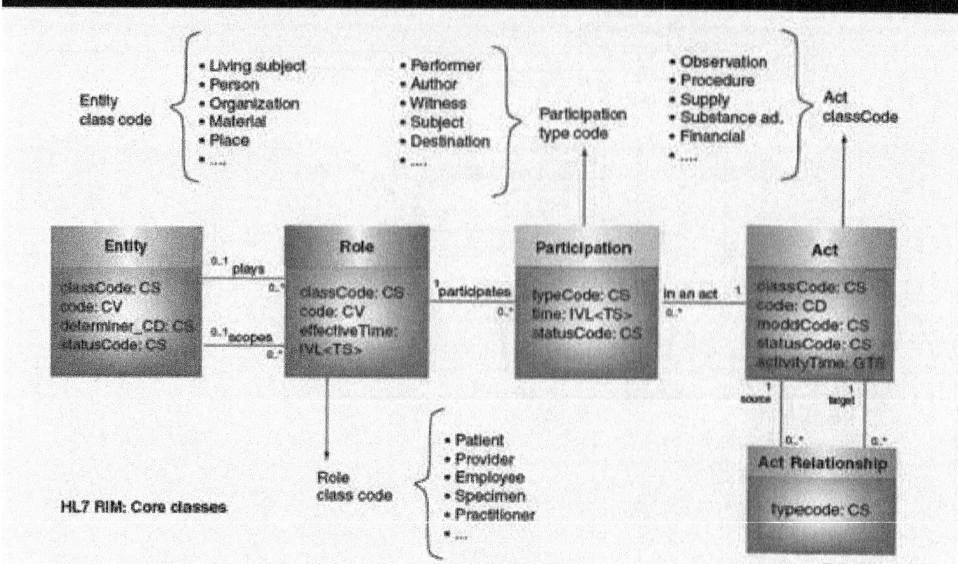

An entity plays a role that participates in an action. The RIM backbone is a generic set of classes from which health specific classes are derived, for example, observation, procedure and substance administration are all subclasses of act.
CD: Concept descriptor; CS: Coded simple value; CV: Coded value; GTS: General Timing Specification; HL7: Health Level Seven; IVL: Interval; RIM: Reference Information Model; TS: Point in time.

Fig. : Example of a Generic data model.

Generic data models are generalizations of conventional data models. They define standardized general relation types, together with the kinds of things that may be related by such a relation type. The definition of generic data model is similar to the definition of a natural language. For example, a generic data model may define relation types such as a 'classification relation', being a binary relation between an individual thing and a kind of thing (a class) and a 'part-whole relation', being a binary relation between two things, one with the role of part, the other with the role of whole, regardless the kind of things that are related.

Given an extensible list of classes, this allows the classification of any individual thing and to specify part-whole relations for any individual object. By standardization of an extensible list of relation types, a generic data model enables the expression of an unlimited number of kinds of facts and will approach the capabilities of natural languages. Conventional data models, on the other hand, have a fixed and limited domain scope, because the instantiation (usage) of such a model only allows expressions of kinds of facts that are predefined in the model.

Semantic Data Modelling

The logical data structure of a DBMS, whether hierarchical, network, or relational, cannot totally satisfy the requirements for a conceptual definition of data because it is limited in scope and biased toward the implementation strategy employed by the DBMS.

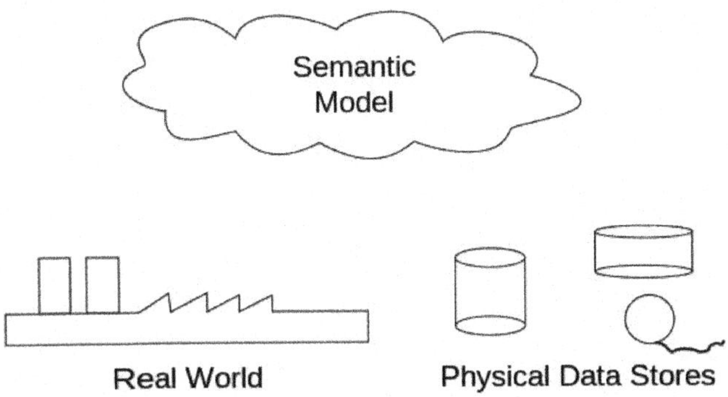

Fig. : Semantic data models.

Therefore, the need to define data from a conceptual view has led to the development of semantic data modelling techniques. That is, techniques to define the meaning of data within the context of its interrelationships with other data. As illustrated in the figure the real world, in terms of resources, ideas, events, etc., are symbolically defined within physical data stores. A semantic data model is an abstraction which defines how the stored symbols relate to the real world. Thus, the model must be a true representation of the real world.

A semantic data model can be used to serve many purposes, such as :

- Planning of data resources
- Building of shareable databases
- Evaluation of vendor software
- Integration of existing databases.

The overall goal of semantic data models is to capture more meaning of data by integrating relational concepts with more powerful abstraction concepts known from the Artificial Intelligence field. The idea is to provide high level modelling primitives as integral part of a data model in order to facilitate the representation of real world situations.

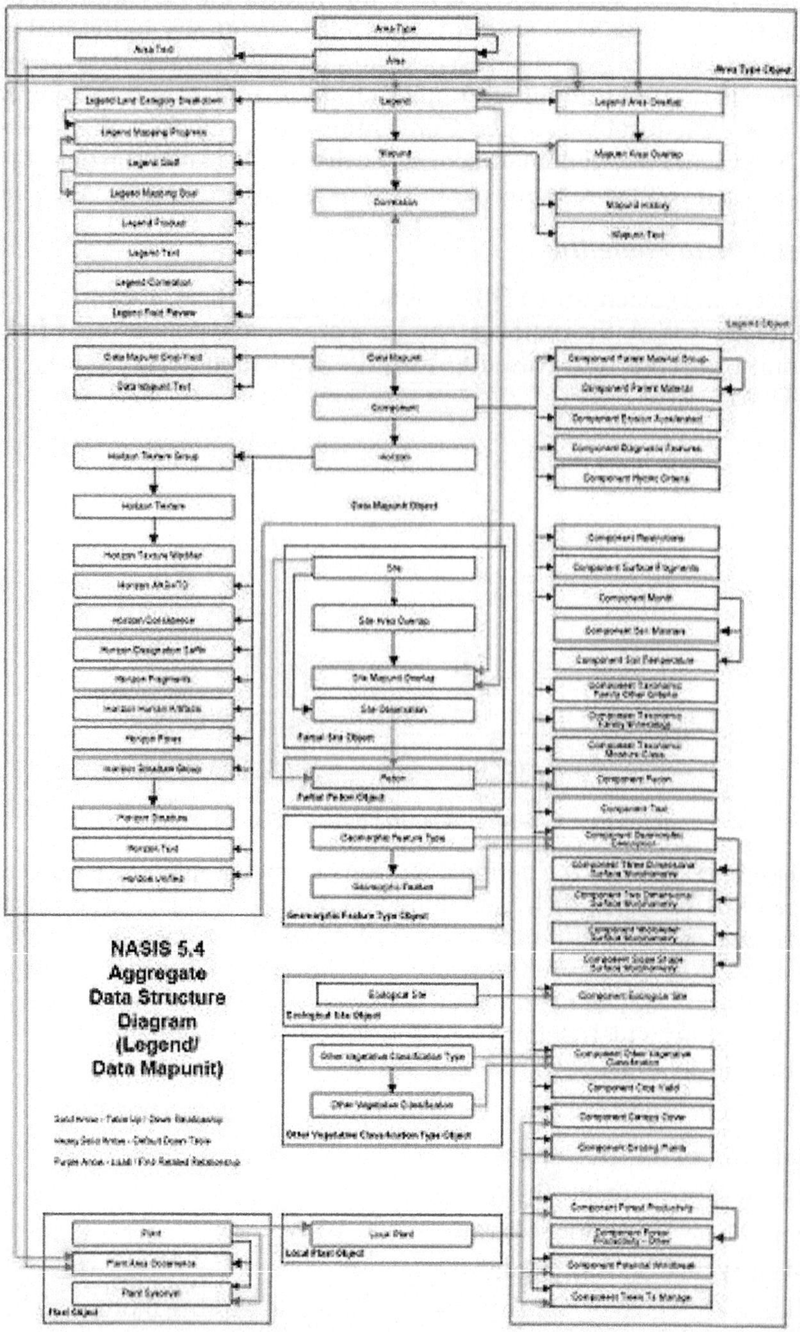

Fig. : Example of a Data Structure Diagram.

DATA STRUCTURE DIAGRAM

Data Structure Diagram (DSD) is a diagram of the conceptual data model which documents the entities and their relationships, as well as the constraints that connect to them.

The basic graphic notation elements of DSDs are boxes which represent entities.The arrow symbol represents relationships. Data structure diagrams are most useful for documenting complex data entities.

Overview

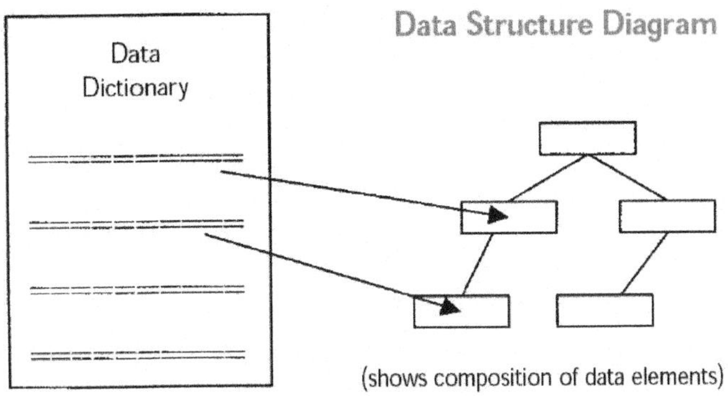

Fig. : Data Structure Diagram.

Data Structure Diagram is a diagram type that is used to depict the structure of data elements in the data dictionary. The data structure diagram is a graphical alternative to the composition specifications within such data dictionary entries.

The data structure diagrams is a predecessor of the entity-relationship model (E-R model). In DSDs, attributes are specified inside the entity boxes rather than outside of them, while relationships are drawn as boxes composed of attributes which specify the constraints that bind entities together. DSDs differ from the E-R model in that the E-R model focuses on the relationships between different entities, whereas DSDs focus on the relationships of the elements within an entity.

There are several styles for representing data structure diagrams, with the notable difference in the manner of defining cardinality. The choices are between arrow heads, inverted arrow heads (crow's feet), or numerical representation of the cardinality.

Bachman Diagram

A Bachman diagram is a certain type of data structure diagram, and is used to design the data with a network or relational "logical" model, separating the data model from the way the data is stored in the system. The model is named after database pioneer Charles Bachman, and mostly used in computer software design.

In a relational model, a relation is the cohesion of attributes that are fully and not transitive functional dependent of every key in that relation. The coupling between the relations is based on accordant attributes. For every relation, a rectangle has to be drawn and every coupling is illustrated by a line that connects the relations. On the edge of each line, arrows indicate the cardinality. We have 1-to-n, 1-to-1 and n-to-n. The latter has to be avoided and must be replaced by two 1-to-n couplings.

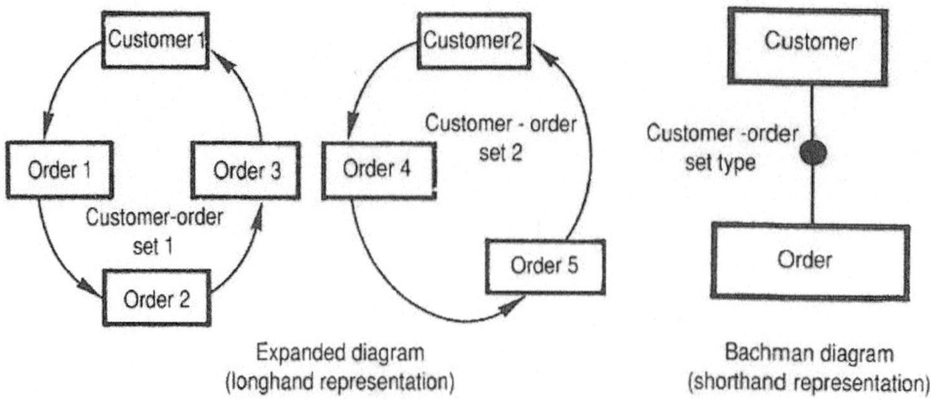

Fig. : Illustration of set type using a Bachman diagram.

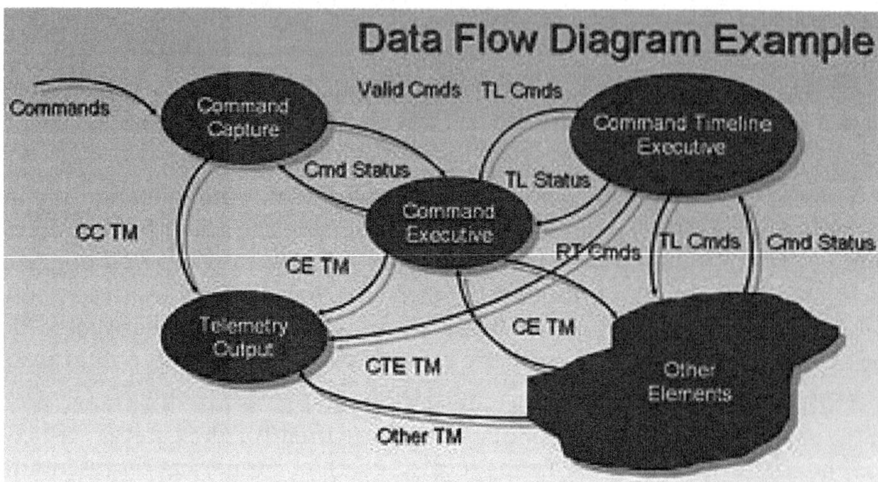

Fig. : Data flow diagram example.

DATA FLOW DIAGRAM

A **data flow diagram (DFD)** is a graphical representation of the "flow" of data through an information system, modelling its *process* aspects. Often they are a preliminary step used to create an overview of the system which can later be elaborated. DFDs can also be used for the visualization of data processing.

A DFD shows what kind of information will be input to and output from the system, where the data will come from and go to, and where the data will be stored. It does not show information about the timing of processes, or information about whether processes will operate in sequence or in parallel (which is shown on a flowchart).

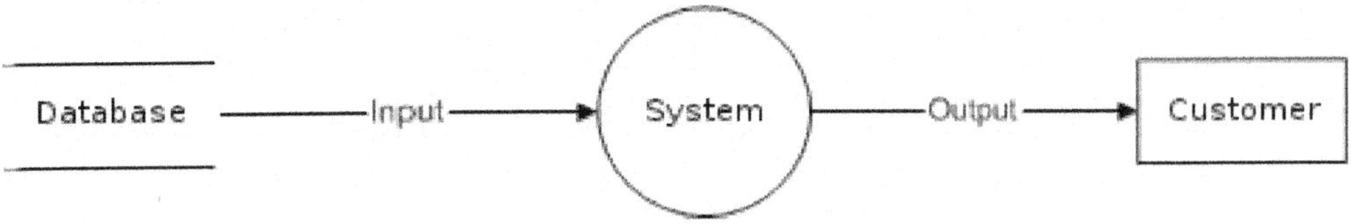

Data flow diagram example.

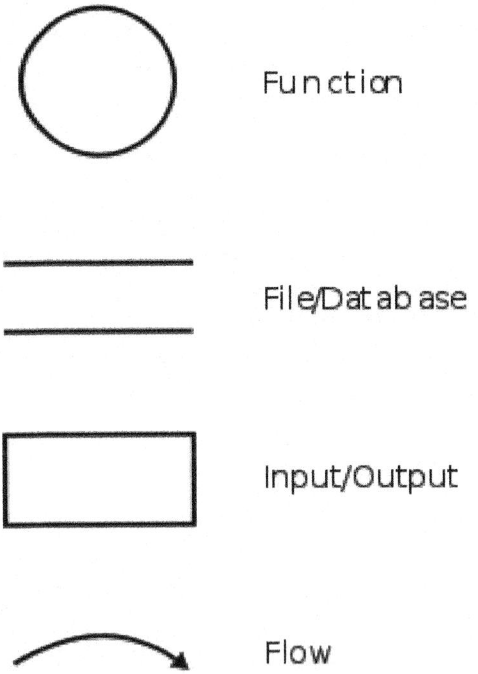

Fig. : Data flow diagram-Yourdon/DeMarco notation.

Theory

It is common practice to draw the context-level data flow diagram first, which shows the interaction between the system and external agents which act as data sources and data sinks. This helps to create an accurate drawing in the context diagram. The system's interactions with the outside world are modelled purely in terms of data flows across the *system boundary*. The context diagram shows the entire system as a single process, and gives no clues as to its internal organization.

This context-level DFD is next "exploded", to produce a Level 1 DFD that shows some of the detail of the system being modelled. The Level 1 DFD shows how the system is divided into sub-systems (processes), each of which deals with one or more of the data flows to or from an external agent, and which together provide all of the functionality of the system as a whole. It also identifies internal data stores that must be present in order for the system to do its job, and shows the flow of data between the various parts of the system.

Data flow diagrams were proposed by Larry Constantine, the original developer of structured design, based on Martin and Estrin's "data flow graph" model of computation.

Data flow diagrams are one of the three essential perspectives of the structured-systems analysis and design method SSADM. The sponsor of a project and the end users will need to be briefed and consulted throughout all stages of a system's evolution. With a data flow diagram, users are able to visualize how the system will operate, what the system will accomplish, and how the system will be implemented.

The old system's dataflow diagrams can be drawn up and compared with the new system's data flow diagrams to draw comparisons to implement a more efficient system. Data flow diagrams can be used to provide the end user with a physical idea of where the data they input ultimately has an effect upon the structure of the whole system from order to dispatch to report. How any system is developed can be determined through a data flow diagram model.

In the course of developing a set of *levelled* data flow diagrams the analyst/designers is forced to address how the system may be decomposed into component sub-systems, and to identify the transaction data in the data model.

Data flow diagrams can be used in both Analysis and Design phase of the SDLC.

There are different notations to draw data flow diagrams (Yourdon & Coad and Gane & Sarson), defining different visual representations for processes, data stores, data flow, and external entities.

INFORMATION MODEL

An **information model** in software engineering is a representation of concepts and the relationships, constraints, rules, and operations to specify data semantics for a chosen domain of discourse. Typically it specifies relations between kinds of things, but may also include relations with individual things. It can provide sharable, stable, and organized structure of information requirements or knowledge for the domain context.

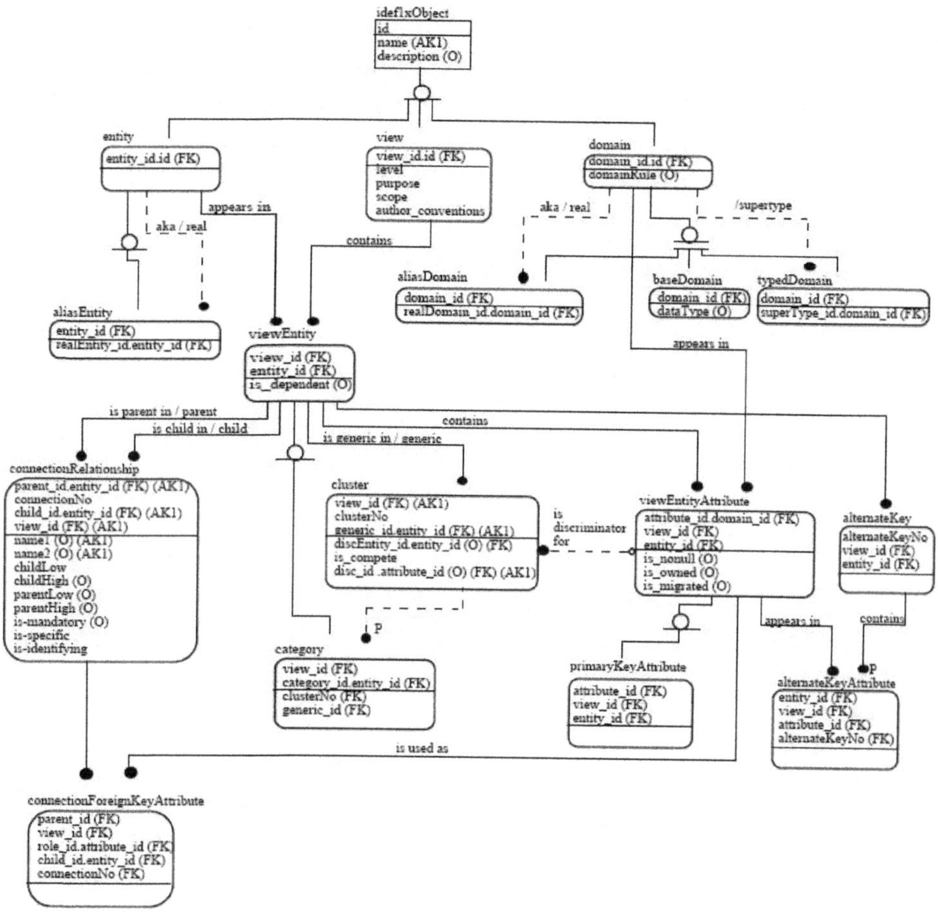

Fig. : An IDEF1X Diagram, an example of an Integration Definition for Information Modelling.

Overview

The term *information model* in general is used for models of individual things, such as facilities, buildings, process plants, etc. In those cases the concept is specialised to facility information model, building information model, plant information model, etc. Such an information model is an integration of a model of the facility with the data and documents about the facility.

Within the field of software engineering and data modelling an information model is usually an abstract, formal representation of entity types that may include their properties, relationships and the operations that can be performed on them. The entity types in the model may be kinds of real-world objects, such as devices in a network, or occurrences, or they may themselves be abstract, such as for the entities used in a billing system. Typically, they are used to model a constrained domain that can be described by a closed set of entity types, properties, relationships and operations.

An information model provides formalism to the description of a problem domain without constraining how that description is mapped to an actual implementation in software. There may be many mappings of the information model. Such mappings are called data models, irrespective of whether they are object models (*e.g.* using UML), entity relationship models or XML schemas.

Information Modelling Languages

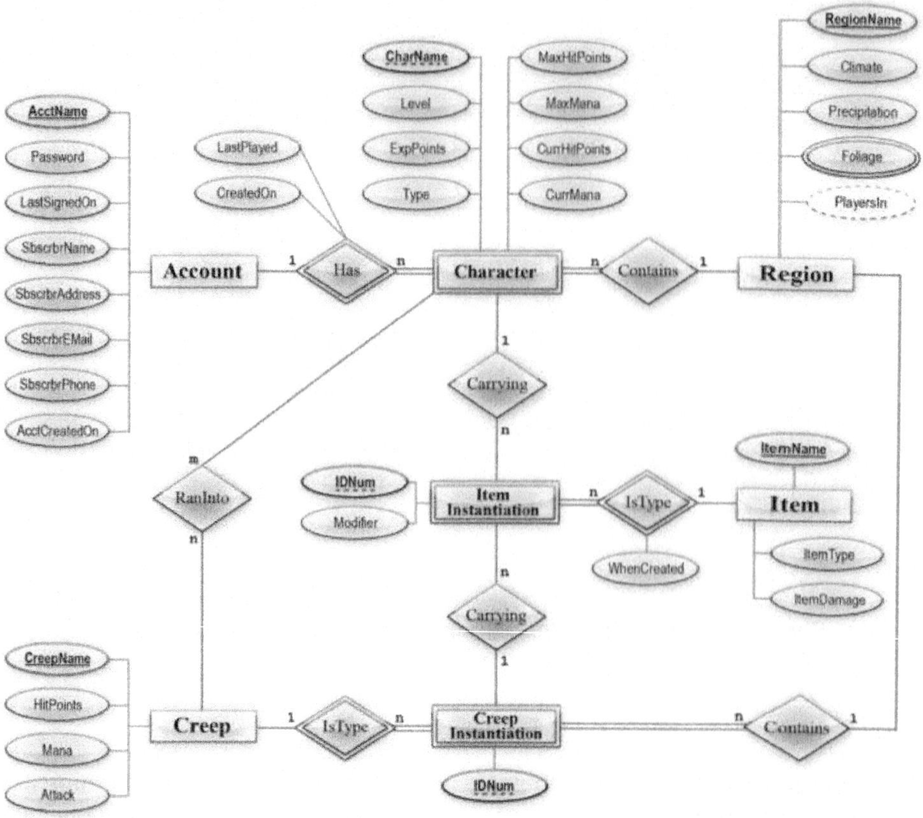

Fig. : A sample ER diagram.

In 1976, an entity-relationship (ER) graphic notation was introduced by Peter Chen. He stressed that it was a "semantic" modelling technique and independent of any database modelling techniques such as Hierarchical, CODASYL, Relational etc. Since then, languages for information models have continued to evolve. Some examples are the Integrated Definition Language 1 Extended (IDEF1X), the EXPRESS language and the Unified Modelling Language (UML).

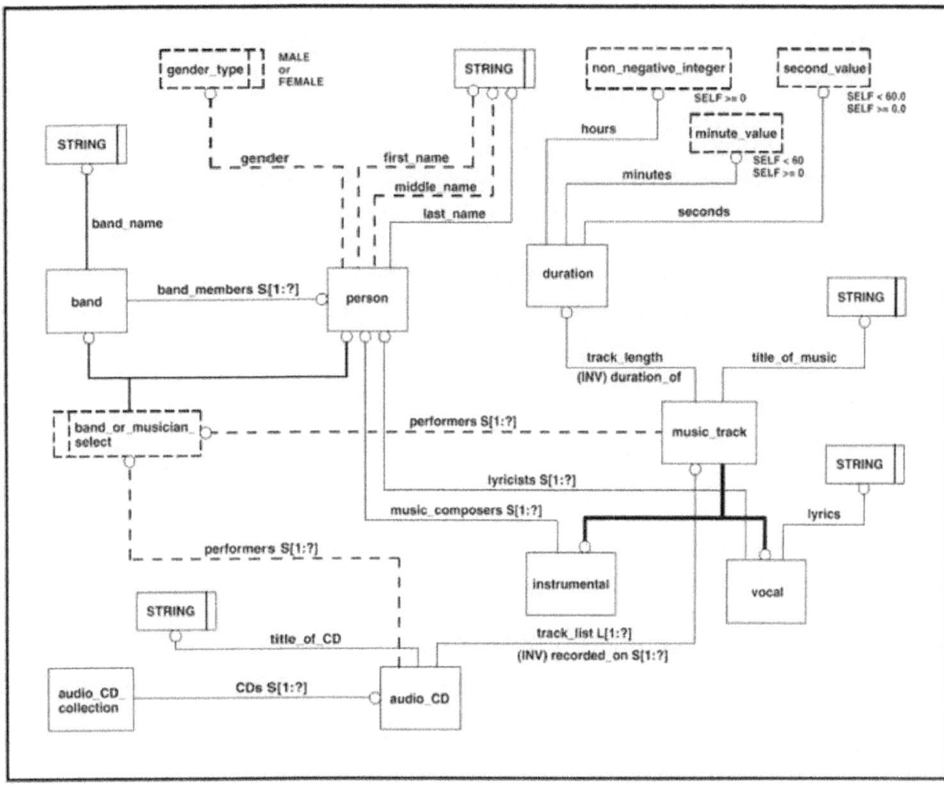

Fig. : Database requirements for a CD collection in EXPRESS-G notation.

Research by contemporaries of Peter Chen such as J.R. Abrial and G.M Nijssen led to today's Fact Oriented Modelling languages which are based on linguistic propositions rather than on "entities". FOM Tools can be used to generate an ER model which means that the modeler can avoid the time-consuming and error prone practice of manual normalization. Object-Role Modelling language (ORM) and Fully Communication Oriented Information Modelling (FCO-IM) are both research results, based upon earlier research.

In the 1980s there were several approaches to extend Chen's Entity Relationship Model. Also important in this decade is REMORA by Colette Rolland.

The ICAM Definition (IDEF) Language was developed from the U.S. Air Force ICAM Program during the 1976 to 1982 timeframe. The objective of the ICAM Program, according to Lee, was to increase manufacturing productivity through the systematic application of computer technology. IDEF includes three different modelling methods : IDEF0, IDEF1, and IDEF2 for producing a functional model, an information model, and a dynamic model respectively. IDEF1X is an extended version of IDEF1. The language is in the public domain. It is a graphical representation and is designed using the ER approach and the relational theory. It is used to represent the "real world" in terms of entities, attributes, and relationships between entities. Normalization is enforced by KEY Structures and KEY Migration. The language identifies property groupings (Aggregation) to form complete entity definitions.

EXPRESS was created as ISO 10303–11 for formally specifying information requirements of product data model. It is part of a suite of standards informally known as the Standard for the Exchange of Product model data (STEP). It was first introduced in the early 1990s. The language, according to Lee, is a textual representation. In addition, a graphical sub-set of EXPRESS called EXPRESS-G is available. EXPRESS is based on programming languages and the O-O paradigm. A number of languages have contributed to EXPRESS. In particular, Ada, Algol, C, C++, Euler, Modula-2, Pascal, PL/1, and SQL. EXPRESS consists of language elements that allow an unambiguous object definition and specification of constraints on the objects defined. It uses SCHEMA declaration to provide partitioning and it supports specification of data properties, constraints, and operations.

UML is a modelling language for specifying, visualizing, constructing, and documenting the artifacts, rather than processes, of software systems. It was conceived originally by Grady Booch, James Rumbaugh, and Ivar Jacobson. UML was approved by the Object Management Group (OMG) as a standard in 1997. The language, according to Lee, is non-proprietary and is available to the public. It is a graphical representation. The language is based on the objected-oriented paradigm. UML contains notations and rules and is designed to represent data requirements in terms of O-O diagrams. UML organizes a model in a number of views that present different aspects of a system. The contents of a view are described in diagrams that are graphs with model elements. A diagram contains model elements that represent common O-O concepts such as classes, objects, messages, and relationships among these concepts.

IDEF1X, EXPRESS, and UML all can be used to create a conceptual model and, according to Lee, each has its own characteristics. Although some may lead to a natural usage (*e.g.*, implementation), one is not necessarily better than another. In practice, it may require more than one language to develop all information models when an application is complex. In fact, the modelling practice is often more important than the language chosen.

Information models can also be expressed in formalized natural languages, such as Gellish. Gellish, which has natural language variants Gellish Formal English, Gellish Formal Dutch, etc., is an information representation language or modelling language that is defined in the Gellish smart Dictionary—Taxonomy, which has the form of a Taxonomy/Ontology. A Gellish Database is not only suitable to store information models, but also knowledge models, requirements models and dictionaries, taxonomies and ontologies. Information models in Gellish English use Gellish Formal English expressions. For example, a geographic information model might consist of a number of Gellish Formal English expressions, such as :

```
-the Eiffel tower <is located in> Paris
-Paris <is classified as a> city
```

whereas information requirements and knowledge can expressed for example as follows :

```
-tower <shall be located in a> geographical area
-city <is a kind of> geographical area
```

Such Gellish expressions use names of concepts (such as 'city') and relation types (such as <is located in> and <is classified as a>) that should be selected from the Gellish Formal English Dictionary—Taxonomy (or of your own domain dictionary). The Gellish English Dictionary—Taxonomy enables the creation of semantically rich information models, because the dictionary contains definitions of more than 40000 concepts, including more than 600 standard relation types. Thus, an information model in Gellish consists of a collection of Gellish expressions that use those phrases and dictionary concepts to express facts or make statements, queries and answers.

Standard Sets of Information Models

The Distributed Management Task Force (DMTF) provides a standard set of information models for various enterprise domains under the general title of the Common Information Model (CIM). Specific information models are derived from CIM for particular management domains.

The Tele Management Forum (TMF) has defined an advanced model for the Telecommunication domain (the Shared Information/Data model, or SID) as another. This includes views from the business, service and resource domains within the Telecommunication industry. The TMF has established a set of principles that an OSS integration should adopt, along with a set of models that provide standardized approaches.

The models interact with the information model (the Shared Information/ Data Model, or SID), via a process model (the enhanced Telecom Operation Map, or eTOM) and a life cycle model.

OBJECT MODEL

*In computing, **object model** has two related but distinct meanings :*

1. The properties of objects in general in a specific computer programming language, technology, notation or methodology that uses them. Examples are the object models of *Java*, the *Component Object Model (COM)*, or *Object-Modelling Technique (OMT)*. Such object models are usually defined using concepts such as class, generic function, message, inheritance, polymorphism, and encapsulation. There is an extensive literature on formalized object models as a sub-set of the formal semantics of programming languages.
2. A collection of objects or classes through which a program can examine and manipulate some specific parts of its world. In other words, the object-oriented interface to some service or system. Such an interface is said to be the *object model of* the represented service or system. For example, the Document

Object Model (DOM) [1] is a collection of objects that represent a page in a web browser, used by script programs to examine and dynamically change the page. There is a Microsoft Excel object model [2] for controlling Microsoft Excel from another program, and the ASCOM Telescope Driver [3] is an object model for controlling an astronomical telescope.

An object model consists of the following important features :

Object Reference : Objects can be accessed via object references. To invoke a method in an object, the object reference and method name are given, together with any arguments.

Interfaces : An interface provides a definition of the signature of a set of methods without specifying their implementation. An object will provide a particular interface if its class contains code that implement the method of that interface. An interface also defines types that can be used to declare the type of variables or parameters and return values of methods.

Actions : An action in object-oriented programming (OOP) is initiated by an object invoking a method in another object. An invocation can include additional information needed to carry out the method. The receiver executes the appropriate method and then returns control to the invoking object, sometimes supplying a result.

Exceptions : Programs can encounter various errors and unexpected conditions of varying seriousness. During the execution of the method many different problems may be discovered. Exceptions provide a clean way to deal with error conditions without complicating the code. A block of code may be defined to throw an exception whenever particular unexpected conditions or errors arise. This means that control passes to another block of code that catches the exception.

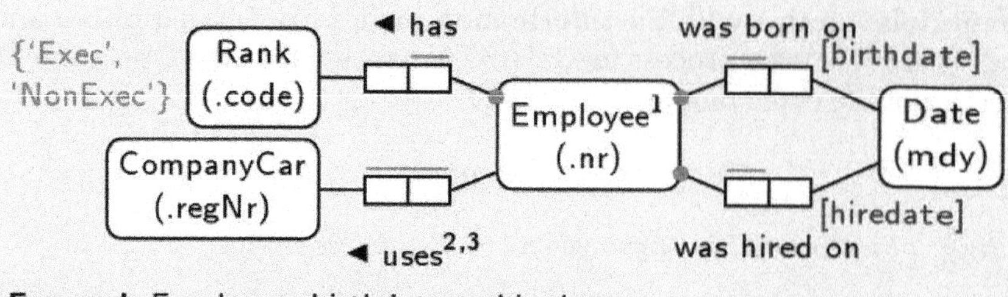

[1] **For each** Employee, birthdate < hiredate.

[2] **Each** Employee **who** has Rank 'NonExec' uses **at most one** CompanyCar.

[3] **Each** Employee **who** has Rank 'Exec' uses **some** CompanyCar.

Fig. : Example of an ORM2 diagram.

OBJECT-ROLE MODELLING

Object-role modelling (ORM) is used to model the semantics of a universe of discourse. ORM is often used for data modelling and software engineering.

An object-role model uses graphical symbols that are based on first order predicate logic and set theory to enable the modeler to create an unambiguous definition of an arbitrary universe of discourse.

The term "object-role model" was coined in the 1970s and ORM based tools have been used for more than 30 years—principally for data modelling. More recently ORM has been used to model business rules, XML-Schemas, data warehouses, requirements engineering and web forms.

History

The roots of ORM can be traced to research into semantic modelling for information systems in Europe during the 1970s. There were many pioneers and this short summary does not by any means mention them all. An early contribution came in 1973 when Michael Senko wrote about "data structuring" in the IBM Systems Journal. In 1974 Jean-Raymond Abrial contributed an article about "Data Semantics". In June 1975, Eckhard Falkenberg's doctoral thesis was published and in 1976 one of Falkenberg's papers mentions the term "object-role model".

G.M. Nijssen made fundamental contributions by introducing the "circle-box" notation for object types and roles, and by formulating the first version of the conceptual schema design procedure. Robert Meersman extended the approach by adding subtyping, and introducing the first truly conceptual query language.

Object role modelling also evolved from the *Natural language Information Analysis Method,* a methodology that was initially developed by the academic researcher, G.M. Nijssen in the Netherlands (Europe) in the mid-1970s and his research team at the Control Data Corporation Research Laboratory in Belgium, and later at the University of Queensland, Australia in the 1980s. The acronym **NIAM** originally stood for "Nijssen's Information Analysis Methodology", and later generalised to "Natural language Information Analysis Methodology" and *Binary Relationship Modelling* since G. M. Nijssen was only one of many people involved in the development of the method.

In 1989 Terry Halpin completed his PhD thesis on ORM, providing the first full formalization of the approach and incorporating several extensions.

Also in 1989, Terry Halpin and G.M. Nijssen co-authored the book "Conceptual Schema and Relational Database Design" and several joint papers, providing the first formalization of object-role modelling. Since then Dr. Terry Halpin has authored six books and over 160 technical papers.

A graphical NIAM design tool which included the ability to generate database-creation scripts for Oracle, DB2 and DBQ was developed in the early 1990s in Paris. It was originally named Genesys and was marketed successfully in France and later Canada. It could also handle ER diagram design. It was ported to SCO Unix, SunOs, DEC 3151's and Windows 3.0 platforms, and was later migrated to succeeding Microsoft operating systems, utilising XVT for cross operating system graphical portability. The tool was renamed OORIANE and is currently being used for large data warehouse and SOA projects.

Also evolving from NIAM is "*Fully Communication Oriented Information Modelling*" FCO-IM. It distinguishes itself from traditional ORM in that it takes a strict communication-oriented perspective. Rather than attempting to model the domain and its essential concepts, it models the communication in this domain (universe of discourse). Another important difference is that it does this on instance level, deriving type level and object/fact level during analysis.

Another recent development is the use of ORM in combination with standardised relation types with associated roles and a standard machine-readable dictionary and taxonomy of concepts as are provided in the Gellish English dictionary. Standardisation of relation types (fact types), roles and concepts enables increased possibilities for model integration and model reuse.

Concepts

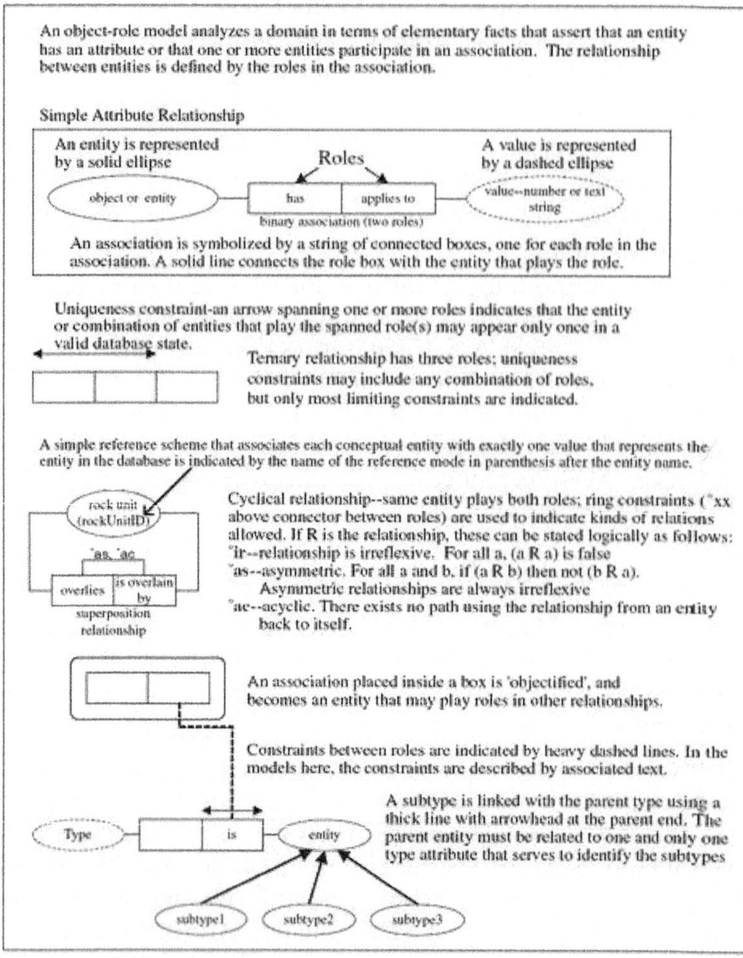

Fig. : Overview of object-role model notation, Stephen M. Richard.

Facts

Object-role models are based on elementary facts, and expressed in diagrams that can be verbalised into natural language. A fact is a proposition such as "John Smith was hired on 5 January, 1995" or "Mary Jones was hired on 3 March, 2010".

With ORM, propositions such as these, are abstracted into "fact types" for example "Person was hired on Date" and the individual propositions are regarded as sample data. The difference between a "fact" and an "elementary fact" is that an elementary fact cannot be simplified without loss of meaning. This "fact-based" approach facilitates modelling, transforming, and querying information from any domain.

Attribute-free

ORM is attribute-free : unlike models in the entity relationship (ER) and Unified Modelling Language (UML) methods, ORM treats all elementary facts as relationships and so treats decisions for grouping facts into structures (*e.g.* attribute-based entity types, classes, relation schemes, XML schemas) as implementation concerns irrelevant to semantics. By avoiding attributes in the base model, ORM improves semantic stability and enables verbalization into natural language.

Fact-based Modelling

Fact-based modelling includes procedures for mapping facts to attribute-based structures, such as those of ER or UML.

Fact-based textual representations are based on formal sub-sets of native languages. ORM proponents argue that ORM models are easier to understand by people without a technical education. For example, proponents argue that object-role models are easier to understand than declarative languages such as Object Constraint Language (OCL) and other graphical languages such as UML class models. Fact-based graphical notations are more expressive than those of ER and UML. An object-role model can be automatically mapped to relational and deductive databases (such as datalog).

ORM 2 Graphical Notation

ORM2 is the latest generation of object-role modelling. The main objectives for the ORM 2 graphical notation are :

- More compact display of ORM models without compromising clarity
- Improved internationalization (*e.g.* avoid English language symbols)
- Simplified drawing rules to facilitate creation of a graphical editor
- Extended use of views for selectively displaying/suppressing detail
- Support for new features (*e.g.* role path delineation, closure aspects, modalities)

Design Procedure

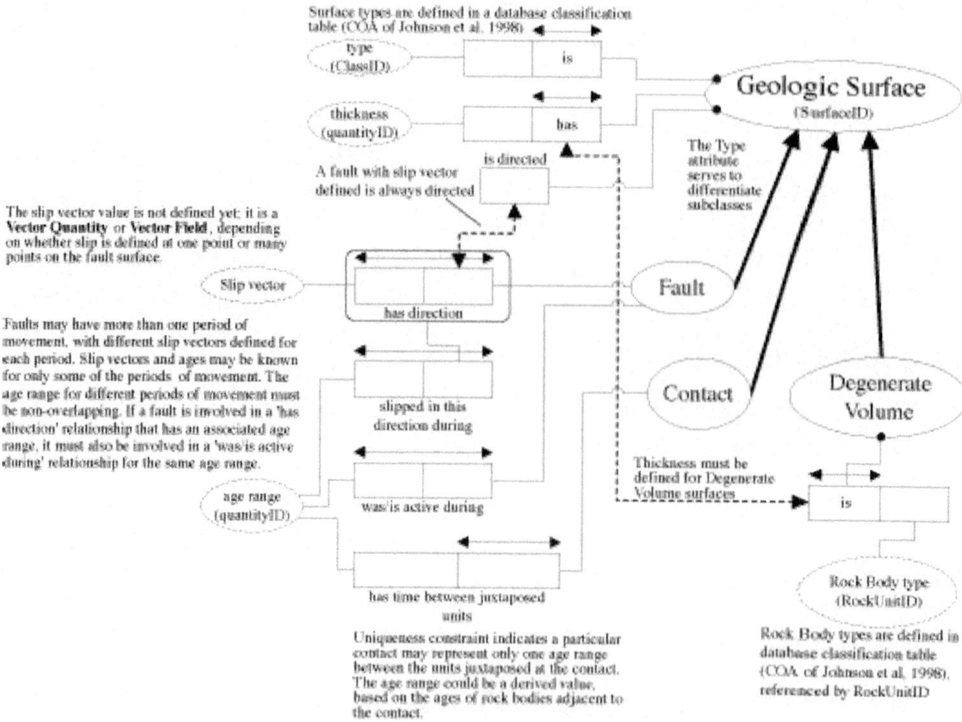

Fig. : Example of the application of Object Role Modelling in a "Schema for Geologic Surface", Stephen M. Richard.

System development typically involves several stages such as : feasibility study; requirements analysis; conceptual design of data and operations; logical design; external design; prototyping; internal design and implementation; testing and validation; and maintenance. The seven steps of the conceptual schema design procedure are :

1. Transform familiar information examples into elementary facts, and apply quality checks.
2. Draw the fact types, and apply a population check.
3. Check for entity types that should be combined, and note any arithmetic derivations.
4. Add uniqueness constraints, and check arity of fact types.
5. Add mandatory role constraints, and check for logical derivations.
6. Add value, set comparison and subtyping constraints.
7. Add other constraints and perform final checks.

ORM's conceptual schema design procedure focuses on the analysis and design of data.

Tools

ORMLite

ORMLite is a free, open-source modelling tool that supports ORM 2 notation. It was created as a self-learning environment to help popularize ORM. It can verbalize facts and generate relational models. It is written in Python and so it is multi-platform. Version 0.13b was released in November 2012.

NORMA

Dr. Terry Halpin led the initial development of the NORMA ORM modelling tool whilst he was a Professor at Neumont University. This is now supported via The ORM Foundation, a UK-based non-profit organization dedicated to the promotion of the object-role modelling.

NORMA (Natural ORM Architect for Visual Studio) is a free and open source plug-in to Microsoft Visual Studio 2005, Visual Studio 2008, Visual Studio 2010 and Visual Studio 2012. NORMA supports second generation ORM(ORM 2), and can be used to generate code from and object-role model to several implementation targets, such as major database engines, object-oriented code, and XML schema. Examples include :

Database engines	Microsoft Sql Server, Oracle, DB2, MySQL, PostgreSQL, etc.
Programming languages	LINQ to SQL, PLiX (Programming Language in XML) and PHP
Other	XML schemas (XSD)

Advantages of NORMA include :

- Accepts text input and automatically generates graphics.
- Validates common constraints and completeness as the model is entered.
- Provides simultaneous narrative and graphic versions of all models.
- Provides multiple vies of a model (ORM graphics, Relational, E-R and text narrative).
- Automatic navigation from error message to graphic view of the error.
- The four views (ORM, Relational, E-R and narrative) provide useful access. for modelers. However, the narrative view of the model is easier to read by those who are not familiar with ORM, Relational or E-R diagrams.
- Can reverse-engineer a physical database (partial).
- Narrative view uses hyperlinks for full cross-referencing.
- Graphical model has no fixed bounds.

Dr. Terry Halpin's 2008 book, *Information Modelling and Relational Databases, Second Edition* "...uses the notation of ORM 2 (second generation ORM), as supported by the NORMA (Natural ORM Architect) tool...", and "(...) At the time of writing, the Natural ORM Architect (NORMA) tool provides the most complete support for the ORM 2 notation discussed in this book." (Preface, xxv).

As of April 2009, the NORMA project delivers frequent releases.

Each new NORMA Community Technical Preview (CTP) is published in the library of The ORM Foundation website.

DogmaModeler

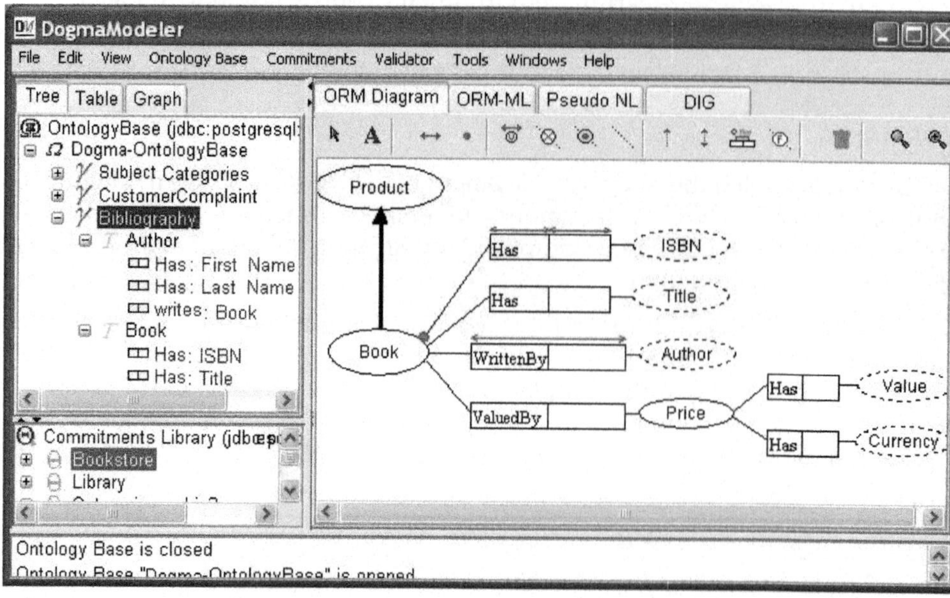

Fig. : DogmaModeler screenshot.

DogmaModeler is a free ontology modelling tool based on Object role modelling. The philosophy of DogmaModeler is to enable non-IT experts to model ontologies with a little or no involvement of an ontology engineer.

This challenge is tackled in DogmaModeler through well-defined methodological principles. The first version of DogmaModeler was developed at the Vrije Universiteit Brussel.

The DogmaModeler Project shows no activity since its creation in 2006, and the source code for the project is not available through that site. The latest version of the program, available at the DogmaModeler website is dated on October 27, 2006.

Since then the project seems to have been continued and expanded into several other tools at the Vrije Universiteit Brussel's Semantics Technology and Applications Research Laboratory (VUB STARLab).

VisioModeler

The former ORM tool known as VisioModeler is an unsupported product from Microsoft Corporation. Models developed in VisioModeler may be exported to Microsoft's current and future ORM solutions.

The early ORM tools such as IAST (Control Data) and RIDL* were followed by InfoDesigner, InfoModeler and VisioModeler.

When Microsoft bought the Visio Corporation, Microsoft extended VisioModeler and made it a component of Microsoft Visual Studio. This was Microsoft's first ORM implementation and it was published in the 2003 Enterprise Architects release of Visual Studio as a component of the tool called "Microsoft Visio for Enterprise Architects (VEA)".

In the same year, a companion "how to" book was published by Morgan Kaufmann entitled "Database Modelling with Microsoft Visio for Enterprise Architects". Microsoft has retained the ORM functionality in the high-end version of Visual Studio 2005 and the Morgan Kaufmann book remains a suitable user guide for both versions.

Visio for Enterprise Architects (VEA)

Microsoft included a powerful ORM and logical database modelling solution within its Visio for Enterprise Architects (VEA) product. The 2005 release of VEA also included some minor upgrades (*e.g.* a driver for SQL Server 2005 was included).

CaseTalk

A modelling tool called CaseTalk based on the ORM-dialect known as Fully Communication Oriented Information Modelling (FCO-IM) is developed and maintained by Oelan in the Netherlands.

Infagon

A freeware ORM tool known as Infagon is available from Mattic software. Infagon is also based on the FCO-IM dialect.

Other tools

- **Doctool** and **CogNIAM** (CogNIAM tools).
- **ActiveFacts** (ORM 2 tools).
- **DogmaStudio** (ORM Ontology tool).
- **Orthogonal Toolbox** (free XML add-on to database modelling COM API for Microsoft's ORM solution).

UNIFIED MODELLING LANGUAGE

Fig. : UML logo.

The **Unified Modelling Language (UML)** is a general-purpose modelling language in the field of software engineering, which is designed to provide a standard way to visualize the design of a system.

It was created and developed by Grady Booch, Ivar Jacobson and James Rumbaugh at Rational Software in the 1990s.

In 1997 it was adopted by the Object Management Group (OMG), and has been managed by this organization ever since. In 2000 the Unified Modelling Language was accepted by the International Organization for Standardization (ISO) as an approved standard. Since then it has been revised to cover the latest revision of UML.

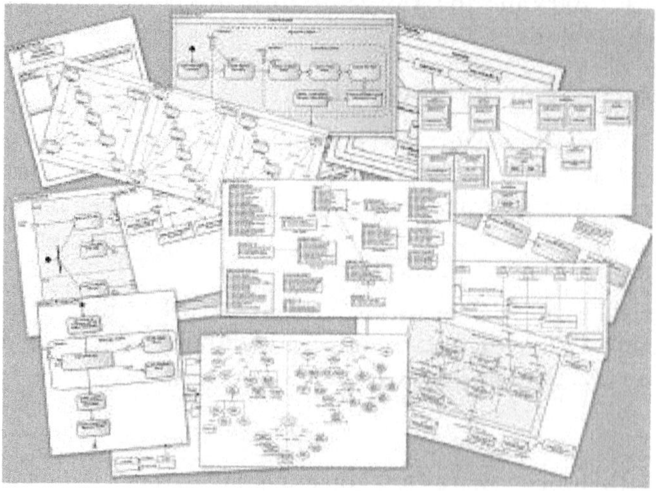

Fig. : A collage of UML diagrams.

Overview

The Unified Modelling Language (UML) offers a way to visualize a system's architectural blueprints in a diagram, including elements such as :

- Any activities (jobs)
- Individual components of the system
  - And how they can interact with other software components.
- How the system will run
- How entities interact with others (components and interfaces)
- External user interface
- How the system is expected to be used.

Although originally intended solely for object-oriented design documentation, the Unified Modelling Language (UML) has been extended to cover a larger set of design documentation and been found useful in many contexts.

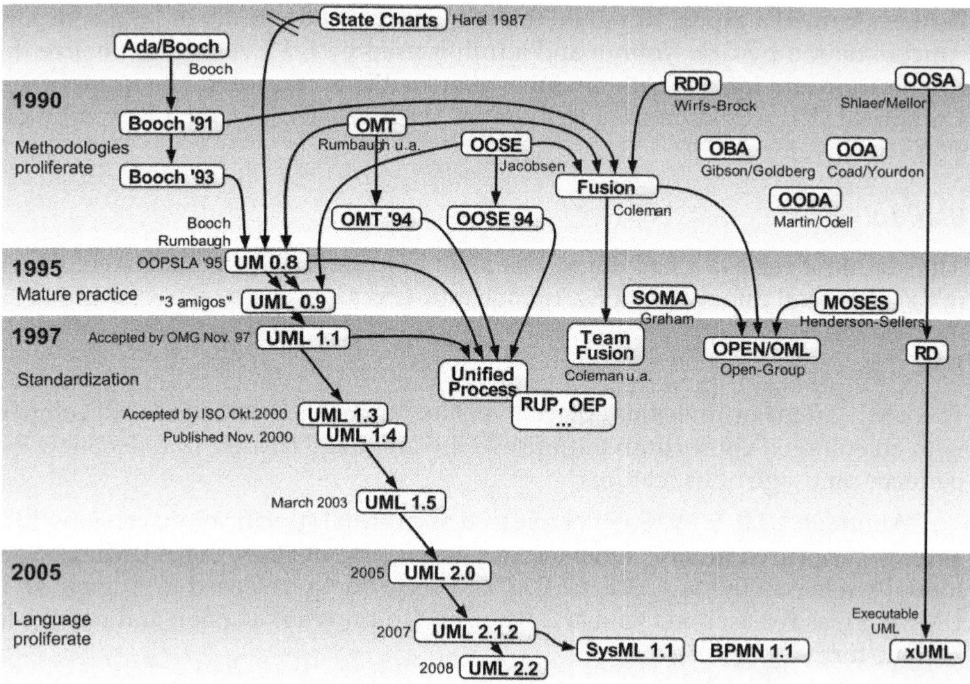

Fig. : History of object-oriented methods and notation.

History

UML has been evolving since the second half of the 1990s and has its roots in the object-oriented methods developed in the late 1980s and early 1990s.

It is originally based on the Booch method, the Object-modelling technique (OMT) and Object-oriented software engineering (OOSE), which it has integrated into a single language.

Before UML 1.x

Rational Software Corporation hired James Rumbaugh from General Electric in 1994 and after that the company became the source for two of the most popular object-oriented modelling approaches of the day : Rumbaugh's Object-modelling technique (OMT) and Grady Booch's method. They were soon assisted in their efforts by Ivar Jacobson, the creator of the object-oriented software engineering (OOSE) method, who joined them at Rational in 1995.

Under the technical leadership of those three, a consortium called the UML Partners was organized in 1996 to complete the *Unified Modelling Language (UML)* specification, and propose it to the Open Modelling Group (OMG) for standardisation. The partnership was also contained additional interested parties (for example HP, DEC, IBM and Microsoft). The UML Partners' UML 1.0 draft was proposed to the OMG in January 1997 by the consortium. During the same month the UML

Partners formed a group, designed to define the exact meaning of language constructs, chaired by Cris Kobryn and administered by Ed Eykholt, to finalize the specification and integrate it with other standardization efforts. The result of this work, UML 1.1, was submitted to the OMG in August 1997 and adopted by the OMG in November 1997.

UML 1.x

After the first release a task force was formed to improve the language, which released several minor revisions, 1.3, 1.4, and 1.5.

UML 2.x

The UML 2.0 major revision replaced version 1.5 in 2005, which was developed with an enlarged consortium to improve the language further to reflect new experience on usage of its features.

Although UML 2.1 was never released as a formal specification, versions 2.1.1 and 2.1.2 appeared in 2007, followed by UML 2.2 in February 2009. UML 2.3 was formally released in May 2010. UML 2.4.1 was formally released in August 2011. UML 2.5 was released in October 2012 as an "In process" version and has yet to become formally released.

There are four parts to the UML 2.x specification :

1. The Superstructure that defines the notation and semantics for diagrams and their model elements.
2. The Infrastructure that defines the core metamodel on which the Superstructure is based.
3. The Object Constraint Language (OCL) for defining rules for model elements.
4. The UML Diagram Interchange that defines how UML 2 diagram layouts are exchanged.

The current versions of these standards follow : UML Superstructure version 2.4.1, UML Infrastructure version 2.4.1, OCL version 2.3.1, and UML Diagram Interchange version 1.0. It continues to be updated and improved by the revision task force, who resolve any issues with the language.

Design/usage

Software Development Methods

UML is not a development method by itself; however, it was designed to be compatible with the leading object-oriented software development methods of its time (for example OMT, Booch method, Objectory).

Modelling

It is important to distinguish between the UML model and the set of diagrams of a system. A diagram is a partial graphic representation of a system's model. The model also contains documentation that drives the model elements and diagrams (such as written use cases).

UML diagrams represent two different views of a system model :

- *Static (or structural) view :* Emphasizes the static structure of the system using objects, attributes, operations and relationships. The structural view includes class diagrams and composite structure diagrams.
- *Dynamic (or behavioural) view :* Emphasizes the dynamic behaviour of the system by showing collaborations among objects and changes to the internal states of objects. This view includes sequence diagrams, activity diagrams and state machine diagrams.

UML models can be exchanged among UML tools by using the XML Metadata Interchange (XMI) interchange format.

Diagrams

UML 2 has many types of diagrams which are divided into two categories. Some types represent *structural* information, and the rest represent general types of *behaviour,* including a few that represent different aspects of *interactions.* These diagrams can be categorized hierarchically as shown in the following class diagram :

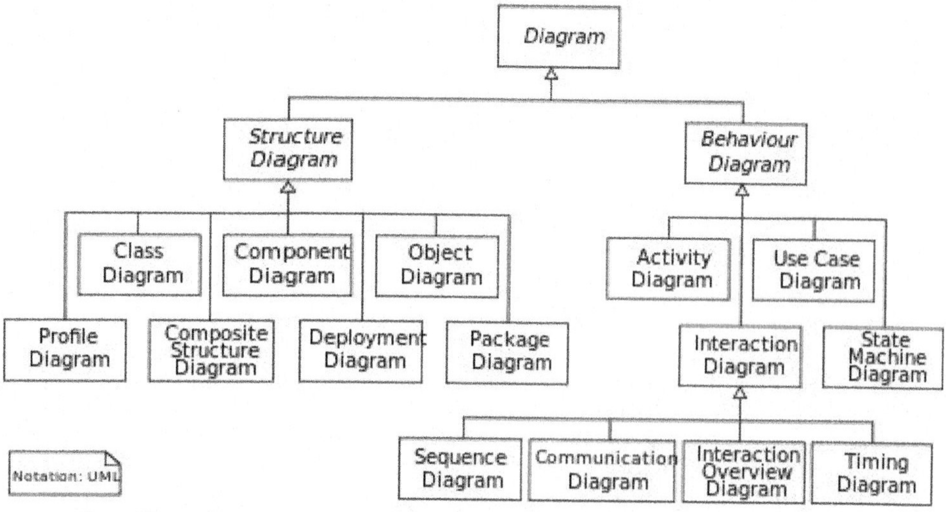

Fig. : These diagrams may all contain comments or notes explaining usage, constraint, or intent.

Structure Diagrams

Structure diagrams emphasize the things that must be present in the system being modelled. Since structure diagrams represent the structure, they are used extensively in documenting the software architecture of software systems. For example, the component diagram which describes how a software system is split up into components and shows the dependencies among these components.

Behaviour Diagrams

Behaviour diagrams emphasize what must happen in the system being modelled. Since behaviour diagrams illustrate the behaviour of a system, they are used extensively to describe the functionality of software systems. As an example, the activity diagram it describes the business and operational step-by-step activities of the components in a system.

Interaction Diagrams

Interaction diagrams, a sub-set of behaviour diagrams, emphasize the flow of control and data among the things in the system being modelled. For example, the sequence diagram which shows how objects communicate with each other in terms of a sequence of messages.

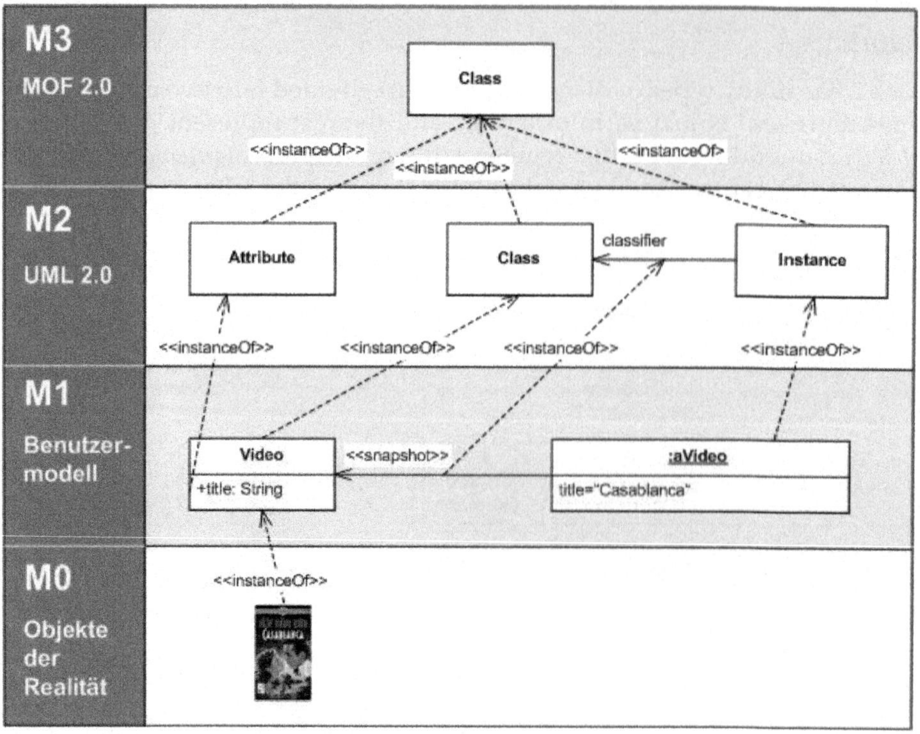

Fig. : Illustration of the Meta-Object Facility.

Meta Modelling

The Object Management Group (OMG) has developed a metamodelling architecture to define the Unified Modelling Language (UML), called the Meta-Object Facility (MOF). The Meta-Object Facility is designed as a four-layered architecture, as shown in the image at right. It provides a meta-meta model at the top layer, called the M3 layer. This M3-model is the language used by Meta-Object Facility to build metamodels, called M2-models.

The most prominent example of a Layer 2 Meta-Object Facility model is the UML metamodel, the model that describes the UML itself. These M2-models describe elements of the M1-layer, and thus M1-models. These would be, for example, models written in UML. The last layer is the M0-layer or data layer. It is used to describe runtime instances of the system.

The meta-model can be extended using a mechanism called stereotyping. This has been criticised as being insufficient/untenable by Brian Henderson-Sellers and Cesar Gonzalez-Perez in "Uses and Abuses of the Stereotype Mechanism in UML 1.x and 2.0".

Adoption

UML has been found useful in many design contexts, so much so that is has become ubiquitous in its field.

It has been treated, at times, as a design silver bullet, which has led to problems in its usage. Misuse of it includes excessive usage of it (design every little part of the system's code with it, which is unnecessary) and assuming that anyone can design anything with it (even those who haven't programmed).

It is seen to be a large language, with many constructs in it. Some feel (including Jacobson) that there are too many and that this hinders the learning (and therefore usage) of it. Efforts are underway to attempt to shrink to the size of the language that needs to be learnt.

Criticisms

Critique of UML 1.x

Linguistic incoherence

The standards have been cited as being ambiguous and inconsistent.

Cardinality notation

As with database Chen, Bachman, and ISO ER diagrams, class models are specified to use "look-across" cardinalities, even though several authors (Merise, Elmasri & Navathe amongst others) prefer same-side or "look-here" for roles and both minimum and maximum cardinalities. Recent researchers (Feinerer, Dullea et. alia) have shown that the "look-across" technique used by UML and ER diagrams is less effective and less coherent when applied to n-ary relationships of order >2.

In Feinerer it says "Problems arise if we operate under the look-across semantics as used for UML associations. Hartmann investigates this situation and shows how and why different transformations fail."

APPLIED DATA-MINING

Cross Industry Standard Process for Data-Mining

CRISP-DM stands for Cross Industry Standard Process for Data-mining. It is a data-mining process model that describes commonly used approaches that expert data miners use to tackle problems. Polls conducted in 2002, 2004, and 2007 show that it is the leading methodology used by data miners. The only other data-mining standard named in these polls was SEMMA. However, 3-4 times as many people reported using CRISP-DM. A review and critique of data-mining process models in 2009 called the CRISP-DM the "de facto standard for developing data-mining and knowledge discovery projects." Other reviews of CRISP-DM and data-mining process models include Kurgan and Musilek's 2006 review, and Azevedo and Santos' 2008 comparison of CRISP-DM and SEMMA.

Major Phases

CRISP-DM breaks the process of data-mining into six major phases.

The sequence of the phases is not strict and moving back and forth between different phases is always required. The arrows in the process diagram indicate the most important and frequent dependencies between phases. The outer circle in the diagram symbolizes the cyclic nature of data-mining itself. A data-mining process continues after a solution has been deployed. The lessons learned during the process can trigger new, often more focused business questions and subsequent data-mining processes will benefit from the experiences of previous ones.

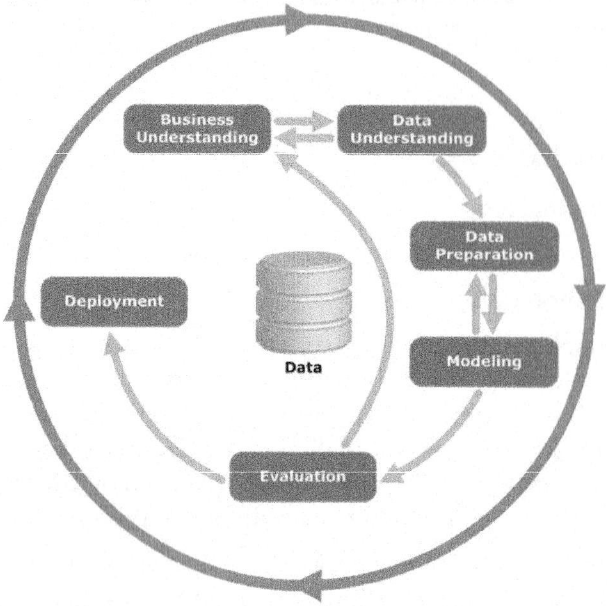

Fig. : Process diagram showing the relationship between the different phases of CRISP-DM.

- *Business Understanding* : This initial phase focuses on understanding the project objectives and requirements from a business perspective, and then converting this knowledge into a data-mining problem definition, and a preliminary plan designed to achieve the objectives.
- *Data Understanding* : The data understanding phase starts with an initial data collection and proceeds with activities in order to get familiar with the data, to identify data quality problems, to discover first insights into the data, or to detect interesting sub-sets to form hypotheses for hidden information.
- *Data Preparation* : The data preparation phase covers all activities to construct the final dataset (data that will be fed into the modelling tool(s)) from the initial raw data. Data preparation tasks are likely to be performed multiple times, and not in any prescribed order. Tasks include table, record, and attribute selection as well as transformation and cleaning of data for modelling tools.
- *Modelling* : In this phase, various modelling techniques are selected and applied, and their parameters are calibrated to optimal values. Typically, there are several techniques for the same data-mining problem type. Some techniques have specific requirements on the form of data. Therefore, stepping back to the data preparation phase is often needed.
- *Evaluation* : At this stage in the project you have built a model (or models) that appear to have high quality, from a data analysis perspective. Before proceeding to final deployment of the model, it is important to more thoroughly evaluate the model, and review the steps executed to construct the model, to be certain it properly achieves the business objectives. A key objective is to determine if there is some important business issue that has not been sufficiently considered. At the end of this phase, a decision on the use of the data-mining results should be reached.
- *Deployment* : Creation of the model is generally not the end of the project. Even if the purpose of the model is to increase knowledge of the data, the knowledge gained will need to be organized and presented in a way that the customer can use it. Depending on the requirements, the deployment phase can be as simple as generating a report or as complex as implementing a repeatable data-mining process. In many cases it will be the customer, not the data analyst, who will carry out the deployment steps. However, even if the analyst will not carry out the deployment effort it is important for the customer to understand up front the actions which will need to be carried out in order to actually make use of the created models.

History

CRISP-DM was conceived in 1996 : In 1997 it got underway as a European Union project under the ESPRIT funding initiative. The project was led by five companies : SPSS, Teradata, Daimler AG, NCR Corporation and OHRA, an insurance company.

This core consortium brought different experiences to the project : ISL, later acquired and merged into SPSS Inc. The computer giant NCR Corporation produced the

Teradata data warehouse and its own data-mining software. Daimler-Benz had a significant data-mining team. OHRA was just starting to explore the potential use of data-mining.

The first version of the methodology was presented at the 4th CRISP-DM SIG Workshop in Brussels in March 1999, and published as a step-by-step data-mining guide later that year.

Between 2006 and 2008 a CRISP-DM 2.0 SIG was formed and there were discussions about updating the CRISP-DM process model. The current status of these efforts is not known. However, the original crisp-dm.org website cited in the reviews, and the CRISP-DM 2.0 SIG website are both no longer active.

While many non-IBM data-mining practitioners use CRISP-DM, IBM is the primary corporation that currently embraces the CRISP-DM process model. It makes some of the old CRISP-DM documents available for download and it has incorporated it into its SPSS Modeler product.

CUSTOMER ANALYTICS

Customer analytics is a process by which data from customer behaviour is used to help make key business decisions via market segmentation and predictive analytics. This information is used by businesses for direct marketing, site selection, and customer relationship management. Marketing provides services in order to satisfy customers. With that in mind, the productive system is considered from its beginning at the production level, to the end of the cycle at the consumer. Customer analytics plays a very important role in the prediction of customer behaviour today.

Uses

Retail

Although until recently over 90% of retailers had limited visibility on their customers, with increasing investments in loyalty programs, customer tracking solutions and market research, this industry started increasing use of customer analytics in decisions ranging from product, promotion, price and distribution management.

Finance

Banks, insurance companies and pension funds make use of customer analytics in understanding customer lifetime value, identifying below-zero customers which are estimated to be around 30% of customer base, increasing cross-sales, managing customer attrition as well as migrating customers to lower cost channels in a targeted manner.

Community

Municipalities utilize customer analytics in an effort to lure retailers to their cities. Using psychographic variables, communities can be segmented based on attributes like personality, values, interests, and lifestyle. Using this information, communities can approach retailers that match their community's profile.

Customer Relationship Management

Analytical Customer Relationship Management, commonly abbreviated as CRM, enables measurement of and prediction from customer data to provide a 360° view of the client.

Predicting Customer Behaviour

Forecasting buying habits and lifestyle preferences is a process of data-mining and analysis. This information consists of many aspects like credit card purchases, magazine subscriptions, loyalty card membership, surveys, and voter registration. Using these categories, profiles can be created for any organization's most profitable customers. When many of these potential customers are aggregated in a single area it indicates a fertile location for the business to situate. Using a drive time analysis, it is also possible to predict how far a given customer will drive to a particular location. Combining these sources of information, a dollar value can be placed on each household within a trade area detailing the likelihood that household will be worth to a company. Through customer analytics, companies can make decisions with confidence because every decision is based on facts and objective Data.

Data-mining

There are two types of categories of data-mining. Predictive models use previous customer interactions to predict future events while segmentation techniques are used to place customers with similar behaviours and attributes into distinct groups. This grouping can help marketers to optimize their campaign management and targeting processes.

Future

By continuing to improve customer prediction techniques it will become a necessity rather than a convenient commodity for businesses to use customer analytics. With this valuable information there is an opportunity to fine-tune retail operations and store manager decisions. Rapid decision-making will increase in speed and effectiveness in the future as tools and information become more easily accessible. The possibilities are still emerging, but applications in political races, jury selection, and developing clinical trial communities are areas that customer analytics could be used in the future.

Chapter 21

BOOSTING

Boosting is a machine learning meta-algorithm for reducing bias in supervised learning. Boosting is based on the question posed by Kearns : Can a set of **weak learners** create a single **strong learner**? A weak learner is defined to be a classifier which is only slightly correlated with the true classification (it can label examples better than random guessing). In contrast, a strong learner is a classifier that is arbitrarily well-correlated with the true classification.

Schapire's affirmative answer to Kearns' question has had significant ramifications in machine learning and statistics, most notably leading to the development of boosting.

When first introduced, the *hypothesis boosting problem* simply referred to the process of turning a weak learner into a strong learner. "Informally, [the hypothesis boosting] problem asks whether an efficient learning algorithm [...] that outputs a hypothesis whose performance is only slightly better than random guessing [*i.e.* a weak learner] implies the existence of an efficient algorithm that outputs an hypothesis of arbitrary accuracy [*i.e.* a strong learner]." Algorithms that achieve hypothesis boosting quickly became simply known as "boosting". Freund and Schapire's arcing (Adapt[at]ive Re-sampling and Combining), as a general technique, is more or less synonymous with boosting.

BOOSTING ALGORITHMS

While boosting is not algorithmically constrained, most boosting algorithms consist of iteratively learning weak classifiers with respect to a distribution and adding them to a final strong classifier. When they are added, they are typically weighted in some way that is usually related to the weak learners' accuracy. After a weak learner is added, the data is reweighted : examples that are misclassified gain weight and examples that are classified correctly lose weight (some boosting algorithms actually decrease the weight of repeatedly misclassified examples, *e.g.*, boost by majority and BrownBoost). Thus, future weak learners focus more on the examples that previous weak learners misclassified.

There are many boosting algorithms. The original ones, proposed by Robert Schapire (a recursive majority gate formulation) and Yoav Freund (boost by majority), were not adaptive and could not take full advantage of the weak learners. They won the prestigious Gödel Prize for their work on AdaBoost, a boosting algorithm that they developed together. Only algorithms that are provable boosting algorithms in the probably approximately correct learning formulation are called boosting algorithms. Other algorithms that are similar in spirit to boosting algorithms are sometimes called "leveraging algorithms", although they are also sometimes incorrectly called boosting algorithms.

EXAMPLES OF BOOSTING ALGORITHMS

The main variation between many boosting algorithms is their method of weighting training data points and hypotheses. AdaBoost is very popular and perhaps the most significant historically as it was the first algorithm that could adapt to the weak learners. However, there are many more recent algorithms such as LPBoost, TotalBoost, BrownBoost, MadaBoost, LogitBoost, and others. Many boosting algorithms fit into the Any Boost framework, which shows that boosting performs gradient descent in function space using a convex cost function.

Boosting algorithms are used in Computer Vision, where individual classifiers detecting contrast changes can be combined to identify Facial Features.

CRITICISM

In 2008 Phillip Long (at Google) and Rocco A. Servedio (Columbia University) published a paper at the 25th International Conference for Machine Learning suggesting that many of these algorithms are probably flawed. They conclude that "convex potential boosters cannot withstand random classification noise," thus making the applicability of such algorithms for real world, noisy data-sets questionable. The paper shows that if any non-zero fraction of the training data is mis-labelled, the boosting algorithm tries extremely hard to correctly classify these training examples, and fails to produce a model with accuracy better than 1/2. This result does not apply to branching program based boosters but does apply to AdaBoost, LogitBoost, and others.

BOOTSTRAP AGGREGATING

Bootstrap aggregating (bagging) is a machine learning ensemble meta-algorithm designed to improve the stability and accuracy of machine learning algorithms used in statistical classification and regression. It also reduces variance and helps to avoid overfitting. Although it is usually applied to decision tree methods, it can be used with any type of method. Bagging is a special case of the model averaging approach.

Description of the Technique

Given a standard training set D of size n, bagging generates m new training sets D_i, each of size n', by sampling from D uniformly and with replacement. By sampling with replacement, some observations may be repeated in each D_i. If $n'=n$, then for large n the set D_i is expected to have the fraction (1-1/e) (≈63.2%) of the unique examples of D, the rest being duplicates. This kind of sample is known as a bootstrap sample. The m models are fitted using the above m bootstrap samples and combined by averaging the output (for regression) or voting (for classification).

Bagging leads to "improvements for unstable procedures" (Breiman, 1996), which include, for example, neural nets, classification and regression trees, and sub-set selection in linear regression (Breiman, 1994). An interesting application of bagging showing improvement in preimage learning is provided here. On the other hand, it can mildly degrade the performance of stable methods such as K-nearest neighbours (Breiman, 1996).

Example : Ozone Data

To illustrate the basic principles of bagging, below is an analysis on the relationship between ozone and temperature (data from Rousseeuw and Leroy, available at classic data sets, analysis done in R).

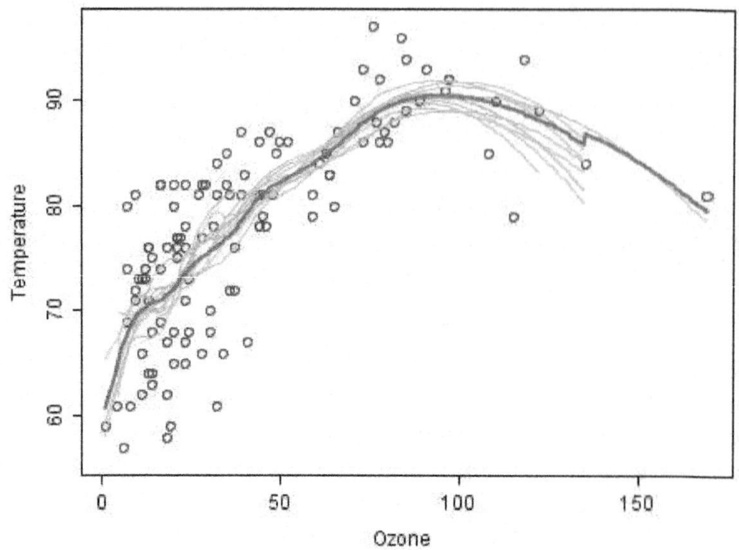

The relationship between temperature and ozone in this data-set is apparently non-linear, based on the scatter plot. To mathematically describe this relationship, LOESS smoothers (with span 0.5) are used. Instead of building a single smoother from the complete data-set, 100 bootstrap samples of the data were drawn. Each sample is different from the original data-set, yet resembles it in distribution and variability. For each bootstrap sample, a LOESS smoother was fit. Predictions from these 100 smoothers were then made across the range of the data. The first 10 predicted smooth fits appear as grey lines in the figure below. The lines are clearly very *wiggly* and they overfit the data—a result of the span being too low.

But taking the average of 100 smoothers, each fitted to a sub-set of the original data-set, we arrive at one bagged predictor (red line). Clearly, the mean is more stable and there is less overfit.

Bagging for Nearest Neighbour Classifiers

It is well known that the risk of a 1 nearest neighbour (1NN) classifier is at most twice the risk of the Bayes classifier, but there are no guarantees that this classifier will be consistent. By careful choice of the size of the resamples, bagging can lead to substantial improvements of the performance of the 1NN classifier. By taking a large number of resamples of the data of size n', the bagged nearest neighbour classifier will be consistent provided $n' \to \infty$ diverges but $n'/n \to 0$ as the sample size $n \to \infty$.

Under infinite simulation, the bagged nearest neighbour classifier can be viewed as a weighted nearest neighbour classifier. Suppose that the feature space is d dimensional and denote by $C^{bnn}_{n,n'}$ the bagged nearest neighbour classifier based on a training set of size n, with resamples of size n'. In the infinite sampling case, under certain regularity conditions on the class distributions, the excess risk has the following asymptotic expansion

$$R_R\left(C^{bnn}_{n,n'}\right) - R_R(C^{Bayes}) = \left(B_1 \frac{n'}{n} + B_2 \frac{1}{(n')^{4/d}}\right)\{1 + o(1)\},$$

for some constants B_1 and B_2. The optimal choice of n', that balances the two terms in the asymptotic expansion, is given by $n' = Bn^{d/(d+4)}$ for some constant B.

History

Bagging (Bootstrap aggregating) was proposed by Leo Breiman in 1994 to improve the classification by combining classifications of randomly generated training sets.

Chapter 22

A Hybrid PCA-CART-MARS-Based Prognostic Approach of the Remaining Useful Life for Aircraft Engines

Fernando Sánchez Lasheras [1,†,*], Paulino José García Nieto [2,†], Francisco Javier de Cos Juez [3,†], Ricardo Mayo Bayón [4,†] and Victor Manuel González Suárez [4,†]

1 Department of Construction and Manufacturing Engineering, University of Oviedo, Gijón 33204, Spain

2 Department of Mathematics, University of Oviedo, Oviedo 33007, Spain; E-Mail: lato@orion.ciencias.uniovi.es

3 Department of Mining Engineering and Exploitation, University of Oviedo, Oviedo 33004, Spain; E-Mail: fjcos@uniovi.es

4 Department of Electrical Engineering, University of Oviedo, Gijón 33204, Spain; E-Mails: rmayo@uniovi.es (R.M.B.); vmsuarez@uniovi.es (V.M.G.S.)

† These authors contributed equally to this work.

* Author to whom correspondence should be addressed; E-Mail: sanchezfernando @uniovi.es; Tel.: +34-984-833-135; Fax: +34-985-522-828.

ABSTRACT

Prognostics is an engineering discipline that predicts the future health of a system. In this research work, a data-driven approach for prognostics is proposed. Indeed, the present paper describes a data-driven hybrid model for the successful prediction of the remaining useful life of aircraft engines. The approach combines the multivariate adaptive regression splines (MARS) technique with the principal component analysis (PCA), dendrograms and classification and regression trees (CARTs). Elements extracted from sensor signals are used to train this hybrid model, representing different levels of health for aircraft engines. In this way, this hybrid algorithm is used to predict the trends of these elements. Based on this fitting, one can determine the future health state of a system and estimate its remaining useful life (RUL) with accuracy. To evaluate the proposed approach, a test was carried out using aircraft engine signals collected from physical sensors (temperature, pressure, speed, fuel flow, *etc.*). Simulation results show that the PCA-CART-MARS-based approach can forecast faults long before they occur and can predict the RUL. The proposed hybrid model presents as its main advantage the fact that it does not require information about the previous operation states of the input variables of the engine. The performance of this model was compared with those obtained by other benchmark models (multivariate linear regression and artificial neural networks) also applied in recent years for the modeling of remaining useful life. Therefore, the PCA-CART-MARS-based approach is very promising in the field of prognostics of the RUL for aircraft engines.

Keywords

Prognostics; aircraft engine; remaining useful life; principal component analysis (PCA); dendrogram; classification and regression trees (CART); multivariate adaptive regression splines (MARS)

1. INTRODUCTION

In this section an introduction to prognostics and health management (PHM) is performed. Also, the aircraft sensors are briefly described and the functioning of the aircraft engine is explained. PHM is now an engineering discipline with a growing importance. PHM combines [1] detection and analysis of environmental, operational and performance-related parameters to assess the health of a product and predict its remaining useful life (RUL). The current state of the art of the PHM [2] presents three main research lines:

- *Data-driven approach:* The data-driven models for the prediction of the health or RUL of any system uses machine learning and pattern recognition techniques [3-5]. The main characteristic of this kind of models is that they do not require any previous knowledge about the device under analysis as the prediction is made according to the available information without taking into account the operating principles of the system.
- *Model-based approach:* in this approach the prognostic models are based on an understanding of the physical process and interrelationships among the different components and subsystems of the device [6]. The model-based-approach not only includes system modeling but would also apply physics of failure modeling [1] or any similar methodology.
- *Hybrid approaches:* These kinds of models try to make the most of both data-driven and model based approaches [1]. A hybrid model combines both data-driven methodologies with the knowledge of the device under study. Please do not confuse PHM hybrid approach with hybrid data-driven models,

which are models that combine different pattern-recognition and machine learning models in order to predict either the RUL or the status of the system under study.

Although the two traditional approaches to PHM have been model-based or data driven, some of the latest research in the PHM discipline combines the strength of the data-driven prognostic methods with the physical models of the system under study, achieving remarkable results [7] in spite of this, data driven approaches like the one presented in this research are also shown to be performing well [8].

The methodology presented in this paper can be considered as data-driven and proposes a model that combines well-known pattern recognition and machine learning methodologies in a hybrid model that is trained in order to be able to predict the RUL of the aircraft engine under study taking into account the current status of some of its variables, but without requiring information about the previous status of the variables of the engine. In other words, the concept of hybrid algorithm does not refer to an algorithm that combines both data-driven methodologies with the understanding of the system under study but rather that it refers to a hybrid algorithm with the meaning that it is usually employed in the field of machine learning and pattern recognition [9,10]. The model proposed hybridizes the following techniques: Principal Component Analysis (PCA), Dendrogram, Regression Trees and Multivariate Adaptive Regression Splines (MARS). The advantages of the proposed algorithm are twofold; on the one hand a reduced number of variables is employed for its training, which makes the training of the final models more efficient and their results easier to replicate, while on the other hand the final MARS models that are trained are not for the complete state of space but for a subset which increases the accuracy of predictions over an individual method. Although there have been previous papers that employ data-driven models based on hybrid algorithms [2,11], the methodology proposed in this paper is new and performs well when compared with previous models used with the same database.

In the present research, the concept of RUL [12] must be understood to be the number of time units remaining that the equipment has before it reaches its limit of operative safety. In recent years a vast number of articles related to different approaches to the concept of remaining useful life based on statistical concepts have been written. In general, the remaining useful life of any device can be considered as a random variable [12].

The Aircraft Engine Sensors

Aircraft engine monitoring systems are used to check the health of the aircraft engines in order to avoid costly repairs by means of preventive maintenance. Engine monitoring systems involve using sensors placed in various locations in an aircraft engine to gather information about its performance. The sensors provide real-time information to pilots on the operation of the engines and also capture data for analysis of the performance of the engine over time. The information obtained provides information about the engine status. The main kinds of sensors present in an aircraft engine are the following:

- *Temperature sensors:* These provide output readings proportional to temperature. The measurement of temperature can be accomplished using thermocouple or resistance temperature device (RTD) techniques depending upon system interface, temperature, or accuracy considerations.
- *Pressure sensors:* Currently the most common pressure sensors are those formed by two separate absolute pressure-sensing capsules. One senses system pressure and the other atmospheric pressure. In these devices, the output signal is the difference between the signals of both sensors.
- *Sensors of speed (rpm):* These kinds of sensors are usually installed directly on the low-pressure compressor rotor in order to determine the revolutions at the low-pressure section. They are also installed in the casing of driving mechanisms, as this is the most suitable place for the measurement of the speed of the high-pressure compressor section.
- *Sensors of fuel flow:* These are usually vane sensors located downstream behind the fuel filter and the pump measuring the amount of fuel going to the engine. The importance of the fuel flow sensor is remarkable as it can immediately detect any engine abnormality involving an increase in fuel consumption.

The diagram in Figure 1 shows the main elements of the engine model proposed in the present research. Air gets into the low pressure compressor through the fan passing afterwards to the high pressure compressor. The air is then heated in the combustor where it is mixed with fuel and ignited. Combustion of the

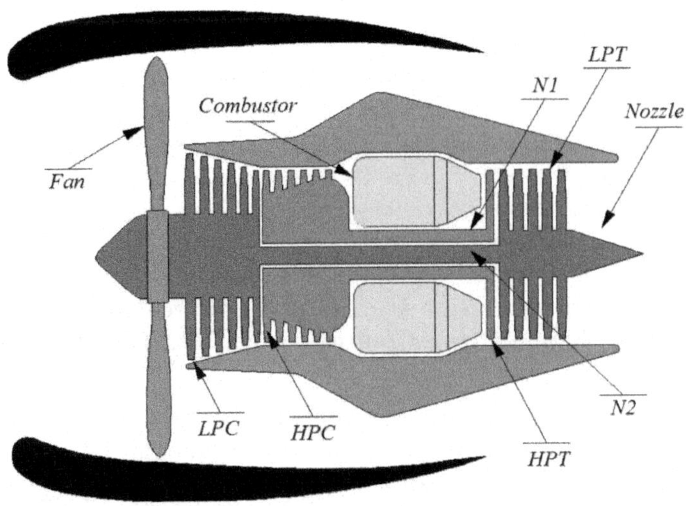

Figure 1. Simplified diagram of the engine simulated [13] (LPC: Low pressure compressor, HPC: High pressure compressor, LPT: Low pressure turbine, HPT: High pressure turbine, N1: Turbine axis and N2: Turbine shaft).

fuel increases the HPC discharge air velocity to drive the high and low pressure turbines. This engine model is based on a low frequency, transient, performance model of a high-pressure ratio, dual-spool, low by-pass, variable cycle, turbofan engine with a digital controller. The controller update rate is 50 Hz, and the component level model balances the mass/energy equations of the system at a rate of 2500 Hz. These capabilities allow the user to verify the performance of the control algorithms and its validation in a generic engine model. The MAPSS program has versions for both civil and military application. A complete explanation of the full C-MAPSS subsystem hierarchy can be found in the C-MAPSS User's Guide [13]. The military version is able to perform realistic simulations within the standard Full Authority Digital Engine Controllers (FADEC) [14] and it will be used in the present research work.

2. MATERIALS AND METHODS

The materials and methods section explains the variables included in the database used for this research and how they were obtained. This section also describes the hybrid algorithm that is the original contribution of this research and all the statistical procedures that are part of it.

2.1. The Database

The present research uses data corresponding to an aircraft engine of 90,000 lb thrust class. The data includes working conditions with heights from sea-level conditions up to heights of 40,000 feet and temperatures ranging from -51 °C to 39 °C. This data was obtained through a software application known as Modular Aero-Propulsion System Simulation (MAPSS). This is a simulation environment for aeronautical turbines which allows access to a range of monitoring parameters controlling the operating state of the system through a graphical user interface. This provides the user with a simulation environment which is able to develop advanced algorithms for control and diagnosis that can be tested on a generic aircraft engine simulator. MAPSS is able to generate linear state space models from which the user can create linear controls. Furthermore, MAPSS is able to perform simulations of transient states. These capabilities allow the user to verify the performance of the control algorithms and its validation in a generic engine model [14]. The MAPSS program has versions for both civil and military application. The military version is able to perform realistic simulations within the Full Authority Digital Engine Controllers (FADEC) [14] standard and will be used for the present research.

The database contains input parameters that simulate the effects of faults and deterioration in any of the rotating components of the engine (fan, low pressure compressor or LPC, high pressure compressor or HPC, high pressure turbine or HPT, and low pressure turbine or LPT). The data belongs to a fleet of similar aircraft with the same engines. Each engine starts with a different health status, but initially all are within the operational range and the data set convers from that moment to the engine failure. Each record of the engine state is formed by a set of 24 variables. Three of these are operational settings, while the other 21 represent values for measurements of engine performance which are contaminated with noise. Also, the remaining useful life of the engines for each record is known. Please note that the length of the RUL is a theoretical time unit that would be considered in either hours or cycles. This information has been introduced in the manuscript, in the section that describes the database. A complete list of variables is presented in Table 1. This database will be used for the training and validation of the hybrid model proposed in this paper. All the engines considered in this study had the same failure mode, namely, HPC degradation. To simulate HPC degradation, HPC flow and efficiency modifiers were used [12].

Table 1. MAPSS input variables of the simulated engine to simulate the remaining useful life (RUL).

Input Variables	Symbol	Sensor
Total temperature at fan inlet (°R)	T2	Sensor.Measurement1
Total temperature at LPC outlet (°R)	T24	Sensor.Measurement2
Total temperature at HPC outlet (°R)	T30	Sensor.Measurement3
Total temperature at LPT outlet (°R)	T50	Sensor.Measurement4
Pressure at fan inlet (psia)	P2	Sensor.Measurement5
Total pressure in bypass-duct (psia)	P15	Sensor.Measurement6
Total pressure at HPC outlet (psia)	P30	Sensor.Measurement7
Physical fan speed (rpm)	Nf	Sensor.Measurement8
Physical core speed (rpm)	Nc	Sensor.Measurement9
Engine pressure ratio (P50/P2)	Epr	Sensor.Measurement10
Static pressure at HPC outlet (psia)	Ps30	Sensor.Measurement11
Ratio of fuel flow to Ps30 (pps/psi)	phi	Sensor.Measurement12
Corrected fan speed (rpm)	NRf	Sensor.Measurement13
Corrected core speed (rpm)	NRc	Sensor.Measurement14
Bypass ratio	BPR	Sensor.Measurement15
Burner fuel-air ratio	farB	Sensor.Measurement16
Bleed enthalpy	htBleed	Sensor.Measurement17
Demanded fan speed (rpm)	Nf_dmd	Sensor.Measurement18
Demanded corrected fan speed (rpm)	PCNfR_dmd	Sensor.Measurement19
HPT coolant bleed (lbm/s)	W31	Sensor.Measurement20
LPT coolant bleed (lbm/s)	W32	Sensor.Measurement21

2.2. The Proposed Algorithm

The model proposed in this research is a hybrid model. In recent years, hybrid methods in machine learning have attracted a great deal of attention from the scientific community. There has been an increasing tendency to use of hybrid models due to the good performance which these kinds of models have given, not only in the forecast of RUL [15,16] but also in many other different applications [17,18]. In other words, it may be observed from the comparison between methods and hybridizations that the performance of the hybrid method is superior [19,20].

The steps followed in the proposed hybrid algorithm are shown in Figure 2. The steps of this algorithm are as follows:

- *Dimensional reduction by means of Principal Components Analysis (PCA):* In the first stage, a dimensional reduction by means of PCA was performed. These variables, which represented up to 90% of the total data variability, were held in order to be used in the next step of the process. Also, a rotated factor pattern of the variables retained was created using the varimax method [21,22].
- *Study of selected variables similarity by means of a dendrogram:* The remaining variables were classified by means of a dendrogram in order to find those which are most similar to the output variable, which in this case is the remaining useful life.
- *Calculation of the regression tree:* Using those variables that are most similar to the output variable, a regression tree was calculated.
- *Calculation of a multivariate adaptive regression splines (MARS) model for each one of the leaf nodes of the regression tree:* A MARS model, using as input variables all those considered of relevance by the dendrogram, was calculated for each of the leaf nodes of the regression tree.

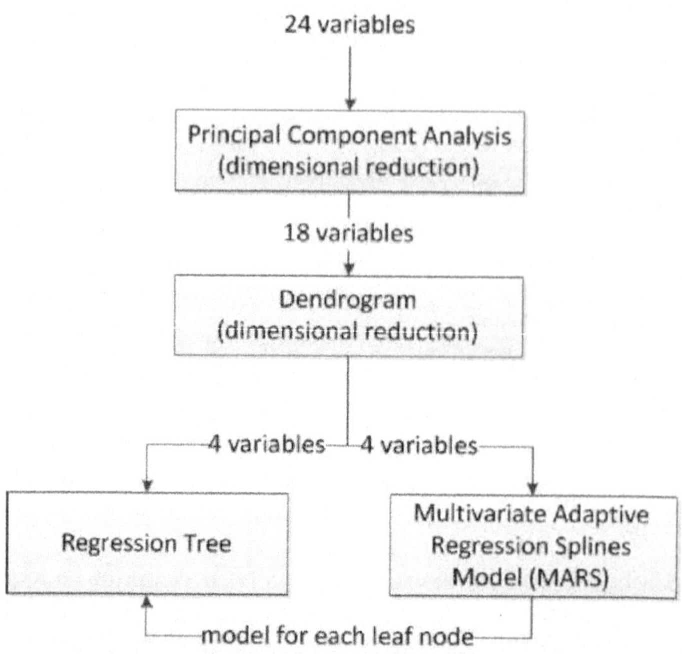

Figure 2. Flowchart of the proposed algorithm.

The contribution of each of the techniques included in the hybrid model is as follows: the two first methods, Principal Components Analysis and the Dendrogram are used for dimensional reduction. In the case of the PCA, it is used as a dimensional reduction tool able to eliminate those variables that contribute the least to data variance, while the dendrogram makes it possible to remove those sensors and operational settings which have a weak relationship with the RUL. The aim of the regression tree in the framework of the proposed hybrid algorithm is to create different subsets of input variables values according to their RUL. Finally, a different MARS model was trained for each of these subsets. The advantages of the proposed algorithm are twofold; on the one hand a reduced number of variables is employed for its training, which makes it more efficient, and on the other hand the final MARS models that are trained are not for the complete state of space, but for a subset which increases the accuracy of predictions over an individual method. As far as the authors know, and according to revision articles [15,16], this research presents the first application of both dendrogram and decision trees to the C-MAPSS database. The next section describes one by one all the techniques that are used in this algorithm.

Finally, the performance of the proposed hybrid model, obtained by following all the steps detailed above, was also compared with the performance obtained by two other models that have been used in recent years for the calculation of remaining useful life: multivariate linear regression [14] and back-propagation artificial neural networks [14]. In these comparisons two classical methods, as the root mean square error (RMSE) and mean absolute error (MAE), have been used because they are commonly used in literature; as the information provided by these metrics is not satisfactory for prognostics another three more suitable metrics [23] have also employed: accuracy, steadiness and risk level. Accuracy is a bias measure of the prognostic algorithm, which is defined as the relative error of the RUL prediction. The steadiness measures the volatility of the RUL prediction and is computed as the square root of the RUL predicted in function of time. In the present research it has been considered a delta of 10 time units. The smaller the value of the steadiness the more stable the predictions [24]. The risk level is the probability of obtaining a RUL estimate smaller than the true RUL.

2.3. Statistical Procedures Applied in the Algorithm

2.3.1. Dimensional Reduction: Principal Components Analysis

The most traditional, well-known and efficient method for reducing data dimensionality by transformation is a technique called principal components analysis (PCA), which involves transforming the original attributes or variables $x_1, x_2, \ldots, x_m$ into another set $f_1, f_2, \ldots, f_p$, *where* $p \leq m$ .The PCA method was independently introduced by Pearson in 1901 [23] and Hotelling in 1933 [24]. The use of PCA as a pre-processing tool is extremely well-documented in literature and is commonly accepted. There are several recent papers that used it in this way [25-27]. From a geometrical point of view, this process can be seen as a change of axes in the representation (projection) [21]. From a mathematic point of view can be expressed as:

$$\underset{(m\times 1)}{\bar{f}} = \underset{(m\times m)}{\begin{bmatrix} \bar{a}_1^T \\ \bar{a}_2^T \\ \cdot \\ \cdot \\ \cdot \\ \bar{a}_m^T \end{bmatrix}} \underset{(m\times 1)}{\bar{x}} = \underset{(m\times m)}{A^T} \underset{(m\times 1)}{\bar{x}}$$

In other words, it is necessary to get a m x m coefficients matrix which, multiplied by each vector of the original attributes, transform it into another vector in a new dimensional space. The Principal Components Analysis assumes that it is required that the variance of the new attributes be ordered according to their value, from larger to smaller. It can be demonstrated that the calculus of the result consists of only the calculus of the covariance matrix [23]:

$$\underset{(m\times m)}{S} = \frac{1}{n-1} \sum_{i=1}^{n} \underset{(m\times 1)}{\left(u_i\right)} \cdot \underset{(1\times m)}{\left(u_i\right)^T} \tag{2}$$

where n is the number of examples in the data set, m is the number of attributes and the vectors of differences $\bar{u}_i$ are calculated subtracting each example i (attributes vector x_1, x_2,..., x_m) with the average value of all the examples [23]:

$$\underset{(m\times 1)}{\bar{u}_i} = \underset{(m\times 1)}{\bar{x}_i} - \underset{(m\times 1)}{\bar{m}} \quad \text{where} \quad \underset{(m\times 1)}{\bar{m}} = \frac{1}{n} \sum_{i=1}^{n} \underset{(m\times 1)}{\bar{x}_i} \tag{3}$$

After the calculus of the covariance matrix S, the coefficient vectors are obtained by calculating the eigenvalues of the covariance matrix:

$$e_1 \geq e_2 \geq \ldots \geq e_m \geq\ = 0 \tag{4}$$

There are different criteria for determining how many dimensions to preserve. For example, in this case, as many dimensions as would be necessary to add up to at least 90% of the variance of the original data will be chosen. Another criterion, for example, chooses all those dimensions whose variance is greater than the average of the eigenvalues. In our case a quite restrictive criterion was chosen, as PCA is used as a first stage tool for pruning variables. It involved adding at least as many dimensions as would be necessary to achieve 90% of the variance of the original data.

2.3.2. Dendrograms

A dendrogram is a kind of tree diagram frequently used to illustrate the arrangement of clusters obtained after a hierarchical clustering process [21]. The top row of nodes represent individual observations or data, while the remaining nodes represent the groups to which the data belongs and the arrows, the distances or differences. To perform this grouping, it is usual to use the Euclidean distance as the distance metric. In the present research the metric referred to was employed.

The Euclidean distance between two points $(x_{i1}, x_{i2}, ..., x_{ip})$ and $(x_{j1}, x_{j2}, ..., x_{jp})$ in a p dimensions space can be defined as follows:

$$d_{ij} = \sqrt{(x_{i1} - x_{j1})^2 + (x_{i2} - x_{j2})^2 + ... + (x_{ip} - x_{jp})^2} \tag{5}$$

In the case of the present research, and before the application of the Euclidean distance, variables are normalized in order to avoid problems related to their different scales.

2.3.3. Decision Trees

A decision tree is a set of terms in a hierarchical structure arranged in such a way that the final decision shall be determined following the conditions that are present in each branch of the tree, from its roots to its leaves [21,28] One of the great advantages of decision trees is that, in its general form, the possible options from a given condition are exclusive. This allows a situation to be analysed and, following the decision tree properly, action to be taken or one decision to be made. In the case of the present research, an implementation of recursive regression partitioning trees was employed [29]

2.3.4. Multivariate Adaptive Regression Splines Models (MARS)

The multivariate adaptive regression models (MARS) is a technique of classification or multivariate nonparametric regression introduced by Friedman [29-35] Its main purpose is to predict the values of a continuous dependent variable $y(n \times 1)$ from a set of independent explanatory variables $\mathbf{X}(n \times p)$ The MARS model can be represented as:

$$y = f(X) + e \tag{6}$$

where f is the sum of the base functions and e is an error vector of dimensions $(n \times 1)$.

One of the main advantages of the MARS model is that it does not require any *a priori* assumptions about the functional relationship between the dependent and independent variables. This relationship is modelled by a subset of coefficients and piecewise linear splines (basic functions) The basis functions span the space of q-th order spline approximation, fitting the coefficients of the basis function expansion to the data by ordinary least-squares. In the univariate case $(n = 1)$ with $K + 1$ regions delineated by K points on the real line (knots), one such basis is represented by the function [29-31]:

$$[-(x-t)]_+^q = \begin{cases} (t-x)^q & if\ x < t \\ 0 & otherwise \end{cases} \tag{7}$$

$$[+(x-t)]_+^q = \begin{cases} (t-x)^q & if\ x \geq t \\ 0 & otherwise \end{cases} \tag{8}$$

where $q(\geq 0)$ is the power of the basis functions. The MARS model of a dependent variable y with M basis functions (terms) can be described as follows [21,27,28]:

$$\hat{y} = \hat{f}_M(\mathbf{x}) = c_0 + \sum_{m=1}^{M} c_m B_m(\mathbf{x}) \tag{9}$$

where $\hat{y}$ is the dependent variable predicted by the MARS model, c_0 is a constant, B_m (x) is the m-th basis function and c_m is the coefficient of the m-th basis function. Both variables entered in the model as the positions of the nodes for each individual variable must be optimized. For each data set **X** containing n objects and p explanatory variables, there are $N = n \times p$ pairs of basis functions given by Equations (2) and (3), with the locations of the nodes $x_{ij}(i = 1,2,...,n; j = 1,2,...,p)$. The final model is built using a two-step model. Firstly, in order to select the consecutive base pairs of the model functions, a method is implemented of two steps forward each time [29-31]. This progressive selection of basis functions leads to a very complex and overfitted model. Such a model, although it fits the data well, has poor predictive ability for new objects. To improve prediction, redundant base functions are removed one at a time using a regression procedure. To determine which basis functions will be included in the model, MARS uses the generalized cross validation *(GCV)* [29-35]. Thus, the *GCV*is the RMS residual error divided by a penalty parameter, which depends on the complexity of the model. *GCV* criterion is defined as follows [29-35]:

$$GCV(M) = \frac{\frac{1}{n}\sum_{i=1}^{n}\left(y_i - \hat{f}_M(\mathbf{x}_i)\right)^2}{\left(1 - C(M)/n\right)^2} \tag{10}$$

where $C(M)$ is the penalty due to the complexity, which increases with the number of basis functions, and the model is defined as [29-31]:

$$C(M) = (M+1) + dM \tag{11}$$

where M is the number of basis functions in Equation (7), and the parameter d is a penalty for each base function included in the model. In the present study, the parameter d is equal to 2, and the maximum number of tracer interaction type base functions is restricted to 3.

3. RESULTS AND DISCUSSION

This section presents the results obtained from the application of the data-driven hybrid algorithm proposed in this paper. These results are discussed and the performance compared with that obtained using different techniques for the same database. Finally, the advantages of this new algorithm are also mentioned.

The database used for evaluating the performance of the hybrid model consisted of a total of 16,609 observations, corresponding to a total of 100 different aircraft engines. Besides the identification number of the aircraft engine to which the data belongs, each observation contains a total of 25 variables: the remaining useful life, together with the values for three operation settings and a total of 21 measurements belonging to different sensors located in the aircraft engine. Table 2 summarizes the descriptive statistics of the database. In order to perform the training of all the models, information was used from 80 out of the 100 aircraft engines available, chosen at random. Afterwards, the validation of the models was made using the data from the 20 remaining aircraft engines. In other words, the engine set was randomly divided into two subsets: one with 80 engines and the other with the 20 remaining engines. The information subset with 80 engines was used for the training and the other 20 for the validation. The validation process consisted of applying the prediction model to the values of the variables in the other 20 models in order to predict their remaining useful life. This sequence of random division of the data available on the basis of training data and validation was repeated five times, and the predicted results were compared with the real values of the remaining useful life.

Table 2. Descriptive statistics of all the variables of the database.

Input Variables	Mean	Standard Deviation
Remaining useful life	108.808	68.881
Operational setting 1	-8.870×10^{-6}	0.003
Operational setting 2	2.350×10^{-6}	0.003
Operational setting 3	100.000	10^{-6}
Total temperature at fan inlet (°R)	518.670	10^{-6}
Total temperature at LPC outlet (°R)	642.681	0.500
Total temperature at HPC outlet (°R)	1590.523	6.131
Total temperature at LPT outlet (°R)	1408.934	9.000
Pressure at fan inlet (psia)	14.620	10^{-6}
Total pressure in bypass-duct (psia)	21.609	0.001
Total pressure at HPC outlet (psia)	553.368	0.885
Physical fan speed (rpm)	2388.097	0.070
Physical core speed (rpm)	9065.243	22.082
Engine pressure ratio (P50/P2)	1.300	10^{-6}
Static pressure at HPC outlet (psia)	47.5412	0.267
Ratio of fuel flow to Ps30 (pps/psi)	521.414	0.738
Corrected fan speed (rpm)	2388.096	0.719
Corrected core speed (rpm)	8143.753	19.076
Bypass ratio	8.442146	0.038
Burner fuel-air ratio	0.0300	10^{-6}
Bleed enthalpy	393.212	1.549
Demanded fan speed (rpm)	2388.000	10^{-6}
Demanded corrected fan speed (rpm)	100.000	10^{-6}
HPT coolant bleed (lbm/s)	38.8163	0.181
LPT coolant bleed (lbm/s)	23.279	0.108

First of all, a multivariate linear regression model was performed using the remaining useful life as the output variable and the rest of the available variables as input variables. The model obtained had a R^2 value of 0.5846 (the average value of five replicates where the model was trained with 5 different training subsets). The application of this model to the multivariate linear regression validation data subset provided a result for the RMSE of 44.9307 (the average of five repetitions), a MAE of 33.6063 (the mean of the five repetitions) and an accuracy of 0.5635 (the average of five repetitions). Afterwards, a back-propagation neural network model with one hidden layer was trained, obtaining a RMSE of 42.3311 (the average of five repetitions), a MAE of 29.6364 (the mean of the five replicates) and an accuracy of 0.4996 (the average of five repetitions). The aim of the training and validation of both models was to obtain two benchmark references and compare the hybrid model presented in this research. In order to make it easy to compare the results obtained by means of the different techniques, the five training and validation subsets employed for all the models were the same.

Afterwards, the proposed hybrid algorithm was applied, producing the following results:

- *Dimensional reduction by means of principal component analysis:* The PCA algorithm was fed with 24 variables: the three operational settings and the 21 sensor measurements. A total of seven variables were discarded, and thus the total input variables subset was reduced from 24 to 17, containing a 91.2% of the variability of the data. This percentage of variability was explained by the seven *principal components* which were retained. A rotated pattern of variables was created using varimax method. Variables and factor loadings are shown in Table 3. Please note that h^2 represents the communality estimate and measures the ratio per unit of variance in an observed variable accounted for by the retained components. When the research was conducted, the use of a larger number of dimensions was checked. Using eight dimensions increased the variability contained to 92.1% and two variables more were added to the remaining variables subset, passing from 17 to 19. Nevertheless, the use of eight principal components was ruled out as in the next stage of the algorithm neither of the two new variables included were preserved when similarities were calculated by means of the dendrogram. The steps of the algorithm after the PCA are not performed in the space of the principal components as the next step of the algorithm is a dendrogram, in which similarities among variables and the RUL are analyzed; and although the model developed in this paper is data-driven, it is worthwhile to have the information corresponding to the original variables rather than the transformed ones, in order to be able to analyze the variables that take part in the model. Please also note that the variable subset obtained after this stage was also used for the training of both the linear regression and neural network models previously presented without any remarkable improvement in their results.

Table 3. Rotated pattern of variables with loadings and estimated communality.

Variable	PC1	PC2	PC3	PC4	PC5	PC6	PC7	h^2
Op.Set1	0.01	−0.01	1	0.01	0	0	0	1
Op.Set2	0.01	−0.01	0.01	1	0	0	0	1
Sensor.Measurement2	0.7	0.11	0.01	0	0.1	0.09	0.69	0.99
Sensor.Measurement3	0.63	0.16	0	0	0.75	0.09	0.09	1
Sensor.Measurement4	0.87	0.12	0.01	0.01	0.14	0.14	0.12	0.82
Sensor.Measurement7	−0.87	−0.02	0	−0.01	−0.14	−0.14	−0.12	0.81
Sensor.Measurement9	0.19	0.97	0	0	0.06	0.06	0.04	0.98
Sensor.Measurement11	0.89	0.08	0.01	0	0.14	0.14	0.14	0.86
Sensor.Measurement12	−0.89	−0.01	0	0	−0.15	−0.14	−0.14	0.85
Sensor.Measurement14	0.07	0.99	0	0	0.04	0.04	0.02	0.98
Sensor.Measurement15	0.85	0.13	0	0.01	0.09	0.09	0.08	0.76
Sensor.Measurement17	0.67	0.17	0	0	0.1	0.71	0.09	0.99
Sensor.Measurement20	−0.88	−0.13	0	0	−0.05	−0.01	0	0.79
Sensor.Measurement21	−0.85	−0.13	−0.01	0	−0.09	−0.05	−0.05	0.76

- *Making a dendrogram with the selected variables:* Figure 3 shows the dendrogram obtained with the 17 variables selected by the PCA plus the RUL. In this dendrogram, what can clearly be seen is the degree of similarity between variables, which will be decisive in choosing those variables that are related to the remaining useful lifetime (RUL) variable. The dendrogram shown in Figure 3 corresponds to one of the five repetitions. All the dendrograms obtained were very similar. The variables selected corresponded to four sensors: Sensor, Measurement7, Sensor,Measurement12, Sensor,Measurement20 and Sensor,Measurement21. As can be observed in Figure 3, these four variables have the largest degree of similarity to the remaining useful life variable. Therefore, in this step of the algorithm the number of input variables was reduced from 18(17 plus the RUL) to 4.

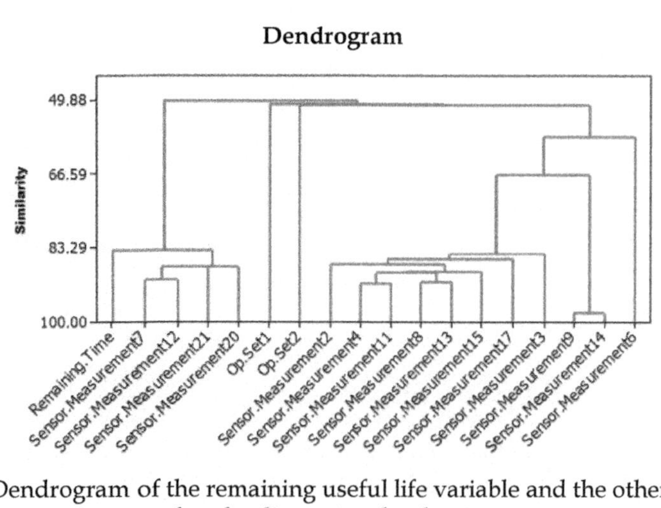

Figure 3. Dendrogram of the remaining useful life variable and the other variables after the dimensional reduction.

- *Calculating a regression tree using variables that are more similar to the variable remaining useful life:* For this problem, the four variables in the dendrogram that showed a level of similarity to the remaining useful life (called Remaining,Time in Figure 3) over 83.29% were used as input variables for a regression tree in which the remaining useful life is the output variable. The regression tree obtained is shown in Figure 3, and used to determine the cut-off points corresponding to the input variables. As can be observed in Figure 4, although four input variables were proposed for the regression tree model, only three are used for the cut-off points. The variables referred to are the following: Sensor,Measurement7, Sensor,Measurement12 and Sensor,Measurement21. The numbers named as N in the leaf nodes of the regression tree represent the number of cases that are in said node and that will therefore be used for the training of the corresponding MARS model.
- *Training a MARS model:* after the performance of the regression tree, a total of five different MARS models were trained. Each of these models corresponded to one of the end nodes of the regression tree. Each of the MARS models was trained with the part of the training subset that satisfied the conditions of their corresponding branches of the tree, using as dependent variables the four obtained as the outcome of the dendrogram (Sensor.Measurement7, Sensor. Measurement12, Sensor.Measurement12 and Sensor.Measurement21).

After the training of the MARS models referred to in the last stage of the algorithm, the resulting hybrid model was applied to the validation data subset, resulting in RMSE and MAE values of 36.0836 and 21.5624 respectively (the mean values of five replications of the algorithm), with an average value for the accuracy of 0.3123 (five replications). The values of steadiness and risk levels were also calculated for the hybrid and the two benchmark models. The results of the

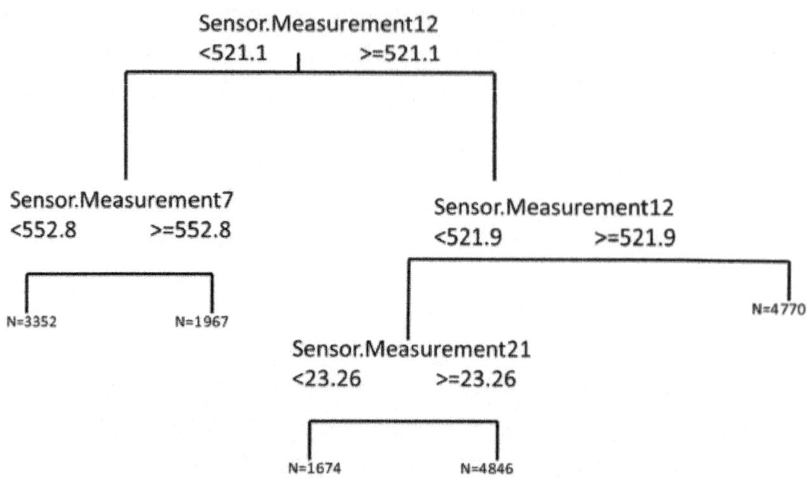

Figure 4. Regression tree of the remaining useful life variable using as input variables Sensor.Measurement7, Sensor.Measurement12, Sensor.Measurement20 and Sensor.Measurement21.

steadiness are presented in Figure 5. As can be observed the steadiness of the hybrid model is better than that obtained by both the linear and RNA models for almost any of the RUL values considered. Figure 6 shows the risk level of each model. In this case, the probability of obtaining a RUL estimate smaller than the true RUL is higher with the linear and hybrid models than with the hybrid one. In spite of this result, which would seem to be a better performance from both linear and hybrid models, we would like to remark that the predictions of both models are quite far from the real RUL when compared with the hybrid model. In other words, both models underestimate the RUL, but this underestimation is too far from the real value to be useful.

According to the results of the regression tree, the three most relevant variables for the determination of the RUL of the aircraft engine are Sensor.Measurement7 (total pressure of the HPC outlet), Sensor.Measurement12 (ratio of fuel flow to PS30) and Sensor.Measurement21 (LPT coolant bleed). The turbine compressor is required to have a constant outlet pressure despite the fluctuating turbine load and fluctuating supply. According to the results, variations in the total pressure of the HPC outlet are important in order to predict the engine's RUL. The presence of the ratio of fuel low to PS30 as one of the remaining input variables allows us to connect the RUL of the aircraft engines with this variable and also indirectly with the gross shaft power of the engine and the possible power loss that would be found due to degradation during the engine operation.

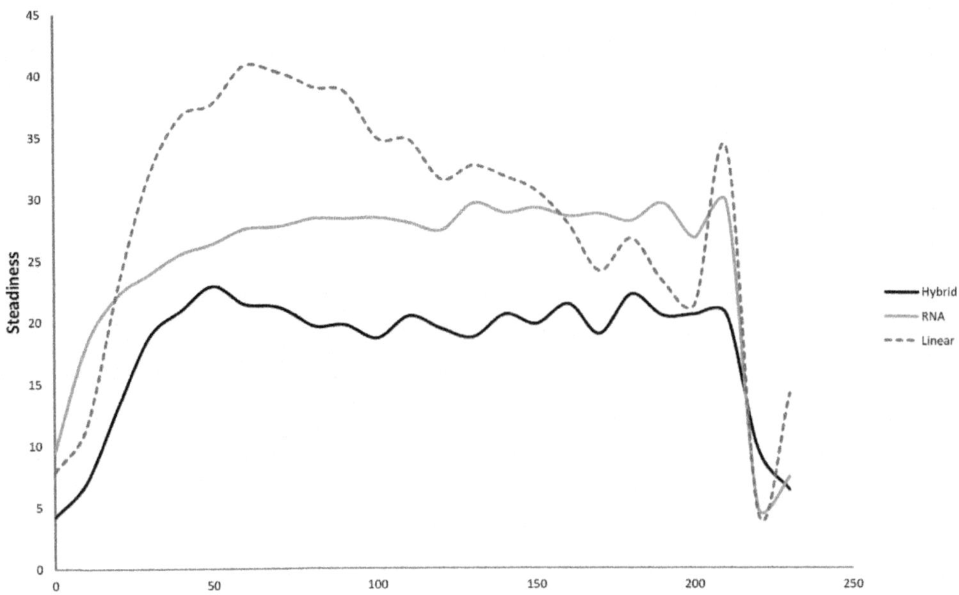

Figure 5. Steadiness of the hybrid model compared with the two benchmark techniques.

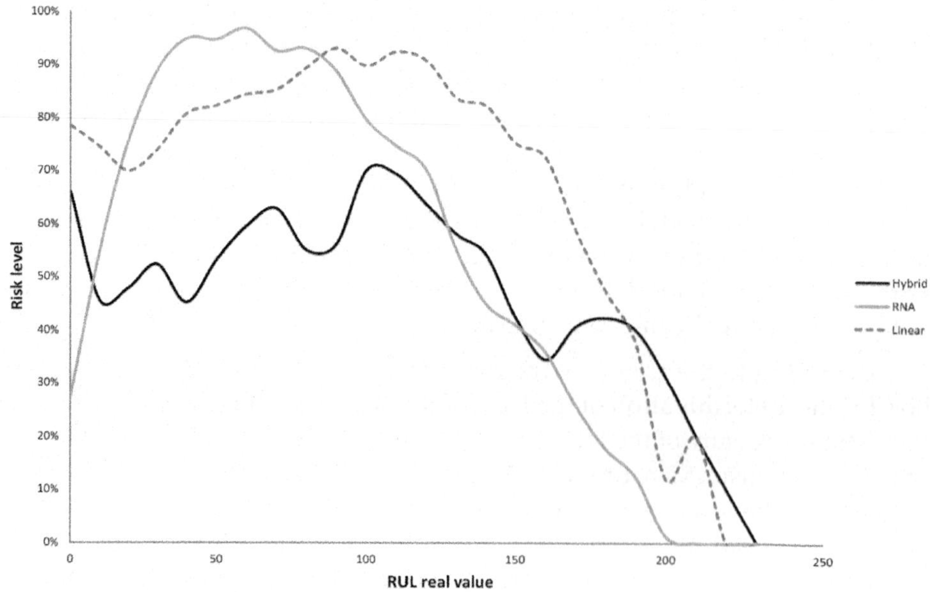

Figure 6. Risk level of the hybrid model compared with the two benchmark techniques.

From a design point of view, the geometry of a coolant channel of a LP blade is relatively simple in comparison with that of HP blade in spite of this, flow variations in during the operation of the aircraft engine, have been shown to be linked to a decrease in the RUL of the aircraft engine. The results obtained in the five replications were statistically equivalent when compared by means of a one-way ANOVA test. The range of dispersion of differences between real and predicted

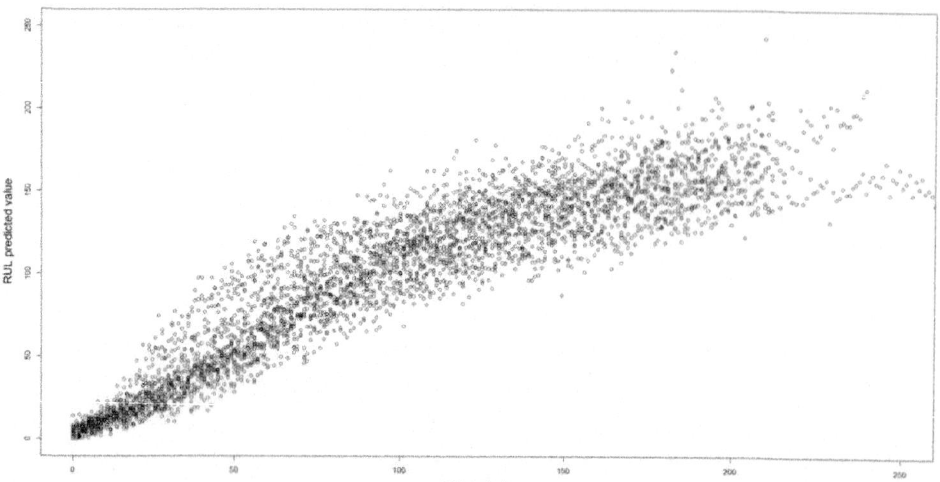

Figure 7. Remaining useful life of one of the validation subsets *versus* the remaining useful life calculated by the hybrid model.

values changes significantly from one aircraft engine to another. The same pattern of behavior was found in the regression and neural network models. In spite of this, the RMSE results obtained by means of the artificial neural network was 14.76% greater than those obtained by means of the hybrid model. In the case of the MAE differences the figure was 27.24% more than with the hybrid method. Figure 7 shows the remaining useful life of one of the validation subsets *versus* the remaining useful life calculated by means of the hybrid model. As may be observed, there is a visual correlation that corresponds to a Pearson's correlation coefficient of 0.9167.

Table 4 shows the confusion matrix of the hybrid, multivariate linear regression and back-propagation neural network models. As the RUL is a continuous variable, a cut-off point was defined in order to determine which aircraft engines were close to the end of their useful life. As it was not possible to establish objective engineering criteria, it was considered that those situations in which the actual RUL of the engine was below 15% of the maximum RUL value considered in the study (312 time units) meant that the unit was close to the end of its useful life. According to the values of this table, the overall misclassification rate of the hybrid model is 5.72%, while the rate for the linear regression and neural networks models were respectively 8.43% and 6.91%. In the case of sensitivity, the highest value was achieved by the hybrid model with a 92.20%, followed by the neural network model with 82.55% and the linear regression model with 65.64%. Finally, in the case of the specificity, the highest value is for the linear regression model with 99.67%, followed by the neural network with 96.38% and the hybrid model with 94.85%. Please note that in this problem the most important parameter is the sensitivity, as it shows the ability of the model to detect engines close to the end of their useful life. Finally, and due to the arbitrary selection of the cut-off point, we would like also to point out that in the case of having considered a cut-off point corresponding to 10% of the maximum remaining useful life of the validation data subset, the sensitivity and specificity values of the hybrid model would have been 86.57% and 97.23% respectively and in the case of the 20% they would be 94.81% for sensitivity and 91.19% for specificity.

Table 4. Confusion matrix of the hybrid, multivariate linear regression and back-propagation neural network models.

		Hybrid		Linear Regression		Neural Network	
		Predicted					
		Negative	Positive	Negative	Positive	Negative	Positive
Actual	Negative	2946	66	3002	10	2903	109
	Positive	160	780	323	617	164	776

According to the report published by Ramasso and Saxena [36] in 2014, up to that time more than seventy publications had developed prognostic algorithms using the C-MAPSS datasets. The paper reported the 10 best models, from the authors' point of view that had been applied to the C-MAPSS data sets. Three of them correspond to an unpublished master's thesis and two unpublished doctoral dissertations. When compared with those models, the one presented in the current research outperformed the correlation score of 7 out of 10 [37-43] and the MAE of 3 out of 10 [41-43]. Please note that on the one hand the training and validation sets used in the different articles are heterogeneous and that on the other hand, no detailed information of their results is available (real RUL values *versus* estimated RUL values for all the referred works). It must also be taken into account that MAE results should be treated with care, as the reliability of this parameter is partially linked to the sample size [44] which varies from one research to another. In the case of the three studies that would appear to outperform the result obtained in this research, the solutions proposed by Heimes [45] and Peel [46] require the use of different multilayer perceptron neural networks, with both models using the 24 variables as inputs for the model. In the case of the research by Heimes, the multilayer perceptron neural network has 24 neurons in the input layer and two hidden layer, one with three neurons and another with two. This model requires a minimum RUL length of 127 cycles. Such a demanding requirement concerning RUL length is not necessary for our model. In the case of the model proposed by Peels, not only did it use 24 input variables, but it was also an ensemble model, composed of two multilayer perceptron neural networks and a Kalman filter. Finally, the approach of Wang [14] is more similar to our research, but requires a total of 7 input variables.

As may be noticed, the algorithm proposed in the present paper performs a dimensional reduction by means of PCA, and afterwards a study of the similarity of the remaining variables with the output variable Remaining Time is performed. Despite the similarity of both methods, previous research [22,47] has demonstrated that the use of both techniques together, PCA with varimax rotation and hierarchical clustering by means of a dendrogram, are more effective than the use of either one of them alone. This fact is confirmed by the results of the present papers although no comparison with the use of only one of these techniques is presented as it is not the main purpose of this research. Similar methodologies have also been used in other research [48] with satisfactory results.

Finally, from the authors' point of view, it would be also worth highlighting that although it is possible to find in the literature better results for predictions performed over the same database, the main advantage of the present approach is that it does not require the use of the previous engine states unlike in other models [49], which in some cases makes it necessary to have a record length of even 500 consecutive data [50] in order to detect the fault with a reasonable confidence. In other words, the output variable, the RUL, is calculated at any given moment knowing only the values of the input variables in the same moment. Therefore it is not necessary to store previous values of these variables. Of course, the value of any input variable at any given moment would be conditioned by the previous operational conditions and degradation status of the engine, but such information is not directly required as a direct input for the model.

4. CONCLUSIONS

In this paper, a hybrid prognostics approach which integrates the MARS method and PCA plus CART techniques is presented. It is a well-known fact that fault prognosis is an essential and complex technology within health management and CBM. Testing of the proposed MARS-PCA-CART-based prognostic approach was carried out using sensor data from aircraft engines. The simulation results showed that the new approach is efficient in predicting the remaining useful life (RUL) for the aircraft engines, as it improved the results of the benchmark technique. The proposed hybrid model improves upon the performance of classical predictive models used for comparison (linear multivariate regression model and neural network model with back propagation) [12,51-54]. It has also been stated in literature on predictive maintenance that one of the most important issues in prognostic technology [55] is the estimation of the risk of failure and the RUL of a component, given the component's age and its past and current operating conditions. In the case of the model proposed in this paper, when used for the prediction of the RUL, it does not require information about the previous status of the variables of the engine: it only requires information regarding the current situation of said variables. Please note that this does not mean that the model is not affected by the previous conditions of the engine, but that the only input variables required for the prediction belong to the current temporal moment and not to any previous status. However, of course, the values of the input variables at each temporal moment will certainly be affected by the previous operative conditions of the engine, and also by the previous engine status. In other words, this model provides two main advantages: firstly, system robustness against possible failures in the memory registers, as well as a reduction in the complexity of actual implementation, which is linked to fewer variables to monitor in real time. Please note that in the end, the number of variables required for the prediction of remaining useful life is a total of four. In other words, the algorithm presented in this paper is suitable for its application to any other system we have historical records for regarding its variables, and where the information about the RUL for each data set is available. This algorithm is particularly recommended if the initial number of variables implied is large enough to require a previous dimensional reduction and there are non-linear relationships of the input variables with the RUL. According to the research of Sikorska *et al.* [16], the model developed in the present research would be considered as a Level 1 model in the prognostics category.

As a future line of research, the authors propose the performance of models taking into account the values of the input variables at earlier time points, in order to determine the remaining useful life of the aircraft engines. From the authors' point of view, these models should provide a better fit, but suffer from the need to have in memory the values at previous time points of the variables included in the model. These models would be of interest only if they could be trained with fewer variables than the model that does not require information of the previous state. Finally, this future line of research should be completed with a test on real data as the simulations performed in this research were made on simulation data.

ACKNOWLEDGMENTS

This paper has been funded by the Government of the Principality of Asturias through funds from the Programme of Science, Technology and Innovation (PCTI) of Asturias 2006-2009, co-financed by 80% within the priority Focus 1 of the Operational Programme FEDER of the Principality of Asturias 2007-2013 (Research project FC-11-PC10-19). Finally, we would like to thank Anthony Ashworth for his revision of English grammar and spelling of the manuscript.

AUTHOR CONTRIBUTIONS

Fernando Sánchez Lasheras and Francisco Javier de Cos Juez conceived and designed the experiments; Fernando Sanchez Lasheras, Ricardo Mayo Bayón and Paulino José García Nieto performed the experiments; Victor Manuel González Suárez, Francisco Javier de Cos Juez and Paulino José García Nieto analyzed the data; Fernando Sanchez Lasheras, Paulino Jose Garcia Nieto, Francisco Javier de Cos Juez, Ricardo Mayo Bayon and Victor Manuel González Suárez wrote the paper.

CONFLICTS OF INTEREST

The authors declare no conflict of interest

REFERENCES

1. Cheng, S.; Azarian, M.H.; Pecht, M.G. Sensor systems for prognostics and health management. *Sensors* **2010,** *10,* 5774-5797.
2. Pecht, M.; Jaai, R. A prognostics and health management roadmap for information and electronics-rich systems. *Microelectron. Reliab.* **2010,** 50, 317-323.
3. Pecht, M. *Prognostics and Health Management of Electronics;* Wiley-Interscience: New York, NY, USA, 2008.
4. Vichare, N.; Pecht, M. Prognostics and Health Management of Electronics. *IEEE Trans. Compon. Packag. Technol.* **2006,** *29,* 222-229.
5. Schwabacher, M.; Goebel, K. A Survey of Artificial Intelligence for Prognostics. In Proceedings of the 2007 AAAI Fall Symposium: AI for Prognostics, Arlington, VA, USA, 8-11 November 2007; pp. 9-11.
6. Liu, J.; Djurdjanovic, D.; Marko, K.; Ni, J.A. Divide and Conquer Approach to Anomaly Detection, Localization and Diagnosis. *Mech. Syst. Signal Process.* **2009,** *23,* 2488-2499.
7. Liu, J.; Wang, M.; Ma, F.; Yang, Y.B.; Yang, C.S. A data-model-fusion prognostic framework for dynamic system state forecasting. *Eng. Appl. Artif. Int.* **2012,** 25, 814-823.
8. Bai, G.; Wang, P.; Hu, C.; Pecht, M. A generic model-free approach for lithium-ion battery health management. *Appl. Energ.* **2014,** *135,* 247-260.
9. Sanchez-Lasheras, F.; de Andres, J.; Lorca, P.; de Cos Juez, F.J. A hybrid device for the solution of sampling bias problems in the forecasting of firms' bankruptcy. *Expert Syst. Appl.* **2012,** *8,* 7512-7523.
10. Garcia Nieto, P.J.; Alonso Fernandez, J.R.; de Cos Juez, F.J.; Sanchez Lasheras, F.; Diaz Muniz, C. Hybrid modelling based on support vector regression with genetic algorithms in forecast-

ing the cyanotoxins presence in the Trasona reservoir (Northern Spain). *Environ. Res.* **2013,** 122, 1-10.

11. Xi, Z.; Jing, R.; Wang, P.; Hu, C. A copula-based sampling method for data-driven prognostics. *Reliab. Eng. Syst. Safe* **2014,** *132,* 72-82.
12. Si, X.-S.; Wang, W.; Hu, C.-H.; Zhou, D.-H. Remaining Useful Life Estimation—A Review on the Statistical Data Driven Approaches. *Eur. J. Oper. Res.* **2011,** 213, 1-14.
13. Dean, F.; de Castro, J.; Litt, J. *User's Guide for the Commercial Modular Aero-Propulsion System Simulation (C-MAPSS), NASA/ARL;* Technical Manual TM2007-215026; NASA Center for AeroSpace Information: Cleveland, OH, USA, 2007.
14. Wang, T.; Yu, J.; Siegel, D.; Lee, J. A similarity-based prognostics approach for remaining useful life estimation of engineered systems. In Proceedings of the IEEE International Conference on Prognostics and Health Management (PMH 2008), Denver, CO, USA, 6-9 October 2008; pp. 1-6.
15. Sikorska, J.Z.; Hodkiewicz, M.; Ma, L. Prognostic modelling options for remaining useful life estimation by industry. *Mech. Sys. Signal Process.* **2011,** 25, 1803-1836.
16. Jardine, A.K.S.; Lin, D.; Banjevic, D. A review on machinery diagnostics and prognostics implementing condition-based maintenance. *Mech. Syst. Signal Process.* **2006,** *20,* 1483-1510.
17. Wu, C.H. Behavior-based spam detection using a hybrid method of rule-based techniques and neural networks. *Expert Syst. Appl.* **2009,** *36,* 4321-4330.
18. Bi, Y.; Bell, D.; Wang, H.; Guo, G.; Greer K. Combining Multiple Classifiers Using Dempster's Rule of Combination for Text Categorization. *Lect. Notes Artif. Int.* **2004,** *3131,* 127-138.
19. Cho, S.B.; Lee, J.H. Learning Neural Network Ensemble for Practical Text Classification. In *Intelligent Data Engineering and Automated Learning;* Lecture Notes in Computer Science; Springer: Berlin/Heidelberg, Germany, 2003; Volume 2690, pp. 1032-1036.
20. Wong, W.K.; Guo, Z.X. A hybrid intelligent model for medium-term sales forecasting in fashion retail supply chains using extreme learning machine and harmony search algorithm. *Int. J. Prod. Econ.* **2010,** 128, 614-624.
21. Hastie, T.; Tibshirani, R.; Friedman, J. *The Elements of Statistical Learning;* Springer: New York, NY, USA, 2003.
22. Grice, J.W. Computing and evaluating factor scores. *Psychol. Methods* **2001,** 6, 430-450.
23. Orchard, M.; Tobar, F.; Vachtsevanos, G. Outer feedback correction loops in particle filtering-based prognostic algorithms: Statistical performance comparison. *Stud. Inform. Control* **2009,** 18, 295-304.
24. Hu, Y.; Baraldi, P.; di Maio, F.; Zio, E. A particle filtering and kernel smoothing-based approach for new design component prognostics. *Rel. Eng. Syst. Safe* **2015,** *134,* 19-31.
25. Pearson, K. On lines and planes of closest fit to systems of points in space. *Philos. Mag.* **1901,** 2, 559-572.
26. Hotelling, H. Analysis of a Complex of Statistical Variables into Principal Components. *J. Educ. Psychol.* **1933,** 24, 417-441 and 498-520.
27. Ismail, S.; Shabri, A. Stream Flow Forecasting using Principal Component Analysis and Least Square Support Vector Machine. *J. Appl. Sci. Agric.* **2014,** 9, 170-180.
28. Guhathakurta, P.; Rajeevan, M.; Thapliyal, V. Forecasting Indian summer monsoon rainfall by a Principal Component Neural Network model. *Meteorol. Atmos. Phys.* **1999,** 71, 255-266.
29. Hu, T.; Wu, F.; Zhang, X. Rainfall-Runoff Modelling using Principal Component Analysis and Neural Network. *Nord. Hydrol.* **2007,** *38,* 235-248.
30. Therneau, T.M.; Atkinson, B.; Ripley, B. *Rpart: Recursive Partitioning;* R package Version 4.1-1; 2013.
31. Friedman, J.H. Multivariate Adaptive Regression Splines (with discussion). *Ann. Stat.* **1991,** 19, 1-141.

32. Sekulic, S.S.; Kowalski, B.R. MARS: A Tutorial. *J. Chemometr.* **1992,** 6, 199-216.

33. Friedman, J.H.; Roosen, C.B. An Introduction to Multivariate Adaptive Regression Splines. *Stat. Methods Med. Res.* **1995,** 4, 197-217.

34. Garcia Nieto, P.J.; Alonso Fernandez, J.R.; Sanchez Lasheras, F.; de Cos Juez, F.J.; Diaz Muniz, C. A New improved study of cyanotoxins presence from experimental cyanobacteria concentrations in the Trasona reservoir (Northern Spain) using the MARS technique. *Sci. Total Environ.* **2012,** *430,* 88-92.

35. Garcia Nieto, P.J.; Sanchez Lasheras, F.; de Cos Juez, F.J.; Alonso Fernandez, J.R. Study of Cyanotoxins Presence from Experimental Cyanobacteria Concentrations Using a New Data Mining Methodology Based on Multivariate Adaptive Regression Splines in Trasona Reservoir (Northern Spain). *J. Hazard. Mater.* **2011,** 195, 414-421.

36. Ramasso, E.; Saxena, A. Performance Benchmarking and Analysis of Prognostic Methods for CMAPSS Datasets. *Int. J. Progn. Health Manag.* **2014,** 5, 1-15.

37. Coble, J.; Hines, J. Prognostic algorithm categorization with PHM challenge application. In Proceedings of International Conference on prognostics and health management, Denver, CO, USA, 6-9 October 2008; IEEE: New York, NY, USA, 2008.

38. Siegel, D. Evaluation of Health Assessment Techniques for Rotating Machinery. Master's Thesis, Division of Research and Advanced Studies of the University of Cincinnati, Cincinnati, OH, USA, 2009.

39. Ramasso, E. Contribution of belief functions to hidden Markov models with an application to fault diagnosis. In Proceedings of the International Workshop on Machine Learning for Signal Processing, Grenoble, France, 1-4 September 2009; IEEE: New York, NY, USA, 2009.

40. Wang, T. Trajectory Similarity Based Prediction for Remaining Useful Life Estimation. Ph.D. Thesis, University of Cincinnati, Cincinnati, OH, USA, 2010.

41. Riad, A.; Elminir, H.; Elattar, H. Evaluation of neural networks in the subject of prognostics as compared to linear regression model. *Int. J. Eng. Technol.* **2010,** *10,* 52-58.

42. Abbas, M. System Level Health Assessment Of Complex Engineered Processes. Ph.D. Thesis, Georgia Institute of Technology, Atlanta, GA, USA, 2010.

43. Ramasso, E.; Gouriveau, R. Prognostics in switching systems: Evidential Markovian classification of real-time neuro-fuzzy predictions. In Proceedings of the International Conference on Prognostics and Health Management, Macao, China, 12-14 January 2010; IEEE: New York, NY, USA, 2010.

44. Chai, T.; Draxler, R.R. Root mean square error (RMSE) or mean absolute error (MAE)?—Arguments against avoiding RMSE in the literature. *Geosci. Model Dev.* **2014,** 7, 1247-1250.

45. Heimes, F. Recurrent neural networks for remaining useful life estimation. In Proceedings of the International Conference on Prognostics and Health Management, Denver, CO, USA, 6-9 October 2008; IEEE: New York, NY, USA, 2008.

46. Peel, L. Data driven prognostics using a Kalman filter ensemble of neural network models. In Proceedings of the International Conference on Prognostics and Health Management, Denver, CO, USA, 6-9 October 2008; IEEE: New York, NY, USA, 2008.

47. Yidana, S.M.; Ophori, D.; Banoeng-Yakubo, B. A multivariate statistical analysis of surface water chemistry data—Tha Ankobra Basin, Ghana. *J. Environ. Manag.* **2008,** *86,* 80-87.

48. Lewis, P.D.; Parry, J.M. An exploratory analysis of multiple mutation spectra. *Mutat. Res.* **2002,** 518, 163-180.

49. Sarkar, S.; Mukherjee, K.; Sarkar, S.; Ray, A. Symbolic Transient Time-series Analysis for Fault Detection in Aircraft Gas Turbine Engines. In Proceedings of the 2012 American Control Conference, Montreal, QC, Canada, 27-29 June 2012; Institute of Electrical and Electronics Engineers Inc.: New York, NY, USA, 2012; pp. 5132-5137.

50. Liu, K.; Gebraeel, N.Z.; Shi, J. A Data-Level Fusion Model for Developing Composite health Indices for Degradation Modeling and Prognostic Analysis. *IEEE Trans. Autom. Sci. Eng.* **2013,** *10,* 652-664.

51. Zio, E.; di Maio, F. A Data-driven Fuzzy Approach for Predicting the Remaining Useful Life in Dynamic Failure Scenarios of a Nuclear System. *Reliab. Eng. Syst. Safe* **2010,** 95, 49-57.

52. Ahmadzadeh, F.; Lundberg, J. Remaining Useful Life Prediction of Grinding Mill Liners Using an Artificial Neural Network. *Miner. Eng.* **2013,** 53, 1-8.

53. Guzman, D.; de Cos Juez, F.J.; Myers, R.; Guesalaga, A.; Sanchez Lasheras, F. Modeling a MEMS deformable mirror using non-parametric estimation techniques. *Opt. Express* **2010,** *18,* 21356-21369.

54. Garcia Nieto, P.J.; Martinez Torres, J.; de Cos Juez, F.J.; Sanchez Lasheras, F. Using multivariate adaptive regression splines and multilayer perceptron networks to evaluate paper manufactured using *Eucalyptus globulus. Appl. Math. Comput.* **2012,** *219,* 755-763.

55. Lee, J.; Ni, J.; Djurdjanovic, D.; Qiu, H.; Liao, H. Intelligent prognostics tools and e-maintenance. *Comput. Ind.* **2006,** *57,* 476-489.

This page left intentionally blank.

INDEX

D

E

L

M

N

O

P

Q

R

S

T